Intentional Interviewing and Counseling

ABOUT THE AUTHORS

Allen E. Ivey is Distinguished University Professor (Emeritus), University of Massachusetts, Amherst and Professor, Counseling, University of South Florida, Tampa. He is president of Microtraining Associates, an educational publishing firm. He received his undergraduate degree (Phi Beta Kappa) from Stanford University in psychology with a minor in social work. After a Fulbright year studying social work at the University of Copenhagen, Denmark, he earned his counseling doctorate at Harvard University.

Allen is the author or coauthor of more than 30 books and 200 articles and chapters, translated into 18 languages. A past president and Fellow of the Division of Counseling Psychology of the American Psychological Association, he is a Diplomate of the American Board of Professional Psychology. He is also a Fellow of APA's Society for the Study of Ethnic and Minority Psychology and the Asian American Psychological Association. He received the International Award from the Division of Counseling and Psychotherapy of the Portuguese Psychological Association for his innovative writings on developmental counseling and therapy. Allen's contributions led him to receive the Professional Development Award from the American Counseling Association. He was named "Distinguished Elder" for lifetime contributions to multicultural psychology at the National Multicultural Summit and Conference. His recent work has focused on applying Developmental Counseling and Therapy to the analysis and treatment of severe psychological distress. With Mary, he is writing and conducting workshops on counseling and spirituality.

Mary Bradford Ivey is Vice President of Microtraining Associates and Professor of Counseling, University of South Florida, Tampa. She is a former counselor in the Amherst, Massachusetts, schools and has served as visiting professor at the University of Massachusetts, Amherst; University of Hawai'i, Manoa; and Flinders University, South Australia. Mary's undergraduate degree in social work and education is from Gustavus Adolphus College, and she has a master's degree in counseling from the University of Wisconsin. She earned her doctorate in organizational development at the University of Massachusetts, Amherst.

Mary is the author or coauthor of 11 books, translated into multiple languages, and several articles and chapters. She is a Nationally Certified Counselor (NCC) and a licensed mental health counselor (LMHC); she has held a certificate in school counseling. She has presented workshops and keynote lectures with Allen throughout the world, including Australia, New Zealand, Japan, China, Israel, Greece, Sweden, Canada, Great Britain, Portugal, and Germany. She is also known for her work in promoting and explaining development guidance and counseling in the United States and abroad. Mary received national recognition when her elementary counseling program at the Fort River School was named one of the 10 best in the nation at the Christa McAuliffe Conference. She is one of the first 15 honored Fellows of the American Counseling Association and is also a recipient of the American Counseling Association's O'Hana Award for her work in multicultural counseling in the schools.

SIXTH EDITION

Intentional Interviewing and Counseling

Facilitating Client Development in a Multicultural Society

Allen E. Ivey

University of Massachusetts, Amherst

Mary Bradford Ivey

Microtraining Associates, Inc.

Australia • Brazil • Canada • Mexico • Singapore • Spain
United Kingdom • United States

THOMSON

BROOKS/COLE

Intentional Interviewing and Counseling: Facilitating Client Development in a Multicultural Society, Sixth Edition
Allen E. Ivey and Mary Bradford Ivey

Senior Acquisitions Editor: *Marquita Flemming*
Assistant Editor: *Alma Dea Michelena*
Editorial Assistant: *Sheila Walsh*
Technology Project Manager: *Inna Fedoseyeva*
Executive Marketing Manager: *Caroline Concilla*
Marketing Assistant: *Rebecca Weisman*
Senior Marketing Communications Manager: *Tami Strang*
Project Manager, Editorial Production: *Christine Sosa*
Creative Director: *Rob Hugel*
Art Director: *Vernon Boes*

Print Buyer: *Barbara Britton*
Permissions Editor: *Joohee Lee*
Production Service: *Scratchgravel Publishing Services*
Copy Editor: *Patterson Lamb*
Illustrator: *Scratchgravel Publishing Services*
Cover Designer: *Katherine Minerva*
Cover Image: *Katherine Minerva*
Cover Printer: *Phoenix Color Corp*
Compositor: *International Typesetting and Composition*
Printer: *Courier Corporation/Westford*

Library of Congress Control Number: 2005936430

Student Edition: ISBN 0-495-00619-X

Thomson Higher Education
10 Davis Drive
Belmont, CA 94002-3098
USA

For more information about our products, contact us at:
Thomson Learning Academic Resource Center
1-800-423-0563

For permission to use material from this text or product, submit a request online at **http://www.thomsonrights.com.**
Any additional questions about permissions can be submitted by e-mail to **thomsonrights@thomson.com.**

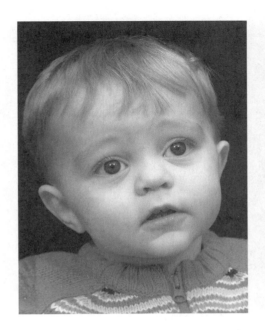 

To Ryan Quirk and Aidan Robey,
our latest editions/additions

CONTENTS

SECTION III   *Helping Clients Generate New Stories That Lead
to Action: Influencing Skills and Strategies*   **259**

Chapter 15   Determining Personal Style and Future Theoretical/Practical Integration   482

Appendix   Glossary of Terms for Brain Areas Discussed in Research Portions of This Book   497

LIST OF BOXES

PREFACE

We live in a world that seems continually in crisis. Terrorism and war, religious and political strife, and economic change provide a backdrop for the issues clients bring to us. Clients come for help with issues ranging from posttraumatic stress and depression to decisions about how to raise children and whether to change jobs.

Problems, concerns, and challenges are the stuff of counseling and therapy. This book seeks to help students meet the challenges of clients through providing a solid, well-researched foundation in the skills of the interview. We firmly believe that problem resolution and client change can best be affected through a wellness and positive psychological approach that provides a "center" for clients. Clients who know their strengths and positive assets are better prepared to dig deep within themselves and meet the challenges of life. The microskills approach has long been based on client strengths and resources.

"Culture counts." U.S. Surgeon General David Satcher spoke of the central importance of multicultural awareness in his 2001 report on mental health. He commented that racial and ethnic minorities face "striking disparities" in access to adequate mental health care. If we are to make a difference in the lives of our clients, it is vital that we be aware of individual and multicultural differences and see that services are provided equitably. We honestly believe that in our new era, interviewers, counselors, and therapists must become active in helping the world realize that for all of us "to get along is essential." Multicultural and social justice issues have now become a practical and ethical imperative.

Students are encouraged to develop a portfolio of competencies as they work through this book. By the time the student is halfway through the book, he or she should be able to complete a full interview using only listening skills. Accountability and demonstrated performance are now essential in our field. The Ivey Taxonomy in Chapter 8 shows students how they can anticipate specific results when they use interviewing skills. Ideally, all students will complete at least one full interview and analyze their work and its effect on clients by the completion of the course.

We all face new challenges, and with this observation we welcome you to the sixth edition of *Intentional Interviewing and Counseling: Facilitating Client Development in a Multicultural Society*. Textbooks are always a work in progress, and we have been influenced by instructors and by students to make constant changes. We look forward to feedback and ideas from you to help this book grow in the future.

Features New to the Sixth Edition

Intentional Interviewing and Counseling: Facilitating Client Development in a Multicultural World, Sixth Edition, continues the tradition of accountability and multicultural understanding of earlier editions, but we have increased our emphasis in both key areas. Among the many changes in this new edition, you will find the following:

▲ Wellness concepts have joined ethics and multicultural competence to form the base of the Microskills Hierarchy. A new Chapter 2 brings a special focus on these foundational issues.

▲ Motivational interviewing, key skills, and strategies have been added to Chapter 14. Students will be able to conduct a beginning interview using this method. Those who complete this text should be able to work through five methods of counseling: decisional, person-centered, assertiveness training, solution-oriented brief therapy, and motivational interviewing. With some groups, it may be wise to select one or two of these theories as you teach the text and recommend that the students use the rest as reference material later in their practicum and professional lives.

▲ Discernment of life's mission and direction has been added to Chapter 11 on reflection of meaning. Students will be able to examine their own life goals and also guide clients who need to focus their lives and find direction.

▲ Each chapter now begins with an interactive case study and students are asked to provide their own thoughts on how to help the client work through these issues. At the end of the chapter, Allen and Mary provide their thoughts on the case.

▲ Recent research in neuropsychology has supported many of the concepts of this book, and brain science findings have been added to the research boxes in most chapters. It is becoming apparent that effective counseling not only affects cognitive growth but also impacts the brain's development. We believe that students need to start thinking about this new area. The research material is designed to stimulate awareness; we are not asking students to learn the structure of the brain. No exercises or questions about this new area have been added to the book, the CD-ROM, or the instructor's guide.

▲ Research summaries presented throughout the book have been updated with a new emphasis on "research evidence that you can use." The microskills of this book are informed by over 450 data-based studies. Dr. Thomas Daniels reviews this research in detail on the optional accompanying CD-ROM.

▲ We have made a small, but potentially useful addition in the practice exercises. Professor Gary Miller of the University of South Carolina believes that we should introduce the ideas of supervision much earlier in the curriculum. He uses microskills practice sessions as an introduction to supervision. We have coined the word *microsupervision* for those instructors who want to start instruction early. But this addition is not such that practice sessions should change except for a shift in perspective.

▲ The popular case commentaries by Weijun Zhang and Azara Santiago-Rivera are now complemented by observations from national and international experts.

> James Lanier points out key differences among the words "problem, concern, issues, and challenges," stating that a problem-centered language may not be appropriate for all clients (Chapter 2).
>
> Mark Pope provides a comprehensive view of multiculturalism (Chapter 2).
>
> Courtland Lee discusses issues around counseling teens and shows how questions can help their development (Chapter 4).

Derald Wing Sue presents his analysis of the meaning of Whiteness and its varying definitions (Chapter 7).

Kathryn Quirk, a graduate student, provides an alternative perspective on empathy and culture (Chapter 8).

Owen Hargie from Northern Ireland speaks of spiritual conflict and confrontation (Chapter 9).

Bob Manthei from New Zealand presents data helping students understand what's going on in the client's head during counseling (Chapter 13).

Mary Rue Brodhead from Canada discusses how microskills can be used throughout one's career whether one stays in counseling or not (Chapter 15).

Important Issues That Remain Unchanged

Some key competencies and issues have been basic to this book from the beginning. Among many other competencies, students will be able to

▲ Complete a full interview using only listening skills by the time they are halfway through the book.

▲ Use microskills intentionally and be able to predict how clients will respond to their interventions through using the Ivey Taxonomy.

▲ Draw out client stories through the basic listening sequence.

▲ Understand the basics of multicultural counseling and apply these concepts to their own self-understanding and to the interview.

▲ Develop competence in the use of family and community genograms.

▲ Analyze their own interviewing behavior and its effectiveness with clients.

▲ Develop beginning competence in five approaches to the interview: decisional counseling, person-centered, behavioral assertiveness training, brief solution-oriented therapy, and motivational interviewing.

Teaching Aids

The CD-ROM, *Intentional Interviewing and Counseling: Your Interactive Resource,* has been updated with three new video segments, including special emphasis on that key skill, reflection of feeling. A set of video presentations on ethical, multicultural, and wellness issues begins the CD-ROM. You will also find flashcard summaries of each chapter, interactive exercises for skill practice and mastery, automatically scored practice quizzes, a downloadable portfolio, and many web links to supplement each chapter.

We have endeavored to provide choices for our readers by offering the book alone (ISBN 0-495-00620-3) or the book and CD-ROM prepack (ISBN 0-495-00619-X) for those who wish to integrate technology into their classes. We are pleased that our publisher is able to offer the CD-ROM for a nominal additional fee, as we believe that the interactivity and learning potential available through this technology are truly valuable. Although the book stands alone, the addition of the CD-ROM provides a vehicle with which students can test and catalogue their learning in the comprehensive portfolio sections.

The *Instructor's Resource Guide* (ISBN 0-495-09025-5) has been revised in accordance with the changes to the book. There you will find chapter objectives, suggested class procedures, additional discussion of end-of-chapter exercises, and multiple-choice and essay questions. The *Instructor's Resource Guide* also includes basic information on developmental counseling and therapy (DCT) that many professors find useful in beginning skills courses. This material may be duplicated for student use. The IRG is available electronically and can be ordered through your Thomson sales representative or through the Academic Resource Center at 800-423-0563. The test bank is also available on ExamView (ISBN 0-495-13079-6), which allows instructors to manipulate and create tests easily.

A new set of PowerPoint® slide presentations by Esther Dyer is now available on the Book Companion Website at http://counseling.wadsworth.com/ivey6. You may download and print these slides to make overhead transparencies or project them as PowerPoints from your computer.

Videotapes illustrating the microskills are available from Microtraining Associates, 25 Burdette Avenue, Framingham, MA 01702, phone/fax 888-505-5576, and on their web site, www.emicrotraining.com. Additional weblinks for ethics, multicultural, and professional issues can also be found at the web site. A new *Basic Attending Skills* videotape is now available featuring Deryl Bailey and Azara Santiago-Rivera as well as Mary and Allen Ivey and Norma Gluckstern. The *Basic Influencing Skills* video can be obtained for the last half of this book. A new video on microcounseling supervision has recently been released with a supplementary CD-ROM. This should be helpful to students in classifying and working with skills. Those with an orientation to theoretical approaches should find the new skill and strategy videotapes useful. Microtraining Associates, a privately owned company independent of Thomson Brooks/Cole, is known for its wide array of multicultural training videos and now has supplementary materials on most areas of multicultural concern as well as many videos on counseling and therapy skills and strategies. Teaching microskills and social skills to individual clients, peer helpers, and others has become a central part of practice for many professionals. Health maintenance organizations, social work agencies, community helpers, AIDS and cancer volunteers, and many others benefit from the skills training approach.

Alternative Instructional Sequences

The order of the chapters in this book remains basically the same as in the past. However, some instructors will want to reorder chapters to meet their own instructional goals. We have tried to organize the chapters in such a way as to make alternative sequencing easy.

For example, some prefer to teaching questioning *after* the listening skills of encouraging, paraphrasing, and reflection of feeling. They point out that some students have difficulty "going beyond" questions and really listening to clients. This more person-centered approach is certainly effective and a good way to emphasize the importance of active listening.

Another major sequencing issue concerns the placement of confrontation. In our book with Paul Pedersen, *Intentional Group Counseling: A Microskills Approach* (Thomson Brooks/Cole, 2001), we place confrontation skills as the last set of microskills to be learned. We do this because confrontation in groups is particularly complex. We are aware that it can be equally complex with individuals. Thus, we recommend a change in chapter sequencing for those of you who prefer placing confrontation later in the course. Despite this, we discuss confrontation in Chapter 9. We find that the attending, observing, and basic listening skill emphasis of the first half of the book allows effective and early teaching of basic confrontation. The Confrontation Impact Scale is useful in work with focusing, reflection of meaning, and the other influencing skills and strategies.

Chapter 14 presents four methods of counseling and therapy. Not all students will be able to master these in one semester. Many instructors select from these four methods or have groups in each class study a single theory and present it to their classmates. Advanced students will be able to engage in all four by the end of the course.

Another approach to Chapter 14 would be to pair material on person-centered interviews with the first eight chapters and behavioral assertiveness training with Chapter 12 on influencing skills. Brief solution-oriented approaches could be paired with Chapter 4 on questions, particularly if you teach questions after the other listening skills. Motivational interviewing is a variety of decisional counseling and could be paired with Chapter 13.

Each of us needs to shape and adapt textbooks to meet our students' needs and our own approach to teaching. Other sequences of skills can be arranged, and we welcome your feedback on this important and challenging instructional issue.

Acknowledgments

Thomas Daniels, Memorial University, Cornerbrook, has been central to the development of the microskills approach for many years, and we are pleased that his summary of research on over 450 data-based studies is now available on the CD-ROM that accompanies this book. We are appreciative of one of our students, Penny John, for permission to use her interview as an example in Chapter 13. Amanda Russo, a student at Western Kentucky University, also allowed us to share some of her thoughts about the importance of practicing microskills.

Weijun Zhang's writing and commentaries remain central to this book. We also thank Owen Hargie, James Lanier, Courtland Lee, Robert Manthei, Mark Pope, Kathryn Quirk, Azara Santiago-Rivera, Sandra Rigazio-DiGilio, and Derald Wing Sue for their written contributions. Robert Marx and Joseph Litterer were important in the early development of this book. Discussions with Otto Payton and Viktor Frankl have clarified the presentation of reflection of meaning. William Matthews was especially helpful in formulating the five-stage model of the interview. Lia and Zig Kapelis of Flinders University and Adelaide University are thanked for their support and participation while we served as visiting professors in South Australia.

David Rathman, Chief Executive Officer of Aboriginal Affairs, South Australia, has constantly supported and challenged this book, and his influence shows in many ways. Matthew Rigney, also of Aboriginal Affairs, was instrumental in introducing us to new ways of thinking. These two people first showed us that traditional, individualistic ways of thinking are incomplete, and therefore they were critical in the development of the focusing skill with its emphasis on the cultural/environmental context.

The skills and concepts of this book rely on the work of many different individuals over the past 30 years, notably Eugene Oetting, Dean Miller, Cheryl Normington, Richard Haase, Max Uhlemann, and Weston Morrill at Colorado State University, who were there at the inception of the microtraining framework. The following people have been especially important personally and professionally in the growth of microcounseling and microtraining over the years: Bertil Bratt, Norma Gluckstern, Jeanne Phillips, John Moreland, Jerry Authier, David Evans, Margaret Hearn, Lynn Simek-Morgan, Dwight Allen, Paul and Anne Pedersen, Lanette Shizuru, Steve Rollin, Bruce Oldershaw, Oscar Gonçalves, Koji Tamase, and Elizabeth and Thad Robey.

The board of directors of the National Institute of Multicultural Competence— Michael D'Andrea, Judy Daniels, Don C. Locke, Beverly O'Bryant, Thomas Parham, and Derald Wing Sue—are now part of our family. Their support and guidance have become central to our lives. Many of our students at the University of South Florida, Tampa, University of Massachusetts, the University of Hawai'i, Manoa, and Flinders University, South Australia, also contributed in important ways through their reactions, questions, and suggestions.

We are grateful to the following reviewers for their valuable suggestions and comments: Markus P. Bidell, University of New Mexico; Seth Brown, University of Northern Iowa; Matthew R. Draper, Indiana State University; Lonnie Duncan, Western Michigan University; Jerold Gold, Adelphi University; Bonnie Hinkle, Dallas Baptist University; Cheryl McGill, Florence Darlington Technical College; Gary M. Miller, University of South Carolina-Columbia; Nathalie D. Mizelle, San Francisco State University; Julieta Monteiro-Leitner, Southeast Missouri State University; Lorri L. Petties, Valparaiso University; Lesley D. Riley (Student), University of South Carolina-Columbia; Brent Taylor, San Diego State University; and Douglas M. Thorpe, Marymount University.

Machiko Fukuhara, Professor Emeritus, Tokiwa University, and president of the Japanese Association of Microcounseling, has been our friend, colleague, and co-author for many years. Her understanding and guidance have contributed in many direct ways to the clarity of our concepts and to our understanding of multicultural issues. We give special thanks and recognition to this wise partner.

Finally, it is always a pleasure to work with the group at Brooks/Cole, notably Lisa Gebo, Alma Dea Michelena, Sheila Walsh, and their associates. Our manuscript editor, Patterson Lamb, has become an important adviser to us. Anne and Greg Draus of Scratchgravel Publishing Services always do a terrific job. We thank all of the above.

We would be happy to hear from readers with your suggestions and ideas. Please use the form at the back of this book to send us your comments. Feel free to contact us also via e-mail. We appreciate the time that you as a reader are willing to spend with us.

Allen E. Ivey
Mary Bradford Ivey
E-mail: info@emicrotraining.com

Intentional Interviewing
and Counseling

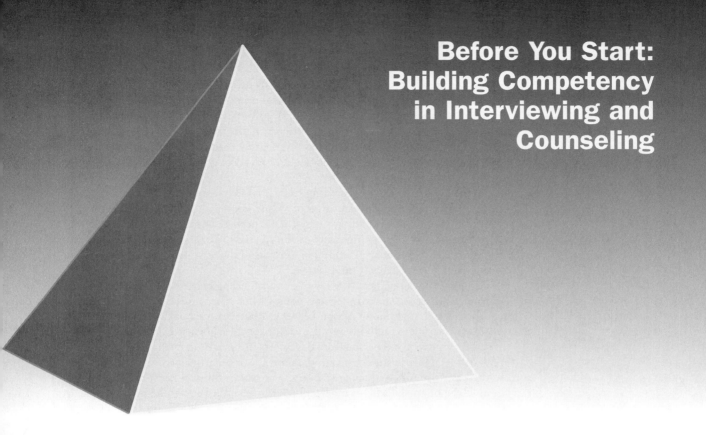

WELCOME!

Allen: My first courses in counseling were fascinating. I liked the theoretical ideas and the information about testing and careers, but what I enjoyed most was the course on theories of counseling. To me, this was the foundation of the whole process.

But then came the second semester and my first real opportunity to practice what I had learned in my field internship. I found myself overwhelmed by the amount of information shared by clients making complicated decisions or facing difficult issues. The theories in the books I had read did not easily apply to real people. How was I to survive and help? Somehow, I made it through, but I know I could have done a better job with those early sessions if I had been more skilled before I started.

Intentional Interviewing and Counseling is designed to teach you specific skills that you can use immediately in the session. The book seeks to "demystify" the art of helping. As you move through your interviews in this book, you will find that each step of the microskills hierarchy presents the specifics of counseling in clear and usable form. Whereas I learned from a "guess and try" framework, the concepts in this book will enable you to enter the reality of counseling with understanding and expertise. You'll encounter many practice exercises that allow you to test out your understanding and competence. Later in this book, you will discover that you can apply these skills with multiple theories of helping.

Mary: I arrived at the University of Wisconsin shortly after Carl Rogers had left, but his influence remained. Rogers came back several times and shared his ideas and his being with us. His impact on all of us was profound. His person-centered theory emphasized positive regard and the ability of clients to solve their own problems. He described our role as counselors as one that focused on listening and reflecting feelings. Asking questions or influencing a client through directives was something we should not do.

My first experience in counseling found me focusing on a reflective approach. While listening seemed critical, it often wasn't enough, especially with less verbal clients. Over time, I learned that many clients needed a more active stance. Gradually, I learned the influencing skills of the microskills hierarchy and when it was appropriate to be reflective or leading. I also found that clients could still make their own decisions, even if I was more active.

Moreover, I was lucky to work with an outstanding behavioral psychologist, Ray Hosford, who passed away much too young from ALS, Lou Gehrig's disease. Ray helped me see that although person-centered Rogerian methods are critical foundations, there are many methods and theories that can help clients. He also helped me see that an interviewer's personal style is highly influential in client growth.

I still recognize listening as fundamental, particularly as manifested when we use the basic listening sequence to hear client stories fully and accurately. The precision of the microskills helps me be a better listener and also to be more flexible in using varying approaches to meet the needs of an extremely diverse set of clients.

Mary and Allen: Welcome! We are delighted to have you join us. We believe that there are multiple ways that we can help clients. Some of you who read this book will become committed to the person-centered approach; others will move toward the cognitive behavioral and psychodynamic orientations. We hope all of you will consider the newer multicultural counseling and therapy. All theories and methods have value, particularly if we match them to client needs. Our own orientation is developmental/integrative for we believe that are several routes toward the "truth" of effective interviewing.

One of your important tasks as a beginning professional is to develop your own system for integration of skills and theories. We suggest that you start immediately identifying your own natural style and positive strengths and then use these as a base throughout this book. Each of us has a natural gift that enables us to reach others and help them achieve their goals. We hope that you will take what we present here and then shape the material to fit your natural style and the needs of those whom you would serve.

WHAT DOES THIS BOOK OFFER FOR YOUR DEVELOPMENT?

Throughout this book, you will be examining intentional interviewing and counseling—an interviewing approach that is concerned with flexibility and competence. You will learn specific skills that will enable you to help others find new ways to understand their thoughts, feelings, and behavior. In addition, you will learn how to facilitate their understanding of the *meaning* of what happens to them and their vision of deeper lifetime goals.

There are many concepts, ideas, and skills presented in this book, but there are some important general goals that you can expect to achieve. Through step-by-step study and practice you will encounter and master specific interviewing skills that will enable you to achieve the following:

▲ Engage in the basic skills of the interview: listening, influencing, and structuring an effective session with individual and multicultural sensitivity.

▲ Conduct a full interview using only listening skills.

▲ Analyze with considerable precision your own natural style of helping and its impact on clients.

▲ Master a basic structure of the interview that can be applied to many different theories. Specifically, you may expect to be able to conduct a decision-making interview, a person-centered interview, a solution-oriented brief counseling session, a behavioral assertiveness training session, and an interview using motivational interviewing. Included in this is active treatment planning.

▲ Integrate positive wellness, ethics, and multicultural issues into your practice.

▲ Generate your own story of the practice of interviewing and counseling by constructing your personal theory about the helping process. As you encounter client uniqueness and cultural complexity, anticipate that your story and theory will be one of constant change, growth, and development.

DEVELOPING COMPETENCE IN THE INTERVIEW

The intentional interviewer, counselor, or therapist is a person with many options for helping others. You, as a developing professional, are also responsible to be competent in interviewing skills and strategies so that you can best serve diverse populations of clients. Self-understanding and practice are basic to this goal.

Build on Your Natural Style of Helping

People enter the counseling, psychology, social work, and other human service fields intending to help others. Perhaps you, like most who enter our field, have been told that you are a good listener. You may have had friends or family come to you for advice or simply to tell you their problems. This is not counseling, but it is helping. You enter this book with some degree of social skills and natural abilities to assist others.

Natural style is defined here as your spontaneous way of working with others to help them achieve their goals. As you work through this book, look at yourself and respect your own natural competence as the foundation for your growth in interviewing skills. However, your natural style may not "work" with everyone—it may be necessary to shift your style to be fully effective. The effective interviewer gradually develops a blend of natural style with learned competencies.

As you begin work with *Intentional Interviewing and Counseling*, it is vital that you maintain a focus on your own abilities as a foundation for growth and further development. Competency Practice Exercise 1 presents an exercise in self-understanding that you may wish to complete before moving further in your reading. *Let us start with your natural expertise.*

Competency Practice Exercise 1:
What Are the Strengths of Your Natural Style?

A good place to start in interviewing and counseling is awareness of yourself as a person of capability. List below specifics in which you are a person of competence. This is your foundation for growth.

When have you helped someone else? Be as specific as possible.

_____

_____

What, specifically, did you do that was helpful?

_____

_____

_____

Be honest—what strengths do you think you bring to a course of study in helping and interviewing skills? (Saying good things about oneself is not always easy, but do it!)

_____

_____

_____

_____

Ask a friend or family member to identify qualities and skills that you naturally have that might make you an effective helper. Record what they say below:

_____

_____

_____

_____

Once our strengths are identified, it is much easier to face up to our personal challenges and limitations. Your base of positive skills and qualities will enable you to develop further as an interviewer or counselor.

Self-Understanding and Emotional Intelligence

Self-understanding and emotional intelligence are essential to interviewing competence and to enhancing your natural style. Knowing oneself is basic to the artistic and skilled interviewer. Self-understanding is the broad concept of knowledge about oneself. Closely related to self-understanding is emotional intelligence. Interviewing and counseling, themselves, can be described as exercises in emotional intelligence. Emotional intelligence was first defined by Peter Salovey and John D. Mayer in 1990 and was first brought to wide attention by Daniel Goleman (1998, 2005).

Several domains of emotional intelligence have been defined. Below you will see how the skills and strategies of this book relate to self-understanding and self-development (Goleman, Boyatzis, & McKee, 2002):

1. *Self-awareness.* Throughout this book you will have continual opportunities to examine yourself and your work with others in the interview. You will have the opportunity to hear your communication style on audiotape and, most likely, you will see yourself in action on videotape as well. What are your strengths and limitations? Do you have a positive image of yourself? Unless you feel good about yourself, you may have struggles helping clients.
2. *Self-regulation.* When an interview is challenging, can you handle your feelings and avoid allowing your "buttons to be pushed" too easily? Do you have self-control, flexibility, and the ability to generate new ideas on the spot? The skills and strategies of this book are designed to provide you with an array of possibilities for dealing with those challenging situations.
3. *Motivating yourself and using your abilities.* Each chapter in this book contains practice exercises designed to help you master interviewing skills and concepts, listen actively to others, and learn how to deal with challenging situations. Understanding a concept is not mastery. The emotionally competent interviewer or counselor is motivated for peak performance. Are you persistent in achieving your goals or do you give up easily?
4. *Empathy.* Empathic understanding and listening skills are intimately intertwined. Are you interested in others, can you be empathic to their concerns, can you pick up small signs of the many emotions your clients will have? Do have an understanding of people different from yourself and can you see their perspective?
5. *Social skills.* Relationship is central to the helping process. How effective have you been in working with others? How competent will you be in establishing rapport, trust, and confidence in the helping interview? Can you listen? Can you influence others? And you will also find that your clients can benefit from training in social skills as part of the counseling process.

Throughout this book, and in the accompanying CD-ROM, we will provide many exercises and skill practice ideas that can serve as avenues for self-exploration and a clearer definition of your own competencies. Empathy and social skills, of course, are central to this book and the interviewing and counseling professions. Self-awareness, self-regulation, and motivation may be described as qualities that are vital for effective work in the helping field.

Practice Leads to Mastery and Competence

There is no end to your development as an interviewer. All of us can get better, no matter how good we are (or think we are). The many skills and concepts of this book will be mastered to full competence only if you work actively with them. Role-plays in class or workshops followed by audiotape or videotape practice will help you develop expertise and mastery of skills. Practice will also be vital in your development of personal self-understanding and emotional intelligence.

And when you practice, seek feedback from as many others as possible on your performance. Learning how others see you and your work is central to professional and personal growth. Feedback has been called the "breakfast of champions."

You can "practice" the skills by going through the skill once and saying to yourself, "That was easy." Or you can really practice by aiming to see if you can get specific and predictable results in the session as a result of your skills. Many students find that several practice sessions are useful with as much feedback as possible. For example, you can practice the skill in a situation in which a friend tells you a positive story that he or she is eager to share. More challenging practice follow-ups might include role-plays in which you work with a variety of difficult issues that you might face—clients who are less verbal, people who are hostile and aggressive, and those with complex concerns.

If you are motivated, you will find that several alternative practice sessions increase your skills, confidence, and competence. Practice will determine where you stand in terms of your abilities, skills, and expertise. Four levels of competence are identified in each chapter in this book:

▲ *Level 1: Identification and classification.* Elementary competence and mastery occurs when you can identify and classify interviewing behavior. You can observe others' behavior on audiotape or videotape and know what they are doing. A quiz or an examination measures your ability to understand.

▲ *Level 2: Basic competence.* This involves being able to perform the skills in an interview, most often a practice role-play. You may, for example, demonstrate in an audiotaped or videotaped session that you can use both open and closed questions—even though you may not use the skills at a high level.

▲ *Level 3: Intentional competence.* You will find that you can use a skill with predictable results. For example, effective use of attending behavior skills increases client talk-time, while the lack of them reduces client conversation. Intentional competence means that you can help clients talk about their issues in specific ways and you can even predict what clients will say if you use a certain skill. But clients are real people and not always predictable. If you act intentionally and the predicted result does not occur, you can move to an alternative skill or strategy which may facilitate client growth in a different way.

▲ *Level 4: Teaching competence.* One way for you to acquire greater mastery is to teach the skill to someone else. Others can profit from your knowledge about interviewing—paraprofessional community volunteers, firefighters and police, or teens in a church, synagogue, or mosque. You will also find that your clients can benefit *in the interview* through direct teaching of skills. The communication skills emphasized in this book are also basic social skills, useful in daily interactions. The microskills are counseling and therapeutic strategies in themselves, useful in enhancing client self-understanding and efficacy.

Vital to effectiveness as an interviewer is a clear sense of how ethical and multicultural understanding relate to daily practice. In addition, ethical and multicultural competence can be considered vital aspects of self-understanding and emotional intelligence.

A FINAL WORD

This book is based on years of clinical experience and research. Over 450 data-based studies affirm the validity of the microskills approach; thus, the approach is the most-researched model of interviewing and counselor training available. Microskills have demonstrated effectiveness in counselor and therapist training. They have been used throughout the world in a multitude of training projects for professionals, paraprofessionals, and community volunteers.

But what can it do for you? Your special challenge is to maintain a focus on yourself and your own natural skills and abilities. We hope that you will study each idea and determine whether it feels comfortable to you, appears correct, and will be useful to you and your clients. There are many ways to be an effective interviewer, counselor, or therapist. We know that what is here "works," but we also believe that it will be valuable only if it fits with your natural way of being. We do suggest that even if a concept doesn't feel right that you try it and see if it fits, as you would try on a new jacket. What seems awkward at first may be a favored method later in your interviewing practice.

The many practice exercises are designed to remind us all that counseling is an art as well as a science. While we can supply ideas and suggestions, it is you who will apply these concepts to the canvas of real people's lives. As you learn skills, you may also find it helpful to think of yourself as an artist who is going to put things together in new ways. You will paint the pictures with your client—not us!

To provide a baseline for examining your natural style, we suggest that you audio- or videorecord an interview as soon as possible. An early recording will help you identify what you are doing and it can be compared with your later work as you progress through the book.

We hope that this book helps you identify and put into practice your own definition of how you want to help others. Your attitude of caring and respect toward yourself and your clients will always be fundamental in successful helping, counseling, and psychotherapy. Recognize and respect your natural uniqueness, thus making it easier to recognize and respect the diverse clients you will meet.

Over the years, we have learned much from student comments and suggestions. Please feel free to contact us with your issues and questions. We would welcome your ideas.

Allen E. Ivey
Mary Bradford Ivey
Info@emicrotraining.com
www.emicrotraining.com

REFERENCES

Goleman, D. (1998). *Working with emotional intelligence.* New York: Bantam.

Goleman, D. (2005). *Emotional intelligence.* New York: Bantam.

Goleman, D., Boyatzis, R., & McKee, A. (2002). *Primal leadership: Learning to lead with emotional intelligence.* Boston: Harvard Business School Press.

Salovey, P., & Mayer, J. (1990). Emotional intelligence. *Imagination, Cognition, and Personality, 9,* 185–211.

SECTION I

Introduction

Listening is the foundation of counseling and interviewing. In the session, we seek to enable clients to tell their stories. Through this narrative exploration, it then becomes possible to help clients rewrite and act on their stories in new ways. Our task is to expand client possibilities for intentional response and action.

You can make a difference in the lives of others. This difference can be for the better . . . or for the worse. Through the individual interview you can enrich your clients or hinder their growth. The first section of this book is oriented to joining clients where they are, hearing them narrate their stories, and reflecting with them on what they have learned. Often listening is enough, but when more is needed, later chapters will provide you with ideas to supplement the listening foundation.

Chapter 1 asks you to start the process of defining yourself as an interviewer and counselor. We suggest that you audio- or videotape an interview with a volunteer client as soon as possible. This taped interview can serve as a baseline in which you demonstrate your natural style of helping. Ask your client for feedback and to discuss the session with you. Pay special attention to what you have done right while also listening to what the volunteer client feels that you might improve. A wellness approach demands that you search for strengths and then use these as a foundation for problem solving and further development.

Cultural intentionality is a major concept underlying this book. A simple but useful summary of this concept in personal terms would be the following:

Cultural intentionality is demonstrated when you are flexible and have many possible responses to any client issue. Though you may have an excellent natural personal style, you are able to build on it by expanding your alternatives for response.

In contrast, lack of intentionality may be manifested by inflexibility, which may result in your using only one or two ways to help clients. These styles of helping may be generally effective, but at times you may find yourself having difficulty in reaching certain people.

Cultural intentionality requires you to engage in constant growth and change, to learn new skills and strategies, and to develop flexibility with ever-changing clients who may come from widely varying multicultural groups. It is also part of the larger concept of multiculturalism discussed in Chapter 2.

Chapter 2, *Ethics, Multicultural Competence, and Wellness,* discusses the foundation of the microskills hierarchy presented on the inside cover. Professional ethical standards are central to the helping professions of counseling, family therapy, human services, psychology, and social work. Chapter 2 presents key aspects of ethical standards important for interviewing and refers you to web sites where the complete standards are available.

We are all multicultural beings as each of us, regardless of our background, comes from an array of multicultural contexts. These cultural contexts may include many dimensions of diversity including, but not restricted to, age, ethnicity/race, gender, geographical location or community, language, sexual orientation, spiritual/religious beliefs, socioeconomic situation, physical ability, and experience with traumatic situations. In this sense, all interviewing and counseling is cross-cultural as each of us comes from a unique cultural background. The terms *diversity, cross-cultural,* and *multicultural* refer to all of these dimensions and will be used interchangeably.

Wellness and the positive psychology movement focus on the strengths that clients bring to the interview. We solve our problems and work through life challenges on what we *can do,* rather than on what we can't do. Yes, we need to listen carefully to client stories and understand their difficulties, but an important part of the counseling process is helping the client search out and understand strengths and resources. Counseling and therapy, by their very nature, are optimistic professions believing that people can change and be fully involved in their own growth.

Chapter 3, *Attending Behavior: Basic to Communication,* presents the first and most basic listening skill of interviewing, counseling, and psychotherapy. We must attend or listen to the client if we are to help. Many beginning helpers inappropriately strive to solve the client's problems in the first 5 minutes of the interview by giving premature advice and suggestions. You may want to set one early goal for yourself as you start working with this book: *Allow your clients to talk.* It may have taken your clients several years to develop the problems before they consult you. It is vital that you listen to them carefully and hear their story *before* you undertake problem solving.

Later portions of this book emphasize action skills of helping such as confrontation, interpretation, and directives. However, virtually all who work in interviewing, counseling, and psychotherapy consider the ability to listen to and enter into the world of the client to be the most important part of effective helping.

We suggest that you start this book with a commitment to yourself and your own natural communication expertise. We recommend that you intentionally enhance your natural style and expand your alternatives for working with clients. Wellness, ethics, and multicultural understanding will serve as guides to enrich and make our natural style more professional. Finally, please pay special attention to the skill of attending behavior. It is like the air we breathe and we can easily miss its importance. When in doubt about what to do in an interview, listen . . . and then listen again!

CHAPTER 1

Toward Intentional Interviewing and Counseling

How can intentional interviewing and counseling help you and your clients?

Major objective	This chapter is designed to identify five key ideas of the microskills approach and show how the step-by-step model of the microskills hierarchy relates to broad concepts of interviewing, counseling, and psychotherapy. Intentional interviewing is designed to facilitate the drawing out of client stories, enabling clients to find new ways of thinking about these stories and finding new ways of acting. It is important that the interviewer have multiple ways of responding to clients in a culturally sensitive fashion with awareness of contextual issues.
Specific objectives	Knowledge and skill in the concepts of this chapter result in the following:

▲ Awareness of development as the aim of interviewing, counseling, and psychotherapy and the story—positive assets—restory—action model.

▲ Ability to define intentionality, cultural intentionality, and intentional competence.

▲ Awareness of the microskills hierarchy and its step-by-step model that encourages you to define your personal style and theory of counseling.

▲ Understanding how important theories of counseling and psychotherapy have varying patterns of microskill usage.

▲ Encouraging you to examine your own natural helping style immediately and use your own skills as a base as you work through this text.

Imagine that you are the interviewer, counselor, or therapist and a new client comes in. She immediately starts talking rapidly with a list of multiple issues. What might you say or do next that could be helpful?

Client: I'm overwhelmed. My husband was let go in the latest downsizing and is impossible to live with. My job is going OK, but I worry about making the next car payment. Our ancient washer broke, flooded our basement, and ruined a box of family photographs. Our daughter came home crying because the kids are teasing her and my mother-in-law is coming to visit next week. What should I do?

How would you respond? What would you say? Write it here.

_____

_____

_____

_____

_____

Compare your response with those of other people. You'll probably find that their responses are different from yours. A key question is *who made the "correct" response?*

The answer, of course, is that there are many possibly useful responses in any interviewing situation. Reflecting the client's emotions can be helpful ("You're swamped and overwhelmed with all that's happening"). Selecting one aspect to focus on can be useful, and then later you can examine other dimensions by asking an open question ("What, specifically, is happening between you and your husband right now?" or "Could you share a bit more about your financial situation?"). You might even say, "I hear the anxiety and tension in your voice. . . . Let's slow down, take a deep breath, and start from the beginning."

Among many other possibilities, you could direct attention to the job loss, the daughter's school problem, or the mother-in-law's forthcoming visit. You might even choose to sit silently and see what happens next.

DEVELOPMENT AS THE AIM OF INTENTIONAL INTERVIEWING: DRAWING OUT CLIENT STORIES

Interviewing and counseling are concerned with client stories. You will hear many different story lines—for example, procrastination and the inability to take action, tales of depression and abuse, and most important, narratives of strength and courage. Your first task is to listen carefully to these stories and learn how clients come to think, feel, and act as they do. On occasion, simply listening carefully and empathetically is enough to produce meaningful change.

You will also want to help clients think through new ways of approaching their stories. Through the conversation that is interviewing and counseling, it is possible to rewrite and rethink old stories in new, more positive and proactive narratives. The result can be deeper awareness of emotional experience, more useful ways of thinking, and new behavioral actions.

Development is the aim of counseling and interviewing. Expect your clients to have enormous capacity for growth. In the middle of negative and deeply troubling stories, one of your tasks is to search for strengths, positives, and power in the client. If you can develop exemplary models from the client's past and present, you are well on the way to combating even the most difficult client situation or story.

One brief example: Imagine that an 8-year-old child comes to you in tears, having been teased by friends. You listen and draw out the story. Through your warmth and caring interest, the child calms down. The child has strengths, and you point out some of them. You may talk about the wisdom of coming to talk about problems with a counselor who cares. You comment on a concrete example of the child's strengths such as verbal or physical ability, a time you noticed the child helping someone else, or perhaps family members who support the child. You may read a short book or tell a story that metaphorically illustrates that problems can be overcome through internal strength.

As a result of listening, developing positive assets and strengths, and gaining new perspectives through storytelling, you and the child are rewriting the event and planning new narratives and action for the future. The basic treatment structure used with the child can be expanded for counseling with adolescents, adults, and families: Listening to the story, finding positive strengths in that story or another life dimension, and rewriting a new narrative for action are what interviewing and counseling are about. In short: *story—positive assets—restory—action.*

As you listen, be careful not to minimize the story. Behind the tears of the child may be a history of abuse or other more serious concerns. So as you listen to stories, simultaneously search for more complex, unsaid stories that may lie behind the initial narrative.

STORY—POSITIVE ASSETS—RESTORY—ACTION

Narrative theory is a relatively new model for understanding counseling and interviewing sessions (Semmier & Williams, 2000; White & Epston, 1990; Monk, Winslade, Crocket, & Epston, 1997). Narrative theory emphasizes storytelling and the generation of new meanings. The concepts of narration, storytelling, and conversation are useful frameworks as we examine skills, strategy, and theory in counseling and interviewing.

The narrative model of *Intentional Interviewing and Counseling* may be described as follows: First we need to hear client stories. We also need to listen for strengths and assets—empowerment, wellness, and positive psychology are an increasingly vital part of both interviewing and counseling (cf. Myers & Sweeney, 2005; Peterson & Seligman, 2004; Seligman, 2002). With an understanding of client issues and personal power, we have a base for change. Restorying is about developing client stories in new directions. The new story often makes action and change possible.

Story

The listening skills described in Section I are basic to learning how clients make sense of their world—the stories clients tell us about their lives, their problems, challenges, and issues. Let us help them tell their stories in their own way.

Positive Assets

It is not enough to listen; it is vital that positive strengths and assets be discovered as part of clients' stories. At times, counseling and interviewing can degenerate into a depressing repetition of negative stories and even whining and complaining. Unless you help develop respect for and empower your clients with a wellness approach, moving in positive directions becomes most difficult. Thus, throughout this book, by using the positive asset search and wellness research and theory (see the next chapter), we will be suggesting ways you can build on client strengths. You will find that a positive base often helps clients deal with more challenging and complex issues in the interview.

Also seek out and listen for stories in which clients describe times when they have succeeded in overcoming obstacles. Listen for and be "curious about their competencies—the heroic stories that reflect their part in surmounting obstacles, initiating action, and maintaining positive change. . . ." (Duncan, Miller, & Sparks, 2004, p. 53). You'll find that the listening and observational skills of Chapters 3 through 8 will help you identify positive stories to balance negative self-perceptions and eventually rewrite difficult stories in a more positive fashion.

Restory

If you understand client stories and strengths, you are prepared to help clients restory—generate new ways to talk about themselves. One important strategy for restorying is provided in Chapter 8, in which you will demonstrate your ability to conduct a full interview using only listening skills. Many times, effective listening to their stories is sufficient to provide clients with the strength and power to develop their own new narratives.

Focusing, confrontation, and the influencing skills presented in Section II are important parts of helping clients generate new stories about their experience. The five-stage interview with its many adaptations can be used to enable clients to find new ways of making meaning.

Awareness of theoretical diversity is also central. Counseling theory provides us with alternative ways to think and talk about client stories. You will profit from learning the vitality and power in counseling theories as diverse as cognitive-behavioral, psychodynamic, existential-humanistic, multicultural counseling and therapy, and others. As you define your own being, you will want to remain open to the multitude of possibilities offered by the professional helping field.

Action

The influencing skills are also about helping clients bring new ways of thinking and being into action. Through the use of action-oriented directives, the skilled use of the interpretation/reframe, or assertiveness training, you may enable clients to take their new ideas into action. Unless something changes in terms of the client's thoughts, feelings, or behaviors outside the interview, we can assume that interviewing and counseling have not been effective.

If your work with this book is successful, you will have developed a solid understanding of foundation skills and strategies, an ability to conduct interviews from several perspectives, and—perhaps most important—you will have begun the process of writing your own narrative, your own personal theory of interviewing and counseling. We hope your personal construction of theory and practice will remain open to constant challenge and growth from your clients and from your professional colleagues.

INTERVIEWING, COUNSELING, AND PSYCHOTHERAPY

The terms *counseling* and *interviewing* are often used interchangeably in this book. Though the overlap is considerable, interviewing may be considered the most basic process used for information gathering, problem solving, and advice giving. Interviewing is usually short term with only one or two sessions. A human service worker may interview a client about financial needs and planning. Managers interview potential employees, and college admissions staff interview students applying for admission. After a major disaster (terrorist bombing, hurricane, or flood) a crisis worker may interview a family about their needs and plans for recovery and then give them advice about what they can do to meet tomorrow's needs.

Counseling is a more intensive and personal process. It is generally concerned with helping people cope with normal problems and opportunities, although these "normal problems" often become quite complex. Though many people who interview may also counsel, counseling is most often associated with the professional fields of social work, guidance, psychology, pastoral counseling, and, to a limited extent, psychiatry. Clients with financial difficulties may need several sessions of counseling to straighten out their situation. The employee or college student facing challenges often needs help in understanding issues and making decisions. After the early interview, the family experiencing a major disaster may need crisis counseling over a longer period of time.

We can clarify the overlapping differences and similarities of interviewing and counseling with some examples. A personnel manager may interview a candidate for a job but in the next hour counsel an employee who is deciding whether to take a new post in a distant town. A school counselor may interview each class member for 10 minutes during a term to check on course selection but also may counsel some of them about personal concerns. A psychologist may interview a person to obtain research data but in the next hour be found counseling a client concerned about an impending divorce. Even in the course of a single contact, a social worker may interview a client to obtain financial data and then move on to counseling about personal relationships.

Many people who interview will find themselves counseling on occasions. Most counselors at some time find themselves interviewing. Both interviewing and counseling may be distinguished from psychotherapy, which is a more intense process, focusing on deep-seated personality or behavioral difficulties. The skills and concepts of intentional interviewing are equally important for the successful conduct of longer-term psychotherapy. Moreover, with the advent of managed care and brief therapy, many psychotherapists will often engage in counseling.

Despite relatively clear differences among interviewing, counseling, and psychotherapy, overlap remains (see Figure 1-1). Effective interviewing at times can help a person make important decisions, and that is therapeutic. And psychotherapists, of

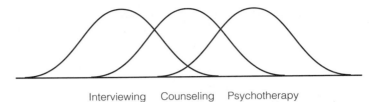

Interviewing   Counseling   Psychotherapy

Figure 1-1   The interrelationship of interviewing, counseling, and psychology

necessity, must interview clients to obtain basic facts and information and will also counsel them on many of life's normal concerns.

INCREASING SKILL AND FLEXIBILITY: INTENTIONALITY AND CULTURAL INTENTIONALITY

There are many ways to facilitate client development. In becoming increasingly competent, you will learn to blend what is natural for you with new interviewing skills and theory. Intentionality asks you to be yourself but also to realize that if you are to reach a wide variety of clients, you will need to be flexible and constantly learn new ways of being in the interview.

Intentionality

Many possibilities could be used with any client, but we urge you to listen first and do problem solving later. Clients come to us with multiple issues and concerns. How you listen and how you respond may say as much about you and your style as it says about the person you are trying to help. One of the goals of this book is to encourage you to look at yourself and your style of listening.

Beginning interviewers are often eager to find the "right" answer for the client. In fact, they are so eager that they often give quick patch-up advice that is inappropriate. For example, it is possible that cultural factors such as ethnicity, race, gender, lifestyle, or religious orientation may have determined your response and interview plan for the client. How ideal it would be to find the perfect empathic response that would unlock the door to the client's world and free the individual for more creative living! However, the tendencies to search for a single "right" response and to move too quickly can be damaging.

Intentional interviewing is concerned not with which single response is correct, but with how many potential responses may be helpful. Intentionality is a core goal of effective interviewing. We can define it as follows:

Intentionality is acting with a sense of capability and deciding from among a range of alternative actions. The intentional individual has more than one action, thought, or behavior to choose from in responding to changing life situations. The intentional individual can generate alternatives in a given situation and approach a problem from different vantage points, using a variety of skills and personal qualities, *adapting styles to suit different individuals and cultures.*

The culturally intentional interviewer remembers a basic rule of helping: If something you try doesn't work, don't try more of the same—try something different!

Cultural Intentionality

One of the critical issues in interviewing is that the same skills may have different effects on people from varying cultural backgrounds. Intentional interviewing requires awareness that racial and ethnic groups may have different patterns of communication. Eye contact patterns differ. For example, in U.S. culture, middle-class patterns call for rather direct eye contact, but in some cultural groups direct eye contact is considered rude and intrusive. Some find the rapid-fire questioning techniques of many North Americans offensive. Spanish-speaking groups have more varied vocal tones and sometimes a more rapid speech rate than do English-speaking people.

It is also important to remember that the word *culture* can be defined in many ways. Religion, class, racial/ethnic background (for example, Irish American and African American), gender, and lifestyle differences, as well as the degree of a client's developmental or physical disability, also represent cultural differences. There is also a youth culture, a culture of those facing imminent death through AIDS or cancer, and a culture of the aging. In effect, any group that differs from the "mainstream" of society can be considered a subculture. All of us at times are thus part of many cultures that require a unique awareness of the group experience. In fact, many people suggest that *diversity* is what constitutes the mainstream. This may suggest that respecting and honoring our differences is what will bring us together as one people.

However, avoid stereotyping. Individuals differ as much as or more than cultures. You will want to attune your responses to the unique human being before you. Lack of intentionality shows in the interview when the helper persists in using only one skill, one definition of the problem, or one theory of interviewing, even when that approach isn't working.

For an overview of key issues in counseling and therapy from a multicultural perspective, the following are suggested: Ponterotto, Casas, Suzuki, and Alexander (2001); Ridley (2005); Ivey, D'Andrea, Ivey, and Simek-Morgan (2002); and Sue and Sue (2002). These, in turn, will lead you to many other references.

Intentional Competence

Closely allied to intentionality is *intentional competence,* which is associated with predictability and the anticipated consequences of what you say in the session. Intentional competence adds another dimension to basic microskills. If you use a specific skill, you may anticipate a specific result. For example, consider the following:

If you ask an open question, the anticipated result is that your client will talk in more depth about the topic. If the question is closed, the client is expected to give a short, directed answer.

If you reflect feelings, expect clients to acknowledge and deal with their feelings more fully.

If you confront client conflict, the consequence is likely to be that the client will seek to resolve these issues.

Even though you can make intentional predictions about what will happen as a result of your skills, it is perhaps more important to remember that clients may vary in their responses to the same skills. You may ask an open question and find that some clients simply do not want to answer completely. You may reflect feelings but learn that clients may not be ready to explore emotions. They may resist your confrontation. In short, the intentional prediction may not always "work."

Intentionality, as discussed earlier, stresses the importance of having more than one action, thought, or behavior to choose from in changing life situations. True intentional competence demands more than predictability; it also requires you to flex, change direction and skills, and be with your client in new ways. Your own natural style, self-understanding, and artistic abilities will be important in the ever-changing world of the interview.

THE MICROSKILLS HIERARCHY

Microskills are communication skill units of the interview that will help you interact more intentionally with a client. They will provide specific alternatives for you to use with different types of clients. Microskills form the foundation of intentional interviewing.

The microskills hierarchy (see Figure 1-2) summarizes the successive steps of intentional interviewing. The skills of the interview rest on a base of ethics, multicultural competence, and wellness. On this foundation lies the first microskill—culturally and individually appropriate *attending behavior,* which includes patterns of eye contact, body language, vocal qualities, and verbal tracking. Attending behavior is the first microskill unit discussed in this text. In your study and practice sessions you will have the opportunity to define this skill further, see attending demonstrated in an interview, read about further implications of this skill, and finally, practice attending yourself.

Once you have mastered attending behavior, you will move up the microskills pyramid to learn about questioning, client observation, paraphrasing, and other basic listening skills. Effective interviewing and eventual skill integration require listening to and understanding the client before you engage in problem solving. You will use the basic listening sequence again and again.

Higher is not necessarily better in this hierarchy. Unless you have developed skills of listening and respect, the upper reaches of the pyramid are meaningless. The foundational skills of attending behavior, client observation, questioning, and reflective listening are critical parts of the practice of even the most experienced professional. Develop your own style of interviewing and counseling, but always with respect for this grounding.

With a solid background in these central skills, you will then learn how to structure a "well-formed interview." You will be able to conduct a complete interview with a verbal client using only listening skills. You will find that the structure of the interview presented here can be adapted and shaped to help you understand and master several alternative methods and theories of counseling.

You will then encounter action skills of interviewing and counseling. The first of these is confrontation, considered basic to client growth and change. The

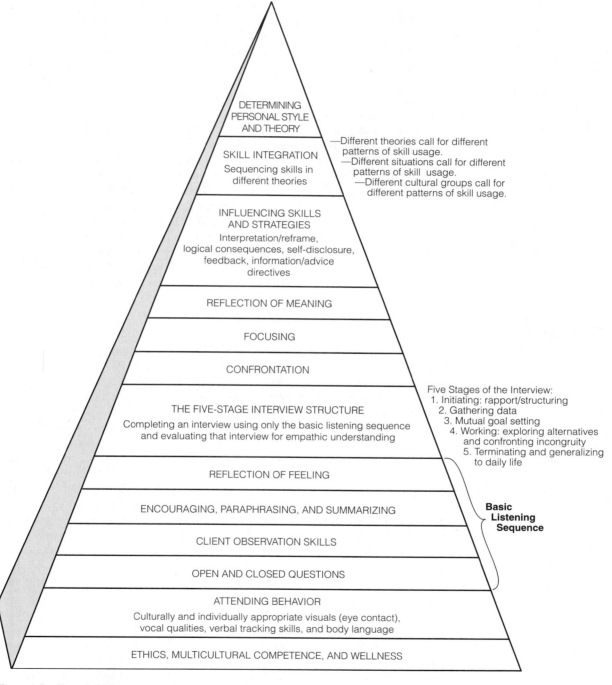

Figure 1-2 The microskills hierarchy: A pyramid for building cultural intentionality. (Copyright © 1982, 1987, 2003, 2007 Allen E. Ivey. Reprinted by permission.)

microskills of focusing, reflection of meaning, and interpersonal influence come next in the hierarchy. In Chapter 12, you will have the opportunity to study and master directives (helping a client to engage in specific actions), interpretation (providing the client with an alternative frame of reference for viewing a concern), and other skills, such as giving feedback and pointing out logical consequences of possible actions.

With a mastery of listening, the ability to conduct an interview using only listening skills, and a command of the advanced skills, you are prepared to consider alternative theories and modes of helping. You will find that these microskills can be organized into different patterns utilized by different theories. For example, if you have mastered the listening skills and the structure of the interview, you have an important beginning in learning Rogerian person-centered theory. Later, as you move on to other theories of counseling and therapy, you will find that the basic skills of this book will be used as a foundation for mastering those as well.

But as you can see from a glance at the apex of the microskills pyramid (determining personal style and theory), it isn't enough just to master skills and theories. You will eventually have to determine your own theory and practice of counseling and interviewing. Interviewers, counselors, and therapists are an independent lot; the vast majority of helpers prefer to develop their own styles, and through eclecticism move toward their own blend of skills and theories.

Theory and Microskills

Many students ask about the theory underlying the microskills approach. To this question there are three responses. The first is that interviewing and counseling are informed by more than 250 theories and that certainly we don't need another one. Thus, as you begin working with this book, we'd prefer that you focus on skills and not emphasize theoretical implications at this time.

At the second level of response, your expertise with the basic skills of microcounseling can be very helpful to you in understanding and practicing multiple theoretical approaches. For example, if you become proficient in attending skills, basic listening skills, and influencing strategies, you have laid the groundwork for developing competence in many different theories ranging from person-centered to cognitive-behavioral to multicultural counseling and therapy. All theories require practitioners to listen to client stories. Influencing skills such as directives and the interpretation/reframe supplement the foundation of listening skills. Although each theory uses these skills, you will find that varying theories require different ways of using them.

Table 1-1 presents the microskill patterns of some key theoretical approaches. The first five listed (decisional, person-centered, cognitive behavioral assertiveness training, solution-oriented, and motivational interviewing) are discussed in Chapters 13 and 14. Other important theories are presented as examples for comparisons. You may note, for example, that virtually all theories give considerable attention to the listening skills. However, the influencing skills vary widely in their use. For example, the interpretation/reframe is important in brief solution-oriented, psychodynamic, and rational-emotive behavioral therapy while the use of this skill would be expected to be infrequent in person centered theory.

Table 1-1 Microskills patterns of differing approaches to the interview

MICROSKILL LEAD	Decisional counseling	Person centered	Cognitive behavioral assertiveness training	Solution oriented	Motivational interviewing	Psychodynamic	Gestalt	Rational-emotive behavioral therapy	Feminist therapy	Business problem solving	Medical diagnostic interview	Traditional teaching	Student-centered teaching	Eclectic/metatheoretical
BASIC LISTENING SKILLS														
Open question	●	○	◒	●	●	◒	●	◒	●	◒	◒	◒	◒	◒
Closed question	◒	○	●	◒	◒	○	◒	◒	◒	◒	◒	◒	●	◒
Encourager	●	◒	◒	◒	●	◒	◒	◒	●	◒	◒	○	◒	◒
Paraphrase	●	●	◒	◒	●	◒	◒	○	◒	◒	◒	○	◒	◒
Reflection of feeling	●	●	◒	◒	●	◒	○	○	◒	◒	◒	○	◒	◒
Summarization	◒	◒	◒	●	●	◒	○	◒	◒	◒	◒	○	●	◒
INFLUENCING SKILLS														
Reflection of meaning	◒	●	○	○	○	◒	◒	○	◒	○	○	○	◒	◒
Interpretation/reframe	◒	○	◒	○	●	●	●	◒	●	◒	◒	○	◒	◒
Logical consequences	◒	○	◒	○	●	○	○	◒	◒	●	◒	●	◒	◒
Self-disclosure	◒	◒	○	○	◒	○	◒	○	◒	○	○	○	◒	◒
Feedback	◒	◒	◒	◒	◒	◒	◒	◒	◒	●	○	◒	●	◒
Advice/information/ and others	◒	○	●	○	◒	○	○	◒	●	●	●	●	◒	◒
Directive	◒	○	●	○	◒	○	●	●	◒	●	●	●	◒	◒
CONFRONTATION (Combined Skill)	◒	◒	◒	◒	●	◒	●	●	●	◒	◒	◒	◒	◒
FOCUS														
Client	●	●	●	●	●	●	●	◒	◒	◒	●	●	●	◒
Main theme/problem	●	○	◒	●	●	◒	◒	◒	◒	●	●	●	●	◒
Others	◒	○	◒	◒	◒	◒	◒	◒	◒	○	○	○	◒	◒
Family	◒	○	◒	◒	◒	◒	○	◒	◒	○	○	○	◒	◒
Mutuality	○	◒	○	◒	◒	○	○	○	◒	○	○	○	◒	◒
Counselor/interviewer	○	◒	○	○	◒	○	◒	◒	●	○	○	○	◒	◒
Cultural/ environmental context	◒	○	◒	◒	○	○	○	○	●	◒	○	○	◒	◒
ISSUE OF MEANING (Topics, key words likely to be attended to and reinforced)	Problem solving	Relationship	Behavior problem solving	Problem solving	Change	Unconscious motivation	Here-and-now behavior	Irrational ideas/logic	Problem as a "women's issue"	Problem solving	Diagnosis of illness	Information/ facts	Student ideas/ info./facts	Varies
AMOUNT OF INTERVIEWER TALK-TIME	Medium	Low	High	Medium	Medium	Low	High	High	Medium	High	High	High	Medium	Varies

LEGEND

● Frequent use of skill
◒ Common use of skill
○ Occasional use of skill

The focus skill is another area where theories differ. Most theories focus their interventions and awareness on the individual client and the client's problem with minimal attention to other dimensions. Feminist and multicultural counseling and therapy, by way of contrast, give considerable attention to cultural/environmental/contextual dimensions as they tend to see societal issues as important as individual concerns.

Consequently, microskills is a foundational theory that can help you understand and work more effectively with many widely varying approaches to facilitating client growth and development. The microskills approach has long been focused on alternative perceptions and constructions of reality.

In a third theoretical view, microskills have been termed the technical skills of a constructivist-developmental theory (Kelly, 1955; Ivey, 2000). What does this mean? In short, the goal of a constructivist approach is to understand how clients experience and make sense of the world. What are their basic story lines and narratives? If you use the basic listening skills effectively, you will learn to enter the world of a very important person—the client whom you seek to help. Rather than seeing the world through your own glasses, you seek to understand the world through the client's perspective—how does that client see, feel, and hear the world? How is he or she constructing the experience, and what sense is the client making of that experience? What strengths can we find in the client that can be used to resolve these issues?

Many clients have developed ineffective constructions or stories about the world through their lifetime developmental history. When you listen carefully you can help clients understand their own way of thinking. They then may begin to change the way they think, feel, and behave in relation to themselves and others. If listening skills are not sufficient, focusing, confrontation, and influencing skills may be used to help clients restory and find new ways of being in the world.

Using microskills as a tool of constructivism and narrative is an approach through which you can gently enter the client's way of seeing the world. Jointly with the client you can search out new ways to envision and construct more effective and comfortable stories for thinking, feeling, and acting. So, from this third frame of reference—the constructivist-developmental approach—the microskills approach can be viewed as a helping theory in its own right.

In terms of multicultural issues, the microskills constructivist approach has considerable merit. Different cultural groups construct stories and make meaning in different ways. The microskills approach has been particularly effective in adaptations to various cultures and has been translated into many languages. It is important that cultural differences be recognized as part of the constructivist process. Thus far, traditional theories of helping tend to give little attention to multicultural or gender issues.

Once you have mastered the entire microskills hierarchy you will be able to use skills and strategies in a variety of settings and will find that you can master the complex theories of counseling more easily, particularly when it comes to applying these theories in the actual interview.

The Microskills Model of Interviewing and Counseling

The single skills microskills model was developed in 1966–1968 by a group at Colorado State University, and in 1974 multicultural differences in communication

Box 1-1   Research Evidence That You Can Use: Validating the Microskills Approach

More than 450 data-based microskills studies have been completed to date. The model has been tested nationally and internationally in over 1,000 clinical and teaching programs in the past 35 years (Daniels & Ivey, 2006; Daniels, Rigazio-DiGilio, & Ivey, 1997; Molen, Hommes, Smit, & Lang, 1995). Microcounseling was the first systematic video-based counseling model to identify specific observable skills (Ivey et al., 1968) and the first emphasizing multicultural issues (Ivey, Gluckstern, & Ivey, 1974/2006). The CD-ROM accompanying this book provides a detailed research review (Daniels, 2007).

Specific research implications and applications of the microskill model will be identified throughout the book. Included will be suggestions of ways you can apply research in your own work. Some of the most important findings relevant to your work with clients include the following:

1. *You can expect results.* Several critical reviews have been conducted and they have all found microtraining to be an effective framework for teaching skills to a wide variety of groups. There are also consistent data attesting to the effectiveness of teaching microskills to clients and patients. You will find that this book will increase your abilities and provide you with more alternatives for action. The skills of the microskills hierarchy have been shown again and again to be clear, useful in the interview, and teachable. You will find that the step-by-step model makes it possible for you to identify more clearly what you are already doing, how you are affecting clients, and how you can extend your abilities.
2. *Practice is essential.* Practicing the skills to mastery (intentional competence) is important if the skills are to be maintained and used after training. If you do not practice or use the skills and strategies of this book, they will disappear over time. Completing the practice exercises included and working actively to generalize this knowledge to daily practice are essential. Use it or lose it!
3. *Multicultural differences are real.* People from different multicultural groups (e.g., ethnicity/race, gender) have different patterns of skill usage. There is a lifelong need for all of us to learn about groups other than our own and adapt skill usage in a culturally appropriate manner without stereotyping any client. Each of us is unique.
4. *Different counseling theories have varying patterns of skill usage.* Expect person-centered counselors to focus almost exclusively on listening skills whereas cognitive-behaviorists use more influencing skills. Microskills expertise should help you define your own theory of choice and how it integrates with your natural style.
5. *If you use a specific microskill, then you can expect a client to respond in predictable ways.* A special value of this approach is that you know what *may* happen in the interview. But each client is different and predictability is not perfect. Cultural intentionality enables you to react to changes in the interview and prepares you for the unexpected.

styles were identified (Ivey, Normington, Miller, Morrill, & Haase, 1968; Ivey, Gluckstern, & Ivey, 1974/2006). The extensive research on the microskills model is summarized in Box 1-1. The CD-ROM accompanying this book provides a comprehensive research report on the model. Multiple approaches and adaptations of the original skills model are now available—see, for example, Brammer and MacDonald,

2002; Hackney and Cormier, 2004; Egan, 2002; and Hill, 2004. These and other authors have utilized the framework, although they give varying emphasis to multi-cultural issues. Carkhuff (2000) also utilizes the single skills model but focuses on qualitative issues—discussed in Chapter 8 on skill integration.

The microskills model is not linked to any theoretical approach but to the premise that the skills are useful in multiple theories and settings (Corey, 2005; Ivey, D'Andrea, Ivey, & Simek-Morgan, 2002). The model for learning microskills is practice oriented in a step-by-step progression and includes the following central dimensions:

1. *Orientation to the skill and how it may be useful in the session.* This is focusing on a single skill, a vital part of the holistic interview.
2. *Observation of the skill in action.* This may be done through reading a transcript of an interview, listening to an audiotape, viewing a video, or observing a live demonstration.
3. *Reading about the skill or hearing a lecture on the main points of effective usage.* Cognitive understanding is vital for skill maintenance.
4. *Practicing the skill in role-plays.* Ideally, skill practice utilizes video, but role-played practice with observers and feedback sheets is also effective.
5. *Self-assessment and plans for generalization.* How can you take this skill into your own daily work in the interview?

Note that the chapters of the book follow this five-step model. You are also encouraged to consider this model as a way of teaching your clients and small groups the skills of this book. The microskills are dimensions of emotional intelligence and social competence. Their use has proven to be an effective counseling and therapeutic technique in itself.

MICROSKILLS PRACTICE, SUPERVISION, AND LIFETIME GROWTH

The microcounseling framework was developed to clarify and ease the transition from the classroom to actual interviewing practice. Once we have learned the basic skills, we are well prepared for improving our counseling and therapy work throughout our professional lives. The interview is a place where we all can improve. We hope you will find that the openness and specificity of the microskill practice sessions will continue as you move on beyond this text to your next steps.

When you move to field placements and internships in schools and community agencies, supervision of your sessions will be central to your learning. Here you will share your work with professionals who can help and guide you toward further competence. In your microskill practice sessions, you actually will be engaging in minisupervision sessions. Your classmates will provide you with feedback, both on your strengths and areas where you might develop further. Microskill practice with "microsupervision" will provide a structure as you develop a habit of sharing your interviewing work with others in an open atmosphere.

Supervision and openness are so important that they should be "introduced in the first class students take and then be intertwined throughout the curriculum by all faculty" (Miller & Dollarhide, in press). Microskills provide a vocabulary and system through which you can identify what you are doing and its effectiveness. Microskills supervision provides a comprehensive framework that can be used in many models of supervision ranging from person-centered through multicultural and psychodynamic (Daniels, Rigazio-DiGilio, & Ivey, 1997; Russell-Chapin & Ivey, 2004).

BRAIN RESEARCH: IMPLICATIONS FOR THE INTERVIEW

Research on the brain over the past decade has reached a state of precision where it now has important implications for interviewers, counselors, and therapists. Neuropsychology can be defined as "the study of relations between brain function and behavior" (Kolb & Whishaw, 2003, p. G16). Perhaps the most important discovery is *neuroplasticity*—the brain develops new neural connections throughout the lifespan and changes in response to new situations or experiences in the environment. "Neuroplasticity can result in the wholesale remodeling of neural networks . . . a brain can rewire itself" (Schwartz & Begley, 2002, p. 16).

What does this mean for you and the helping process? When you interact with clients, both your and your client's brain functioning can be measured through a variety of brain imaging techniques, especially functional magnetic imaging (fMRI). An example of neuroplasticity is that both you and your client may learn, change, and develop new neural connections are a result of your interaction. Successful interviewing and counseling help clients develop new and useful connections. Neuropsychology's research on emotion validates past counseling and therapy theory and research from a new perspective (see especially the research boxes in Chapters 7 and 8).

Perhaps particularly important for you as you build counseling microskills is research cited by Restak (2003, p. 9). This research

> . . . carried out weekly fMRI's on volunteers while they learned a sequence of finger movements. Within three to four weeks of training, [the reseacher] could discern changes in activity patterns in three progressive parts of the brain: the prefrontal cortex, which is responsible for the intention to carry out movements; the supplementary motor cortex, responsible for organizing the sequencing and coordination of the muscles involved in carrying out the action; and the primary motor cortex that gives the "orders" for the movements to take place.

Systematic step-by-step learning, such as that emphasized in this book, is a highly effective learning method used in ballet, music, golf, and many other settings. If there is sufficient practice, new connections in the brain may be expected, and increased ability in demonstrating these skills will appear in areas ranging from finger movements to dance—and from the golf swing to interviewing skills.

However, this is not a book on neuropsychology and brain science. Our purpose here is to acquaint you with a few particularly interesting findings that relate to becoming a more intentional interviewer. While we, of necessity, will identify certain brain areas and their functions by name, please do not consider this book the place to learn brain science and/or brain functioning. A glossary of the brain with the key functions of some areas we discuss will be found in the Appendix.

We have placed some key information about neuropsychology within several of the research boxes that you will find in each chapter. As you go forward to your professional study after this book, you can expect brain research to be an increasingly important part of our field. Consider what you read here a beginning and an opening to a future in which the relationship between behavior and brain functioning will become increasingly clear.

YOUR NATURAL STYLE: AN IMPORTANT AUDIO OR VIDEO EXERCISE

At the beginning of this chapter, you were asked to give your own response to an interviewee experiencing multiple issues. Skills learned through microtraining provide you with additional alternatives for intentional responding to the client. However, these responses must be genuinely your own. If you adopt a response simply because it is recommended, it is likely to be ineffective for both you and your client. Not all parts of the microtraining framework are appropriate for everyone. You have a natural style of communicating, and it is that natural style these concepts should supplement. In effect, learn new skills and be yourself.

Coupled with your natural style will be awareness of, knowledge of, and skills with the clients with whom you work. How do they respond to your natural style? Your style will likely have to change as you encounter diversity among clients. You will find that you work more effectively with some clients than others. Due to her experience, a woman may feel uncomfortable with a male counselor or vice versa. Many clients lack trust with interviewers who come from a different race or ethnicity than their own. You may yourself be less comfortable with teenagers than you are with children or adults. In all these cases, you will need to expand your competence and add new methods and information to your natural style.

You are about to engage in a systematic study of the interviewing process. By the end of the book you will have experienced many ideas for analyzing your interviewing style and skill usage. Along the way, it is helpful to have a record of where you were before you started this training. Your present natural style is a baseline you will want to keep in touch with and honor. It is invaluable to identify who you are and what you do before beginning systematic training.

Before going further, audiorecord or videorecord yourself in a natural interview. Find someone who is willing to role-play a client with a concern, problem, opportunity, or issue. Interview that "client" for at least 15 minutes using your own natural communication style. We believe it is best to do this as soon as possible so that you can obtain an accurate picture of where you are as you start this course. You will want to compare your session with later work that you do as you progress through this text.

When you complete the interview, request that your client fill out the Client Feedback Form at the end of the chapter. In practice sessions, you will find it very helpful if you get immediate feedback from your client. As you practice the microskills of this book, we encourage you to again use the Client Feedback Form. You may even find it helpful to continue the use of this form or some adaptation of it in your work in the helping profession.

Before engaging in that session, please read pages 35–40 and follow the ethical guidelines presented there. Because you are going to record the interview, be sure to ask the role-playing volunteer client the critical question "May I record this interview?" Also inform the client that if he or she wishes, the tape recorder may be turned off at any time. Common sense demands ethical practice and respect for the client.

You can select almost any topic for the interview. A friend or classmate discussing a school or job problem may be appropriate. Many people have worked in some type of helping relationship in the past. If so, they can role-play clients they have had in the past. A useful topic is interpersonal conflict—for example, concerns over family tensions or decisions about a new job opportunity.

Box 1-2   Key Points

Story—positive assets—restory—action	Our first task is to help clients tell their stories. To facilitate development, we need to draw out narratives of their personal assets. With a positive foundation, clients may learn to write new stories with the possibility of new actions.
Interviewing, counseling, and psychotherapy	These are interrelated processes that sometimes overlap. Interviewing may be considered the more basic and is often associated with information gathering and advice giving. Counseling focuses on normal developmental concerns whereas psychotherapy emphasizes treatment of more deep-seated issues.
Intentionality and cultural intentionality	Achieving intentionality is the major goal of this book and a central goal of the interviewing process itself. Intentionality is acting with a sense of capability and deciding from among a range of alternative actions. The intentional individual has more than one action, thought, or behavior to choose from in responding to life situations. The culturally intentional individual can generate alternatives from different vantage points, using a variety of skills and personal qualities within a culturally appropriate framework.
Intentional competence	When you use specific skills in the interview, you can predict what the client is likely to say next. However, each person is different and often will not behave exactly as predicted. With intentional competence you will shift style and change skills to continue the interview smoothly.
Microskills	Microskills are the single communication skill units of the interview (for example, questions, interpretation). They are taught one at a time to ensure mastery of basic interviewing competencies.
Microskills hierarchy	The hierarchy organizes microskills into a systematic framework for the eventual integration of skills into the interview in a natural fashion. The microskills rest on a foundation of ethics, multicultural competence, and wellness. The attending and listening skills are followed by confrontation, focusing influencing skills, and eventual skill integration.
Theory and microskills	All counseling theories use the microskills but in varying patterns with differing goals. Mastery of the skills will facilitate your becoming able to work with many theoretical alternatives. The microskills framework can also be considered a theory in itself in which interviewer and client work together to enable the construction of new stories, accompanied by changes in thought and action.

(continued)

Box 1-2   (continued)

Microskills teaching model	Five steps are used to teach *single skills* of interviewing: (1) orientation; (2) seeing the skill in action; (3) reading and learning about broader uses of the skill; (4) practice; and (5) planning for generalization to daily life. The model will also be useful to teach social skills to clients in the interview.
Research validation	The microskills model has been validated by more than 450 databased studies and over 35 years of clinical practice. The skills can be learned, and they do have an impact on clients, but they must be practiced constantly or they may disappear.
Brain research and neuropsychology	Interviewing and counseling will be increasingly informed by research in this area in the coming years and you will want to keep abreast of new developments. Research relevant to interviewing and counseling will be presented in the research boxes throughout this book. Of particular importance is neuroplasticity. "Neuroplasticity can result in the wholesale remodeling of neural networks . . . a brain can rewrite itself" (Schwartz & Begley, 2002, p. 16). Successful interviewing may be expected to help clients develop new and useful connections.
You, microskills, and the interview	Microskills are useful only if they harmonize with your own natural style in the interview. Before you proceed further with this book, audiorecord or videorecord an interview with a friend or a classmate, and make a transcript of this interview. Later, as you learn more about interview analysis, analyze your behavior in that interview. You'll want to compare it with your performance in an interview some months from now.

COMPETENCY PRACTICE EXERCISE AND SELF-ASSESSMENT

Each chapter concludes with suggestions on how to take ideas from the book into daily practice of interviewing and counseling. This first practice set focuses on two key competency practice exercises.

Individual Practice

Exercise 1: What Is Your Natural Helping Style?

The final pages of this chapter speak to the importance of your recording an interview of 15 minutes or more with a volunteer client. This interview can serve as a baseline for comparison as you work through this book. Again, we recommend that you record this interview before you move too far into the book. In this way, you can best discover aspects of your natural style. Developing a written transcript of this interview for later study and analysis will be helpful. Before you conduct the interview, review the issues around ethical informed consent on page 37.

We believe that each person who works with this book has important natural talents that need to be recognized, reinforced, and developed further as he or she becomes more expert in the helping process. Interviewing and counseling are very personal processes and you need to respect yourself and learn from yourself throughout your career. Be sure to obtain feedback from your client (Box 1-3).

Box 1-3   Client Feedback Form

_____ (Date)

_____   _____
(Name of Interviewer)   (Name of Person Completing Form)

Note: You and your instructor may wish to change and adapt this form to meet the needs of varying clients, agencies, and situations.

1. What one thing stood out for you from this session? What might you remember and take home with you?

2. Did the interviewer listen to you? Did you feel heard? Rate this on a 7-point scale with "1" representing that you felt very much listened to and "7" that you were not heard.

     High   Medium   Low
      7   6   5   4   3   2   1

3. What, if anything, did the interviewer miss that you would have liked to explore today or in another session?

4. What did you find helpful? What did the interviewer do that was right? Be specific. For example, not "You did great," but rather, "You listened to me carefully when I talked about _____."

5. Overall: Rate the quality of this session in terms of its helpfulness.

     High   Medium   Low
      7   6   5   4   3   2   1

Self-Assessment

Review your audio- or videotape and ask yourself and the volunteer client the following questions:

1. What did you do that you think was effective and helpful?
2. What stands out for you from the Client Feedback Form and any other comments the client may have said to you about the session?
3. Can you identify one thing you would like to improve?
4. What strengths do you bring to the study of interviewing? Include the natural skills observed in the session plus personal strengths and qualities that you believe will be helpful in your future growth.

Exercise 2: Diversity, Multiculturalism, and You—Culture Counts!

Cultural and social influences are not the only influences on mental health service and delivery, but they have been historically underestimated—*and they do count.* Cultural differences must be *accounted for* to ensure that minorities, like all Americans, receive mental health care tailored to their needs. (Office of Surgeon General, 1999)

This quotation is from the U.S. Surgeon General's Report entitled *Mental Health, Culture, Race and Ethnicity.* We encourage you to visit the web site for the full report: www.mentalhealth.org/cre/toc.asp.

Professional associations in counseling and psychology have developed guidelines for multicultural proficiency. Several of these guidelines that relate specifically to interviewing practice are presented in Chapter 2, pages 42–43, and you may wish to read these guidelines now. Part of cultural intentionality and multicultural competence is your awareness of yourself as a cultural being and your ability to work empathetically with people different from you.

Consider the list below and the multiple cultural identities we all have as part of our being:

Language	Physical ability/disability
Race/ethnicity	Socioeconomic status
Gender	Age (young, old)
Sexual orientation	Significant life experience (e.g., rape, abuse, cancer, war)
Spirituality	Area of the country, nationality

You may want to add other issues to this list. We are all multicultural beings deeply affected by our cultural and environmental context.

The first of the multicultural guidelines discussed in Box 2-3, Chapter 2 (page 42), explains that awareness of yourself as a cultural being is vital as a beginning. Unless you are aware of yourself as a cultural being, you will have difficulty in developing awareness of others. You need to understand you own cultural background and the differences that may exist between you and those who may come from different cultures. You will be constantly learning about other cultures and thus developing new skills. An important skill is recognizing your limitations and the need in certain cases for referral.

Review the list of dimensions of diversity above and identify yourself from this list as a multicultural being. Add to this list other characteristics that you think are important in defining diversity and multiculturalism.

Then examine yourself for personal preferences and biases. How much experience do you have with people who are different from you? How able are you to work with those who may be different from you? For example, if you are heterosexual, how able are you to work with the gay or lesbian culture? If you are gay or lesbian, how able are you to work with the heterosexual culture? What developmental steps do you need to take to increase your understanding and awareness?

REFERENCES

Brammer, L., & MacDonald, G. (2002). *The helping relationship: Process and skills* (8th ed.). New York: Pearson.

Carkhuff, R. (2000). *The art of helping* (8th ed.). Amherst, MA: Human Resources Development Press.

Corey, G. (2005). *Theory and practice of counseling and psychotherapy* (7th ed.). Belmont, CA: Brooks/Cole–Thomson Learning.

Daniels, T. (2007). A review of research on microcounseling: 1967–present. In A. Ivey & M. Ivey, *Intentional interviewing and counseling: An interactive CD-ROM.* Pacific Grove, CA: Brooks/Cole.

Daniels, T., & Ivey, A. (2006). *Microcounseling* (3rd ed.). Springfield, IL: Thomas.

Daniels, T., Rigazio-DiGilio, S., & Ivey, A. (1997). Microcounseling: A training and supervision paradigm. In E. Watkins (Ed.), *Handbook of psychotherapy supervision.* New York: Wiley.

Duncan, B., Miller, S., & Sparks, J. (2004). *The heroic client.* San Francisco: Jossey-Bass.

Egan, G. (2002). *The skilled helper* (7th ed.). Pacific Grove, CA: Brooks/Cole.

Hackney, H. & Cormier, S. (2004). *The professional counselor* (5th ed.). New York: Pearson.

Hill, C., & O'Brien, K. (2004). *Helping skills.* Washington, DC: American Psychological Association.

Ivey, A. (2000). *Development therapy: Theory into practice.* North Amherst, MA: Microtraining Associates.

Ivey, A., D'Andrea, M., Ivey, M., & Simek-Morgan, L. (2002). *Theories of counseling and psychotherapy: A multicultural perspective* (5th ed.). Boston: Allyn & Bacon.

Ivey, A., Gluckstern, N., & Ivey, M. (2006). *Basic attending skills* [Manuals and videos]. Framingham, MA: Microtraining Associates. (First edition 1974)

Ivey, A., Normington, C., Miller, C., Morrill, W., & Haase, R. (1968). Microcounseling and attending behavior: An approach to pre-practicum counselor training [Monograph]. *Journal of Counseling Psychology, 15,* Part II, 1–12.

Kelly, G. (1955). *The psychology of personal constructs.* New York: Norton.

Kolb, B., & Whishaw, I. (2003). *Fundamentals of human neuropsychology* (5th ed.) New York: Worth.

Miller, G., & Dollarhide, C. (in press). Supervision of preparation and practice of school counselors: Pathways to excellence. *Counselor Education and Supervision.*

Monk, G., Winslade, J., Crocket, K., & Epston, D. (1997) *Narrative theory in practice: The archaeology of hope.* San Francisco: Jossey-Bass.

Myers, J. E., & Sweeney, T. J. (Eds.). (2005). *Counseling for wellness: Theory, research, and practice.* Alexandria, VA: American Counseling Association.

Office of the Surgeon General. (1999). *Mental health, culture, race, and ethnicity.* Washington, DC: Department of Health and Human Services.

Peterson, C., & Seligman, M. (Eds.). (2004). *Character strengths and virtues.* Oxford: Oxford University Press.

Ponterotto, J., Casas, J. M., Suzuki, L., & Alexander, C. (Eds.). (2001). *Handbook of multicultural counseling* (2nd ed.). Thousand Oaks, CA: Sage.

Restak, R. (2003). *The new brain.* New York: Rodale.

Ridley, C. (2005). *Overcoming unintentional racism in counseling and therapy.* Thousand Oaks, CA: Sage.

Russell-Chapin, L., & Ivey, A. (2004). *Your supervised practicum and internship.* Belmont, CA: Brooks/Cole–Thomson Learning.

Schwartz, J., & Begley, S. (2002). *The mind and the brain: Neuroplasticity and the power of mental force.* New York: Regan.

Seligman, M. (2002). *Authentic happiness.* New York: Free Press.

Semmier, P., & Williams, C. (2000). Narrative therapy: A storied context for multicultural counseling. *Journal of Multicultural Counseling and Development, 28,* 51–62.

Sue, D. W., & Sue, D. (2002). *Counseling the culturally different.* New York: Wiley.

van der Molen, H., Hommes, M., Smit, G., & Lang, G. (1995). Two decades of cumulative microtraining in the Netherlands: An overview. *Educational Research and Evaluation: An International Journal on Theory and Practice, 1,* 347–387.

White, M., & Epston, D. (1990). *Narrative means to therapeutic ends.* New York: Norton.

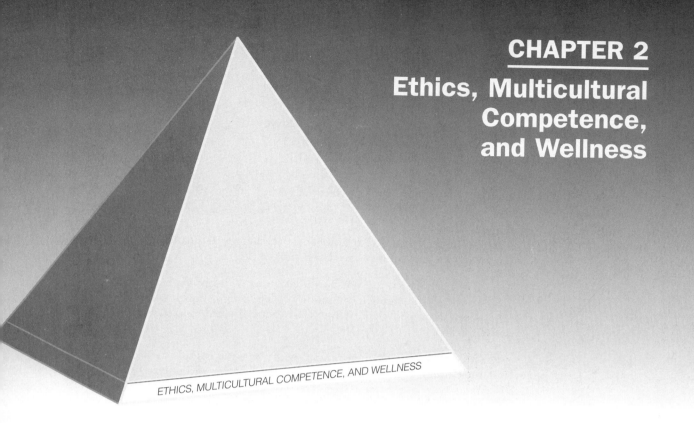

CHAPTER 2
Ethics, Multicultural Competence, and Wellness

How can this chapter help you and your clients?

Major objective	This chapter emphasizes that effective interviews build on professional ethics, multicultural sensitivity, and a positive wellness approach. It is designed to provide specifics for action in the interview.
Specific objectives	Knowledge and skill in the concepts of this chapter result in the following:

- ▲ Understanding of key ethical principles in interviewing, counseling, and psychotherapy.
- ▲ Ability to apply these ethical principles in developing your own informed consent form.
- ▲ Understanding a broad definition of multicultural competence that encourages awareness of multiple cultural identities.
- ▲ Examination of your own multiple cultural identities.
- ▲ Ability to define wellness and positive psychology and then to apply these concepts in an assessment interview.
- ▲ Understanding the distinctions between self and self-in-relation and the importance of placing the client in relation to the environmental context.

I am (and you also)
Derived from family
Embedded in a community
Not isolated from prevailing values
Though having unique experiences
In certain roles and statuses
Taught, socialized, gendered, and sanctioned
Yet with freedom to change myself and society.   (R. Jacobs, *Be an Outrageous Older Woman*, 1991, p. 37. Reprinted by permission of Knowledge, Trends, and Ideas, Manchester, CT.)

Kendra, age 25, enters your office and after some preliminaries in which you establish rapport, she comments on her concerns:

> I'm really upset. I've got a child at home with my mother and I'm trying to work my way through community college and my boss at the nursing home has been hitting on me. I want to leave, but I can't afford to stay in school without this job.

We need to work with all our clients with a sense of ethical practice and awareness of their multicultural backgrounds. Each client we encounter is one of a kind. Kendra's uniqueness stems from her biological background and the way she has lived her life in connection to others. Family, community, and culture have deeply influenced Kendra's value and socialization. Kendra is a person-in-environment, a self-in-relation to others. Our task, as interviewers, counselors, and therapists, is to facilitate her growth within this broad context and perhaps even to encourage her to change society if she wishes.

The final section of this chapter discusses wellness and positive psychology. Even from her brief personal statement, you already have some ideas about Kendra. What are some possible personal wellness strengths you imagine that Kendra might already possess that would lead ultimately to problem resolution?

_____

_____

_____

_____

_____

What are some resources that Kendra might draw on from her surrounding community?

_____

_____

_____

_____

How might you ethically address Kendra's social context and cultural background?

_____

_____

_____

You may wish to compare your thoughts with those of Mary and Allen on pages 61–62.

ETHICS IN THE HELPING PROCESS

All major helping professions throughout the world have codes outlining guidelines for ethical practice. "The codes promote professional empowerment by assisting professionals and professionals-in-training to: (a) keep good practice, (b) protect their clients, (c) safeguard their autonomy, and (d) enhance the profession" (Pack-Brown & Williams, 2003, p. 4). Ethical codes would also be summarized by the words "Do no harm to your clients, treat them responsibly with full awareness of the social context of helping." As interviewers, counselors, and therapists, we are responsible for the client before us and for society as well. At times these responsibilities conflict and here you will especially want to seek more detailed guidance from your supervisor and professional written ethical codes.

Box 2-1 presents Internet sites of some key ethical codes in English-speaking areas of the globe. You will find statements on competence, informed consent, confidentiality, and diversity in all codes. Issues of power and social justice are implicit in all codes, but most explicit in social work and human services. We strongly recommend that you review the complete text of the code for your intended profession.

Competence

Regardless of the human services profession with which you identify, competence is central. Part of personal expertise is recognizing that you can't do it all and that you may need to seek appropriate supervision. Very few of us can work with all clients. Thus, it is important that you be ready to refer clients when you are unable to help them effectively.

The American Counseling Association's (2005) statement on competence brings diversity and professional competence together. Note the emphasis on continuing to learn and expand one's qualifications over time.

> *C.2.a. Boundaries of Competence.* Counselors practice only within the boundaries of their competence, based on their education, training, supervised experience, state and national professional credentials, and appropriate professional experience. Counselors will demonstrate a commitment to gain knowledge, personal awareness, sensitivity, and skills pertinent to working with a diverse client population.

In working with a client such as Kendra, you need to constantly monitor yourself to be sure you are competent to work with her on each issue she presents. For example,

Box 2-1   Professional Ethical Codes With Web Sites

Listed below are some important ethical codes. Web site addresses are correct at the time of printing but can change. For a keyword web search, use the name of the professional association and the words *ethics* or *ethical code*.	
American Association of Marriage and Family Therapy (AAMFT) Code of Ethics	www.aamft.org
American Counseling Association (ACA) Code of Ethics	www.counseling.org
American Psychological Association (APA) Ethical Principles of Psychologists and Code of Conduct	www.apa.org
Australian Psychological Society (APS) Code of Ethics	www.psychology.org.au
British Association for Counselling and Psychotherapy Ethical Framework	www.bacp.co.uk
Canadian Counselling Association (CCA) Codes of Ethics	www.ccacc.ca
National Association of Human Service Education (NAHSE) Ethical Standards of Human Service Professionals	www.nohse.com
National Association of Social Workers (NASW) Code of Ethics	www.naswdc.org
New Zealand Association of Counsellors Inc. (NZAC) Code of Ethics	www.nzac.org.nz
Ethics Updates provides updates on current literature, both popular and professional, that relate to ethics.	www.ethics.acusd.edu

you may be able to help her work out her difficulties with her supervisor at the nursing home, but in the process of this, you discover a more difficult problem in relation to her mother that requires family counseling. You may need to refer her to another counselor for family therapy while continuing to work with her on her individual issues. If Kendra or another client demonstrates severe distress or an issue with which you are not comfortable, referral is essential.

Informed Consent

Informed consent is one of the most important elements in counseling. The counselor tells the client the goals, procedures, benefits, and risks of the counseling process and the client agrees to what has been outlined. Even as a learner, when you

Box 2-2   Sample Practice Contract

The following is a sample contract that you adapt for your own practice sessions with volunteer clients.

Dear Friend,

I am a student in interviewing skills at [insert name of class and college/university]. One of the requirements of this course is that I practice counseling skills with volunteers. I appreciate your willingness to work with me on my class assignments.

You may wish to talk about real concerns that you may have or you may prefer to role-play a problem or issue that does not necessarily relate to you. Please let me know, however, which of these two possibilities you chose.

Here are some important dimensions of our work together:

Confidentiality. As a student, I cannot offer any form of legal confidentiality. You may rest assured, however, that what you tell me in real or role-played situations will remain confidential and remain with me except for the following important exceptions which must be reported as required by state law: 1. A serious issue of harm to yourself; 2. Indications of abusing or neglecting children; 3. Other special conditions as required by our state [insert as appropriate].

Audio- and/or Videotaping. An important part of interviewing training is making a recording and listening to my own work. This may be shared with the supervisor [insert name and phone number of professor or supervisor] and/or students in my class. You'll find that recording does not affect our practice session so long as you and I are comfortable. If you wish, we can turn off the recorder at any time. Recordings or written transcripts of the recording will be destroyed at the end of the course unless I have additional written permission from you.

Boundaries of Competence. As I am beginning as an interviewer, I obviously cannot do counseling and therapy. This is a practice session so that I can learn more about the interview. In fact, I'd appreciate feedback from you as to my performance and what you find helpful.

_____   _____

Volunteer Client   Interviewer

Date _____

enter into role-plays and practice sessions, it is important that you inform your volunteer "clients" of their rights, your own competence, and what is likely to happen. For example, you might say,

> Kendra, I'm taking an interviewing course and I appreciate your being willing to help me out. Obviously, I'm beginning this type of work, so only talk about things that you want to talk about. I'll audiotape the interview, but if you want

me to turn it off, I'll do so immediately and erase it as soon as possible. I'll type out a transcript of this session and share it with you before passing it in to the instructor. I'll take out anything that might identify you personally. Remember, we will stop anytime you wish. Do you have any questions?

Use the sample contract in Box 2-2 as an ethical starting point and eventually develop your own approach to this critical issue. Counseling is an international profession. The Canadian Counselling Association (1999) approach to informed consent is particularly clear.

> *B4. Client Rights and Informed Consent.* When counselling is initiated, and throughout the counselling process as necessary, counsellors inform clients of the purposes, goals, techniques, procedures, limitations, potential risks and benefits of services to be performed, and other such pertinent information. Counsellors make sure that clients understand the implications of diagnosis, fees and fee collection arrangements, record keeping, and limits to confidentiality. Clients have the right to participate in the ongoing counselling plans, to refuse any recommended services, and to be advised of the consequences of such refusal.

The American Psychological Association (2002) stresses that psychologists should inform clients if the interview is to be supervised and provides additional specifics:

> *Standard 10.01* . . . When the therapist is a trainee and the legal responsibility for the treatment provided resides with the supervisor, the client/patient, as part of the informed consent procedure, is informed that the therapist is in training and is being supervised and is given the name of the supervisor.

> *Standard 4.03 Recording.* Before recording the voices or images of individuals to whom they provide services, psychologists obtain permission from all such persons or their legal representatives.

Confidentiality

As a learner or beginning professional, you usually do not have legal confidentiality. Nonetheless, what you hear in class role-plays or what is said to you in a practice session needs to be kept to yourself. Trust is built on your ability to keep confidences. Be aware that each state has varying laws on confidentiality.

The American Counseling Association's Ethical Code (2005) states:

> *Section B: Introduction.* Counselors recognize that trust is the cornerstone of the counseling relationship. Counselors aspire to earn the trust of clients by creating an ongoing partnership, establishing and upholding appropriate boundaries, and maintaining confidentiality. Counselors communicate the parameters of confidentiality in a culturally competent manner.

Professionals encounter *many* challenges to this issue. State law sometimes requires you to inform parents before even counseling a child and states that information from the interview must be shared with the parents. If issues of abuse should appear, you must report this to the authorities. If the client is dangerous to self or others, then rules of confidentiality change. You will want to study this important professional and legal issue in much more detail.

Power

The National Organization for Human Service Education (2000) comments on an important ethical issue that often receives insufficient attention:

> *Statement 6.* Human service professionals are aware that in their relationships with clients power and status are unequal. Therefore, they recognize that dual or multiple relationships may increase the risk of harm to, or exploitation of clients, and may impair professional judgment. . . .

The very act of helping has power implications. The client or helpee starts in a position of lesser power than the counselor. Power differentials occur in a society where privilege goes with skin color, gender, sexual orientation, or other multicultural dimension. Thus, you may find yourself in a situation in which institutional or cultural oppression becomes part of the counseling relationship. Awareness of and openness to talking about these issues is one way to work toward a balance of power in helping sessions. If you are a male counseling a woman, for example, it can be helpful to bring up the gender difference. For example, "How does it feel, being a woman, to talk about this issue to a man?" If your client or you are uncomfortable, it is wise to discuss this issue further. Referral may be necessary at times.

Dual relationships are the problems that can occur when you have more than one relationship with a client. If Kendra is a classmate or friend, you are engaged in a dual relationship in your practice session. This situation may occur if you work in a small town and counsel a member of your church or school community. Dual relationships can become a complex issue in the helping profession and you will want to examine this issue in more detail in the ethical codes.

Social Justice

Is the problem, concern, or challenge "in the client," "in the environment," or in some balance of the two? Is the interviewer's task completed when the session is over? The National Association of Social Workers (1999) suggests that action beyond the interview is needed and includes a major statement on social justice.

> ***Ethical Principle:*** *Social workers challenge social injustice.* Social workers pursue social change, particularly with and on behalf of vulnerable and oppressed individuals and groups of people. Social workers' social change efforts are focused primarily on issues of poverty, unemployment, discrimination, and other forms of social injustice. These activities seek to promote sensitivity to and knowledge about oppression and cultural and ethnic diversity. Social workers strive to ensure access to needed information, services, and resources; equality of opportunity; and meaningful participation in decision making for all people.

As Kendra talks about the unwelcome advances her boss is making toward her at work, the issue of oppression of women has arisen and should be named as such. The social justice perspective requires that you help her understand that the problem is not her fault and you can support her in efforts to change the working environment. At a broader level, you can work outside the interview to promote higher standards in nursing home care.

This book is about skills and strategies for human change. At the same time, we need to remember that our clients live in relationship to the world.

The microskill of focusing stresses that we need to be aware of the cultural/environmental/social context of our clients.

DIVERSITY AND MULTICULTURAL COMPETENCE

The American Counseling Association (2005) focuses the Preamble to their Code of Ethics on diversity as a central ethical issue.

> The American Counseling Association is an educational, scientific, and professional organization whose members work in a variety of settings and serve in multiple capacities. ACA members are dedicated to the enhancement of human development throughout the lifespan. Association members recognize diversity and embrace a cross-cultural approach in support of the worth, dignity, potential, and uniqueness of each individual within their social and cultural contexts.

Box 2-3   National and International Perspectives: Multiculturalism Belongs to All of Us

Mark Pope, Cherokee Nation and Past President of the American Counseling Association

Multiculturalism is a movement that has changed the soul of our profession. It represents a re-integration of our social work roots with our interests and work in individual psychology.

Now, I know that there are some of you out there who are tired of culture and discussions about culture. You are the more conservative elements of us, and you have just had it with multicultural this and multicultural that. And, further, you don't want to hear about the "truth" one more time.

There is another group of you that can't get enough of all this talk about culture, context, and environmental influences. You are part of the more progressive and liberal elements of the profession. You may be a member of a "minority group" or you have become a committed ally. You may see the world in terms of oppressor and oppressed.

Perhaps now you are saying, "good analysis" or alternatively, "he's pathetic" (especially if you disagree with me). I'll admit it is more complex than these brief paragraphs allow, but I think you get my point.

Here are some things that perhaps can join us together for the future:

1. We are all committed to the helping professions and the dignity and value of each individual.
2. The more we understand that we are part of *multiple cultures,* the more we can understand the multicultural frame of reference and enhance individuality.
3. *Multicultural* means just that—many cultures. Racial and ethnic issues have tended to predominate, but diversity also includes gender, sexual orientation, age, geographic location, physical ability, religion/spirituality, socioeconomic status, and other factors.
4. Each of us is a multicultural being and thus all interviewing and counseling involve multicultural issues. It is not a competition as to which multicultural dimension is the most important. It is time to think of a "win/win" approach.

(continued)

The Ethical Standards of Human Service Professionals (National Organization of Human Service Professionals, 2000) include the following three assertions:

Statement 17. Human service professionals provide services without discrimination or preference based on age, ethnicity, culture, race, disability, gender, religion, sexual orientation or socioeconomic status.

Statement 18. Human service professionals are knowledgeable about the cultures and communities within which they practice. They are aware of multiculturalism in society and its impact on the community as well as individuals within the community. They respect individuals and groups, their cultures and beliefs.

Statement 19. Human service professionals are aware of their own cultural backgrounds, beliefs, and values, recognizing the potential for impact and their relationships with others.

Diversity and multiculturalism have become central to the helping professions throughout the world. For example, if Kendra focuses on multicultural issues in

Box 2-3   (continued)

5. We need to address our own issues of prejudice—racism, sexism, ageism, heterosexism, ableism, classism, and others. Without looking at yourself, you cannot see and appreciate the multicultural differences you will encounter.
6. That said, we must always remember that the race issue in Western society is central. Yes, I know that we have made "great progress," but each progressive step we make reminds me how very far we have to go.

All of us have a legacy of prejudice that we need to work against for the liberation of all, including ourselves. This requires constantly examining yourself, honestly and painfully. You are going to make mistakes as you grow multiculturally; but see these errors as an opportunity to grow further.

Avoid saying, "Oh, I'm not prejudiced." We need a little discomfort to move on. If we realize that we have a joint goal in facilitating client development and continue to grow, our lifetime work will make a significant difference in the world.

Mary Ivey comments: Mark says it well. We all have a lot to learn and denying our roots in racism and prejudice is unwise. I was born in Minnesota. When I was a child, multicultural meant which church you went to—Swedish Lutheran, Finnish Lutheran, or Norwegian Lutheran. I did not get an understanding of multicultural issues until graduate school where I soon became involved with the women's movement and worked with the founding of the first women's center at the University of Massachusetts, Amherst. Soon followed Black and White encounter groups and my awareness was even more expanded. Since that time, I continue to learn and grow in multicultural understanding. It takes patience and time—in fact, you will discover that multicultural growth is a lifelong process.

Source: Mark Pope, Elder of the St. Francis River Band of Cherokees and Past-President of the American Counseling Association.

which you are not competent, you may need to refer her to someone else. You also have the responsibility to engage in constant learning to minimize the possibility for the need of referral. Referral to others cannot be an ethical excuse over the long term. You have a responsibility to build your multicultural competence through constant study and supervision.

Multicultural Practice

The American Psychological Association Multicultural Guidelines begin with this statement: "All individuals exist in social, political, historical, and economic contexts and psychologists are increasingly called upon to understand the influences of these contexts on individuals' behavior" (APA, 2002, p. 377). *Multicultural Counseling Competencies* have been developed to provide specifics for culturally sensitive helping (Roysircar, Arredondo, Fuertes, Ponterotto, & Toperek, 2003; Sue et al., 1998). This section reviews some of the most important of these practice guidelines for the beginning professional and you are urged to consult these three major resources for more details and information.

Expect the issue of multicultural competence to become increasingly important to your professional helping career. For example, cultural competency training is now required for medical licensure in New Jersey, while at least four other states have pending legislation with similar bills (Adams, 2005).

In this section, the words *multiculturalism* and *diversity* are defined broadly to include race/ethnicity, gender, sexual orientation, language, spiritual orientation, age, physical ability/disability, socioeconomic status, geographical location, and other factors. The multicultural competencies talk about awareness, knowledge, and skills. They ask you to become aware that specific issues exist, to learn about them, and to develop skills that can be used with clients. Given that each client, as a cultural being, has a life experience different from everyone else's you face a lifetime of multicultural learning.

Awareness of Your Own Assumptions, Values, and Biases

Awareness of yourself as a cultural being is vital as a beginning. Unless you have this awareness, you will have difficulty developing awareness of others. You need to understand your own cultural background and the differences that may exist between you and people from different cultures. You will be constantly learning about other cultures and thus developing new skills. An important skill is recognizing your limitations and the need in certain cases for referral.

Privilege is related to power, discussed in the preceding section. Privilege is power given to people through cultural assumptions and stereotypes. The concept of White privilege was originated by McIntosh (1998) when she pointed out that White people enjoy certain benefits simply by the color of their skin. She spoke of the "invisibility of Whiteness," commenting that White European Americans tend to be unaware of their color or the advantages that come to them because of it. Five examples of White privilege are presented below and you can use your computer search function to access the complete list (keywords: White, privilege, McIntosh):

▲ If I should need to move, I can be pretty sure of renting or purchasing housing in an area that I can afford and in which I would want to live.

Box 2-4 Guidelines for Multicultural Proficiency

The American Counseling Association and the American Psychological Association have developed guidelines of multicultural competence in practice, research, and training (American Psychological Association, 2002; Arredondo et al., 1996; Sue et al., 1998).

The following guidelines are a summary of some key issues related to interviewing and counseling practice. At a later point, you should examine the full statements provided by professional associations.

Guideline 1—Awareness

The intentional interviewer or counselor will make a lifetime commitment to developing increased cultural expertise. Interviewers strive to demonstrate the following:

1. They are aware of themselves as cultural beings, paying special attention to developing awareness of personal preferences and biases that might enhance as well as impede or work against the effective delivery of services.
2. They are aware of how contextual problems outside a person's control affect the way that an individual discusses his or her concerns. For example, external issues such as oppression or discrimination (sexism, racism, failure to recognize disability) may deeply affect a client without his or her conscious awareness. Is the problem "in the individual," "in the environment," or in some balance of the two?

Guideline 2—Knowledge

The intentional interviewer or counselor strives to make a lifelong commitment to learning the multicultural base of practice.

1. Interviewers strive to learn about multicultural groups, their history, and their present concerns as a constantly ongoing process.
2. Interviewers learn about helping processes in non-Western cultures and seek to include them, as appropriate, in their own practice. For example, how can spiritual or community leaders supplement counseling practice?
3. Interviewers learn their own limitations in cultural expertise and seek supervision as necessary. In addition, they learn when and how to refer a client for more appropriate assistance.

Guideline 3—Skills

The intentional interviewer or counselor strives to develop effective multicultural practice in these ways:

1. Through developing skills that are attuned to the unique worldview and culture of a widely varying base of clients. Each skill, strategy, or helping theory is examined for its cultural appropriateness.
2. By respecting the first language of the client and/or ensuring that careful translation is available. Obtain informed consent about the language in which interviewing is to be conducted.
3. By ensuring that contextual and diversity factors such as level of education, socioeconomic status, physical ability/disability, acculturation stress, and others are considered a part of overall treatment planning and action. In addition, clients are assisted in learning how their "individual" concerns are related to these contextual issues.

▲ I can go shopping alone most of the time, pretty well assured that I will not be followed or harassed.

▲ Whether I use checks, credit cards, or cash, I can count on my skin color not to work against the appearance of financial reliability.

▲ I can do well in a challenging situation without being called a credit to my race.

▲ I am never asked to speak for all the people of my racial group.

▲ I can choose blemish cover or bandages in flesh color and have them more or less match my skin.

McIntosh's concepts have been extended to privileges enjoyed by men, the benefits of middle-class economic status, and descriptions of many other groups who hold less power and privilege in our society. Here are five examples from a male privilege checklist (Deutsch, 2002):

▲ My odds of being hired for a job, when competing against female applicants, are probably skewed in my favor. The more prestigious the job, the larger the odds are skewed.

▲ I can be confident that my co-workers won't think I got my job because of my sex even though that might be true.

▲ If I am never promoted, it's not because of my sex.

▲ The odds of my encountering sexual harassment on the job are so low as to be negligible.

▲ I am not taught to fear walking alone after dark in average public spaces.

Following are five example items from the middle-class privilege list (Lui, Pickett, & Ivey, 2004):

▲ I can be assured that I have adequate housing for my family and myself. I have the resources to make choices regarding my medical care.

▲ I can buy not only what I need to have, but also what I want.

▲ When politicians speak of the middle class, I know they are referring to me.

▲ If my child runs into a problem in school, I feel that my concerns as a parent will be heard.

▲ I have enough financial reserves that I can handle a major car problem without a large financial crisis.

Whites, males, heterosexuals, middle-class people, and others who enjoy power and privilege all have the additional privilege of not being aware of their privilege. The physically able see their capacities as "normal" with little awareness that most of them are "temporarily able" until old age or a trauma occurs. When Christians are the dominant religion as in the United States and much of the Western world, they often are not aware of the privileges that exist for them. Similarly, when other religions are the majority in other parts of the world, they also may be unaware of their privilege and power.

We have said that the interviewer is in a power situation and thus enjoys some privileges that the client does not have; for example, counselors and therapists often can decide whether they want to work with a particular client. Clients, in turn, are often expected to work with the person they are referred to. With little knowledge, clients are in a "one down" position and can be subject to abuses of power.

You as the interviewer also face some challenges. If you are of White European descent, male, middle-class, and heterosexual, and the client is female, working class, and of a different race from you, she is less likely to trust you and thus establishing rapport may be more difficult. Your task is to improve your awareness, knowledge, and skills about who you are if you are to work with clients different from you.

Understanding the Worldview of the Culturally Different Client

Worldview is formally defined as the way we interpret humanity and the world. For interviewing purposes, think of worldview as the way your clients see themselves and the world around them. Due to varying multicultural backgrounds, each client views the world differently. The next section of this book on attending, observation, and listening skills is designed to help you learn the skills of listening to the many worldviews of clients.

The multicultural competences stress the importance of our being aware of negative emotional reactions we may have to groups different from us. Our own multicultural backgrounds sometimes taught us to view certain groups from stereotyped frameworks that may be inaccurate. Thus, it is doubly important to listen and learn the worldview of the client and not impose your own way of thinking on the interview. Diversity also means that *each* individual is unique.

We work toward multicultural competence when we develop knowledge of the many cultural groups that we will meet. Some traditional approaches to counseling theory and skills may be inappropriate and ineffective with some groups. We also need to learn about socioeconomic influences on the client and give special attention to how racism, sexism, heterosexism, and other oppressive forces may act on a client's worldview.

You will develop skills in understanding various worldviews through academic study and reading. But perhaps more important, you will be become actively involved in the community where your clients reside. This means attending community events, social and political functions, celebrations and festivals, and—most important— getting to know those who are culturally different from you on a personal basis.

Developing Appropriate Intervention Strategies and Techniques

The multicultural movement in counseling grew out of the dissatisfaction of minorities, women, gays/lesbians, people with disabilities, and other groups who felt that traditional counseling and therapy were not working effectively for them. A classic study found that 50% of minority clients did not return to counseling after the first session (cited in Sue & Sue, 2003). A general theory of multicultural counseling and therapy (MCT) has been developed to address this problem (Sue, Ivey, & Pedersen, 1996). Feminist therapy (Brown, 1996) is an example of a culturally specific approach to new ways of working with women.

Over time, you will want to expand your knowledge and skills in learning traditional theory and strategies as well as newer methods. Counseling and therapy theory is now being adapted with many specifics for using traditional theory in a more culturally respectful manner (Ivey, D'Andrea, Ivey, & Simek-Morgan, 2002).

It is particularly important that you become aware of the history of cultural bias in assessment and testing instruments and the impact of discrimination on clients.

Cultural intentionality—the ability to engage in many and varied verbal and nonverbal helping responses is a basic skill for multicultural work. *Intentional Interviewing and Counseling* seeks to address intentionality through providing you with ideas for multiple responses to your clients. *If your first response doesn't work, you need to be ready with another.* For example, the skill of focusing (Chapter 10) can help clients, who may be blaming themselves for a problem with a classmate or instructor, determine if their issues are actually related to discrimination.

Adapting present methods and theories to be more culturally sensitive, of course, is a major objective for us all. Psychoeducation, the direct teaching of skills and knowledge to clients, can be a useful intervention and this book gives considerable attention to this area. In addition, you may wish to add social action to your skills as an interviewer. Through work in the community and schools, you may be able to prevent some of the problems your clients face.

WELLNESS AND POSITIVE PSYCHOLOGY

This section presents the background of a strength-based approach to the interview. These systems do not deny human problems and difficulties. Rather they seek to present an alternative approach to these problems through a wellness approach. If clients' issues are discussed in an atmosphere of positive psychology and strength, we will enhance our chances for enabling them to work through complex issues.

Positive Psychology: The Search for Strengths

Clients come to us to discuss their problems, their issues, and their concerns. They are talking with us about what is *wrong* with their lives and may even want us to *fix* things for them. There is no question that our role is to enable clients to live their lives more effectively and meaningfully. But an important part of this process of problem solving is helping clients discover their strengths.

Leona Tyler, one of the first women presidents of the American Psychological Association, developed a system of starting counseling based on strengths:

> initial stages of . . . therapy include a process that might be called exploration of resources. The counselor pays little attention to personality weaknesses . . . (and) is most persistent in trying to locate . . . ways of coping with anxiety and stress, already existing resources that may be enlarged and strengthened once their existence is recognized. (Tyler, 1961, p. 213)

Then, once the client's strengths and resources are known, the counselor and the client can better approach complex issues and problems. Tyler's positive ideas have been central to the microcounseling framework since its inception (Ivey & Gluckstern, 1974; Ivey, Gluckstern, & Ivey, 2006). The strength and resource-oriented *story—positive asset—restory—action* model is an elaboration of her ideas. Chapters 13 and 14 of this book discuss four concrete theoretical approaches that all focus on human strengths—Carl Rogers's person-centered theory, decisional counseling, solution-oriented brief therapy, and motivational interviewing.

Box 2-5 National and International Perspectives on Counseling Skills: Problems, Concerns, Issues, and Challenges—How Shall We Talk About the Story?

James E. Lanier, University of Illinois, Springfield

Counseling and therapy historically have tended to focus on client problems. The word *problem* implies difficulty and the necessity of eliminating or solving the problem. Problem may imply deficit. Traditional diagnosis such as that found in the *Diagnostic and Statistical Manual of Mental Disorders-TR* (American Psychiatric Association, 2000) carries the idea of problem a bit further, using the word *disorder* with such terms as *panic disorder, conduct disorder, obsessive-compulsive disorder*, and many other highly specific *disorders*. The way we use these words often defines how clients see themselves.

I'm not fond of problem-oriented language, particularly that word "disorder." I often work with African American youth. If I asked them, "What's your problem?" they likely would reply, "I don't have a problem, but I do have a concern." The word *concern* suggests something we all have all the time. The word also suggests that we can deal with it—often from a more positive standpoint. Defining *concerns* as *problems* or *disorders* leads to placing the blame and responsibility for resolution almost solely on the individual.

Finding a more positive way to discuss client concerns is relevant to all your clients, regardless of their background. *Issue* is another term that can be used instead of *problem*. This further removes the pathology from the person and tends to put the person in a situational context. It may be a more empowering word for some clients. Carrying this idea further, *challenge* may be defined as a call to our strengths. Some might even talk about *an opening for change*.

Beyond that, the concepts of wellness and the positive asset search make good sense for the youth with whom I have worked. Change is most easily made from a position of strength—criticism and problem-oriented language can weaken. However, don't be afraid to challenge people to grow. Confrontation can help your clients grow in positive ways.

Allen Ivey comments: Language is central to counseling and interviewing. You may want to explore ways of listening to your clients' stories. Are their narratives *problems*? Do they have *disorders*? Have they expressed *concerns* in the story? What *issues* do they face both internally and externally? How is the story different if constructed as a *challenge*? Does the experience provide *an opening for change*? Can we help them *restory a more positive frame* for their lives?

Recently, the term *positive psychology* has become important and the field has developed an extensive body of knowledge and research supporting the importance of a strengths-based approach, also oriented to moral character (Pedersen & Seligman, 2004; Snyder & Lopez, 2002). Seligman (2002, p. 1) states that psychology has overemphasized the disease model—"We've become too preoccupied with repairing damage when our focus should be on building strength and resilience." Positive psychology brings together a long tradition of emphasis on positives within counseling, human services, psychology, and social work.

In the case of Kendra, presented at the beginning of this chapter, we see a client distressed by several possible issues—living at home, caring for her child while she works, the stresses of community college, and facing sexual harassment in a needed job (which likely does not pay a living wage). The positive approach reminds us to look for Kendra's personal strengths and environmental resources, but the question remains, "how?" A wellness assessment is one route toward finding strengths, positive assets, and resources for later problem solving.

Wellness: A Practical Model

The counseling profession's *wellness* approach is a way of life oriented toward optimal health and well-being, in which body, mind, and spirit are integrated by the individual so that he or she may live life more fully within the human and natural community (Myers, Sweeney, & Witmer, 2000). The wellness model speaks specifically to a holistic view of wellness and was developed through extensive empirical research (Myers & Sweeney, 2004, 2005; Sweeney & Myers, 2005). An integrative view of the wellness model is available in Ivey, Ivey, Myers, and Sweeney (2005).

A Contextual/Holistic View of Wellness

The wellness model is holistic (Myers & Sweeney, 2005a) and refers to a *self-in-relation,* the *person-in-community,* and *individual-in-social context* (Baker Miller, Stiver, & Hooks, 1998; Ogbonnya, 1994; Sweeney & Myers, 2005b). All parts are related to the whole, thus the *self is actually indivisible* and self is developed in connection to others. Any change in any part of the self affects the total individual and others as well. In addition, major changes in contextual factors may be as important or more important than individual change. We should not forget that individuals can change the surrounding context.

Figure 2-1 presents the Wellness Model, which provides us with a useful map of possible places to search for strengths and resources in the client. Important in the wellness model is the environment and client social contexts. When considering any client, we need to be aware of family, neighborhood, and community for it is here that the client will most often find supports and, of course, the source of many difficulties as well. With Kendra, it will be useful to know more about the strengths in this family. We can also look to people and groups in her community for other supports.

Institutions affect us all, but we seldom think of them. The nursing home setting is one obvious example, as is her community college. What other supports does the community college offer? Is she involved in a church and what resources do they have to help her? Government and business programs sometimes are there as helpers (and sometimes as hindrances). World events can affect Kendra, particularly if she has a friend serving overseas in the armed forces. Time and key generational events can be a factor in how a client thinks about the world. Those who experienced the Vietnam period and the 60s may have a different worldview from those born later. In future years, expect to find a significant difference between those who experienced the Iraq war as an adult and younger people coming up who do not have these vivid memories.

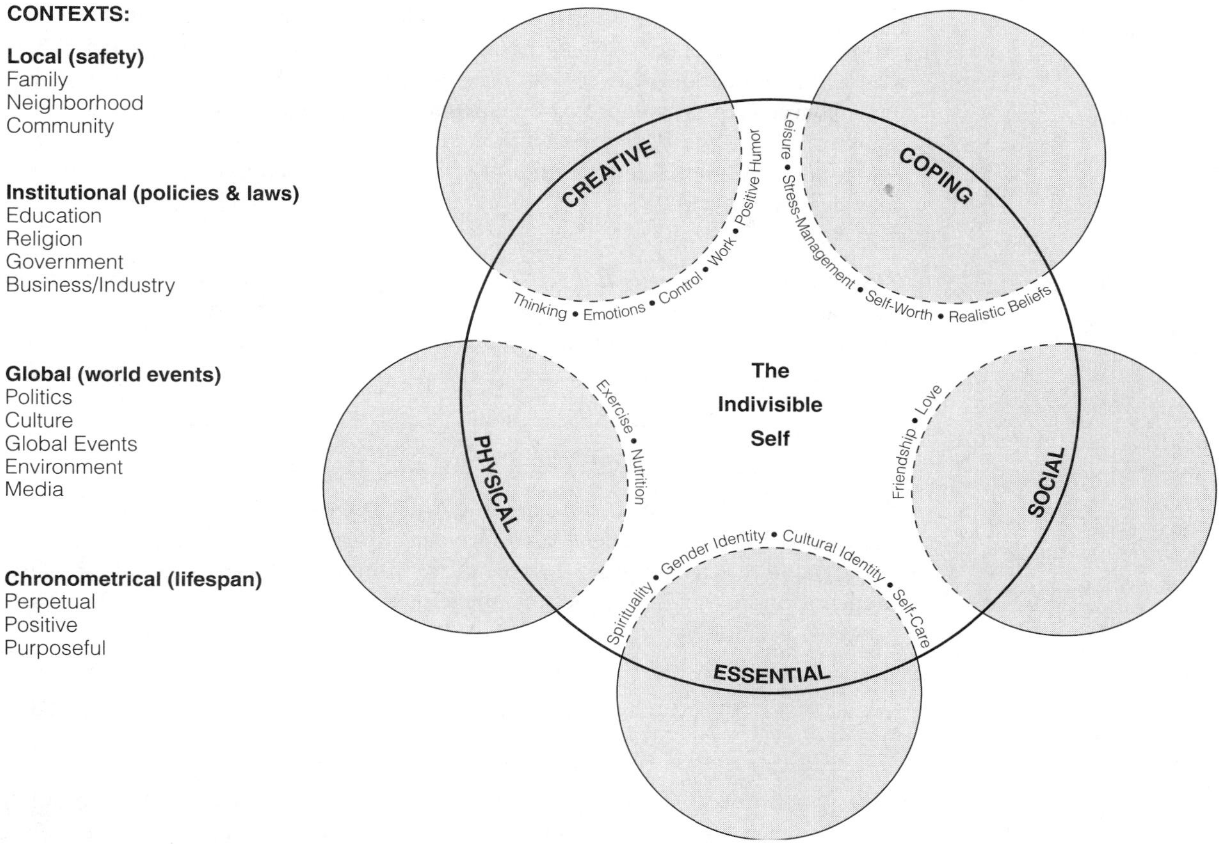
CONTEXTS:
Local (safety)
Family
Neighborhood
Community
Institutional (policies & laws)
Education
Religion
Government
Business/Industry
Global (world events)
Politics
Culture
Global Events
Environment
Media
Chronometrical (lifespan)
Perpetual
Positive
Purposeful
CREATIVE
Thinking • Emotions • Control • Work • Positive Humor
COPING
Leisure • Stress-Management • Self-Worth • Realistic Beliefs
PHYSICAL
Exercise • Nutrition
The Indivisible Self
SOCIAL
Friendship • Love
ESSENTIAL
Spirituality • Gender Identity • Cultural Identity • Self-Care

Figure 2-1 The indivisible self: An evidence-based model of wellness. (Copyright T. J. Sweeney & J. E. Myers, 2004. Reprinted with permission.)

Wellness Assessment: Identifying Client Strengths

This section presents 17 dimensions of wellness basic to optimal health. These factors, organized into five sections, have direct implications for helping clients become aware of their capacities and strengths. The wellness assessment, based on extensive research, shows us how to take empirical research into practice skills useful in the interview (Myers & Sweeney, 2004, 2005; Sweeney & Myers, 2005a).

Seventeen dimensions are a lot to deal with, but if you first focus on yourself and your own wellness strengths, you will have an initial understanding of the power of the positive in the interview. Later, you can work through this wellness assessment with a volunteer client. You will not ordinarily conduct a full wellness assessment as presented here, but—as appropriate to the situation—you will want to use portions of the wellness assessment with most clients. And in longer-term counseling situations, a full session given to wellness assessment can prove extremely valuable in the long run.

Let's focus on you as we first consider wellness assessment. We suggest that you write down your specific strengths relating to each dimension presented below. Later, you can conduct a wellness assessment by going through the questions following with a real or volunteer client. You may wish to photocopy these pages and share them with the interviewee. Throughout the questioning process, try to draw out concrete examples and specifics of strengths available to you or the client you may be working with.

Dimension 1—The essential self.   *Spirituality.* What strengths and supports do you gain from your spiritual/religious orientation? Be as specific as possible. How could you draw on this resource when faced with life challenges?

_____

_____

_____

Gender identity. Two dimensions are important here. First, what strengths can you draw on as a female or male? Do you have positive models who help you live your life? Second, what strengths do you draw from your sexual orientation? What models support you as a gay person, a heterosexual, a lesbian, a bisexual, or a transgendered person?

_____

_____

_____

Cultural identity. What strengths do you derive from your multicultural background, including race, ethnicity, and other dimensions of diversity? Who are your heroes?

_____

_____

_____

Self-care. This refers to how well people take care of themselves—cleanliness, healthy habits, avoidance of drugs, safety habits (wearing seat belts) are all examples that lead to a longer life. Those clients who do not engage in self-care may be depressed or have other serious issues. How are you doing in this area? What are your strengths?

_____

_____

_____

Dimension 2—The social self.   *Friendship.* This focuses on your ability to be a friend and have friends and have healthy long-term relationships. We are people in connection and not meant to be alone. It takes time to nurture relationships. Give examples of yourself as a social being and think of the positive feelings that come from a good relationship with friends.

_____

_____

_____

Love. Caring for special people such as family and a loved one results in intimacy, trust, and mutual sharing. Give examples of family members and role models from whom you can gain some support. Sexual intimacy and sharing with a close partner are key areas of wellness. What strengths from the past and present can you identify?

_____

_____

_____

Dimension 3—The coping self.   *Leisure.* This is a topic that interviewing and counseling often forget, but it is essential to wellness. People who take time off to enjoy things can return to the world of realism and stress more effectively. What leisure time activities do you enjoy? Equally important—*do you take time to do them?* When was the last time and how did it feel?

_____

_____

_____

Stress-management. Life in the modern world provides us with endless opportunities to become "stressed out." What do you do when you encounter stress? What works best for you and do you remember to use these strategies? Give at least one example when you managed stress well.

_____

_____

_____

Self-worth. Self-esteem—feeling good about yourself—is needed for personal comfort and effective living. We need to accept imperfections as well as acknowledge our strengths. How do you evaluate yourself on this important dimension?

_____

_____

_____

Realistic beliefs. Seeing the situation *as it is* is one portion of mental health. *Facts are friendly,* even when they are uncomfortable. On the other hand, some of us have come to believe negatively about our capabilities and ourselves; these unrealistic beliefs need to be challenged. How good are you at seeing what *is* in a realistic fashion? Give specific examples when you have made adequate and realistic self-assessments.

_____

_____

_____

Dimension 4—The creative self.   *Thinking.* This refers to effective problem solving and overall personal adjustment. What are some things that have gone well for you in the past? The present? What might you anticipate in a positive way for the future?

_____

_____

_____

Emotions. Ability to experience emotion appropriately is vital to a healthy lifestyle. When have you expressed emotion appropriately with a good result? Can you allow yourself to feel positive emotions? Be specific.

_____

_____

_____

Control. People who feel in control of their lives see themselves as making a difference—they are in charge of their own "space." When have you been able to control difficult situations in a positive way? How are you in control of your own destiny? Again, provide positive, concrete examples.

_____

_____

_____

Work. We all need to work to sustain ourselves and to give meaning to our lives. What work habits do you have that are particularly strong? What do you enjoy about work? Give examples of each.

_____

_____

_____

Positive humor. A sense of humor certainly helps an individual cope with the world. Can you laugh easily? What things occur to you around this area? What is fun?

_____

_____

_____

Dimension 5—The physical self.   *Exercise.* These last two areas need to become a more important part of our interview practice. If the client is not getting good nutrition and sufficient exercise, the mind is not going to work as well as it could. One useful treatment for clients who may be depressed is helping them get their bodies moving. What is your exercise routine? What works for you? If your physical activity is limited at present, what are you going to do about it?

_____

_____

_____

Nutrition. This is another area that needs more attention from interviewers and counselors. What are your eating habits? Especially, what strengths have you developed in your understanding and ability to apply concepts of effective nutrition?

_____

_____

_____

_____

Intentional Wellness Plan

Assessment is not enough. We need to develop an *Intentional Wellness Plan* for ourselves and also with our clients. The first step is a concrete assessment of wellness strengths. The second step is an honest appraisal of areas where you and your clients need to examine what could be done to develop a healthier lifestyle. Effective wellness assessment provides a summary of strengths and areas where improvements might be made (Myers & Sweeney, 2005a).

It is particularly important not to overwhelm the client with the many things that could be done to improve overall wellness. Work with the client to select one or two items from the wellness assessment as a start and negotiate a contract for action and follow-up. Check with your client on a regular basis to see how the plan is working.

Briefly, place here your early ideas for a personal wellness plan for yourself.

_____

_____

_____

_____

This wellness plan can become part of one session or a major portion of a longer-term interviewing and counseling plan. You can facilitate the entire process of interviewing, counseling, and therapy with portions of the wellness approach. The growing interest in positive psychology further supports these practices in a variety of settings and with clients of all ages.

Box 2-6   Research That You Can Use: Wellness

	Wellness has been studied in various populations; this research has involved many cross-cultural and cross-national studies involving participants of all ages, resulting in a database containing information on more than 12,000 individuals (Myers & Sweeney, 2005a). Among the major findings have been that wellness measures are associated with general psychological well-being, a positive body consciousness as contrasted with body shame, healthy love styles, job satisfaction, ethnic identity, and acculturation.
	The essential conceptualizations gleaned from these studies thus far affirm that (a) wellness is indivisible—that is, positive choices for living well in one area of life have implications for other areas of a person's physical, mental, emotional, and spiritual well-being; (b) wellness over the lifespan is equally important regardless of gender, culture, race, or geographical location; and (c) wellness is measurable in meaningful ways suitable for counseling and other educational interventions.
Wellness and neuropsychology	Brain research supports the wellness approach. The holistic right hemisphere is associated more with positive emotions such as happiness and joy while the left hemisphere and amgydala (deep in the brain) more often with negative feelings. In depression and deep sadness, brain scans reveal that the positive areas are less active (Davidson, Pizzagalli, Nitschke, & Putnam, 2002). Happiness involves physical pleasure, the absence of negative emotions, and positive meanings (Carter, 1999; Davidson, 2001). In effect positive thoughts and action can help override fear, anger, and sadness. This finding is directly parallel to the wellness research cited above.
	What, specifically does this mean for your practice? When a client focuses solely on problems and negative emotions, we can help the client through a wellness approach that strengthens and nourishes the individual. In this way we can help clients "build a tolerance for negative emotions and gradually acquire a knack for generating positive ones" (Damasio, 2003, p. 275). So, the wellness approach is not just "window dressing." This strengthening can be both psychological (reminding clients of positive experiences and personal strengths) and physical (exercise and sports, nutrition, and adequate sleep). A base of strengths facilitates problem solving and working through the many complex issues we all face.

SUMMARY: INTEGRATING WELLNESS, ETHICS, AND MULTICULTURAL PRACTICE

As we review this chapter, let us return to Kendra, our mythical client presented at the beginning of this chapter. Her IDENTITY has been formed through multiple relations in her family, community, and broader society. These relationships have socialized her by placing her in certain roles and statuses. Kendra is a person-in-community, a self-in-relation to others. Yet she still has freedom to change herself and society and our task is to facilitate that process.

We have suggested that Kendra's issues (college, child, living at home with mother, working, possible sexual harassment, and financial problems) can best be addressed if we include dimensions of positive psychology and wellness in our work with her. While it might be helpful to conduct a full wellness assessment, this will not always be possible. What you can do is search constantly for strengths that you can then use to help her address the challenges she faces.

Key ethical issues in the interview include making sure that you are competent to work with her, obtaining appropriate informed consent, preserving confidentiality, and using counselor power responsibly. The possible sexual harassment needs to be explored and Kendra will need your support as she moves to a decision here. At this moment in her life, Kendra appears to have real financial needs. Those who have a lower income clearly do not have the same possibilities and privileges as those who are more economically stable.

Every client we meet has a unique multicultural background and we constantly need to develop and improve our awareness, knowledge, and skills in many areas. The areas discussed in this chapter have been ethnicity/race, gender, sexual orientation, spirituality, ability/disability, socioeconomic status, language, and age. Kendra at 25 has differing needs and life experience than if she were 45. If she is White, she has access to some privileges despite her economic situation. If she is a Person of Color, she may face discrimination for race as well as for gender.

The challenge of multiculturalism is a broad one. The route toward multicultural competence first lies in understanding yourself as a multicultural being. Armed with this beginning, you are better prepared to learn how to work with people who are different from you. The task is humbling, but one we should take on with joy and enthusiasm for we constantly will learn how difference enriches all our lives. It would not be very interesting if we all had the same experiences, behaved the same, and had similar values.

COMPETENCY PRACTICE EXERCISES AND SELF-ASSESSMENT

Intentional interviewing and counseling is achieved through practice and experience. It will be enhanced by your own self-awareness, emotional competence, and ability to observe yourself, thus learning and growing in skills.

The competency-practice exercises on the following pages are designed to provide you with learning opportunities in three areas:

1. *Individual practice.* A short series of exercises is provided to give you an opportunity to practice the concepts.
2. *Group practice.* Practice alone can be helpful, but working with others in role-played interviews or discussions is where the most useful learning occurs. Here you can obtain precise feedback on your interviewing style. And if videotapes or audiotapes are used with these practice sessions, you'll find that seeing yourself as others see you is a powerful experience.
3. *Self-assessment.* You are the person who will use the skills. We'd like you to look at yourself as an interviewer and counselor via some additional exercises.

Individual Practice

Exercise 1: Review an Ethical Code

Select the ethical code from Box 2-1 that is most relevant to your interests and review it in more detail. Then, visit the ethical code of another country or another helping profession and note similarities and differences on competence, informed consent, confidentiality, social justice, and diversity. What is your own position on these issues? Write your observations and comments in a journal or on the CD-ROM.

Exercise 2: You as a Multicultural Being, Your Self-in-Relation

Again, we are all multicultural beings, although many of us are unaware of that fact. Please take a moment to review the following dimensions of multiculturalism and identify yourself. This is not an exhaustive list, so you should feel free to add your own additional dimensions. After you have finished this process, indicate those areas where people typically have privilege, and usually along with that privilege some implicit power over others. How privileged are you?

Ethnicity/race	Spirituality/religion	Language
Gender	Ability/disability	Age
Sexual orientation	Socioeconomic status	Other factors of importance

As you review the above, what are your thoughts about self-in-relation, person-in-community?

Exercise 3: Personal Wellness Assessment

Review the Wellness Model in Figure 2-1 and use it as a personal positive asset search. What strengths and resources do you find in your context and in your own personal wellness?

Strengths and resources from your context

Local	*Institutional*	*Global (World Events)*
Family	Education	Politics
Neighborhood	Religion	Culture
Community	Government	Global Events
	Business/Industry	Environment
	Media	

Personal strengths (the indivisible self)

Essential	*Coping*	*Physical*	*Social*	*Creative*
Spirituality	Realistic beliefs	Exercise	Friendship	Thinking
Gender identity	Stress management	Nutrition	Love	Emotions
Cultural identity	Self-worth		Control	
Self-care	Leisure		Positive humor	
			Work	

Group Practice

Exercise 4: Conduct a Wellness Assessment and Develop a Wellness Plan

Now that you have engaged in a wellness assessment for yourself, meet with three of your class members and engage in a wellness assessment with one of them. Conclude this practice with discussion of a plan for the future. The third person will be an observer and provide comments and give feedback on the process. We recommend that your volunteer client fill out the Client Feedback Form from Chapter 1. Alternatively, do this as a homework assignment with a volunteer.

Exercise 5: Develop an Informed Consent Form

Box 2-2 presents a sample informed consent form, or practice contract. With your small group, develop your own informed consent form that is appropriate for your particular school situation and for your state or commonwealth.

Exercise 6: Exploring Multicultural Competence

Again in a small group situation, review the major concepts of multicultural competence presented here. Please maintain awareness that this is a very brief summary and does not cover the full richness of this topic.

▲ Awareness of your own assumptions, values, and biases including dimensions of privilege
▲ Understanding the worldview of the culturally different client
▲ Developing appropriate intervention strategies and techniques

As your group reviews these issues, what goals for learning can you establish as you work through this and further study in interviewing, counseling, and psychotherapy?

Self-Assessment

How would you assess your understanding and competencies in the concepts of this chapter? We are asking you to assess yourself and your ability to use the ideas presented here.

Exercise 1: Journal Assessment

Reflecting on yourself as a future interviewer, counselor, or psychotherapist via a written journal can be a helpful way to review what you have learned, evaluate your understanding, and think ahead to the future. Here are three questions that you may wish to consider.

1. What stood out for you personally in the section on ethics? What one thing did you consider most important? Ideas of social justice and action in the community are considered by some a controversial topic. What are your thoughts?
2. How comfortable are you with ideas of diversity and working with people different from you? Can you recognize yourself as a multicultural person full of many dimensions of diversity?
3. Wellness and positive psychology have been stressed as a useful part of the counseling and therapy interview. At the same time, relatively little attention has been given so far to the very real problems that clients bring to us. While many difficult issues will be covered throughout this text, what are your personal thoughts at this moment on wellness and positive psychology? How comfortable are you with this approach?

Exercise 2: Self-Evaluation of Chapter Competencies

Use the following as a checklist to evaluate your present level of mastery. Check those dimensions that you currently feel able to do. Those that remain unchecked can serve as future goals. *Do not expect to attain intentional competence on every dimension as you work through this book.* You will find, however, that you will improve your competencies with repetition and practice.

Level 1: Identification and classification. You will need this minimal level of mastery for those coming examinations.

❑ Ability to define and discuss the key aspects of ethics as they relate to the interview: competence, informed consent, confidentiality, power, and social justice

❑ Ability to define and discuss the three dimensions of multicultural competence: awareness of your own assumptions, values, and biases; understanding the worldview of the culturally different client; developing appropriate intervention strategies and techniques

❑ Ability to define and discuss positive psychology and wellness

❑ Ability to define and discuss the contextual factors of the wellness model

❑ Ability to define and discuss the five personal dimensions of the wellness model: essential self, coping self, social self, creative self, and physical self

Level 2: Basic competence. Here you are asked to perform the basic skills in a more practical context such as self-evaluation or an actual interview. This is an initial level of competence and can be built on and improved throughout your use of this text.

❑ Ability to write an informed consent form

❑ Ability to define myself as a multicultural being

❑ Ability to evaluate my own wellness profile, both personal and contextual

❑ Ability to take another person through a wellness assessment

Levels 3 and 4, intentional competence and teaching competence, will not be presented in this chapter. You will encounter them in Chapter 3 on attending behavior.

DETERMINING YOUR OWN STYLE AND THEORY: CRITICAL SELF-REFLECTION ON ETHICS, MULTICULTURAL COMPETENCE, AND WELLNESS

Regardless of what any text on interviewing, counseling, and psychotherapy says, the fact remains that it is *you* who will decide whether to implement the ideas, suggestions, and concepts. What single idea stood out for you among all those presented in this chapter, in class, or through informal learning? What stands out for you is likely to be important as a guide toward your next steps. What points in this chapter struck you as most important? What might you do differently? How might you use ideas in this chapter to begin the process of establishing your own style and theory? Write your ideas here.

_____

_____

_____

REFERENCES

Adams, D. (2005). Cultural competency now law in New Jersey. Amednews.com, April 25, http://www.ama-assn.org/amednews/2005/04/25/prl20425.htm.

American Counseling Association. (2005). *Code of ethics and standards of practice.* Alexandria, VA: Author.

American Psychiatric Association. (2000). *Diagnostic and statistical manual of mental disorders, Text Revision.* Washington, DC: Author.

American Psychological Association. (2002). *Ethical principles of psychologists and code of conduct.* Washington, DC: Author.

Arredondo, P., Toporek, R., Brown, S., Jones, J., Locke, D. C., Sanchez, J., & Stadler, H. (1996). Operationalization of the multicultural counseling competencies. *Journal of Multicultural Counseling and Development, 24,* 42–78.

Baker Miller, J., Stiver, I., & Hooks, T. (Eds.). (1998). *The healing connection: Women in relationships in therapy and life.* Boston: Beacon.

Brown, L. (1996). *Subversive dialogues: Theory in feminist therapy.* New York: Basic Books.

Canadian Counselling Association. (1999). *Code of ethics.* Ottawa, Ontario: Author.

Carter, R. (1999). *Mapping the mind.* Berkeley, CA: University of California Press.

Damasio, A. (2003). *Looking for Spinoza: Joy, sorrow, and the feeling brain.* New York: Harcourt.

Davidson, R., Pizzagalli, D., Nitschke, J., & Putnam, K. (2002). Depression: Perspectives from affective neuroscience. *Annual Review of Psychology, 53,* 545–574.

Davidson, R. (2001). The neural circuitry of emotion and affective style: Prefrontal cortex and amygdala contributions. *Social Science Information, 40.*

Deutsch, B. (2002). The male privilege checklist. *Expository Magazine, 2*(2), http://www.expositorymagazine.net/maleprivilege_checklist.htm.

Goleman, D. (1995). *Emotional intelligence: Why it can matter more than IQ.* New York: Bantam.

Goleman, D. (1998). *Working with emotional intelligence.* New York: Bantam.

Ivey, A., D'Andrea, M., Ivey, M., & Simek-Morgan, L. (2002). *Theories of counseling and psychotherapy: A multicultural perspective* (5th ed.). Boston: Allyn & Bacon.

Ivey, A., & Gluckstern, N. (1974). *Basic attending skills.* North Amherst, MA: Microtraining Associates.

Ivey, A., Gluckstern, N., & Ivey, M. (2006). *Basic attending skills* (4th ed.). Framingham, MA: Microtraining Associates.

Ivey, A., Ivey, M., Myers, J., & Sweeney, T. (2005). *Developmental counseling and therapy: Promoting wellness over the lifespan.* Boston: Lahaska/Houghton-Mifflin.

Liu, W., Pickett, T., & Ivey, A. (2004). Middle-class privilege: Entitlement, social class bias, and implications for training and practice. Iowa City: University of Iowa. Manuscript submitted for publication.

McIntosh, P. (1988). *White privilege and male privilege: A personal account of coming to see correspondences through work in women's studies.* Wellesley, MA: Wellesley College Center for Research on Women.

Myers, J. E., & Sweeney, T. J. (Eds.). (2005a). *Counseling for wellness: Theory, research, and practice.* Alexandria, VA: American Counseling Association.

Myers, J. E., & Sweeney, T. J. (2005b). The indivisible self: An evidence-based model of wellness. *Journal of Individual Psychology, 63*(3).

Myers, J. E., Sweeney, T. J., & Witmer, M. (2000). Counseling for wellness: A holistic model for treatment planning. *Journal of Counseling and Development, 78,* 251–266.

National Association of Social Workers. (1999). *Code of ethics.* Washington, DC: Author.

National Organization of Human Service Professionals. (2000). Ethical standards of human service professionals. *Human Service Education, 20,* 61–68.

Ogbonnya, O. (1994). Person as community: An African understanding of the person as an intrapsychic community. *Journal of Black Psychology, 20,* 75–87.

Pack-Brown, S., & Williams, C. (2003). *Ethics in a multicultural context.* Thousand Oaks, CA: Sage.

Pedersen, C., & Seligman, M. (2004). *Character, strengths, and virtues: A handbook and classification.* Oxford: Oxford University Press.

Roysicar, G., Arredondo, P., Fuertes, J., Ponterotto, J., & Toperek, R. (2003). *Multicultural competencies, 2003.* Washington, DC: Association for Multicultural Counseling and Development.

Seligman, M. (2002). *Authentic happiness.* New York: Free Press.

Sue, D. W., Carter, R. T., Casas, J., Fouad, N., Ivey, A., Jensen, M., LaFromboise, T., Manese, J., Ponterotto, J., & Vazquez-Nutall, E. (1998) *Multicultural counseling competencies.* Thousand Oaks, CA: Sage.

Sue, D. W., Ivey, A., & Pedersen, P. (1996). *A theory of multicultural counseling and therapy.* Pacific Grove, CA: Brooks/Cole.

Sue, D. W., & Sue, D. (2003). *Counseling the culturally diverse.* New York: Wiley.

Snyder, C., & Lopez, S. (2002). *Handbook of positive psychology.* Oxford: Oxford University Press.

Sweeney, T. J., & Myers, J. E. (2005). Optimizing human development: A new paradigm for helping. In A. Ivey, M. B. Ivey, J. E. Myers, & T. J. Sweeney (Eds.), *Developmental strategies for helpers* (2nd ed., pp. 39–68). Amherst, MA: Microtraining.

Tyler, L. (1961). *The work of the counselor* (2nd ed.). East Norwalk, CT: Appleton & Lange.

ALLEN AND MARY'S THOUGHTS ABOUT KENDRA

Kendra presents her central concern as dealing with sexual harassment on the job, complicated by her need to finance her education. A quick solution is to quit the job, but then the loss of income while searching for another can be a serious issue.

Beyond that, how are things going with her mother and child? Single parents lead a complex life.

Personal resources.   While we need to address Kendra's problems, we also need to be fully aware of her strengths and also make sure that Kendra is conscious of them as well. As part of getting to know Kendra, we'd discuss several of her positive assets, particularly her ability to balance work, parenting, and school. Any client who does all this should be recognized as a person of strength, full of personal wellness assets. Further discussion will bring out other strengths. The skills she has in all these areas will combine to help her find a suitable solution to the current issue.

Community resources.   The most obvious resource that Kendra has may likely be found in her family and her mother's support. Kendra's child is also an important emotional support, even though likely challenging at times. *Exploration of resources* may reveal friends and the church, mosque, or synagogue as a place of emotional and physical support. There may be scholarships and other support options available at the college. Finally, let us hope that there is someone in her work setting who can serve as a friend and advocate.

Social context and cultural background.   We believe that individual counseling is best conducted when we are able to see the client in social and cultural context. All the major helping professions now stress the importance of culture, even making it an ethical matter. There are two issues that we'd like you to consider at this time, although more could be commented on. First is our own awareness of multicultural issues and context that might affect Kendra. We need to be aware that Kendra is a woman, a single parent, from a specific ethnic/racial background (which is unstated in this case). These three factors (plus others, of course) help us understand Kendra more completely. Our awareness and sensitivity to possible culture issues is vital, even if they are never discussed. Second, as appropriate, we need to help Kendra understand the broader context of her issues. Most clearly, harassment tends to be a women's issue and issues of sexual discrimination may need to be discussed. Other multicultural factors can be brought into the session as the interview progresses. What is a solution that is personally satisfying to Kendra? This may require a social justice approach and may involve helping her develop action plans to change the work-setting rules. Or it may simply be recognizing the nature of the cultural context as Kendra makes a personal decision.

CHAPTER 3

Attending Behavior: Basic to Communication

ATTENDING BEHAVIOR

ETHICS, MULTICULTURAL COMPETENCE, AND WELLNESS

How can attending behavior be used to help your clients?

Major functions	Attending behavior encourages client talk. You will want to use attending behavior to help clients tell their stories and to reduce your own talk-time. Conversely, the lack of attending behavior can also serve a useful function. Through nonattention you can help others talk less about topics that are destructive or nonproductive. You will also find that teaching clients listening skills is a useful treatment supplement. Shy teenagers, depressed clients focused on negative stories, and the businessperson who fails to listen to others are three examples of clients who may benefit from this form of social skills training.
Secondary functions	Knowledge and skill in attending result in the following:

- ▲ Communicating to the client that you are interested in what he or she is saying.
- ▲ Increasing your awareness of the client's pattern of focusing on certain topics.
- ▲ Modifying your patterns of attending to establish rapport with each individual. Different people and different cultural groups sometimes have different patterns of attending.
- ▲ Having some recourse when you are lost or confused in the interview. Even the most advanced professional doesn't always know what is happening. When you don't know what to do, attend!

INTRODUCTION: THE BASICS OF LISTENING

Foremost in any interview or counseling situation is the ability to make contact with another human being. We make this contact through listening and talking as well as by nonverbal means. *Listening* to other people is most critical, as it enables them to continue to talk and explore. Effective attending behavior may be considered the foundation skill of this book. If you are to develop a trusting relationship with a client, the skills of attending are essential.

How can we define effective listening more precisely? A simple warm-up exercise may help. It is best if you work with someone else, but you can do it in your imagination, especially if you recall times in your life when you felt you weren't heard. Think of times when someone failed to listen to you. Perhaps a family member or friend failed to hear your concerns, a teacher or employer misunderstood your actions and unfairly blamed you, or you had that all-too-familiar experience of calling a helpline telephone tree and never getting the answer you needed. Each of these represents the importance of listening and the frustration that happens when we are not heard.

One of the best ways to understand good quality listening is to experience the opposite—poor listening. Find a partner to role-play an interview in which one of you does not listen. If no partner is available, think back on some of the bad listening experiences you have gone through. Spend about 3 minutes role-playing the poor interview (or recalling the specifics of the ineffective session). The emphasis here is on what went wrong; you should feel free to exaggerate in order to underline the many things that can cause a session to go poorly.

List here some behaviors, traits, and qualities that are characteristic of the ineffective interviewer or counselor who fails to listen.

_____

_____

_____

_____

This exercise is often humorous. You may recall with laughter the bored interviewer, the teacher or family member who misinterpreted what you said, or the ineffective person on the helpline tree. Yet as you remember your own experience of not having your story heard, your strongest memory may be of a feeling of emptiness, disappointment, and perhaps even anger. Think of your frustration with the insensitivity of the interviewer. Lists of ineffective interviewing behaviors sometimes run to 30 items or more. If you are to be effective and competent, do the opposite of the ineffective counselor.

Obviously, you can't learn all the qualities and skills immediately. Rather, it is best to learn them step by step. The first step, the skill of attending behavior, consists of four central dimensions and is critical to all other helping skills. To communicate that you are indeed listening or attending to the client, you need the "three V's" of attending plus attentive body language ("+ B").[*]

---

[*] We thank Norma Block Gluckstern for the three V's acronym.

Attending behavior concepts were first introduced to the helping field by Ivey, Normington, Miller, Morrill, and Haase (1968). Cultural variations in microskills usage were first identified by Ivey and Gluckstern (1974) and have been central in the model since that time. However, you will still find some authors, researchers, and the popular press discussing attending skills without attention to cultural differences. These skills are infinitely more effective when they are used within an individually and culturally appropriate framework. Let us first examine these dimensions, as they will often appear in the European-North American context.

1. *Visual/eye contact.* If you are going to talk to people, look at them.
2. *Vocal qualities.* Your vocal tone and speech rate also indicate clearly how you feel about another person. Think of how many ways you can say "I am really interested in what you have to say" just by altering your vocal tone and speech rate.
3. *Verbal tracking.* The client has come to you with a topic of concern. Don't change the subject; stick with the client's story.
4. *Attentive and authentic body language.* Clients know you are interested if you face them squarely and lean slightly forward, have an expressive face, and use facilitative, encouraging gestures. In short, allow yourself to be yourself—authenticity in attending is essential.

The three V's + B have one goal in common—to reduce interviewer talk-time and provide clients with an opportunity to tell the story as concretely and with as much detail as needed. In addition, these skills will help you enable clients to think about the meaning of their stories.

Moreover, you will find it helpful to note your clients' patterns of attending. As such, you can learn a good deal about your clients through noting the topics to which they attend and those that they may feel less comfortable talking about. You will also find that clients from varying racial and ethnic backgrounds will differ in their patterns of listening to you. Use client observation skills (Chapter 5) to help you adapt your style to meet the needs of the unique person before you.

EXAMPLE INTERVIEWS: DO I WANT TO BECOME A COUNSELOR?

The client in this case is a first-semester junior exploring career choice. Like many reading this book, he is considering the helping professions as an alternative.

The first interview segment, presented below, is part of a career counseling session. Jared has already stated that he is considering psychology as a major and counseling as a field. The first illustration is designed as a particularly ineffective example so that it provides a sharp contrast with the second example. Note how patterns of visual contact, vocal qualities, failure to maintain verbal following, and poor body language can disrupt a session.

Box 3-1 Attending Behavior and People With Disabilities

Attending behaviors (visuals, vocals, verbals, and body language) all may require modification if you are working with people who are disabled. It is your role to learn their unique ways of thinking and being, for these clients will vary extensively in the way they deal with their issues. Focus on the person, not on the disability. For example, think of a person with hearing loss rather than "a hearing impaired client," a person with AIDS rather than "an AIDS victim," a person with a physical disability rather than "a physically handicapped individual." So-called handicaps are often societal and environmental rather than personal.

People who may have limited vision or be blind

Eye contact is so central to the sighted that initially you may find it very demanding to work with clients who are blind or partially sighted. Some may not face you directly when they speak. Expect clients with limited vision to be more aware of your vocal tone. People who are blind from birth may have unique patterns of body language. At times, it may be helpful to teach them the skills of attending nonverbally even if they cannot see their listener. This may help them communicate more easily with the sighted.

People who may have hearing loss or be deaf

An important beginning is to realize that some people who are deaf do not consider themselves impaired in any way. Many of this group were born deaf and have their own language (signing) and their own culture, a culture that often excludes those who hear. You are unlikely to work with this type of client unless you are skilled in sign language and are trusted among the deaf community.

You may counsel a deaf person through an interpreter. This experience will not prove to be positive without some training in the use of an interpreter in addition to a basic understanding of deaf culture. Too many counselors speak to the interpreter instead of to the client, and often use phrases such as "tell him . . ." This certainly will cut off the client. Also, eye contact is vital, whether in direct counseling with a deaf client or while using an interpreter.

Those who have moderate to severe hearing loss will benefit if you extensively paraphrase their words to ensure that you have heard them correctly. Speak in a natural way, but not fast. Speaking more loudly is often ineffective, as ear mechanisms often do not equalize for loud sounds as they once did. In turn, teaching those with hearing loss to paraphrase what others say to them can be helpful to them in communicating with others.

People with physical disabilities

First, each person is unique. We cannot place people with physical disability in any one group. Consider the differences among the following: a person who uses a wheelchair, an individual with cerebral palsy, one who has Parkinson's disease, one who has lost a limb, or a client who is physically disfigured by a serious burn. They all may have the common problem of lack of societal understanding and support, but you must work with each individual from her or his own perspective. Their body language and speaking style will vary. What is important is to attend to each one as a complete person.

People who are temporarily able

Most people reading this section will have experience with some but not all of the issues above. We suggest that you consider yourself one of the many who are *temporarily able*. Age and life experience will bring most of you some variation of the challenges above. For older individuals, the issues discussed here may become the norm rather than the exception. Approach all clients with humility and respect.

In both cases, the interviewer, Jerome, is reviewing a form outlining key aspects of career choice.

Negative Example

Interviewer and Client Conversation	Process Comments
1. *Jerome:* The next thing on my questionnaire is your job history. Tell me a little bit about it, will you?	The vocal tone is overly casual, almost uninterested. The interviewer looks at the form, not at the client. He is slouching in a chair.
2. *Jared:* Well, I guess the job that, uh . . .	Client appears a bit hesitant and unsure as to what to say or do. There is a slight stammer.
3. *Jerome:* Hold it! There's the phone. (Long pause while Jerome talks to a colleague.) Uh, okay, okay, where were we?	The interview is for the client. Avoid phone and other interruptions during the session. Such behavior shows disrespect for Jared. And if a break is necessary, remember what was occurring just before you were called out.
4. *Jared:* The job that really comes to mind is my work as a camp counselor during my senior year of high school. I really liked counseling kids and . . .	The client's eyes brighten and he leans forward as he starts to talk about something he likes.
5. *Jerome:* (interrupts) Oh, yeah, I did a camp counselor job myself. It was at Camp Itasca in Minnesota. I wasn't that crazy about it, though. But I did manage to have fun. What else have you been doing?	When he is interrupted, Jared looks up with surprise, then casts his eye downward, as if to realize that he is no longer the focal point of attention. It would have been better if Jerome had not shifted the focus but rather had asked Jared for some specifics of what he liked about his work as a camp counselor.
6. *Jared:* Ah . . . well, I . . . ah . . . wanted to tell you a little bit more about this counseling job, I . . .	Jared is interested in this topic and tries again, but notice the speech hesitations.
7. *Jerome:* (interrupts) I got that down on the form already, so tell me about something else. A lot of counseling types have done peer counseling in school. Have you?	Most of us have experienced "form-centered" interviewing. It is not a pleasant experience and could be contrasted with the ideas of person-centered counseling. The leading closed question puts Jared on the spot.
8. *Jared:* Ah . . . no, I didn't.	The topic jump now has Jared at a complete standstill. He looks puzzled and confused.
9. *Jerome:* Many effective counselors were peer helpers in school . . .	Jerome is now working on his agenda. One senses that he is not fond of his job and would rather be doing something else.
10. *Jared:* (interrupts) No, I lived in a small town. I never learned about peer counseling until I got to college.	Jared is now interrupting and sits forward and talks a little more loudly. His anger is beginning to show.

11. *Jerome:* Too bad, it might have helped.	Jerome looks away and dismisses Jared's thoughts and feelings.
12. *Jared:* I only had 75 kids in my graduating class, but I know that counseling is what . . .	Jared is valiantly trying to focus the conversation on himself and his interest in counseling.
13. *Jerome:* (interrupts) Hmmm, that sounds like something I ought to write down. You came from a small school.	Jerome does remember he has a job to do. But he is looking intently at his form and still does not look at Jared.
14. *Jared:* One person that I always admired was my school counselor. He really got along with the kids.	Jared starts to look a bit more enthusiastic, but again is interrupted.
15. *Jerome:* My school counselor wasn't so great. I think it was just a job for him.	Jerome returns the focus to Jerome and Jared sits back in discouragement and disarray.

At one level, this interview is extreme, but it does illustrate the many different ineffective things an interviewer can do. Jerome clearly had poor visuals, vocals, verbal following, and body language. At another level, perhaps this session is not so rare. Most likely you have encountered supervisors or bureaucrats who have treated you in this way. Some clients you encounter will have similar poor listening skills.

Positive Example

Let us give Jerome another chance. What differences do you note in this second session?

Interviewer and Client Conversation	Process Comments
20. *Jerome:* Jared, so far, I've heard that you are interested in counseling as a career. You've liked your psychology courses and you find friends come to you to talk about their problems. In this next phase of the session, I'd like to review a form with you that may help us plan together. The next thing on this form is something about your job history. Could you tell me a little bit about the jobs you've had in the past? Would that be OK?	Jerome personalizes the interview by using the client's name. He first summarizes what he has heard so far and tells the client what is going to happen next. He asks an open question to obtain information from Jared's point of view. He leans forward slightly, maintains good eye contact, and uses a warm and friendly vocal tone.
21. *Jared:* Sure, the job that comes to mind is one that I had in my senior year of high school and during my first two years of college as a camp counselor at the YMCA.	Jared's vocal tone is confident and relaxed. He seems eager to explore this area.

(continued)

Interviewer and Client Conversation	Process Comments
22. *Jerome:* Sounds like you really liked it. Tell me more.	Jerome catches the feeling under Jared's enthusiastic words and encourages him to continue.
23. *Jared:* Well, I was . . . as a kid, people always asked me what I wanted to be when I grew up. I had no idea, but when I worked at this camp, I realized that I truly liked working with kids and was good at it.	Jared smiles and continues.
24. *Jerome:* Uh-huh.	Continued good eye contact and an encourager indicate to Jared that he should go on.
25. *Jared:* I like to work with people, and perhaps that is why I am thinking of majoring in psychology.	As Jared shares his story, we begin to see the relationship between past jobs he has held and possible future majors in college and job future.
26. *Jerome:* I see. So you really liked helping kids in the camp. It even makes you think that is the career direction you might like to head in. Could you give me a specific example of what happened at the camp that you particularly liked?	Jerome summarizes what Jared has said to this point. As Jared is speaking somewhat generally, it is useful to ask for specific examples. Note especially that Jerome has given the lead to his client and uses careful attending and verbal following skills.
27. *Jared:* Well, I seemed to be able to get kids in the group to work well together—there were kids from lots of different racial groups with differing amounts of money in their backgrounds. Yet I seemed to be able to organize them in a way that built on the strengths of everybody. And we resolved conflicts. I think I helped them feel pretty good about themselves.	Identifying client strengths can serve as a basis for both decision making and problem solving.
28. *Jerome:* Sounds like you really feel good about your work there.	Jerome focuses on Jared's underlying feeling tone.
29. *Jared:* Oh, I was delighted. I felt my cabin was the most cohesive group in the camp sometimes. I feel real good because I helped them and we had lots of fun at the same time. I learned how to play the ukulele and we had some great singing sessions. I thought, you know, that this could be a profession for me and that would be great . . . if I could do something like that year round.	Jared continues with enthusiasm and verifies good feelings about himself. If you catch the feeling tone of your clients, they will often acknowledge these feelings and you will learn about them in more depth.
30. *Jerome:* So, continuing to help others the year round may be important. What are some things along that line you've thought of that you might want to do?	Jerome picks up Jared's enthusiasm, paraphrases Jared's last important words, and then asks an open question to encourage more exploration on the topic.

31. *Jared:* Well, I thought I might want to become a counselor so that I could help people with real personal problems. Kids came to me a lot to talk about their issues with their families and friends. I might even want to be a camp director. I know I want to work with people, but am not exactly sure how.	Throughout all this discussion Jerome and Jared have a comfortable relaxed relationship with a solid demonstration of the three V's + B.
32. *Jerome:* So, Jared, it seems clear that working with people is something you like and one of your real strengths. People skills are important in many jobs. Could we shift gears here for a moment? I know something now about what you like, but could you tell me perhaps about a job that you didn't like so well?	Jared seems to have explored his positive associations fairly completely and is ready to move on. Jerome has found that comparing positive job experiences with those less pleasant can be helpful.
33. *Jared:* During the other months of the year, I had a paper route where I lived. It was so boring and I hated it when the weather was bad. And the papers sometimes were so heavy. Sometimes they were so thick, I couldn't throw them to the upper floors. The pay wasn't all that good, but that was the only job I could find.	Here we learn some of Jared's values and how he approaches work.
34. *Jerome:* Uh-huh, it was hard, but the only job you could find. Sounds like boring is not something you want in a job.	Jerome follows Jared's conversation, paraphrasing portions to indicate that he is following what has been said.
35. *Jared:* Right, I thought of quitting many times, but I hate to be thought of as a quitter. I was determined not to quit.	If you attend to what a client has said, he or she often will smile and say "right." Small signals such as this indicate that your attending skills are helping the client examine herself or himself.
36. *Jerome:* Uh-huh.	Jerome nods his head showing interest.
37. *Jared:* I had a couple friends who had the same job. They'd deliver from, you know, August to October, and as soon as it got cold, they'd quit and just live off the allowance their parents gave them. I decided I was going to tough it out through the winter, and the guy who was the boss down there, he told me in May when I finally quit to go to camp, he said, "Jared, you're a good kid, you know, you stuck through the winter." That meant a lot to me.	An important part of interviewing and counseling is encouraging clients to share positive stories and experiences. On this basis of strength, we can best encounter our problems.

(continued)

Interviewer and Client Conversation	Process Comments
38. *Jerome:* So, to wind up here at this particular point—we've talked about a job you liked and one that you didn't. Can you put anything together in the two jobs might be useful for you to note and us to remember? Your sticking through the winter, that may even tie in with some of the stuff we talked about earlier in camp.	Jerome invites Jared to think back and reflect on his two jobs. What can Jared learn about himself from his own storytelling?
39. *Jared:* Well, I didn't like the paper route, but it sure felt good when I discovered that I wasn't a quitter and the boss said those good things. And, I felt good about myself when I saw the kids doing so well as I worked with them. I guess I have some natural talents. I seem to be able to work with people and maybe I'll be able to hang with them when the going gets tough.	Jared is making his own connections and conclusions rather than having them supplied by the interviewer.
40. *Jerome:* So, Jared, putting it together, it sounds as if you have some real strengths that will be useful in working with people and you have that important quality of persistence—you can hang in when it gets rough. These are some real positives that you can draw on as we start to look toward career choice.	Jerome rephrases Jared's comment and summarizes the entire interviewing section emphasizing strengths that can be brought to bear on future issues.
41. *Jared:* (pause) That's right. I never thought of it that way before. Even my work as a paperboy might be valuable in making my mind up for sure. I certainly know that boring and tedious are not for me.	Jared sits back for a moment. Jerome's use of attending skills has focused on Jared's perspectives and Jared has supplied the strengths and ideas that will help him come to a clearer career choice.

Although the focus in this second session was on attending to Jared and seeking to draw out his career story, Jerome used a common strategy that is helpful in many different settings. He first sought to bring out a positive story that almost inevitably brings out personal strengths in the client. He then turned to a more problematic area with the potential to discover weakness or areas in which the client might need further growth. In this case, Jared brought out another strength. But we also learn that Jared is the type of person who is unlikely to enjoy routine work that does not offer an opportunity to interact with people. Negative job experiences can be helpful in planning for a positive career future.

The strategy Jerome used is termed *compare and contrast interviewing*. A positive experience is identified in some detail and then the negative is explored. As the two are contrasted, clients often reveal important things about themselves. The identification of positive assets and wellness strengths is particularly important.

Compare and contrast interviewing can also be effective in interpersonal counseling. For example, your client may have difficulties in a significant relationship. He or she may tell you about many problems and the situation may sound totally hopeless. If you ask for a past positive experience in the relationship, you may find strengths on which to build a more positive future for the client.

INSTRUCTIONAL READING: GETTING SPECIFIC ABOUT LISTENING

A frequent tendency of the beginning counselor or interviewer is to try to solve the client's difficulties in the first 5 minutes. Think about it—the client most likely developed her or his concern over a period of time. It is critical that you slow down, relax, and attend to client narratives. *Listen before you leap!*

Attending and giving clients "talk-time" demonstrates that you truly want to hear their story and major concerns. And, it is also important to note and recognize that your clients will have varying styles of attending and interacting with you. The following discussion outlines the specifics of attending behavior and its many individual and cultural variations.

The Three V's + B: Visuals, Vocals, Verbals, and Body Language

Visual/Eye Contact

Cultural differences in eye contact abound. Direct eye contact is considered a sign of interest in European-North American middle-class culture. However, even in that culture a person often maintains more eye contact while listening and less while talking. Furthermore, when a client is uncomfortable talking about a topic, it may at times be better to avoid eye contact. Research indicates, moreover, that some African Americans in the United States may have reverse patterns; that is, they may look more when talking and slightly less when listening. Among some traditional Native American and Latin groups, eye contact by the young is a sign of disrespect. Imagine the problems this may cause the teacher or counselor who says to a youth, "Look at me!" when this directly contradicts basic cultural values. Some cultural groups (for instance, certain traditional Native American, Inuit, or Aboriginal Australian groups) generally avoid eye contact, especially when talking about serious subjects.

Not only do you want to look at clients, you'll also want to notice breaks in eye contact. Clients often tend to look away when thinking carefully or discussing topics that particularly distress them. You may find yourself avoiding eye contact while discussing certain topics. There are counselors who say their clients talk about "nothing but sex" and others who say their clients never bring up the topic. Through eye-contact breaks, both types of counselors indicate to their clients whether the topic is appropriate.

When an issue is especially interesting to clients, you may find that their pupils tend to dilate. On the other hand, when the topic is uncomfortable or boring, their pupils may contract. If you have a chance to observe your face carefully on videotape, you'll note that you as a counselor or interviewer indicate to your clients your degree of interest in the same subtle fashion.

Vocal Qualities

Your voice is an instrument that communicates much of the feeling you have toward another person or situation. Changes in its pitch, volume, or speech rate convey the same things that changes in eye contact or body language do.

Keep in mind that different people are likely to respond to your voice differently. Try the following exercise with a group of three or more people.*

Ask all members of the group to close their eyes while you talk to them. Talk in your normal tone of voice on any subject of interest to you. As you talk to the group, ask them to notice your vocal qualities. How do they react to your tone, your volume, your speech rate, and perhaps even your regional or ethnic accent? Continue talking for 2 or 3 minutes. Then ask the group to give you feedback on your voice. Summarize what you learn here.

_____

_____

_____

This exercise often reveals a point that is central to the entire concept of attending. *People differ in their reactions to the same stimulus.* Some people find one voice interesting, whereas others find that same voice boring; still others may consider it warm and caring. This exercise and others like it reveal again and again that people differ, and that what is successful with one person or client may not work with another.

Accent is a particularly good example of how people will react differently to the same voice. What are your reactions to the following accents—Australian, BBC English, Canadian, French, Pakistani, New England U.S., Southern U.S.? Obviously we need to avoid stereotyping people because their accents are different from ours.

Verbal underlining is another useful concept. As you consider the way you tell a story, you will find yourself giving louder volume and increased vocal emphasis to certain words and short phrases. Clients of course do the same. The key words a person underlines via volume and emphasis are often concepts of particular importance.

Awareness of your voice and of the changes in others' vocal qualities will enhance your skill in attending to their stories. Again, note the timing of vocal changes, as they may indicate comfort or discomfort depending on the cultural affiliation of a client. Speech hesitations and breaks are another signal, often indicating confusion or stress. Clearing one's throat may indicate that words are not coming easily.

Verbal Tracking

Staying with your client's topic is critical in verbal tracking. Encourage the full elaboration of the narrative. Just as people make sudden shifts in nonverbal communication, they change topics when they aren't comfortable. And cultural differences may appear as well. In middle-class U.S. communication, direct tracking is most appropriate, but in some Asian cultures such direct verbal follow-up may be considered rude and intrusive.

---

* This exercise was developed by Robert Marx, School of Management, University of Massachusetts.

Selective attention is a type of verbal tracking that counselors and interviewers need to be especially aware of. We tend to listen to some things and ignore others. Over time we have developed patterns of listening that enable us to hear some topics more clearly than others. For example, take the following statement, in which the client presents several issues.

Client: (speaks slowly, seems to be sad and depressed) I'm so fouled up right now. The first term went well and I passed all my courses. But this term, I am really having trouble with chemistry. It's hard to get around the lab in my wheelchair and I still don't have a textbook yet. (An angry spark appears in her eyes and she clenches her fist.) By the time I got to the bookstore, they were all gone. It takes a long time to get to that class because the elevator is on the wrong side of the building for me. (Looks down at floor) Almost as bad, my car broke down and I missed two days of school because I couldn't get there. (The sad look returns to her eyes) In high school, I had lots of friends, but somehow I just don't fit in here. It seems that I just sit and study, sit and study. Some days it just doesn't seem worth the effort.

There are obviously several different directions in which the interview could go. You can't talk about everything at once. List those several directions here. To which one(s) would you selectively attend? After you have responded please turn to page 90 and read one professional's thoughts about this case.

_____

_____

_____

_____

_____

Another way to respond would be to reflect the main theme of the client's story; for example, "You must feel like you're being hit from all directions." But there are no correct answers. In such a situation different interviewers would place emphasis on different issues. Some interviewers consistently listen attentively to only a few key topics while ignoring other possibilities. Be alert to your own potential patterning of responses. It is important that no issue get lost, but it is equally important not to attack everything at once, as confusion will result.

The concept of verbal tracking may be most helpful to the beginning interviewer or to the experienced interviewer who is lost or puzzled about what to say next in response to a client. *Relax,* take whatever the client has said in the immediate or near past, and direct attention to that through a question or brief comment. You don't need to introduce a new topic. Build on the client's topics, and you will come to know the client very well over time.

Attentive and Authentic Body Language

The anthropologist Edward Hall once examined film clips of Native Americans of the Southwest and of European-North Americans and found more than 20 different variations in the way they walked. Just as cultural differences in eye contact exist, body language patterns differ.

A comfortable conversational distance for many North Americans is slightly more than arm's length, and the English prefer even greater distances. Many Latin people often prefer half that distance, and those from the Middle East may talk practically eyeball to eyeball. As a result, the slightly forward leaning we recommend for attending behavior is not going to be appropriate all the time. A natural, relaxed body style that is your own is most likely to be effective, but be prepared to adapt and flex according to the individual with whom you are talking.

Just as shifts in eye contact tell us about potentially uncomfortable issues for clients, so do changes in body language. A person may move forward when interested and away when bored or frightened. As you talk, notice people's movements in relation to you. How do you affect them? Note your patterns in the interview. When do you change body posture markedly? Are there patterns of which you need to be aware?

Authenticity is vital to the way you communicate. Whether you use visual, vocal, or verbal tracking or attentive body language, it is vital that you be a real person in a real relationship. Take on the skills, be aware and respectful of cultural differences, but also make sure that you are yourself—for your authentic personhood is a vital presence in the helping relationship.

The Value of Nonattention

There are times when it is inappropriate to attend to client statements. For example, a client may talk insistently about the same topic over and over again. A depressed client may want to give the most complete description of how and why the world is wrong. Many clients want to talk about only negative things. In such cases, intentional nonattending may be useful. Through failure to maintain eye contact, subtle shifts in body posture, vocal tone, and deliberate jumps to more positive topics, you can facilitate the interview process.

The most skilled counselors and interviewers use attending skills to open and close client talk, thus making the most effective use of limited time in the interview.

The Usefulness of Silence

Counseling and interviewing are *talking* professions. But sometimes the most useful thing you can do as a helper is to support your client silently. As a counselor, particularly as a beginner, you may find it hard to sit and wait for clients to think through what they want to say. Your client may be in tears, and you may want to give support through your words. However, the best support may be simply being with the person and not saying a word.

For a beginning interviewer, silence can be frightening. After all, doesn't counseling mean talking about issues and solving problems verbally? Perhaps this is true for many of us, but others like to "sit on" their issues, mull them over, and think carefully before responding. The first thing to do when you feel uncomfortable with silence is to look at your client. If the client appears comfortable, draw from his or her ease and join in the silence break. If the client seems as disquieted by the silence as you are, then rely on your attending skills and ask a question or make a comment about something relevant that has been mentioned earlier in the session.

Box 3-2 National and International Perspectives on Counseling Skills: Use With Care—
Culturally Incorrect Attending Can Be Rude

Weijun Zhang, Management Consultant, Shanghai, China

The visiting counselor from North America got his first exposure to cross-cultural counseling differences at one of the counseling centers in Shanghai. His client was a female college student. I was invited to serve as an interpreter. As the session went on I noticed that the client seemed increasingly uncomfortable. What had happened? Since I was translating, I took the liberty of modifying what was said to fit each other's culture, and I had confidence in my ability to do so. I could not figure out what was wrong until the session was over and I reviewed the videotape with the counselor and some of my colleagues. The counselor had noticed the same problem and wanted to understand what was going on. What we found amazed us all.

First, the counselor's way of looking at the client—his eye contact—was improper. When two Chinese talk to one another, we use much less eye contact, especially when it is with a person of the opposite sex. The counselor's gaze at the Chinese woman could have been considered rude or seductive in Chinese culture.

Although his nods were acceptable, they were too frequent by Chinese standards. The student client, probably believing one good nod deserved another, nodded in harmony with the counselor. That unusual head bobbing must have contributed to the student's discomfort.

The counselor would mutter "uh-huh" when there was a pause in the woman's speech. While "uh-huh" is a good minimal encouragement in North America, it happens to convey a kind of arrogance in China. A self-respecting Chinese would say *er* (oh), or *shi* (yes) to show he or she is listening. How could the woman feel comfortable when she thought she was being slighted?

He shook her hand and touched her shoulder. I told our respected visiting counselor afterward, "If you don't care about the details, simply remember this rule of thumb: in China, a man is not supposed to touch any part of a woman's body unless she seems to be above 65 years old and displays difficulty in moving around."

"Though I have worked in the field for more than 20 years, I am still a layperson here in a different culture," the counselor commented as we finished our discussion.

Mary Ivey comments: Weijun has shared with us an excellent example of the need for multicultural awareness if we are to develop authentic counseling relationships. You may not work in China, but you will work with individuals from varying backgrounds whose style of communication will be different from your own. As you develop as a counselor or interviewer, recall the need to adapt not only attending skills but also all other theoretical and practical aspects of the field. What works for you in one interview may not in the next. Be ready for a career of constant learning and challenge.

Finally, remember the obvious: clients can't talk while you do. Examine your interviews from time to time for amount of talk-time. Who talks the most, you or your client? With less verbal young children, however, it may be wise for you to talk slightly more, tell stories, and so on, to help them verbalize.

Box 3-3   Research Evidence That You Can Use: Attending Behavior

Empirical research on attending behavior has been extensive over the years (see Daniels and Ivey, 2006, and the CD-ROM accompanying this book for a comprehensive report). Hill and O'Brien (1999, p. 90) present a review of the literature and conclude that "smiling, a body orientation directly facing the client, a forward trunk lean, both vertical and horizontal arm movements, and a medium distance of about 55 in. between the helper and client are all generally helpful nonverbal behaviors."

On the other hand, Nwachuku and Ivey (1992) tested a program of culture-specific training and found that variations to meet cultural differences were essential. Stewart, Jackson, Neil, Jo, Nehring, and Grondin (1998) found that White counselors' perception of their expressed empathy was not in accord with the African American males' perception of counselor-expressed empathy. The way an interviewer can "be with" a client tends to vary among cultures.

Bensing (1999) provides a thought-provoking quote from her extensive review of the literature: "Simply looking at the patient has proven to be very important . . . and even silence can be very therapeutic, at least when it is used effectively" (p. 295). Among other findings, she noted that U.S. physicians are more detached and task-oriented than are Dutch physicians, who are rated as warmer and more involved. We can expect variations among both individuals and cultures in their style of attending.

The teaching of social skills has been successful with a variety of populations. Early work demonstrated the value of teaching attending behavior to hospitalized patients (Donk, 1972; Ivey, 1973). A study of adult schizophrenics showed that teaching social skills with special attention to attending behavior was successful and that patients maintained the skills over a 2-month period (Hunter, 1984). Shy, withdrawn people (avoidant personality) profited from social skills training in a well-controlled study by van der Molen (2006). Despite these and many other successes, the counseling and therapy fields still give relatively little attention to this psychoeducational mode of treatment (Sanderson, 2002).

Attention and neuropsychology

Attention is not just a psychological concept—it is measurable through brain imaging. When a person attends to a stimulus (e.g., the client's story) many areas of the brain of both interviewer and client become involved (Posner, 2004). Two key factors in attention are arousal and focus. Arousal involves the reticular activating system, at the brain's core, which transmits stimuli to the cortex and activates neurons firing throughout many areas. By giving attention to the client, both interviewer and client thought processes have been activated. Selective attention is a critical aspect of listening—"Focus is brought about by . . . a part of the thalamus, which operates rather like a spotlight, turning to shine on the stimulus" (Carter, 1999, p. 186). If you do not attend to the client's key issues, this focus is lost.

Just as attending behavior underlies all the microskills of this book, so do attention and selective attention provide the physiological foundation for personal growth. By listening and selective attention, both our brain and the client's are reacting and changing.

Social Skills Training, Attending Behavior, and Microskills

Social skills training is a specific set of strategies oriented to teaching clients basic life skills. Rather than talking through issues, social skills training involves educational methods to *teach* clients an array of interpersonal skills and behaviors. These skills include a wide range of behaviors, such as dating behaviors, drug-refusal skills, assertiveness, mediation, and job-interviewing procedures. Virtually all interpersonal actions can be taught through social skills training.

An important domain of interpersonal competence is that of social skills. Listening carefully, building bonds with others, and using interpersonal influence are all social skills. Thus, while attending behavior is considered groundwork for effective helping, it is also a social skill on which all of us can improve. Regardless of the type of concern we have, we want someone who will listen carefully. Great communicators are known for their rhetoric and ability to influence. However, great interviewers and counselors are sought because of their ability to listen to client stories.

Training as treatment is a term that summarizes the method and goal of social skills training. The microtraining format of selecting specific skill dimensions for education has become basic to most social skills programs since its inception (Ivey, 1971). As you extend the counseling and interviewing dimensions to skills training itself, think of the following steps: (a) negotiate a skill area for learning with the client; (b) discuss the specific and concrete behaviors involved in the skill, sometimes presenting them in written form as well; (c) practice the skill with the client in a role-play in the interview or group counseling session; and (d) plan for generalization of the skill to daily life.

Social skills training is a cognitive-behavioral method to help clients communicate more effectively with others. You will find that most of the microskills emphasized in this book are useful in educating clients in life skills. Two forms of social skills training will be emphasized here—how to teach your clients communication skills informally and how to teach larger groups these ideas more systematically.

Shortly after the first work in identifying interviewing microskills, Allen Ivey was working with a first-year college student who suffered a mild depression and complained about the lack of friends. Allen asked the student what he talked about with those in his dormitory. The student responded by continuing his list of complaints and worries. With further probing, the student acknowledged that he spent most of his time with others talking about himself and his difficulties. It was easy to see that potential friends would avoid him. We all tend to move away from those who talk negatively and stay away from those who talk only about themselves and fail to listen to us.

On the spot, Allen talked to the student about attending behavior and its possible rewards. The three V's + B were presented, and the importance of gaining trust and respect from others by listening was emphasized. Allen suggested that the student might profit from actively listening to those around him rather than talking only about himself. The student expressed interest in learning these skills, and a practice session was initiated there in the interview. First, negative attending was practiced, and the student was able to see how his lack of listening might contribute to isolation in the dormitory. Then positive attending was practiced, and the student discovered that he could listen.

Transfer of training to the dormitory was then practiced. Allen and the client discussed specifics of selecting someone with whom to try these skills. When the student returned the following week, he had a big smile and reported that he had found his first friend at the university. Moreover, he discovered an important side effect: "I feel less sad and depressed. First, I don't feel so alone and helpless. The second thing I noticed was that when I am attending to someone else, I am not thinking about myself and then I feel better." One frequent characteristic of depressed clients is that they reinforce pessimistic thoughts about themselves by constant negative self-statements, either silently to themselves or through talking to others constantly about their problems.

When you are attending to someone else, it becomes much more difficult to think negatively about yourself. Instruction in attending behavior is one of the foundations of social skills training. Many types of clients can benefit from learning and practicing these skills. Van der Molen (1984, 2006) used attending behavior and other skills of this book in a highly successful program in which he taught shy people to become more socially outgoing. As just one other example, children diagnosed with attention deficit disorder (ADD) who receive skills training are less disruptive (Pfiffner & McBurnett, 1997). Effective skills training can help children learn better in the classroom.

As you work through this book, think about how educating clients in the various microskills can be beneficial. Although attending behavior is often the foundation, other skills of this book are useful in treatment as well. Later we will explore how to teach attending and other microskills to groups. Social skills training is one system for freeing your clients for intentional responding. Though it is not the total answer, it will be useful for you with many clients over the years.

USING ATTENDING IN CHALLENGING SITUATIONS

Some beginning interviewers and counselors may think that attending skills are simple and obvious. They may be anxious to move to the "hard stuff." You also may have wondered how attending behavior can be useful if you plan to work with challenging clients in schools, community mental health centers, or hospitals. The following is designed to illustrate from our personal experience some things that we and others have observed about attending over time.

Mary: Attending is natural to me and the basic listening sequence has always been central to my work with children, but I do find at times that my natural style is not enough, particularly when I work with distracted or acting-out children. Even with all my experience with children, like other professionals, sometimes I am at a loss as to what to do next. After some analysis, I found that if I moved back to my foundation in attending skills and focused carefully on visuals, vocals, verbal following, and body language, I could regain contact with even the most troubled child. Similarly, in challenging situations with parents, I have at times found myself returning to a focus on attending behavior, later adding the basic listening sequence and other skills. Conscious attending has helped me many times in involved situations. Attending is not a simple set of skills.

I train older students to work as school peer mediators and another group to be peer tutors for younger children. I have found that using the exercise at the beginning of this chapter on poor attending and then contrasting it with good attending works well as an introductory exercise. I then teach attending skills and the basic listening sequence to my student groups.

Allen: One of my most powerful experiences occurred when I first worked at the Veterans Administration with schizophrenic patients who talked in a stream of consciousness "word salad." I found that if I maintained good attending skills and focused on the exact words they were saying, they soon were able to talk in a normal, linear fashion. I also found that teaching communication skills with video and video feedback to even with the most troubled patients was effective. Sometimes this was sufficient treatment by itself to move them out of the hospital. Depressed psychiatric inpatients, in particular, responded well to social skills training (Ivey, 1973, 1991). However, I did find that highly distressed patients could learn only one of the four central dimensions at a time, as trying to grasp all four dimensions of attending was confusing for them. Thus, I would start with visual/eye contact and later move on to other attending skills.

Like Mary, when the going gets rough, I find that it helps me to return to basic attending skills and a very serious effort to follow what the client is saying as precisely as possible. In short, when in doubt, attend. It often works!

SUMMARY: BECOMING A SAMURAI*

Japanese masters of the swords learn their skills through a complex set of highly detailed training exercises. The process of masterful sword work is broken down into specific components that are studied carefully, one at a time. In this process of mastery, the naturally skilled person often suffers and finds handling the sword awkward. The skilled individual may even find his or her performance worsening during the practice of single skills. *Being aware of what one is doing can interfere with coordination and smoothness in the early stages.*

Once the individual skills are practiced and learned to perfection, the samurai retire to a mountaintop to meditate. They deliberately forget what they have learned. When they return they find the distinct skills have been naturally integrated into their style or way of being. The samurai then seldom have to think about skills at all: They have become samurai masters.

Restak (2003) summarizes the work of Ericsson (2001), who examined the world's best golfers, all of whom worked on specific skills to improve their game:

> The key is to increase one's control over each component of the performance. [Ericsson] came to this conclusion after studying the practice habits of the 10 most highly regarded golfers of the 20th century. Each exhibited intense concentration, leading to heightened awareness of each of the many components of their actions, coupled with an ability to adapt in ways that led to higher levels of control. In order to alter subtle aspects of their performance, *many of them studied videotapes of past tournaments.* (Italics ours)

Intentional and relaxed performance of skills and strategies takes time and practice. You may have found discomfort in practicing the single skill of attending. Later you may find the same problem with other skills. This happens to both the beginner and the advanced counselor. Improving and studying our natural skills often

---

* We are indebted to Lanette Shizuru, University of Hawai'i, Monoa, for the example of the samurai.

result in a temporary and sometimes frustrating decrease in competence, just as they do for the samurai.

Consider driving. When you first sat at the wheel, you had to coordinate many tasks, particularly if you drove a car with a stick shift. The clutch, the gas pedal, the steering wheel, and the gear ratios had to be coordinated smoothly with what you saw through the windshield. When you gave primary attention to the skills of driving, you might have lost sight of where you were going. But practice and experience soon led you to forget the specific skills, and you were able to coordinate them automatically and give full attention to the world beyond the windshield. The mastery of single skills led you to achieve your objectives.

Learn the skills of this book, but allow yourself time for meditation and/or integrating these ideas into your own natural authentic being.

There is no need to talk about yourself or give long answers when you attend to someone else. Give clients ownership of "air-time." After all, it is their story we need to hear. Your main responsibility as a helper is to assist others in finding their own answers. You'll be surprised how able they are to do this if you are willing to attend.

Box 3-4   Key Points

Why?	The four attending behaviors all have one goal in common: to reduce interviewer talk-time while providing the client with an opportunity to examine issues and tell her or his story. You can't learn about the other person or the problem while you are doing the talking! Lack of attending may also be used to stop needless client talk at any time during the interview. *But always use attending with individual and cultural sensitivity.*
What?	Attending behavior consists of four simple but critical dimensions (3 V's + B):
	1. *Visual/eye contact.* If you are going to talk to people, look at them.
	2. Vocal qualities. Your vocal tone and speech rate indicate much of how you feel about another person. Think of how you feel about another person. Think of how many ways you can say "I am really interested in what you have to say" just by your vocal tone and speech rate.
	3. *Verbal tracking.* The client has come to you with a topic of interest; don't change the subject. Keep to the topic initiated by the client. If you change the topic, be aware that you have and realize the purpose of your change.
	4. *Attentive body language.* In general, clients know you are interested in them if you face them squarely and lean slightly forward, have an expressive face, and use facilitative, encouraging gestures. Be authentic and sensitive to differences.
How?	Attending is easiest if you focus your attention on the client rather than on yourself. Note what the client is talking about, ask questions, and make comments that relate to your client's topics. For example:
Client:	I'm so confused. I can't decide between a major in chemistry, psychology, or language.
Interviewer:	(nonattending) Tell me about your hobbies. What do you like to do? *or* What are your grades?
Interviewer:	(attending) Tell me more. *or* You feel confused? *or* Could you tell me a little about how each subject interests you? *or* Opportunities in chemistry are promising now. Could you explore that field a bit more? *or* How would you like to go about making your decision?

(continued)

Box 3-4 (continued)

	Note that all attending responses follow the client's verbal statement. Each might lead the client in a very different direction. Interviewers need to be aware of their patterns of selective attention and how they may unconsciously direct the interview. Finally, nonattending responses may sometimes be helpful in discouraging certain client talk and focusing the interview.
With whom?	Attending is vital in all human interactions, be they counseling, medical interviews, or business decision meetings. It is also important to note that different individuals and cultural groups may have different patterns of listening to you. Some may maintain distinctly different patterns of eye contact, for example, and consider the direct gaze rude and intrusive. Consider the possibility of teaching attending behavior to clients. Social skills training may be useful for those who are shy or depressed and for many others in distress. The business person who is overbearing and talks constantly can also benefit from learning how to attend.
And?	A simple but often helpful rule for interviewing is to *attend* when you become lost or confused about what to do. Simply ask the client to comment further on something just said or mentioned earlier in the interview.

COMPETENCY PRACTICE EXERCISES AND SELF-ASSESSMENT

Intentional interviewing and counseling is achieved through practice and experience. It will be enhanced by your own self-awareness, emotional competence, and ability to observe yourself, thus learning and growing in skills.

The competency-practice exercises on the following pages are designed to provide you with opportunities in three areas:

1. *Individual practice.* A short series of exercises is provided to give you an opportunity to practice by yourself aspects of attending behavior.
2. *Group practice.* Practice alone can be helpful, but working with others in role-played interviews is where the most useful learning occurs. Here you can obtain precise feedback on your interviewing style. And if videotape (or audiotape) is used with these practice sessions, you'll find that seeing yourself as others see you is a powerful experience.
3. *Self-assessment.* You are the person who will use the skills. We'd like you to look at yourself as an interviewer and counselor via some additional exercises.

Individual Practice

Exercise 1: Generating Alternative Attending Statements

A client comes to you with the following story as the interview begins:

> Our relationship is getting worse. We used to do so well and somehow seemed to strengthen each other. The parents on both sides were always opposed to our getting together; and lately Carlena is listening to her folks. She thinks I'm too rigid and can't let loose. And I just got in trouble at work. Sleeping is becoming difficult.

Some clients begin their interviews with a long list of problems. Write three things below that you might say in response to this client (remembering each might lead the client in a different direction). As a thought question, "How would you plan to work with these multiple issues over time?"

1. _____

2. _____

3. _____

Exercise 2: Identifying Topic Jumps

Review the two Jerome and Jared interviews demonstrating negative and positive attending behavior sessions. Identify specific places in both segments in which Jerome, the interviewer, changed topic. List each topic jump here and then compare them to our list on page 90.

Negative session: _____

Positive session: _____

Exercise 3: Behavioral Counts, an Evaluation and Research Strategy

This exercise introduces you to early research in microskills. After training in attending behavior, we have found that beginning interviewers have fewer eye contact breaks, speech hesitations, and topic jumps as well as a smaller number of distracting body gestures or movements.

Observe an interview. This could be a role-played counseling session, a television talk show, or simply an interaction between friends and family. Use the following form to count specific behaviors. Specifically, make a mark for each instance of less effective attending.

As you work through this exercise, recall multicultural and individual differences that can occur in the interview. Your observations and interpretations of what you observe need to be moderated with sensitivity to diversity.

_____ Visuals—number of eye-contact breaks

_____ Vocals—number of speech hesitations or uses of disruptive vocal tone

_____ Verbal following—number of topic jumps

_____ Body language—number of distracting movements or gestures

Group Practice

The following instructions are designed for groups of four but may be adapted for use with pairs, trios, and groups up to five or six. Ideally, each group will have access to video- or audiorecorders. However, use of careful observation (see Box 3-5) and the feedback sheets in the book (see Box 3-6) can provide enough structure for a successful practice session without the benefit of recording equipment.

Step 1: Divide into practice groups.   Get acquainted with each other informally before you go further.

Box 3-5   Guidelines for Effective Feedback

	To see ourselves as others see us. To hear how others hear us. And to be touched as we touch others . . . These are the goals of effective feedback.
Feedback	Feedback is one of the skill units of the basic attending and influencing skills developed in this book; it is discussed in more detail in Chapter 12. However, if you are to help others grow and develop in this program, you must provide feedback to them now on their use of skills in practice sessions. Here are some guidelines for effective feedback:
Guidelines	▲ *The person receiving the feedback is in charge.* Let the interviewer in the practice sessions determine how much or little feedback is wanted. ▲ *Feedback includes strengths,* particularly in the early phases of the program. If negative feedback is requested by the interviewer, add positive dimensions as well. People grow from strength, not from weakness. ▲ *Feedback is most helpful when it is concrete and specific.* Not "Your attending skills were good" but "You maintained eye contact throughout except for breaking it once when the client seemed uncomfortable." Make your feedback factual, specific, and observable. ▲ *Corrective feedback should be relatively nonjudgmental.* Feedback often turns into evaluation. Stick to the facts and specifics, though the word *relatively* recognizes that judgment inevitably will appear in many different types of feedback. Avoid the words *good* and *bad* and their variations. ▲ *Feedback should be lean and precise.* It does little good to suggest that a person change 15 things. Select one to three things the interviewer actually might be able to change in a short time. You'll have opportunities to make other suggestions later.

Step 2: Select a group leader.   The leader's task is to ensure that the group follows the specific steps of the practice session. It often proves helpful if the least experienced group member serves as leader first. Group members then tend to be supportive.

Step 3: Assign roles for the first practice session.

▲ *Client.* The first role-play client will be cooperative and have a story or issue to present, talk freely about the topic, and not give the interviewer a difficult time.

▲ *Interviewer.* The interviewer will demonstrate a natural style of attending behavior with the client and practice the basic skills.

▲ *Observer 1.* The first observer will fill out the feedback form (Box 3-6) detailing some aspects of the interviewer's attending behavior. Observation of these practice videos could also be called "microsupervision" in that you are helping the interviewer understand what he or she is doing and the effectiveness of the brief session. Later, when you are working as a professional helper, it is vital that you continue to share your work with colleagues through verbal report or audio/videotape. Supervision is a vital part of effective interviewing and counseling.

▲ *Observer 2.* The second observer will time the session, start and stop any equipment, and fill out a second observation sheet as time permits.

Box 3-6   Feedback Form: Attending Behavior

_____ (Date)

_____   _____
(Name of Interviewer)   (Name of Person Completing Form)

Instructions: Provide written feedback that is specific and observable, nonjudgmental, and support-ive. As an alternative, use behavioral counts as shown in Exercise 3.

1. *Visual/eye contact.* Facilitative? Staring? Avoiding? Sensitive to the individual client? At what points, if any, did the interviewer break contact? Facilitatively? Disruptively?

2. *Vocal qualities.* Vocal tone? Speech rate? Volume? Accent? Points at which these changed in response to client actions? Number of major changes or speech hesitations?

3. *Verbal tracking and selective attention.* Was the client able to tell the story? Stay on topic? Number of major topic jumps? Did shifts seem to indicate interviewer interest patterns? Did the interviewer demonstrate selective attention in pursuing one issue rather than another? Did the client have the majority of the talk-time?

4. *Attentive body language.* Leaning? Gestures? Facial expression? At what points, if any, did the interviewer shift position or show a marked change in body language? Number of facilitative body language movements? Was the session authentic?

5. *Specific positive aspects of the interview.*

6. *Discussion question:* What areas of diversity do the interviewer and client represent? How does this affect the session?

Step 4: Plan. The interviewer states her or his goals clearly, and the members of the group should take time to plan the role-play. The interviewer helps to open and to facilitate the client's talking about the story or concern. An increased percentage of client talk-time on a single topic will indicate a successful session. The interviewer may also plan to close off client talk and then open it again. It may help to keep the conversation going if he or she elicits both positive and negative comments; this also results in a deeper understanding of the client. The more concrete the plan, the greater is the likelihood of success.

The suggested topic for the attending practice session is "Why I want to be a counselor or interviewer." The client talks about her or his desire to do such work while the interviewer demonstrates attending skills. Other possible topics for the session include the following:

▲ A job you have liked, and a job that you didn't (or don't) enjoy.
▲ A positive experience you have had that led to a new learning about yourself.
▲ As part of the planning process, try at least one of the role-played interviews in which the client portrays a person who is physically challenged. The varying aspects of visuals, vocals, verbals, and body language may help sensitize you to varying individual and multicultural communication patterns.

The topics and role-plays are most effective if you talk about something meaningful to you. You will also find it helpful if everyone in the group works on the same topic as roles are rotated. In that way you can compare styles and learn from one another more easily.

While the interviewer and the interviewee plan, the two observers preview the feedback sheets and plan their own practice sessions to follow.

Step 5: Conduct a 3-minute practice session using attending skills. The interviewer practices the skills of attending, the client talks about the current work setting or other selected topic, and the two observers fill out the feedback sheets. Do not go beyond 3 minutes. If possible, record the interview.

Step 6: Review the practice session and provide feedback to the interviewer for 12 minutes. As a first step in feedback, the role-play client gives her or his impressions of the session and completes the Client Feedback Form of Chapter 1. This may be followed by interviewer self-assessment and comments by the two observers. As part of the review, ask yourselves the key question, "Did the interviewer achieve her or his own planning objective?" This is critical for assessing the level of mastery obtained.

On the feedback form, note both verbal and nonverbal behaviors and the different effects they had on the client and observers. Giving useful, *specific* feedback is particularly critical. Note the suggestions for feedback in Box 3-5.

Finally, as you review the audio- or videotapes of the interview, *start and stop the tape periodically.* Replay key interactions. Only in this way can you fully profit from the recording media. Just sitting and watching television is not enough; use media actively.

Step 7: Rotate roles. Everyone should have a chance to serve as interviewer, client, and observer. Divide your time equally!

Some general reminders. It is not necessary to compress a complete interview into 3 minutes. Behave as if you expected the session to last a longer time, and the timer can break in after 3 minutes. The purpose of the role-play sessions is to observe skills in action. Thus, you should attempt to practice skills, not solve problems. Clients have often taken years to develop their interests and concerns, so do not expect to solve one of these problems or obtain the full story in a 3-minute role-play session. Written feedback, if carefully presented, is an invaluable part of a program of interview skill development.

Self-Assessment

Skills and strategies are vital in interviewing and counseling, but they are not enough. No two interviewers are the same—you as a person will integrate the skills in your own way. Artistic and integrated interviewing becomes part of the natural style of the person, but all of us need to examine ourselves and what we are doing. We also need feedback on our performance from others and that is where group practice, supervision, and conversations with our peers are important.

Exercise 1: What Is Your Natural Style of Listening and Attending?

An important part of self-awareness is reflecting on yourself as a listener and your natural style of attending. Attending to others is fundamental to empathic understanding, one of the central dimensions of emotional intelligence. How able are you to listen to and be with others?

Following are several alternatives to help you think about your abilities as a listening person. Some are thought questions to which you can respond in journal form. Others are action oriented.

❑ What led you to a course in interviewing skills? Are you a "people-person"? Have you had friends come to you to share their concerns and problems? Do you like to listen to others? What are your motivations?

❑ How comfortable are you with ideas of diversity and working with people different from you? Can you recognize yourself as a multicultural person full of many dimensions of diversity? Include issues of physical challenge as an important part of multiculturalism and diversity.

❑ View yourself on videotape or listen to an audiotape of an interview. One of the best ways to examine your natural style of listening is to observe the video or audiotape you made as you started this book. Again, if you have not made that first video, now is the time! What do you observe in an informal viewing of yourself?

❑ View the video or listen to the audiotape again. This time, use the observation suggestions of the Feedback Form: Attending Behavior, or Exercise 3 of Individual Practice: Behavioral Counts. If you use the behavioral count system, you may use that as a baseline for comparison with later interviews as you work through this book.

Exercise 2: Self-Evaluation of Attending Skills Competencies

Use the following as a checklist to evaluate your present level of mastery. Check those dimensions that you currently feel able to do. Those that remain unchecked can serve as future goals. *Do not expect to attain intentional competence on every dimension as you work through this book.* You will find, however, that you will improve your competencies with repetition and practice.

Level 1: Identification and classification. You will need this minimal level of mastery for those coming examinations!

❑ Ability to define the 3V's + B.
❑ Ability to discuss issues in diversity that occur in relation to attending behavior.
❑ Ability to write attending statements in response to written client statements.
❑ Ability to make behavior counts of the 3V's + B while observing interviews.

Level 2: Basic competence.   We suggest that you aim for this level of competence before moving on to the next skill area in this book.

❑   Ability to demonstrate culturally appropriate visuals/eye contact, vocal qualities, verbal following, and body language in a role-played interview.
❑   Ability to increase client talk-time while reducing your own.
❑   Ability to stay on a client's topic without introducing any new topics of your own.

Level 3: Intentional competence.   In the early stages of this book, strive for basic competence and work toward intentional competence later. Experience with the microskills model is cumulative and you will find yourself mastering intentional competencies with greater ease as you gain more practice.

❑   Ability to understand and manage your own pattern of selective attention.
❑   Ability to change your attending style to meet client individual and cultural differences.
❑   Ability to note topics that clients particularly attend to and topics that they may avoid.
❑   Ability to use attending skills with more challenging clients.
❑   Ability through attention and inattention to help clients move from negative, self-defeating conversation to more positive and useful topics. Conversely, this also includes helping clients who are avoiding issues to talk about them in more depth.

Level 4: Teaching competence.   Clearly, you are not expected to teach attending behavior at this time, but the opportunity may present itself. You may be asked to work with a small group of children or adolescents who need help with social skills. You may present a class in a church group or the dormitory. You may find a client who would benefit from direct instruction in attending behavior.

❑   Ability to teach clients in a helping session about the social skill of attending behavior. You may either tell clients about the skill or you may practice a role-play with them.
❑   Ability to teach small groups the skills of attending behavior.

DETERMINING YOUR OWN STYLE AND THEORY: CRITICAL SELF-REFLECTION ON ATTENDING BEHAVIOR

This chapter has focused on the importance of listening as the foundation of effective interviewing practice: When in doubt as to what to do—listen, listen, listen! Individual and cultural differences are central—visual, vocal, verbal, and body language styles vary. Avoid stereotyping any group.

Regardless of what any text on interviewing, counseling, and psychotherapy says, the fact remains that it is YOU who will decide whether to implement the ideas, suggestions, and concepts. What single idea stood out for you among all those presented in this chapter, in class, or through informal learning? What stands out for you is likely to be important as a guide toward your next steps. What points in this chapter struck you as most important? What might you do differently? How might you use ideas in this chapter to begin the process of establishing your own style and theory? Write your ideas here.

_____

_____

_____

_____

REFERENCES

Bensing, J. (1992). Instrumental and affective aspects of physician behavior. *Medical Care, 30,* 283–298.

Bensing, J. (1999). *Doctor-patient communication and the quality of care.* Utrecht, the Netherlands: Nivel.

Bleeker, J. (1990). Effects of a social skills training program of mildly mentally retarded adolescents. Lisse, the Netherlands: Swets & Zitlinger.

Carter, R. (1999). *Mapping the mind.* Berkeley: University of California Press.

Daniels, T., & Ivey, A. (2006). *Microcounseling* (3rd ed.). Springfield, IL: Thomas.

Donk, L. (1972). Attending behavior in mental patients. *Dissertation Abstracts International, 33* (Ord. No. 72-22 569).

Ericsson, K. (2001). The path to expert golf performance: Insights from the masters on how to improve performance by deliberate practice. In R. Thomas (Ed.), *Optimising performance in golf* (pp. 21–55). Brisbane: Australia Academic Press.

Hill, C., & O'Brien, K. (1999). *Helping skills.* Washington, DC: American Psychological Association.

Hunter. W. (1984). *Teaching schizophrenics communication skills: A comparative analysis of two microcounseling learning environments.* Unpublished dissertation, University of Massachusetts, Amherst.

Ivey, A. (1971). *Microcounseling: Innovations in interviewing training.* Springfield, IL: Thomas.

Ivey, A. (1973). Media therapy: Educational change planning for psychiatric patients. *Journal of Counseling Psychology, 20,* 338–343.

Ivey, A. (1991). *Media therapy reconsidered.* Paper presented at the Veterans Administration Conference, Orlando, FL, October.

Ivey, A., & Gluckstern, N. (1974). *Basic attending skills.* North Amherst, MA: Microtraining Associates.

Ivey, A., Normington, N., Miller, D., Morrill, W., & Haase, R. (1968). Microcounseling and attending behavior: An approach to pre-practicum counselor training. *Journal of Counseling Psychology, 15,* 1–12.

Nwachuku, U., & Ivey, A. (1992). Teaching culture-specific counseling use in microtraining technology. *International Journal for the Advancement of Counseling, 15,* 151–161.

Pfiffner, L., & McBurnett, K. (1997). Social skills training with parent generalization: Treatment effects for children with attention deficit disorder. *Journal of Consulting and Clinical Psychology, 65,* 749–757.

Posner M. (Ed.). (2004). *Cognitive neuropsychology of attention.* New York: Guilford.

Restak, R. (2003). *The new brain.* New York: Rodale.

Sanderson, W. (2002). Are evidence-based psychological interventions practiced by clinicians in the field? Editorial column in *Medscape Mental Health* 7(1).

Stewart, R., Jackson, A., Neil, D., Jo, H., Hill, M., & Baden, A. (1998). *White counselor trainees: Is there multicultural counseling competence without formal training?* Paper presented at the American Counseling Association World Conference, Indianapolis, IN. (ERIC Document No.: ED 419118)

van der Molen, H. (1984). *Aan verlegenheid valt iets te doen: Een cursus in plaats van therapie* [How to deal with shyness: A course instead of therapy]. Deventer: Van Loghum Slaterus.

van der Molen, H. (2006). Social skills training and shyness. In T. Daniels & A. Ivey (Eds.), *Microcounseling* (3rd ed.). Springfield, IL: Thomas.

COMMENTS ON INDIVIDUAL PRACTICE, EXERCISE 2

Jerome clearly topic jumped at 3, 5, 7, and 15. Needless to say, his attending skills were not all that effective at 9 and 11—he was continuing to talk on his own topic! At 13, he finally does attend to Jared.

All the responses in the second segment represent verbal following. At 32, we see an attending summary followed by a topic shift initiated by Jerome and designed to expand Jared's career thoughts.

RESPONSE TO CLIENT ON PAGE 73

Dr. Howard Busby of Gallaudet College has commented on this case (2002, personal e-mail communication to Allen Ivey, January 24). He points out that the problem could be related to the disability, the problems at school, or both. The "wheelchair might be the problem as much as how this client is dealing with it. The client obviously has a disability (terminology), but it does not have to be disabling unless the client makes it so. Although I am categorized as having a disability due to deafness, I have never allowed it to be disabling.

"My interpretation of the client is that this depression, on the surface, could be the result of mobility restriction. However, there are other factors that might have caused the depression, even if there were no need for a wheelchair. It is easy for the client to blame problems on the disability and thus distract from personal issues. The issue could be a poor campus environment, learned helplessness on the part of the client, or a combination of these and other multiple factors."

Allen Ivey comments: All of us would do well to consider Dr. Busby's comments. He is pointing out to us the importance of taking a broad and comprehensive view of all cases. We must avoid stereotyping, but we must also be sensitive to cultural difference.

Hearing Client Stories: How to Organize an Interview

Attending behavior is basic to all the communication skills of the microskills hierarchy. Without individually and culturally appropriate attending behavior, there can be no interviewing, counseling, or psychotherapy.

This section adds to attending skills by presenting the *basic listening sequence* that will enable you to elicit the major facts and feelings pertinent to a client's concern. Through the skills of questioning, encouraging, paraphrasing, reflecting feelings, and summarizing, you will learn how to draw out your clients and understand the way they think about their stories.

Questioning skills open this section, but in addition to obtaining verbal information about the client, you need to observe what occurs nonverbally. Chapter 5 on observation skills gives you the opportunity to practice noting your clients' verbal and nonverbal behavior. You are also asked to observe your own nonverbal reactions in the session.

The remainder of the basic listening sequence is then presented. Chapter 6 examines the clarifying skills of paraphrasing, encouraging, and summarizing. Reflecting client feelings is the topic of Chapter 7.

Once you have mastered observation skills and the basic listening sequence, you are prepared to conduct a full, well-formed interview, comprising five stages. You will be able to conduct this interview using only listening and observation skills. Furthermore, it is important that you be able to evaluate your interviews and those of others for level of empathic understanding. It is vital not only to listen but also to listen empathetically.

This section, then, has ambitious goals. By the time you have completed Chapter 8, you'll have attained several major objectives, enabling you to move on to the influencing skills of interpersonal change, growth, and development. At an intentional level of competence, you may aim to accomplish the following in this section:

1. Master the basic listening sequence, enabling the client to tell the story. In addition, draw out key facts and feelings related to client issues.

2. Observe clients' reactions to your skill usage and modify your skills and attending behaviors to complement their uniqueness.
3. Conduct a complete interview using only listening and observing skills.
4. Evaluate that interview for its level of empathy; in effect, examine yourself and your ability to communicate warmth, positive regard, and other, more subjective dimensions of interviewing and counseling.

When you've accomplished these tasks, you may find that your clients have a surprising ability to solve their own problems, issues, concerns, or challenges. Furthermore, you may also gain a sense of confidence in your own ability as an interviewer or counselor. The underlying theme of this book's first eight chapters, then, is "When in doubt, listen!"

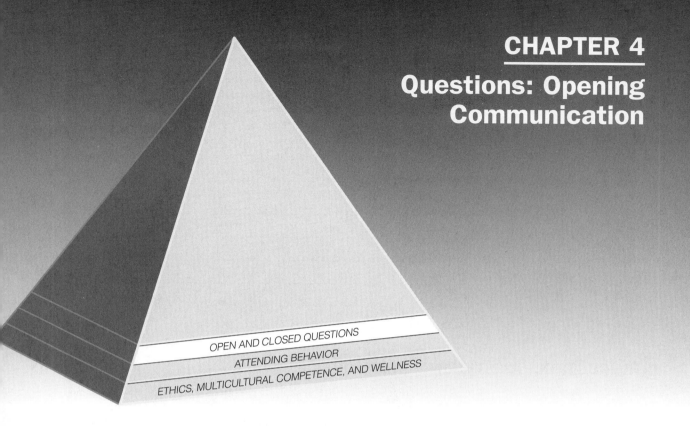

CHAPTER 4

Questions: Opening Communication

OPEN AND CLOSED QUESTIONS

ATTENDING BEHAVIOR

ETHICS, MULTICULTURAL COMPETENCE, AND WELLNESS

How can questions help you and your clients?

Major function	If you use open questions effectively, you can expect the client to talk more freely and openly. Closed questions will elicit shorter responses and may provide you with information and specifics.
	Like attending behavior, questions can encourage or discourage client talk. With questions, however, the leadership comes mainly from the interviewer. The client is often talking within your frame of reference. Questions potentially can take away from client self-direction.
Secondary functions	Knowledge and skill in questioning result in the following:

▲ Bringing out additional specifics of the client's world and enriching his or her story.

▲ Enabling you to make an effective assessment of a client's concern or issue.

▲ Guiding the manner in which a client talks about an issue. For example, *what* questions often lead to talk about facts, *how* questions to feelings or process, and *why* questions to reasons.

▲ Helping to open or close client talk according to the individual needs of the interview.

Benjamin is in his junior year of high school, in the middle third of his class. In this school, each student must be interviewed about plans after graduation—work, the armed forces, or college. You are the high school counselor and have called him in to check on his plans after graduation. His grades are average and he is not particularly verbal and talkative, but is known as a "nice boy."

What are some questions that you could use to draw him out and help him think ahead to the future? And what might be a problem if you ask too many questions?

_____

_____

_____

_____

_____

You may want to compare your questions with our thoughts on page 120 at the end of this chapter.

INTRODUCTION: QUESTIONING QUESTIONS

Although attending behavior is the skill and action foundation of the microskills hierarchy, it is questioning that provides a systematic framework for directing the interview. Questions help an interview begin and move along smoothly. They open up new areas for discussion, assist in pinpointing and clarifying issues, and aid in clients' self-exploration.

Questions are an essential component in many theories and styles of helping, particularly cognitive-behavioral counseling, brief counseling, and much of career decision-making work. The employment counselor, the social worker conducting an assessment interview, and the high school guidance counselor helping a student work on college admissions all need to use questions.

This chapter focuses on two key styles of questioning—open and closed questions:

Open questions are those that can't be answered in a few words. They encourage others to talk and provide you with maximum information. Typically, open questions begin with *what, how, why,* or *could:* For example, "Could you tell me what brings you here today?" You will find these helpful as they can facilitate deeper exploration of client issues.

Closed questions can be answered in a few words or sentences. They have the advantage of focusing the interview and obtaining information, but the burden of guiding the talk remains on the interviewer. Closed questions often begin with *is, are,* or *do:* for example "Are you living with your family?" Used judiciously, they enable you to obtain important specifics.

Some theorists and many practitioners raise important issues around the use of questioning, however. Certainly, excessive use takes the focus from the client and gives too much power to the interviewer. Your central task in this chapter is to find your own balance in using questions in the interview.

Key Issues Around Questions

Why do some people object to questions? Take a minute to recall and explore some of your own experiences with questions in the past. Perhaps you had a teacher or a parent who used questions in a manner that resulted in your feeling uncomfortable or even attacked. Write here one of your negative experiences with questions and the feelings and thoughts the questioning process produced in you.

My difficult personal experience with questions was as follows:

_____

_____

_____

The thoughts and feelings this experience produced in me were these:

_____

_____

_____

People often respond to this exercise by describing situations in which they were put on the spot or grilled by someone. They may associate questions with anger and guilt. Many of us have had negative experiences with questions. Furthermore, questions may be used to direct and control client talk. School discipline and legal disputes typically use questions to control the person being interviewed. If your objective is to enable clients to find their own way, questions may inhibit your reaching that goal, particularly if they are used ineffectively. It is for these reasons that some helping authorities, particularly those who are humanistically oriented, object to questions in the interview. Additionally, in many non-Western cultures, questions are inappropriate and may be considered offensive or overly intrusive.

Nevertheless, questions remain a fact of life in our culture. We encounter them everywhere. The physician or nurse, the salesperson, the government official, and many others find questioning clients basic to their profession. Most counseling theories espouse using questions extensively. Cognitive-behavioral and brief counseling, for example, use many questions. The issue, then, is how to question wisely and intentionally.

The goal of this section is to explore some aspects of questions and, eventually, to determine their place in your communication skills repertoire. The basic focus of this chapter on questioning skills is on open and closed questions. Used carefully, questioning is a valuable skill.

Sometimes Questions Are Essential—"What Else?"

Clients do not always provide you with important information, and sometimes the only way to get at missing data is by asking questions. For example, the client may talk about being depressed and unable to act. As a helper, you could listen to the story carefully but still miss important underlying issues relating to the depression. The open question "What important things are happening in your life right now or with your family?" might bring out the fact that a separation or divorce is near at hand, that a job has been lost, or that there is some other important dimension

Box 4-1 Research Evidence That You Can Use: Questions

Research results can be a guide or a confirmation for what types of questions work best in certain situations or with particular theoretical orientations. Not too surprising is confirmation that open questions produce longer client responses than closed ones (Daniels & Ivey, 2006; Tamase, Torisu, & Ikawa, 1991). Sternberg and others (1996) found this to be true also with children. Research shows that different theoretical orientations to helping have widely varying usage of questions. Effective use of open questions can help interviewers avoid leading the child to false statements, as has sometimes happened in forensic work. Hill (1999, p. 113) summarized research that compared reflecting feelings and responding to questions oriented to feelings and commented that "clients were most likely to talk about their feelings when they were directly asked about them." Tamase (1991) reported similar results.

Research on group leadership found that person-centered leaders tend to use very few questions whereas 40% of the leads of a problem-solving group leader were questions (Ivey, Pedersen, & Ivey, 2001; Sherrard, 1973). You will find that implementing cognitive behavioral therapy (CBT) or brief counseling will require extensive use of questions whereas the person-centered model uses very few. Your decision about which theory of helping you use most frequently will be important in your use of this skill.

Questions and Neuropsychology

Questions are often a good route to help a client discuss issues from the past residing in long-term memory, lodged primarily in the temporal lobe cortex (Kolb & Wishaw, 2003). However, questions that lead clients too much can result in their constructing stories of things that never happened. In a classic study, Loftus (1997) found that false memories could be brought out by simply reminding people of things that never happened. In other studies, brain scans revealed that false memories activate different memories from those that are true (Schachter, 1997). Needless to say, a counselor or therapist may not know whether the memories are true or false. Be careful of putting your ideas into the client's head via probing questions.

underlying the concern. What you first interpreted as a classical clinical depression becomes modified by what is occurring in the client's life, and treatment takes a different direction.

An incident in Allen's life illustrates the importance of questions. His father became blind after open heart surgery. Was that a result of the surgery? No, rather, it was because the physicians failed to ask the basic open question "Is anything else happening physically or emotionally in your life at this time?" If that question had been asked, the physicians would have discovered that Allen's father had developed severe and unusual headaches the week before surgery was scheduled, and they could have diagnosed an eye infection that is easily treatable with medication.

In counseling, a client may speak of tension, anxiety, and sleeplessness. You listen carefully and believe the problem can be resolved by helping the client relax and plan changes in her work schedule. However, you ask the client, "What else is going on in your life?" Having developed trust in you because of your careful listening and interest, the client finally opens up and shares a story of sexual harassment. At this point, the goals of the session change.

Useful questions from the helper than can provide more complete data include the following:

"Looking back at what we've been talking about, *what else* might be added? As you think about this session, what might we have missed? You may even have thought about something and not said it."

"Could you tell me a bit about what occurs to you at this moment?" (This question often provides surprising and helpful new information.)

"*What else* might a friend or family member add to what you've said? From a _____ perspective (insert ethnicity, race, sexual preference, religious, or other dimension), how could your situation be viewed?" (These questions change the focus and help clients see their issues in a broader, network-based context of friends, family, and culture.)

"Have we missed anything?"

EXAMPLE INTERVIEW: CONFLICT AT WORK

Virtually all of us have experienced conflict on the job. Angry, difficult customers, insensitive supervisors, lazy colleagues, and challenges from those whom we may supervise ourselves all give us concern. In the following set of transcripts, we see an employee assistance counselor, Jamila, meeting with Kelly, a junior manager who has a conflict with Peter. The first session illustrates how closed questions can bring out specific facts but then can sometimes end in leading the client in the counselor's direction, even to the point of putting the counselor's ideas into the client's mind.

You are very likely to work with clients who have similar interpersonal issues on the job wherever you may practice. The following case example focuses on the single skill of questioning as a way to bring out client stories. At times this can be helpful, but do not forget the dangers of using too many questions in the session.

Closed-Question Example

Interviewer and Client Conversation	Process Comments
1. *Jamila:* Hi, Kelly. What's happening with you today?	Jamila has talked with Kelly once in the past about difficulties she has had in her early experiences as a manager supervising others for the first time. She begins the session with an open question that could also be seen as a standard social greeting.
2. *Kelly:* Well, I'm having problems with Peter again.	Kelly and Jamila have a good relationship and so Kelly starts out easily. Not all clients are so ready to discuss their issues. More time for developing rapport, trust, and relationship will be necessary for many, even if they are returning for further conversations.

(continued)

Interviewer and Client Conversation	Process Comments
3. *Jamila:* Is he arguing with you?	Closed question, Jamila is already defining the issues without finding out what Kelly is thinking and feeling. Jamila, incidentally, appears to be demonstrating good attending skills. She looks like she is interested and listening.
4. *Kelly:* (hesitates) Not really, he's so difficult to work with.	Kelly sits back in discouragement.
5. *Jamila:* Is he getting his work in on time?	We are now going to see Jamila try to diagnose the problem with Peter by asking a series of closed questions.
6. *Kelly:* No, that's not the issue. He's even early.	Having worked with Jamila before, Kelly sits back and waits for the interviewer to take the lead.
7. *Jamila:* Is his work decent? Does he do a good job?	Jamila is starting to grill Kelly.
8. *Kelly:* That's one of the problems, his work is excellent and always there on time. I can't criticize what he does.	
9. *Jamila:* (hesitates) Is he getting along with others on your team?	Many people who rely on closed questions suddenly find themselves having run out of what to say. They then search for another closed question that may be further off the mark. Jamila's body tenses and she frowns as she thinks of what to ask.
10. *Kelly:* Well, he likes to go off with Daniel and they laugh in the corner. It makes me nervous. He ignores the rest of the staff—it isn't just me.	
11. *Jamila:* So, it's you we need to work on. Is that right?	Jamila has been searching for an individual to blame. She finds her—Kelly. Jamila relaxes a little as she thinks she is on to something.
12. *Kelly:* (hesitates and stammers) . . . Well, I suppose so . . . I . . . I . . . really hope you can help me work it out.	Kelly looks to Jamila as the expert who will find the answers for her. While she dislikes taking blame for the situation, she is also anxious to please and too readily accepts the interviewer's diagnosis.
13. *Jamila:* Does Peter follow your orders and instructions?	Another closed question and Jamila again appears the confident, assured interviewer.
14. *Kelly:* Most of the time, but not as much as I like.	Kelly sits waiting as if a pupil of Jamila.
15. *Jamila:* Are you clear in the way you tell Peter what you want?	Again, the question stems from Jamila's idea of the problem and gives no attention to Kelly's worldview.

16. *Kelly:* (more hesitation) . . . Well . . . I'm not sure. Perhaps that's part of the problem.	Kelly is starting to internalize the difficulty with Peter as "her problem." The issues are now being defined by the interviewer.
17. *Jamila:* Well, I've got some ideas that might help. . . . Let me share them now.	Jamila starts to give advice and direction, but has never heard the issues discussed from Kelly's frame of reference.

Closed questions can quickly bring out data, but they can overwhelm clients and can be used as evidence to force them to agree with the interviewer's idea of what is happening. We'd like to think that the session above is extreme and never happens in real life anywhere. Sad to say, encounters such as this are common in daily life and sometimes even occur in interviewing and counseling sessions. There is a power differential between clients and counselors. It is possible that an interviewer who fails to listen can impose inappropriate decisions on a client.

Open-Question Example

Jamila demonstrates better verbal following. She again uses questions as the main way to draw out Kelly's story, but this time through open questions we learn Kelly's story rather than the one Jamila imposes with closed questions. The interview is for the client, not for Jamila. The session has been condensed so that it focuses on questions. What is presented here actually took about 20 minutes. Again, this is the second meeting for Jamila and Kelly in the employee assistance office.

Interviewer and Client Conversation	Process Comments
18. *Jamila:* Hi, Kelly. What's happening with you today?	Jamila uses the same easy beginning as in the closed question example. She has excellent attending skills and is good at relationship building.
19. *Kelly:* Well, I'm having problems with Peter again.	Kelly responds in the same way as in the first—demonstration.
20. *Jamila:* More problems? Could you share more with me about what's been happening lately?	Open question beginning with "could." Potentially a "could" question may be responded to as a closed question and answered with "yes" or "no." But in the United States, Canada, and other English-speaking countries, it functions more often as an open question. It has a particularly helpful function in that the word "could" provides some control to the client.
21. *Kelly:* This last week Peter has been doing something new. He goes off in the corner with Daniel and the two of them start laughing. He's ignoring most of our staff, and he's been getting under my skin even more lately. And in the middle of all this, his work is fine, on-time and near perfect. But he is so impossible to deal with. It's beginning to tear me apart.	The intentional prediction is that Kelly will respond with information. She provides an overview of the situation and shares how it is affecting her. We are getting much of the same data as in the closed question example, but Kelly is supplying it in her own way. We are hearing Kelly's story.

(continued)

Interviewer and Client Conversation	Process Comments
22. *Jamila:* I hear you. Peter is getting even more difficult and seems to be affecting your team as well. It's really stressing you out and you look upset. Is that pretty much how you are feeling about things?	Jamila summarizes what has been said and acknowledges Kelly's emotions. The closed question at the end is termed a perception check or checkout. Periodically checking with your client to see whether you have heard correctly can help you in two important ways: (a) It communicates to clients that you are listening and encourages them to continue; (b) if you are wrong in your summary of what you have heard, the client can clarify things for you. At times, clients give us a lot of data and we need to ensure through a summary that we hear them accurately.
23. *Kelly:* That's right. I really need to calm down.	
24. *Jamila:* Let's change the pace a bit. Could you give me a specific example of an exchange you had with Peter last week that didn't work well?	Jamila recognizes that Kelly does not want to explore her emotions in depth and asks for a concrete example. Specific illustrations of client issues are often helpful in understanding what is really occurring.
25. *Kelly:* Last week, I asked him to review a report that Anne from bookkeeping had completed for us. It's pretty important that our team understand what's going on. He kind of looked at me when I gave the report to him as if he were thinking, "Who are you to tell *me* what to do?" But he sat down and did it that day. Friday, at the staff meeting, I asked him to summarize the report for everyone. In front of the whole group, he started with a joke about me and that I don't understand numbers so he had to review this report for me. Daniel laughed, but the rest of the staff just sat there. He was obviously trying to get me. I just ignored it. But that's typical of what he does. And he even put Anne down and presented her report as routine and not very interesting.	Specific illustrations of client issues are often helpful in understanding what is really occurring. Single examples can be representative of recurring problems or difficulties. The concrete specifics provided by one or two detailed stories help us better understand what is really going on. At 6, Kelly said, "Peter is getting even more difficult and seems to be affecting the team as well." Although this is important information, it is abstract and the specifics of the narrative are missing. Now that Jamila has heard the specifics, she is better prepared to be helpful.
26. *Jamila:* And, Kelly, how did you feel when he said that?	Jamila follows up with a "how" question oriented to emotions.
27. *Kelly:* I actually blushed and felt embarrassed. But I was so furious. First with him and then with myself for not being able to come up with a response to him.	Kelly outlines her several emotions.
28. *Jamila:* You're furious. Kelly, why do you imagine he is doing that to you?	Will the "why" question lead to the discovery of reasons?

29. *Kelly:* (hesitates) Really, I don't know why. I've tried to be helpful to him.	The intentional prediction did not result in the expected response. This is, of course, not unusual. It is likely too soon for Kelly to know why. This illustrates a common problem with "why" questions.
30. *Jamila:* Gender is sometimes an issue in situations such as this. Men do put women down at times. Could you explore that possibility?	Jamila supplies her own hunch as to what is going on. This is definitely a leading question and needs to be presented carefully. But instead of putting her ideas out as truth, she offers them tentatively with a "could" question and presents her reframing of the situation as "possibility."
31. *Kelly:* Jamila, it makes sense. I've halfway thought of it, but I didn't really want to acknowledge the possibility. It is clear that Peter has taken Daniel away from the team. Until Peter came aboard, we worked together beautifully. (pause) Yes, it makes sense for me. Peter is trying to get in with my supervisor. I see him going up to him all the time. He talks to my two female staff members and our secretary in a demeaning way. I think he's out to take care of himself. He's a great talent, but he *is* difficult. Somehow, I'd like to keep him on the team, but how?	In the first example, Kelly felt her failure to work with Peter was her fault. With Jamila's help, she is beginning to obtain a broader perspective on the difficulties. Kelly is able to identify several situations indicating that Peter's ambition and sexist behavior are issues that need to be addressed.
32. *Jamila:* So, it's becoming clear that you are not the problem. The question now is how are we going to handle this issue. You want a working team and you want Peter to be part of it. Let's talk about it further and then, if it makes sense, we can explore the possibility of assertiveness training as a way to deal with Peter. But, before that, I'd like to review some of your strengths. What do you bring to this situation with Peter that will be helpful to you?	Jamila provides support for Kelly's new frame of reference. She provides ideas for where the interview can go next, but suggests that before this happens, time needs to be spent on the positive asset search and find wellness strengths to build on. Kelly can best resolve these issues if she works from a base of her resources and capabilities.
33. *Kelly:* Good question. I suppose first of all, I really do know more about this area than Peter. He is new to it. I need to remind myself of that. Then, I worked through an issue with Jonathan two years ago when he was a colleague of mine. Jon kept hassling me and finally, I just sat down and had it out with him. He was fine after that. I know my team respects me; they come to me for advice all the time. Daniel was fine until Peter came aboard.	Kelly smiles for the first time. She has had sufficient support from Jamila that she can readily come up with her strengths. However, don't expect it always to be that easy. Remember, this session is condensed. It is easy for clients to return to their weaknesses and ignore their assets.

(continued)

Interviewer and Client Conversation	Process Comments
34. *Jamila:* Kelly, so far we have focused mainly on your difficulties. But let's stop a moment and look at your strengths. First of all, we both know you have done good work with your team over several months now and you have a good relationship with them, so you have their support. I've noticed that even though Peter gets to you, even here you manage to hold your cool. You do lots of things right. What else could we include as your resources and positive strengths as we look at this situation?	Jamila first summarizes Kelly's strengths and gives her positive feedback. She then asks the "What else . . . ?" question so that Kelly can describe some of her strengths herself. This is the wellness approach.
35. *Kelly:* (hesitates) . . . Well, I did work out difficult problems before by talking directly when I was challenged, but Peter is really something. My friends say I am right and can handle it. I certainly have more experience in this area than Peter. . . . Yes, I think you're right. Perhaps I need more confidence in myself.	Clients who come to us to talk over their problems are often surprised to find that the interviewer has also been listening for signs of wellness and positive assets. These strengths can be most helpful to the client in finding resolution and ways to cope with challenge. The change of pace surprises Kelly, but she starts to think more carefully about herself and her strengths.
36. *Jamila:* Could you tell me more about one of those challenges you met in the past? Could you give me a specific example?	The first "could" question opens the topic and asks permission to explore positives. The second searches for concrete specifics. Out of this situation, we expect to find specific skills that later can be applied to the interpersonal challenge.

In this condensed excerpt, we see several important points. The first is that Kelly has been given more talk-time and room to explore what is happening. The questions oriented to a specific example particularly helped clarify what was going on. We also see that question stems such as why, how, and could have some predictability in producing client responses but never a perfect prediction. The positive asset search is a particularly important part of successful questioning. We can best help resolve issues by emphasizing strengths.

INSTRUCTIONAL READING: MAKING QUESTIONS WORK FOR YOU

Following are several issues around the use of questioning techniques and strategies. Questions can be facilitative or they can be so intrusive that clients want to close up and say nothing. Use the ideas presented here to help you define your own position around questioning and how it fits with your natural style.

Eight Major Issues Around Questions

1. Questions Help Begin the Interview

With verbal clients and a comfortable relationship, the open question facilitates free discussion and leaves plenty of room to talk. Here are some examples:

"What would you like to talk about today?"
"Could you tell me what you'd like to see me about?"
"How have things been since we last talked together?"
"The last time we met we talked about your plan to face up and talk with your partner about sexual issues. How did it go this week?"

The first three open questions provide considerable room, in that the client can talk about virtually anything. The last question is open but provides some focus for the session, building on material from the preceding week.

Such open questions may be more than a nontalkative client can handle, however. It may be best in such situations to start the session with more informal conversation—for example, focusing on the weather, a positive part of last week's session, or a current event that you know is of interest to the client. As the client becomes comfortable, you can then turn to the issues for this session.

2. Open Questions Help Elaborate and Enrich the Client's Story

A beginning interviewer often asks one or two questions and then wonders what to do next. Even more experienced interviewers at times find themselves hard-pressed to know what to do. An open question on some topic the client presented earlier in the interview helps the session start again and keep it moving:

"Could you tell me more about that?"
"How did you feel when that happened?"
"Given what you've said, what would be your ideal solution to the problem?"
"What might we have missed so far?"
"*What else* comes to your mind?"

3. Questions Help Bring Out Concrete Specifics of the Client's World

If there is one single open question that appears to be useful in practice sessions and in most theoretical persuasions, it is the one that aims for concreteness and specifics in the client's situation. The model question "Could you give me a specific example?" is the most useful open question available to any interviewer. Many clients tend to talk in vague generalities, and specific, concrete examples enrich the interview and provide data for understanding action. Some additional open questions that aim for concreteness and specifics follow:

Client:	Ricardo makes me so mad!
Counselor:	Could you give me a specific example of what Ricardo does?
	What does Ricardo do specifically that brings out your anger?
	What do you mean by "makes me mad?"
	Could you specify what you do before and after Ricardo makes you mad?

Box 4-2 National and International Perspectives on Counseling Skills: Using Questions With Youth at Risk

Courtland Lee, Past President, American Counseling Association, University of Maryland

Malik is a 13-year-old African American male who is in the seventh grade at an urban junior high school. He lives in an apartment complex in a lower middle (working) class neighborhood with his mother and 7-year-old sister. Malik's parents have been divorced since he was 6 and he sees his father very infrequently. His mother works two jobs to hold the family together and she is not able to be there when he and his sister come home from school.

Throughout his elementary school years, Malik was an honor roll student. However, since starting junior high school, his grades have dropped dramatically and he expresses no interest in doing well academically. He spends his days at school in the company of a group of seventh- and eighth-grade boys who are frequently in trouble with school officials.

This case is one that is repeated among many African American early teens. But this problem also occurs among other racial/ethnic groups as well, particularly those who are struggling economically. And the same pattern occurs frequently even in well-off homes. There are many teens at risk for getting in trouble or using drugs.

While still a boy, Malik has been asked to shoulder a man's responsibilities as he must pick up things his mother can't do. Simultaneously, his peer group discounts the importance of academic success and wants to challenge traditional authority. And Malik is making the difficult transition from childhood to manhood without a positive male model.

I've developed a counseling program designed to empower adolescent Black males that focuses on personal and cultural pride. The full program is outlined in my book *Empowering Black Males* (1992) and focuses on the central question, *"What is a strong Black man?"* While this question is designed for group discussion, it is an important one for adolescent males in general, who might be engaging in individual counseling. The idea is to use this question to help the youth redefine in a more positive sense what is means to be strong and powerful. Some of the related questions that I find helpful include these:

(continued)

Closed questions, of course, can bring out specifics as well, but they place more responsibility on the interviewer. However, if the interviewer knows specifically the desired direction of the interview, closed questions such as "Did Ricardo show his anger by striking you?" "Does Ricardo tease you often?" and "Is Ricardo on drugs?" may prove invaluable, because they may encourage clients to say out loud what before they were only hinting at. But even well-directed closed questions may take the initiative away from the client.

4. Questions Are Critical in Assessment

Physicians must diagnose their clients' physical symptoms. Managers may have to assess a problem on the production line. Vocational counselors need to assess a

Box 4-2   (continued)

▲ What makes a man strong?
▲ Who are some strong Black men that you know personally? What makes these men strong?
▲ Do you think that you are strong? Why?
▲ What makes a strong body?
▲ Is abuse of your body a sign of strength?
▲ Who are some African heroes or elders that are important to you? What did they do that made them strong?
▲ How is education strength?
▲ What is a strong Black man?
▲ What does a strong Black man do that makes a difference for his people?
▲ What can you do to make a difference?

Needless to say, you can't ask an African American adolescent or a youth of any color these questions unless you and he are in a positive and open relationship. Developing sufficient trust so that you can ask these challenging questions may take time. You may have to get out of your office and into the school and community to become a person of trust.

My hope for you as a professional counselor is that you will have a positive attitude when you encounter challenging adolescents. They are seeking models for a successful life and you may become one of those models yourself. I hope you think about establishing group programs to facilitate development and that you'll use some of these ideas here with adolescents to help move them toward a more positive track.

Allen Ivey comments: As counselors, one of our most important responsibilities is prevention. Looking for strengths in family, community, and culture can be vital. Mary and I had the good fortune to work with elementary students in "cultural identity groups." The children in our groups focused on learning about themselves as cultural beings. We brought in family members by giving students take-home questions about family and cultural history. These were shared in the group and students learned pride about their background as well as important information from other children that led to cross-cultural understanding. One of our most powerful questions as part of a unit on oppression was "What have you done to stand up for right?" We still recall one student who had seen the local bus driver pass by a Person of Color without picking him up. The 10-year-old went up to the driver and challenged him. Follow-up 3 years later in high school found that these students still recalled the importance of standing tall against oppression.

client's career history. Questions are the meat of effective diagnosis and assessment. George Kelly, the personality theorist, has suggested for general problem diagnosis the following set of questions, which roughly follow the *who, what, when, where, how, why* of newspaper reporters:

Who is the client? What is the client's personal background? Who else may be involved?

What is the client's problem? What is happening? What are the specific details of the situation?

When does the concern occur? When did it begin? What immediately preceded the occurrence of the issue?

Where does the problem occur? In what environments and situations?
How does the client react to the challenge? How does the client feel about it?
Why does the problem occur?

Needless to say, the *who, what, when, where, how, why* series of questions also provides the interviewer with a ready system for helping the client elaborate or be more specific about an issue at any time during a session. To these, again, we suggest adding the *what else* question to encourage openness.

5. The First Word of Certain Open Questions Partially Determines What the Client Will Say Next

Often, but not always, using key-question stems results in predictable outcomes.

What questions most often lead to facts. "What happened?" "What are you going to do?"

How questions often lead to a discussion about processes or sequences or to feelings. "How could that be explained?" "How do you feel about that?"

Why questions most often lead to a discussion of reasons. "Why did you allow that to happen?" "Why do you think that is so?" Finding reasons can be helpful but also can lead into sidetracks. Remember: Many clients associate *why* with a past experience of being grilled.

Could, would, and *can* questions are considered maximally open and contain some of the advantages of closed questions in that the client is free to say "No, I don't want to talk about that." *Could* questions reflect less control and command than others. "Could you tell me more about your situation?" "*Would* you give me a specific example?" "*Can* you tell me what you'd like to talk about today?"

6. Questions Have Potential Problems

Questions can have immense value in the interview, but we must not forget their potential problems. Among them are the following:

Bombardment/grilling.   Too many questions will tend to put many clients on the defensive. They may also give too much control to the interviewer.

Multiple questions.   Interviewers may confuse their clients by throwing out several questions at once. This is another form of bombardment, although at times it may be helpful to some clients as the client can select which question to answer.

Questions as statements.   Some interviewers may use questions as a way to sell their own points of view. "Don't you think it would be helpful if you studied more?" "What do you think of trying relaxation exercises instead of what you are doing now?" This form of question, just like multiple questions, can be helpful at times. Awareness of the nature of such questions, however, may allow you to consider alternative and more direct routes of reaching the client. A useful rule of thumb is that if you are going to make a statement, it is best not to frame it as a question.

Questions and cultural differences. The rapid-fire North American questioning style is often received less favorably in other societies. If you are working with a member of a cultural group for whom questions may be inappropriate, be aware that excessive use of questions sometimes results in distrust of the counselor.

In addition, research has revealed that men tend to ask more questions than women. Some interpret this as an indication that men are using questions to control their conversations. (If questions feel uncomfortable for you in your experience, remember you need to be aware that questions can still be helpful to many of your clients.)

Why *questions.* As children, most of us experienced some form of the "Why did you do that?" question. *Why* questions often put interviewees on the defensive and cause discomfort. This same discomfort can be produced by any question that evokes a sense of being attacked.

Questions and control. The person who asks the questions is usually in control of the interview. He or she is determining who talks about what, when the talk will occur, and under what conditions it will occur. At times, questions can be helpful in bringing out-of-control interviews under control and direction. At the same time, questions can be used unfairly and intrusively for the interviewer's gain rather than the client's. And excessive use of questions can destroy the relationship built by use of attending skills.

7. In Cross-Cultural Situations, Questions Can Promote Distrust

If your life background and experience are in relative synchrony with those of the client, you may find that you can use questions immediately and freely. On the other hand, if you come from a different cultural background, your questions may be met by distrust and only grudgingly answered. Questions place the power in the interviewer. A barrage of questions to a poor client if you, the interviewer, are clearly middle class may cause the client not to come back for any more interviews. If you are African American or White and working with an Asian American or a Latino/a, an extreme questioning style can produce mistrust. If the ethnicities are reversed, the same problem could occur.

Allen was conducting research and teaching in South Australia with Aboriginal social workers. He was seeking to understand their culture and their special needs for training. Allen is naturally inquisitive and sometimes asks many questions. Nonetheless, the relationship between him and the group seemed to be going well. But one day, Matt Rigney, whom Allen felt particularly close to, took him aside and gave some very useful corrective feedback:

> You White fellas! . . . Always asking questions! Let me tell you what goes on in my mind when a White person asks me a question. First, my culture considers many questions rude. But, I know you and that's what you do. But, this is what goes on in my mind when you ask me a question. First, I wonder if I can trust you enough to give you an honest answer. Then, I realize that the question you asked is too complex to be answered in a few words. But I know you want an answer.

So I chew on the question in my mind. Then, you know what? Before I can answer the first question, you've moved on to the next question!

Allen was lucky that he had developed enough trust and rapport that Matt was willing to share his perceptions. Many People of Color have said that the Australian Aboriginal feedback represents how they felt about many interactions with White people. Moreover, disabled individuals, gays/lesbians/bisexuals, spirituality conservative persons, and many others may be distrustful of the interviewer who uses questions too much if they are from a different group. Finally, *anyone* who is quizzed too much may feel the same way!

8. Questions Can Be Used to Help Clients Search for Positive Assets and Patterns of Wellness

Stories presented in the helping interview are often negative and full of problems and difficulties. People grow from strength, not from weakness. Carl Rogers, the founder of client-centered therapy, was always able to find something positive in the interview. He considered positive regard and respect for the client essential for future growth.

The positive asset search or a wellness review are concrete ways to approach positive regard and respect for the client. As you listen to the client, constantly search for strengths and positives. Then, share your observations. Of course, you do not want to become overly optimistic and minimize the seriousness of the client's situation. However, it is increasingly clear that if you listen only to the sad and negative parts of the client's story, progress and change will be slow and painful (Peterson & Seligman, 2004; White & Epston, 1990).

As a general rule, we suggest that you use the basic listening sequence to draw out the story in relatively brief form. Then repeat and summarize the story to ensure that you have heard accurately. We also recommend that you obtain at least one concrete specific example of the story to ensure mutual understanding.

As appropriate to the client and situation, begin your search for positive assets and strengths. If you develop with your client a list of strengths and assets, you will find you can draw on them later for resolution of concerns and problems. Naturally, do not push strengths against client wishes, as this may appear to minimize her or his concerns. However, seek to make a positive approach part of the interview and later treatment plan.

Some specific, concrete examples of how to engage in a positive asset search include the following:

Personal strength inventory. Clients tend to talk about their problems and what they can't do. This puts them "off-balance." We can help them center and feel better about themselves through a strength inventory. What is the client doing right?

As part of any interview, I like to do a strength inventory. Let's spend some time right now identifying some of the positive experiences and strengths that you either have now or have had in the past.

▲ Could you tell me a story about a success you have had sometime in the past? I'd like to hear the concrete details.

▲ Tell me about a time in the past when someone supported you and what he or she did. What are your currently available support systems?

▲ What are some things you have been proud of in the past? Now?

▲ What do you do well or others say you do well?

Cultural/gender/family strength inventory.   Here we move outside the individual and look at context for positive strengths.

▲ Taking your ethnic/racial/spiritual history, can you identify some positive strengths, visual images, and experiences that you have now or have had in the past?

▲ Can you recall a friend or family member of your own gender who represents some type of hero in the way he or she dealt with adversity? What did that person do? Can you develop an image of her or him?

▲ We all have family strengths despite frequent family concerns. Family can include our extended family, our stepfamilies, and even those who have been special to us over time. For example, some people talk about a special teacher, a school custodian, or an older person who was helpful. Could you tell me concretely about them and what they mean to you?

Positive exceptions to the concern.   Searching for times when the problem doesn't occur is often useful. This approach is common in brief counseling. With this information, you can determine what is being done right and encourage more of the same.

Let's focus on the exceptions. When is the problem or concern absent or a little less difficult? Please give me an example of one of those times.

▲ Few problems happen all the time. Could you tell me about a time when it didn't happen? That may give us an idea for a solution.

▲ What is different about this example from the usual?

▲ How did the more positive result occur?

▲ How is that different from the way you usually handle the concern?

Feedback coupled with positive questions.   As you attend to clients' issues and problems, you will often note that they have strengths they are not discussing. If you provide clients with feedback about your positive observations, they may start thinking about themselves in new ways. Jamila says at 34,

> Kelly, so far we have focused mainly on your difficulties. But let's stop a moment and look at your strengths. First, we both know you have done good work with your team over several months now and you have a good relationship with them, so you have their support. I've noticed that even though Peter gets to you, even here you manage to hold your cool. You do lots of things right. *What else* could we include as your resources and positive assets as we look at this situation?

Many clients will be hesitant to say good things about themselves. Your observations and feedback can be helpful to them in developing a new view of themselves.

The *what else* question provides an opportunity for the client to add more strengths and resources to your feedback.

You obviously will not have time for all of the positive asset searches discussed here. But when you focus only on the negative story, you place your clients in a very vulnerable position. On the other hand, do not become oblivious to their concerns, issues, and problems. Do not use the positive asset search to cover up or hide basic issues. Rather, wellness strengths are resources for resolving our concerns. And in longer-term counseling, we recommend a thorough wellness assessment, as in Chapter 2.

Using Open and Closed Questions With Less Verbal Clients

Generally, open questions are much preferred to closed questions in the interview. Yet it must be recognized that open questions require a verbal client, one who is willing to share information, thoughts, and feelings with you. Here are some suggestions that may encourage clients to talk more freely with you.

Build trust at the client's pace.   A central issue with hesitant clients is trust. If the client is required to meet with you or is culturally different from you, he or she may be less willing to talk. At this time, your own natural openness and social skills are particularly important. Trust building and rapport need to come first. With some clients, trust building may take a full session or more. Extensive questioning too early can make trust building a slow process with some clients. Often it is helpful to discuss multicultural differences openly. "I'm wondering how you feel about my being (White/male/heterosexual or vice versa) as we discuss these issues."

Accept some randomness.   Your less verbal client is not likely to give you a clear, linear story of the problem. If he or she lacks trust or is highly emotionally involved in the concern, it may take some time for you to get an accurate understanding. You may need to use a careful balance of closed and open questions to draw out the story and get "bits and pieces" before you can put together a coherent story. Regardless of what you do, keep your language as simple, straightforward, and concrete as possible. With some clients you will find that disclosing your own stories is helpful— but do this with caution.

Search for concrete specifics.   Counselors and therapists talk about the *abstraction ladder*. If you or the client moves too high on the abstraction ladder, things won't make sense to anyone. This is especially so with less verbal or emotionally distraught clients. Constantly seek to find concrete examples and stories. Avoid searching for general themes and patterns. The open question "Could you give me a *concrete, specific* example?" is particularly effective at such times. Abstractions and repeating patterns of behavior, though important, should wait until the basic facts, feelings, and thoughts have been discovered.

Seek short, concrete answers.   After some trust is generated, you might begin by making an observation such as "The teacher said you and she had an argument." Then try asking a concrete open question such as "What did your teacher say (or do)?" If you focus on concrete events and avoid evaluation and opinion in a nonjudgmental fashion, your chances for helping the client talk will be greatly expanded. You'd like to know what happened specifically, what each person said and did, and perhaps the accompanying emotions. Examples of concrete questions focusing on narrow specifics include the following:

"What happened first? What happened next? What was the result?" (This helps you draw out the linear sequence of the story.)

"What did the other person say? What did he or she do? What did you say or do?" (This focuses on observable concrete actions.)

"What happened afterward? What did you do afterward? What did he or she do afterward?" (Sometimes clients are so focused on the event that they don't yet realize it is over. This helps them see results of events.)

"What did you feel or think just before it happened? During? After? What do you think the other person felt?" (This helps focus emotions.)

Note that each of the preceding questions requires relatively short answers. These are open questions that are more focused in orientation and can be balanced with some closed questions. Do not expect your less verbal client to give you full answers to these questions. You may need to ask closed questions to fill in the details and obtain specific information. ("Did he say anything?" "Where was she?" "Is your family angry?" "Did they say 'yes' or 'no'?")

A *leading* closed question is dangerous, particularly with children. In the previous examples, you can see that a long series of closed questions can bring out the story, but it may provide only the client's limited responses to *your* questions rather than what the client really thought or felt. Worse, the client may end up adopting your way of thinking or may simply stop coming to see you.

Questioning and other listening skills. One effective counseling method is to repeat the client's main words by paraphrasing or reflecting feeling (see Chapters 6 and 7) and then to raise the intonation of your voice at the end in a questioning tone. For example:

Client: I was really upset by my parents. They entered my room when I was gone and searched the whole place. They suspect me of taking drugs.

Some open questions might include "Could you tell me what led your parents to the search?" or "What feelings does this bring out in you?" Another way to help the client keep talking is to repeat what you have heard. A paraphrase might be "Your parents entered your room and suspect you of taking drugs?" while a reflection of feeling could be "You sound really *angry*?" These are not questions, but the rising tone of voice at the end of your comment offers the same openness and often will enable your client to talk more deeply, as you have clearly indicated that you have been listening to him or her carefully. This is one of the reasons some humanistic counselors object to questions. Effective rephrasing can often accomplish the same objective.

Working with children. Children, in particular, may require considerable help from the interviewer before they are willing to share at all. With children, a naturally warm, talkative person who *likes and accepts* children will be able to elicit more information more easily. Thus, it helps to begin sessions with children by sharing something fun and interesting. Games, clay, and toys in the counseling room are useful when dealing with children. You will find that children generally like to do something with their hands while they talk; having a child draw a picture during the conversation can often be useful to the child and to the interviewer. And the drawings often reveal what is going on in the child's life. Many open questions are too broad for young children to understand, but be careful with closed questions and avoid leading too much.

Undergirding the preceding suggestions is the reminder that your ultimate goal is to draw out (a) the general picture of what happened; (b) key facts of the situation; (c) emotions; and (d) where appropriate, the reasons for the situation via the *why* question. *But do not expect this information to come in sequential order.* If you have patience and use attending skills well, you will eventually be able to bring out the full story. (Clarifying skills—minimal encouragement and paraphrasing— reflection of feeling, and summarization are skills elaborated on in later chapters that can be used to help you and the client organize the story more fully.)

SUMMARY: MAKING YOUR DECISION ABOUT QUESTIONS

We began this chapter by asking you to think carefully about your personal experience with questions. It is clear that their overuse can damage an interview. On the other hand, questions do facilitate conversation and help ensure that important points are brought in. Questions can help the client bring in missing information. Among such questions are "What else?" "What have we missed so far?" and "Can you think of something important that is occurring in your life right now that you haven't shared with me yet?"

Person-centered theorists and many professionals sincerely argue against the use of any questions at all. They strongly object to the control implications of questions. They point out that careful attending and use of the listening skills can usually bring out major client issues. If you work with someone culturally different from you, a questioning style may develop distrust. In such cases, questions need to be balanced with self-disclosure and listening.

Our position on questions is clear—*we believe in questions, but we also fear overuse.* We are impressed by the brief counselors who seem to use questions more than any other skill but are still able to respect their clients and help them change. On the other hand, we have seen students who have demonstrated excellent attending skills regress to using only questions. Questions can be an easy "fix" but they require listening to the client if they are to be meaningful.

The positive asset search has been a foundation of the microskills program since its beginnings. We believe that Carl Rogers was correct when he focused on positive regard and unconditional acceptance. We have noted again and again that therapy all too often ends in a self-defeating repetition of problems. Questions that bring out strengths and resources often lead clients to specific assets that they can use to help resolve issues and problems.

The most useful chapter summary will be your impressions and decisions. Where do you personally stand on the use of questions?

Box 4-3 Key Points

Why?	Questions help begin the interview, open new areas for discussion, assist in pinpointing and clarifying issues, and assist the client in self-exploration.
What?	Questions can be described as open or closed. *Open questions* are those that can't be answered in a few short words. They encourage others to talk and provide you with maximum information. Typically, open questions begin with *what, how, why,* or *could.* One of the most helpful of all open questions is "Could you give a specific example of . . . ?" *Closed questions* are those that can be answered in a few words or sentences. They have the advantage of focusing the interview and bringing out specifics, but they place the prime responsibility for talk on the interviewer. Closed questions often begin with *is, are,* or *do.* An example is "Where do you live?" It is important to note that a question, open or closed, on a topic of deep interest to the client will often result in extensive talk-time *if* it is important enough. If an interview is flowing well, the distinction between open and closed questions is less important.
How?	A general framework for diagnosis and question asking is provided by the newspaper reporter framework of *who, what, when, where, how, why.* *Who* is the client? What are key personal background factors? Who else is involved? *What* is the issue? What are the specific details of the situation? *When* does the problem occur? What immediately preceded and followed the situation? *Where* does the issue occur? In what environments and situations? *How* does the client react? How does he or she feel about it? *Why* does the problem or concern happen? *What else* is there to add to the story? Have we missed anything? Interviewing is about more than problems. The above set of questions could be asked to discover what events and issues surround a positive situation or accomplishment. Interview training often overemphasizes concerns and difficulties. A positive approach is needed for balance.
With whom?	Questions may turn off some clients. Some cultural groups find North American rapid-fire questions rude and intrusive, particularly if asked before trust is developed. Yet questions are very much a part of Western culture and provide a way to obtain information that many clients find helpful. If you work from a person-centered theory, try to avoid using questions. On the other hand, if you favor theories such as cognitive behavioral or brief counseling, questioning is a necessary skill.
What else?	Emphasizing only negative issues may result in a downward cycle of depression and discouragement. The positive asset search encourages a balance as we examine client issues and concerns. What is the client doing right? What are the exceptions to the problem? What are client personal, family, and cultural/contextual resources?

COMPETENCY PRACTICE EXERCISES AND PORTFOLIO OF COMPETENCE

Individual Practice

Exercise 1: Writing Closed and Open Questions

Select one or more of the following client stories and then write open and closed questions to elicit further information. Can you ask closed questions designed to bring out specifics of the situation? Can you use open questions to facilitate further elaboration of the topic, including the facts, feelings, and possible reasons? What special considerations might be important with each person as you consider age-related multicultural issues?

Jordan (age 15, African American):

I was walking down the hall and three guys came up to me and called me "queer" and pushed me against the wall. They started hitting me, but then a teacher came up.

Alicja (age 35, Polish American):

I've been passed over for a promotion three times now. Each time, it's been a man who has been picked for the next level. I'm getting very angry and suspicious.

Dominique (age 78, French Canadian):

I feel so badly. No one pays any attention to me in this "home." The food is terrible. Everyone is so rude. Sometimes I feel frightened.

Write open questions for one or more of the above. The questions should be designed to bring out broad information, facts, feelings and emotions, and reasons.

Could _____?

What _____?

How _____?

Why _____?

Now generate three closed questions that might bring out useful specifics of the situation.

Do _____?

Are _____?

Where _____?

Finally, write a question designed to obtain concrete examples and details that might make the problem more specific and understandable.

_____

_____

Exercise 2: Observation of Questions in Your Daily Interactions

This chapter has talked about the basic question stems *what, how, why,* and *could,* and how clients respond differently to each. During a conversation with a friend or acquaintance, try these five basic question stems sequentially:

Could you tell me generally what happened?

What are the critical facts?

How do you feel about the situation?
Why do you think it happened?
What else is important? What have we missed?

Record your observations here. Were the predictions of the book fulfilled? Did the person provide you, in order, with (a) a general picture of the situation, (b) the relevant facts, (c) personal feelings about the situation, and (d) background reasons that might be causing the situation?

_____

_____

_____

_____

_____

Group Practice

Two systematic exercises are suggested for practice with questions. The first focuses on the use of open and closed questions, the second on the assessment of a client's concern or problem. The instructional steps for practice are abbreviated from those described in Chapter 3, on attending behavior. As necessary, refer to those instructions for more detail on the steps for systematic practice.

Exercise 1: Systematic Group Practice on Open and Closed Questions

Step 1: Divide into practice groups.
Step 2: Select a group leader.
Step 3: Assign roles for the first practice session.
▲ Client
▲ Interviewer
▲ Observer 1, who uses the Feedback Form (Box 4-4) and leads the microsupervision process. Remember to focus on interviewer strengths as well as areas for improvement.
▲ Observer 2, who runs equipment, keeps time, and also completes the form.

Step 4: Plan. The interviewer should plan to use both open and closed questions. It is important in the practice session that the key *what, how, why,* and *could* questions be used. Add *what else* for enrichment.

Discuss a work challenge. The client may share a present or past interpersonal job conflict. The interviewer first draws out the conflict, then searches for positive assets and strengths.

Suggested alternative topics might include the following:

▲ A friend or family member in conflict
▲ A positive addiction (such as jogging, health food, biking, team sports)
▲ Strengths from spirituality or ethnic/racial background

Step 5: Conduct a 3- to 6-minute practice session using only questions. The interviewer practices open and closed questions and may wish to have handy a list of suggested question stems. The client seeks to be relatively cooperative and talkative but should not respond at such length that the interviewer has only a limited opportunity to ask questions. More time will be needed if you decide on a more challenging topic.

Step 6: Review the practice session and provide feedback to the interviewer for 12 minutes. Remember to stop the audio- or videotape periodically and listen to or view key happenings several times for increased clarity. Generally speaking, it is wise to provide some feedback before reviewing the tape, but this sometimes results in a failure to view or listen to the tape at all.

Step 7: Rotate roles.

Exercise 2: Systematic Practice in Elementary Assessment

This exercise focuses on the use of the newspaper formula (*who, what, when, where, how,* and *why*) to obtain a basic summary of the client's story. The steps of the exercise are identical to the preceding series of seven steps. The Feedback Form (Box 4-4) may be used.

An assessment interview will, of necessity, require more than 3 minutes. The broad questions of the newspaper formula will need amplification by further open and closed questions. The key question "Could you give me a specific example?" will often prove useful. The assessment interview may run 15 minutes or more in a practice session.

Remember the importance of the *what else* question to enlarge and enrich your assessment. Include a detailed examination of the *who, what, when, where, how,* and *why* of the positive asset search to balance the problem emphasis.

Here are some suggested topics for the practice assessment interview:

Work problems, past or present
Difficulty with a present or past academic course
A past illness or experience with a family member who was ill
Views on alcohol, drugs, sexuality
Problem with discrimination, prejudice, sexual harassment, heterosexism, lack of
    attention to disability

Finally, as part of this practice, search for the positives and exceptions to the problem. What was the person doing right?

Box 4-4   Feedback Form: Questions

_____ (Date)

_____   _____
(Name of Interviewer)   (Name of Person Completing Form)

Instructions: List below as completely as possible the questions asked by the interviewer. At
a minimum, indicate the first key words of the question (*what, why, how, do,
are,* and so on). Indicate whether each question was open (O) or closed (C). Use
additional paper as needed.

_____   1. _____

_____   2. _____

_____   3. _____

_____   4. _____

_____   5. _____

_____   6. _____

_____   7. _____

_____   8. _____

_____   9. _____

_____ 10. _____

1. Which questions seemed to provide the most useful client information?

2. Provide specific feedback on the attending skills of the interviewer.

3. Discuss the use of the positive asset search, wellness, and the use of questions.

Portfolio of Competence

Determining Your Own Style and Theory, the apex of the microskills hierarchy, can be best accomplished on a base of competence. Each chapter closes with a reflective exercise asking your thoughts and feelings about what has been discussed. By the time you finish this book, you will have a substantial record of your competencies and a good written record as you move toward determining your own style and theory.

Use the following as a checklist to evaluate your present level of mastery. Check those dimensions that you currently feel able to do. Those that remain unchecked can serve as future goals. *Do not expect to attain intentional competence on every dimension as you work through this book.* You will find, however, that you will improve your competencies with repetition and practice.

Level 1: Identification and classification.
❑ Ability to identify and classify open and closed questions.
❑ Ability to discuss, in a preliminary fashion, issues in diversity that occur in relation to questioning.
❑ Ability to write open and closed questions that might predict what a client will say next.

Level 2: Basic competence.   Aim for this level of competence before moving on to the next skill area.

❑ Ability to ask both open and closed questions in a role-played interview.
❑ Ability to obtain longer responses to open questions and shorter responses to closed questions.

Level 3: Intentional competence.   Work toward intentional competence throughout this book. All of us can improve our skills, regardless of where we start.

❑ Ability to use closed questions to obtain necessary facts, without disturbing the client's natural conversation.
❑ Ability to use open questions to help clients elaborate their stories.
❑ Ability to use *could* questions and, as predicted, obtain a general client story. ("Could you tell me generally what happened?" "Could you tell me more?")
❑ Ability to use *what* questions to facilitate discussion of facts.
❑ Ability to use *how* questions to bring out feelings ("How do you feel about that?") and information about process or sequence. ("How did that happen?")
❑ Ability to use *why* questions to bring out client reasons. ("Why do think your spouse/lover responds coldly?")
❑ Ability to bring out client concrete information and specifics. ("Could you give me a specific example?")
❑ Ability to use the newspaper sequence for assessment. (*who, what, where, when, why, how*)

Level 4: Teaching competence.   As stated earlier, do not expect yourself to become skilled in teaching others skills at this point. You may, however, find many clients who benefit from direct instruction in questioning others—particularly those who

talk too much about themselves find this skill important in breaking through their self-absorption.

❑ Ability to teach clients in a helping session the social skill of questioning. You may either tell clients about the skill or you may practice a role-play with them.

❑ Ability to teach small groups the skills of questioning.

DETERMINING YOUR OWN STYLE AND THEORY: CRITICAL SELF-REFLECTION ON QUESTIONING

This chapter has focused on the pluses and minuses of using questions in the session. While we, as authors, obviously feel that questions are an important part of the interviewing process, we have tried to point out that there are those who differ from us. Questions clearly can get in the way of effective relationships in interviewing, counseling, and therapy.

Regardless of what any text on interviewing, counseling, and psychotherapy says, the fact remains that it is YOU who will decide whether to implement the ideas, suggestions, and concepts. What single idea stood out for you among all those presented in this chapter, in class, or through informal learning? What stands out for you is likely to be important as a guide toward your next steps. What are your thoughts on multiculturalism and how it relates to your use of questions? How might you use ideas in this chapter to begin the process of establishing your own style and theory?

_____

_____

_____

REFERENCES

Daniels, T., & Ivey, A. (2006). *Microcounseling* (3rd ed.). Springfield, IL: Thomas.

Hill, C. (1999). *Helping skills.* Washington, DC: American Psychological Association.

Ivey, A., Pedersen P., & Ivey, M. (2001). *Intentional group counseling: A microskills approach.* Pacific Grove, CA: Brooks/Cole.

Kolb, B., & Whishaw, I. (2003). *Fundamentals of human neuropsychology* (5th ed.). New York: Worth.

Loftus, E. (1997). Creating false memories. *Scientific American,* September, 51–55.

Pederson, C., & Seligman, M. (2004). *Character strengths and virtues.* New York: Oxford.

Schacter, D. (1997) *Searching for memory.* New York: Basic Books.

Sherrard, P. (1973). *Predicting group leader/member interaction: The efficacy of the Ivey Taxonomy.* Unpublished doctoral dissertation, University of Massachusetts, Amherst.

Sternberg, K., Lamb, M., Hershkowitz, I., Esplin, P., Redlich, A., & Sunshine, N. (1996). The relationship between investigative utterance types and the informativeness of child witnesses. *Journal of Applied Developmental Psychology, 17,* 439–451.

Tamase, K. (1991). Factors which influence the response to open and closed questions: Intimacy in dyad and listener's self-disclosure. *Japanese Journal of Counseling Science, 24,* 111–122.

Tamase, K., Torisu, K., & Ikawa., J. (1991). Effect of the questioning sequence on the response length in an experimental interview. *Bulletin of Nara University of Education, 40,* 199–211.

White, M., & Epston, D. (1990). *Narrative means to therapeutic ends.* New York: Norton.

ALLEN AND MARY'S THOUGHTS ABOUT BENJAMIN

We would likely start the interview by explaining to Benjamin that we'd like to know what he is thinking about his future after he completes school. We would begin the session with some informal conversation about current school events, or something personal we know about Ben. The first question might be stated something like this: "You'll soon be starting your senior year; what have you been thinking about doing after you graduate?" If this question opens up some tentative ideas, we'd listen to these and ask him for elaboration. If he focuses on indecision between volunteering for the army or entering a local community college or the state university, we'd likely ask him some of the following questions:

"What about each of these appeals to you?"
"Could you tell me about some of your strengths that would help you in the army or college?"
"If you went to college, what might you like to study?"
"How do finances play a role in these decisions?"
"Are there any negatives about any of these possibilities?"
"How do you imagine your ideal life 10 years from now?"

On the other hand, Benjamin just might say to any of these, "I don't know, but I guess I better start thinking about it" and look to you for guidance. You sense a need to review his past likes and dislikes as possible clues to the future.

"What courses have you liked best in high school?"
"What have been some of your activities?"
"Could you tell me about the jobs you've had in the past?"
"Could you tell me about your hobbies and what you do in your spare time?"
"What gets you most excited and involved?"
"What did you do that made you feel most happy in the past year?"

Out of questions such as these, we may see patterns of ability and interest that suggest actions for the future.

If Benjamin is uncomfortable in the counseling office, all of these questions might put him off. He might feel that we are grilling him and perhaps even see us as intruding in his world. Generally speaking, getting this type of important information and organizing it requires the use of questioning. But questions are effective only if you and the client are working together and have a good relationship.

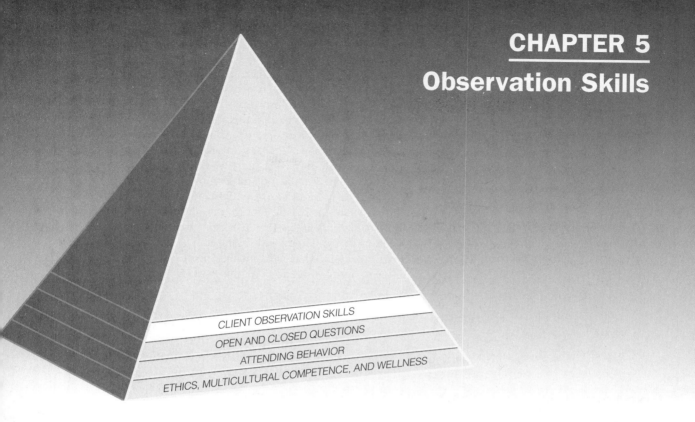

CLIENT OBSERVATION SKILLS

OPEN AND CLOSED QUESTIONS

ATTENDING BEHAVIOR

ETHICS, MULTICULTURAL COMPETENCE, AND WELLNESS

You can observe a lot by watching.
(Yogi Berra)

How can observation skills help you and your clients?

Major function	Your ability to observe what occurs between you and your clients verbally *and* nonverbally is vital to establishing a helping relationship. Observation skills will help you respond appropriately to both individual and multicultural differences. They will also guide you to key issues in the here and now of the interview.
Secondary functions	Knowledge and skill in observation will result in an increased ability to note the following in both yourself and your clients:

▲ Verbal behavior. How do you and your clients use language?
▲ Nonverbal behavior. How do you and your clients behave nonverbally?
▲ Discrepancies and conflict. Much of interviewing is about working through conflict and the inevitable incongruities we face in life.
▲ Styles associated with varying individual and cultural ways of expression. How can you flex intentionally and avoid stereotyping in your observation?

Allen sometimes testifies in cases involving Social Security disability claims. In one case, the "claimant," Horace, was an African American, about 60 years old, who had southern roots. He had a severe back problem from years of arduous lifting in a local factory. This is a common price that is paid in the bodies of men and women who do work that repeats the same motion again and again.

When called to the witness stand, Horace held his hat in his hand, spoke quietly to the judge and took his oath. But as the judge questioned him, his eyes started to wander—all about the room. He did not look at the judge at all. The judge became irritated by what he saw as a lack of attention and spoke sharply to Horace about his lack of respect toward the bench. Horace was obviously embarrassed, but didn't know what to do. In fact, as he became increasingly nervous, he only looked away more. The judge was very unhappy with Horace.

What sense do you make of what is happening here? Is Horace being disrespectful? What would you do in a situation like this if you were sitting in court?

_____

_____

_____

Please turn to page 150 to compare your thoughts with what Allen did.

INTRODUCTION: KEEPING WATCH ON THE INTERVIEW

The skilled interviewer is concerned with facilitating human development, but understanding your clients, their behavior, their thoughts, and their feelings will be central to that effort. It is therefore important that you become skilled in observing client behaviors in your interviewing sessions. This chapter is concerned with sharpening those powers of observation. You will find that the basic skills of attending and questioning continue to be useful in this process.

What should you observe about client behavior in the interview? From your own life experience you are already aware of many things that are important for a counselor or interviewer to notice. Brainstorm from what you already know and make a list.

But there are two people in the relationship. What about you? How are you affecting the client verbally and nonverbally? Looking at your way of being can be equally as important as, or more important than, observing the client. Start by taking a brief inventory here. You might begin by thinking back to your natural style of attending, but expand those self-observations. What is your interpersonal style and how might it affect your relationship with others?

Useful Behaviors to Note in the Client   *My Own Verbal and Nonverbal Styles*

There is an infinite array of nonverbal and verbal behaviors that you can assemble. To these can be added the vast number of theories that provide even more ideas for thinking about interview interaction. In this chapter we shall explore three basic observational issues that can help us develop a foundation for continued growth in understanding the complexity of what goes on between you and the client.

1. *Nonverbal behavior.* What you and your clients say is modified by how you say it through your visuals, vocal tone, and body language—and by many behaviors related to these three. Some authorities say that 85% or more of communication is nonverbal. Regardless of whether this claim is accepted, you need to be constantly aware of the underlying emotional tone often conveyed through nonverbals. *How something is said can sometimes overrule the actual words used by you or your client.*

2. *Verbal behavior.* Verbal following and selective attention were emphasized in Chapter 3 as central to attending behavior. Language is basic to interviewing and counseling and there many ways to consider verbal behavior, ranging from detailed linguistic examination through the differing language systems of varying counseling and therapy theories. This chapter will consider five dimensions useful for direct verbal observation in the session: patterns of selective attention, client key words, "I" statements and "other" statements, and abstract and concrete conversation.

3. *Discrepancies and conflict.* Whether you are helping clients work through problems, deal with issues, encounter challenges, or manage concerns, you will be facilitating the resolution of discrepancies, incongruity, and conflict in their lives. Out of your awareness of verbal and nonverbal behavior will come an increased ability to notice conflicts of many types. These important points will be reviewed again in Chapter 9 on confrontation and conflict management.

This chapter presents observations that can be made about many individuals and cultural groupings. However, you are going to be working with unique individuals. No one individual, regardless of multicultural background, behaves exactly as does the next. Thus, while it is important to learn about general patterns among individuals or groups, it is even more vital that you avoid stereotyping the meaning of any verbal or nonverbal observation.

Acculturation is a fundamental concept of anthropology with significant relevance for observing what is occurring in the interview. The working definition of acculturation for this book is as follows:

> Acculturation is the degree to which an individual has adopted the norms or standard way of behaving in a given culture. Due to the unique family, community, and part of the country in which an individual is raised, no two people will be acculturated to general standards in the same way. In effect, "normative behavior" does not exist in any one individual or even in one area of diversity. Thus, stereotyping individuals or groups needs to be avoided at all costs.

As an example, an African American client raised in a small town in upstate New York in a two-parent family has different acculturation experiences from those of a similar person raised in Los Angeles or East St. Louis. If one were from a single-parent family, the acculturation experiences would change further. If we alter only the ethnic/racial

Box 5-1   National and International Perspectives on Counseling Skills: Can I Trust What I See?

Weijun Zhang

James Harris, an African American professor of education, and I were invited by a national Native American youth leadership organization to give talks in Oklahoma. I attended Dr. Harris's first lecture with about 60 Native American children. Dynamic and humorous, Dr. Harris touched the heartstrings of everyone. But much to my surprise, when the lecture was over, he was very upset. "A complete failure," he said to me with a long face; "they are not interested." "No," I replied, "It was a great success. Don't you see how people loved your lecture?" "No, not at all." After some deliberation, I came to see why he could have such an erroneous impression.

"There were not many facial expressions," Professor Harris said. I said, "You may be right, using African American standards. However, that is not a sign that your audience was not interested. Native American people, in a way, are programmed to restrain their feelings, whether positive or negative, in public; as a result, their facial expressions would be hard to detect. Native American people have always valued restraint of emotion, considering this a sign of maturity and wisdom, as I know. Actually, in terms of emotional expressiveness, African American culture and Native American culture may represent two extremes on a continuum."

"They did not ask a single question, though I repeatedly asked them to," he said. I replied, "Well, they didn't because they respect you." "Come on, you are kidding me." I told him, "Many Native Americans are not accustomed to asking questions in public, probably for the following reasons. (a) If you ask an intelligent question, you will draw attention away from the teacher and onto yourself. That is not an act of modesty and may be seen as showing off. (b) If your question is silly, you will be seen as a laughingstock and lose face. (c) Whether your question is good or bad, one thing is certain: You will disturb the instructor's teaching plan, or you may suggest the teacher is unclear. That goes against the Native American tradition of being respectful to the senior. So you can never expect a Native American audience to be as active as African Americans in asking questions. In today's situation, some kids probably did want to ask you questions,

(continued)

background of this example client to Italian American, Jewish American, or Arab American, the acculturation experience changes again. Many other factors, of course, influence acculturation—religion, economic bracket, and even being the first or second child in a family. Awareness of diversity in life experience is critical if we are to recognize uniqueness and specialness in each individual. If you define yourself as White American, Canadian, or Australian and you think of others as the only people who are multicultural, you need to rethink your awareness. All of us are multicultural beings with varying and singular acculturation experiences.

Finally, consider biculturality and multiculturality. Many of your clients have more than one significant community cultural experience. A Puerto Rican, Mexican, or Cuban American client is likely to be acculturated in both Hispanic and U.S. culture. A Polish Canadian client in Quebec, a Ukrainian Canadian client in Alberta, and an Aboriginal client in Sydney, Australia, may also be expected to represent biculturality. And all Native Americans and Hawai'ians in the United States, Dene and Inuit in Canada, Maori in New Zealand, and Aboriginals in Australia exist in at least two

Box 5-1   (continued)

for you repeatedly encouraged them to. But, unfortunately, they still couldn't do it." He asked why. I said, "You waited only a few seconds for questions before you went on lecturing, which is far from enough. With Native Americans, you have to adopt a longer time frame. European Americans and African Americans may ask questions as soon as you invite them to; American Indians may wait for about 20 to 30 seconds to start to do so. That period of silence is a necessity for them. You might say that Native Americans are true believers in the saying 'Speech is silver, silence is golden.'" He asked, "Well, why didn't you tell me that on the spot, then?" "If it is respectful for those Native American kids not to ask you questions, Dr. Harris," I said, "how could you expect this humble Chinese to be so disrespectful as to come to the stage to correct you?"

The director of the Native American organization, who overheard my conversation with James Harris, approached me with the question "How do you possibly know all this about us Native Americans?" "Well, I don't think it is news for you that Native Americans migrated from Asia some thousands of years ago. You don't mind that your Asian cousins still share your ethnic traits, do you?" The three of us all laughed.

Mary Ivey comments: Differences in communication style abound. Weijun's nonverbal example is important. May I share a verbal style issue that often occurs when we work with children? I was trained in reflective Rogerian person-centered theory, which we now know works best with highly verbal, formal-operational, abstract clients.

Children are usually very concrete, and abstractions often confuse them. I learned rapidly that when I gave a "good" reflection of feeling, children often just stared at me. Children in some ways represent a different culture from that of adults. I soon found out that if I used concrete language and stories, children were easy to reach. All of the listening skills work well with youngsters, but we've got to meet them where they are. Ask yourself if your language style is concrete—can children see, hear, or feel what you say in a very immediate way? Is your story understandable to them?

cultures. There is a culture among people who have experienced cancer, AIDS, war, abuse, and alcoholism. All of these issues and many others deeply affect acculturation.

In short, stereotyping any one individual is not only discriminatory; it is also naïve!

More information will be presented in this chapter than you can absorb or apply immediately, but you will use these ideas throughout the rest of the book and will return to them many times as you practice skills. Moreover, even the most experienced professional interviewer, counselor, or therapist learns more about observing individual and multicultural differences every day.

EXAMPLE INTERVIEW: IS THE ISSUE DIFFICULTY IN STUDYING OR RACIAL HARASSMENT?

Kyle Yellowhorse is a second semester junior business major in a large university in the Northwest. He was raised in a relatively traditional Lakota family on the Rosebud Reservation in South Dakota. Native American Indians are unlikely to come to counseling unless they are referred by others or the interviewer has established herself or himself previously as a person who can be trusted.[*]

Kyle did well in his first 2 years at the university, but during the fall term, his "B" average dropped to barely passing. His professor of marketing has referred him to the campus counseling center. The interviewer in this case is European American.

The following interview, although focused on Native American Indian issues, has many parallels with other native people who have been dispossessed of their land and whose culture has been belittled—Hawai'ians, Aboriginals in Australia, Dene and Inuit in Canada, Maori in New Zealand, and Celtic people in Great Britain. Moreover, the interview in many ways illustrates what might happen in any type of cross-cultural counseling. A European American student meeting an interviewer who is Latina/o, African American, or other Person of Color (or vice versa if the roles of counselor and client are switched) might also have early difficulty in talking and establishing trust.

Interviewer and Client Conversation	Process Comments
1. *Derek:* Kyle, come on in; I'm glad to see you.	Derek walks to the door, he smiles, shakes hands, and has direct eye contact.
2. *Kyle:* Thanks. (pause)	Kyle gives the interviewer eye contact for only a brief moment. He sits down quietly.
3. *Derek:* You're from the Rosebud reservation, I see.	Derek knows that contextual and family issues are often important to Native American Indian clients. Rather than focusing on the individual and seeking "I statements," he realizes that more time may be needed to develop a relationship. Derek's office decorations include artwork from the Native American Indian, African American, and Mexican traditions as well as symbols of his own Irish American heritage.
4. *Kyle:* Yeah. (pause)	While his response is minimal, Kyle notes that Derek relaxes slightly in the chair.
5. *Derek:* There've been some hard times here on campus lately. (pause, but there is still no active response from Kyle) I'm wondering what I might do to help. But first, I know that coming to this office is not always easy. I know Professor Harris asked you to come in because of your grades dropping this last term.	During the fall term, the Native American Indian association on campus had organized several protests against the school mascot—"the fighting Sioux." As a result the campus has been in turmoil with recurring events of racial insults and several fights. Noting Kyle's lack of eye contact, Derek has reduced the amount of direct gaze and he also looks down. Among

[*]Many schools, elementary through university, have a small population of Native American Indians. For example, the Chicago public schools have slightly over 1,000 Native Americans mixed in schools throughout a large system. Most often, the native American Indian population is invisible and you may never know this cultural group exists unless you indicate through your behavior and actions in the community that you are a person who can be trusted and who wishes to know the community.

	traditional people, the lack of eye contact generally indicates respect. At the same time, many clients who are depressed use little eye contact.
6. *Kyle:* Yeah, my grades aren't so good. It's hard to study.	Kyle continues to look down and talks slowly and carefully.
7. *Derek:* I feel honored that you are willing to come in and talk, given all that has happened here. Kyle, I've been upset with all the incidents on campus. I can imagine that they have affected you. But first, how do you feel about being here talking with me, a White counselor?	Derek self-discloses his feelings about campus events. He makes an educated guess as to why Kyle's grades have dropped. As he talks about the campus problems, Kyle looks up directly at him for the first time and Derek notes some fire of anger in his eyes. Kyle nods slightly when Derek says, "I feel honored."
8. *Kyle:* It's been hard. I simply can't study. (pause) Professor Harris asked me to come and see you. I wouldn't have come, but I heard from some friends that you were OK. I guess I'm willing to talk a bit and see what happens.	People who are culturally different from you may not come to your office setting easily. This is where your ability to get into the community is important. In this case, college counseling center staff have been active in the campus community leading discussions and workshops seeking to promote racial understanding. As Kyle talks, Derek notices increased relaxation and senses that the beginning of trust and rapport has occurred. With some clients, reaching this point may take a full interview. Kyle is bicultural in that he has had wide experience in White American culture as well as in his more traditional Lakota family.
9. *Derek:* Thanks, maybe we could get started. What's happening?	We see that Kyle's words and nonverbal actions have changed in the short time that he has been in the session. Derek, for the first time, asks an open question. Questions, if used too early in the session, might have led Kyle to be guarded and say very little.
10. *Kyle:* I'm vice president of the Native American Indian Association—see—and that's been taking a lot of time. Sometimes there are more important things than studying.	Kyle starts slowly and as he gets to the words "more important things," the fire starts to rise in his eyes.
11. *Derek:* More important things?	The restatement encourages the client to elaborate on the critical issue that Derek has observed through Kyle's eyes.
12. *Kyle:* Yeah, like last night, we had a march and demonstration against the Indian mascot. It's so disrespectful and demeaning to have this little Indian cartoon with the big teeth. What does that have to do with education? They talk about	Kyle is now sitting up and talking more rapidly. Anger and frustration show in his body—his fists are clenched and his face shows strain and tension.
	Women, gays, or other minorities who experience disrespect or harassment may feel the same way

(continued)

Interviewer and Client Conversation	Process Comments
"liberal education." I think it's far from liberal; its constricting. But worse, when we got back to the dormitories, the car that belongs to one of our students had all the windows broken out. And inside was a brick with the words "You're next" painted on it.	and demonstrate similar verbal and nonverbal behaviors. The car bashing incident clearly illustrates that the harassing students had a lack of "other" esteem and respect (see page 137).
13. *Derek:* That's news to me. The situation on campus is getting worse. Your leadership of the association is really important and now you face even more challenges.	Derek shares his knowledge of the situation and paraphrases what Kyle has just said. He sits forward in his chair and leans toward Kyle. However, if Kyle were less fully acculturated in White European American culture, the whole tone of the conversation above would be quite different. It probably would have taken longer to establish a relationship, Kyle would have spoken more carefully, and his anger and frustration would likely not have been visible. The interviewer, in turn, would be likely to spend more time on relationship development, use more personal self-disclosure, use less direct eye contact, and—especially—be comfortable with longer periods of silence.
14. *Kyle:* Yeah. (pause) But we're going to manage it. We won't give up. (pause) But—I've been so involved in this campus work that my grades are suffering. I can't help the association if I flunk out.	Kyle feels heard and supported by the interviewer. Being heard allows him to turn to the reasons he came to the counselor's office. He relaxes a bit more and looks to Derek, as if asking for help.
15. *Derek:* Kyle, what I've heard so far is that you've gotten caught up in the many difficulties on campus. As association vice president, it's taken a lot of time—and you are very angry about what's happened. I also understand that you intend to "hang in" and that you believe you can manage it. But now you'd like to talk about managing your academics as well. Have I heard you correctly?	Derek has used his observation skills so that he knows now that Kyle's major objective is to work on improving his grades. He wisely kept questioning to a minimum and used some personal sharing and listening skills to start the interview.
16. *Kyle:* Right! I don't like what's going on, but I also know I have a responsibility to my people back home on the Rosebud Reservation, the Lakota people, and to myself to succeed here. I could talk forever on what's going on here on campus, but first, I've got to get my grades straightened out.	Here we see the importance of self-esteem and self-focus if Kyle is to succeed. But his respect for others is an important part of who he is. He does not see himself as just an individual. He also sees himself as an extension of his group. Kyle has considerable energy and perhaps a need to discuss campus issues, but his vocal tone and body language make it clear that the first topic of importance for him today is staying at the university.

17. *Derek:* OK . . . If you'd like, later we might come back to what's going on. You could tell me a bit about what's happening here with the mascot and all the campus troubles. But now the issue is what's going on with your studies. Could you share what's happening?

Derek observes that Kyle's mission in the interview right now is to work on staying in school. He asks an open question to change the focus of the session to academic issues, but he keeps open the possibility of discussing campus issues later in the interview.

You will find that many interviews start slowly, regardless of the cultural background of the client. Your skill at observing nonverbal and verbal behavior in the here and now and the session will enable you to choose appropriate things to say. One of the reasons Kyle was willing to come to the counseling office was that Derek had been seen regularly on campus at pow-wows as well as other campus multicultural festivals.

Some alternatives that may help you as you begin sessions that don't start easily: patience, a good sense of humor, and a willingness to disclose, share stories, and talk about neutral subjects such as sports or the weather. You will also find an early exploration of positive assets useful at times—"Before we start, I'd like to get to know a little bit more about you. Could you tell me specifically about something from your past that you feel particularly good about?" "What are some of the things you do well?" "What types of things do you like?" As time permits, consider conducting a full wellness review.

To help a child feel comfortable, you may offer a game, a toy, or an opportunity to draw. At times going outside for a walk or to the playground may be important. A quiet basketball court is sometimes a good place to hold a counseling session with an older child or adolescent. You may also find that adults respond favorably to these alternatives. Social workers and school counselors often visit the client's home or work setting.

Kyle's presentation of issues included two major themes—the pervasive campus atmosphere of Native American Indian harassment and Kyle's need to work on his academics. The interviewer, Derek, sought to keep both avenues open and encouraged Kyle to select the one he wanted to talk about. Many clients will introduce more than one issue at the start of a session, and at the beginning it is important to encourage them to define the direction of the session rather than setting the direction yourself. Important issues that you note early on may be brought up later for further discussion after more trust and rapport are developed.

Critical to all of the above is listening to and observing the client and then "going with" the direction the client seems to prefer. Different from Kyle, some clients will present you with a "laundry list" of current issues in their lives. They may jump from topic to topic and seem overwhelmed. In such cases it may be helpful to summarize the multiple concerns and make a list, verbal or even written, and then decide with the client which issue is to be addressed first. If you try to solve all the problems a client has "all at once," frustration and confusion are likely to take over the session.

INSTRUCTIONAL READING

Three organizing principles for understanding interview interaction are stressed in this chapter: nonverbal behavior, verbal behavior, and discrepancies or incongruities. Over time, you will gain considerable skill in drawing out and working with the client's view of the world and how you as a person relate to the client. The best way to learn observation skills may be to observe yourself and your client in a videotaped session.

Nonverbal Behavior

Observing clients' attending behavior patterns is important. Clients may be expected to break eye contact, exhibit bodily movement, and change vocal qualities when they are talking about topics of varying levels of comfort to them. You may observe clients crossing their arms or legs when they want to close off a topic, using rapid alternations of eye contact during periods of confusion, or exhibiting increased stammering or speech hesitations while pursuing difficult topics. This chapter includes a nonverbal behavioral Observation Form in the exercises at the end of the chapter that you may use to increase your awareness of these and other behaviors. And if you watch yourself carefully on tape, you as interviewer will exhibit many of these same behaviors.

Facial Expressions

Facial expressions are particularly important to observe. The brow may furrow, lips may tighten or loosen, flushing may occur, a client may smile at an inappropriate time. Even more careful observation will reveal subtle color changes in the face as blood flow reflects emotional reactions. Breathing may speed up or stop temporarily. The lips may swell and pupils may dilate or contract. These seemingly small responses are important clues to what a client is experiencing; to notice them takes work and practice. You may want to select one or two kinds of facial expressions and study them for a few days in your regular daily interactions and then move on to others as part of a systematic program to heighten your powers of observation.

However, we offer an important caution as you enter the world of observation. It is tempting to interpret the client's world as you see things, possibly forgetting that your interpretation is just that—"your interpretation." Cultural and individual differences modify verbal and nonverbal behavior. It is valuable to observe your client, but always be careful not to stereotype or oversimplify behaviors, thoughts, and feelings.

Body Language

Particularly important are discrepancies in nonverbal behavior. When a client is talking casually about a friend, for example, one hand may be tightly clenched in a fist and the other relaxed and open, possibly indicating mixed feelings toward the friend or something related to the friend.

Hand and arm gestures may give you an indication of how you and the client are organizing things. Random, discrepant gestures may indicate confusion, whereas a person seeking to control or organize things may move hands and arms in straight lines and point with fingers authoritatively. Smooth, flowing gestures, particularly those in harmony with the gestures of others, such as family members, friends, or even you as interviewer, may suggest openness.

Dramatic and interesting patterns of movement exist between people. It is useful to observe the degree of movement harmony among individuals. Often people who are communicating well will "mirror" each other's body language. They may unconsciously sit in identical positions and even make complex hand movements together as if in a ballet. This is termed *movement synchrony.* Other paired movements may not be identical but may be still harmonious, as in *movement complementarity.* For instance, one person talks and the other nods in agreement. You may observe a hand movement at the end of one person's statement that is answered by a related hand movement as the other takes the conversational "ball" and starts talking.

Needless to say, patterns of *movement dissynchrony* should also be observed. Lack of harmony in movement is common between people who disagree markedly or even between those who have subtle conflicts that they may not be aware of.

Echoing is another construct to consider (Willis, 1989). As you observe an interview, note instances when the client makes a body change and then you make a parallel change (usually unconsciously) to join. Other times you may find yourself moving away from the client—your echo is discrepant. This may occur especially when the client says something with which you disagree. You try to maintain neutrality verbally, but your body gives you away. Similarly, you and your client may have congruent postures, but you make a comment that makes the client feel uncomfortable—the client may echo by moving away in response to your words.

You as interviewer will want to observe your degree of body harmony with your clients. How do your movements relate to theirs? Discreetly but deliberately assuming their posture and some of their movements may help you to get in touch with their experience. Many expert counselors and therapists practice "mirroring" their clients. Experience has shown them that matching body language, breathing rates, and key words of the client can heighten their understanding of how the other perceives and experiences the world.

At the same time, genuineness on your part is vital. For example, a practicum student reported difficulty with a client during the session, noting that the client's nonverbal behavior seemed especially unusual. Near the end of the session, the client reported, "I know you guys; you try to mirror my nonverbal behavior. So I keep moving to make it difficult for you." Clearly, you can expect that some of your clients will know as much about observation skills and nonverbal behavior as you do. What should you do in such situations? The key words are *honesty* and *authenticity* on your part. Use the skills and concepts in this book only when they feel honest and real to you. Work in an egalitarian fashion with your client rather than "working on" your client. If you are a genuine person and discuss issues openly in the session, your client will most likely be open with you. In the practicum example, the student worked to generate enough trust with the client that the problem with mirroring was eventually discussed and later sessions became much more productive.

Individual and Multicultural Issues in Nonverbal Behavior

As you engage in observation, recall that each culture has a different style of nonverbal communication. For example, Russians say yes by shaking the head from side to side and no by moving the head up and down. Most Europeans do exactly the opposite. In a similar fashion, we must be careful not to assign our ideas about what is "standard" and appropriate nonverbal communication to our clients.

A study was made of the average number of times friends of different cultural groups touch each other in an hour while talking in a coffee shop. The results showed that English friends did not touch each other at all, French friends touched 110 times, and Puerto Rican friends touched 180 times (cited in Asbell & Wynn, 1991). Clearly, nonverbal styles differ widely from one cultural group to another.

In the European American tradition, people tend to communicate at "arm's length" whereas in some Arabic cultures, speakers prefer to be only 18 inches apart when they talk, a most uncomfortable distance for many Europeans. Arabs who experience the distance of Europeans may interpret such behavior as "cold." Smiling is a sign of warmth in most cultures, but in some situations in Japan, smiling may indicate discomfort. Recall that eye contact may be inappropriate for the traditional Navajo but highly appropriate and expected for a Navajo official who interacts commonly with European American Arizonans. That same official will extend traditional courtesies to Navajos who have less contact with European traditions.

What determines a comfortable interpersonal distance is influenced by multiple factors beyond just cultural background. Hargie, Dickson, and Tourish (2004, p. 45) point out the following:

Gender: women tend to feel more comfortable with closer distances than males
Personality: introverts need more distance than extraverts
Age: children and the young tend to adopt closer distances
Topic of conversation: difficult topics such as sexual worries or personal misbehavior may lead a person to maintain more distance
Personal relationships: friends or couples who are doing well tend to be closer while those who have issues may move apart (expect to see this in couples counseling)

Are you talking to your clients from behind a desk or sitting directly across from them? Or are you willing go to a client's home and perhaps sit on the floor? The importance of the physical setting of the agency and the place where the conversation occurs is important. Consider having several alternative seating arrangements available for the session.

In short, it is appropriate that you begin a lifetime study of nonverbal communication patterns and their variations. In terms of counseling sessions, you will find that changes in style may be as important as, or more important than, finding specific meanings in communication style. Edward Hall's *The Silent Language* (1959) remains a classic. It is a highly useful follow-up to this chapter.

Verbal Behavior

As noted in the chapter introduction, counseling and interviewing theory and practice have an almost infinite array of verbal frameworks within which to examine the interview. The classification of interviewing leads via the microskills is one of them. Four additional useful concepts for interview analysis are presented here: selective attention, key words, concreteness versus abstractions, and "I" statements and "other" statements. We will also discuss some key multicultural issues connected with verbal behavior.

Box 5-2   Research Evidence That You Can Use: Observation

Research on nonverbal behavior has a long and distinguished history. Edward Hall's *The Silent Language* of 1959, is a classic of anthropological and multicultural research and remains the place to start. Early work in nonverbal communication was completed by Paul Ekman, and his 1999 and 2004 summaries of his work are basic to the field. Eye contact and forward trunk lean were found to be highly correlated with ratings of empathy (Sharpley & Guidara, 1993; Sharpley & Sagris, 1995). Hill and O'Brien's review (1999) noted that clients used fewer head nods when they were reacting negatively. Knapp and Hall (2001) is another standard reference.

A classic study by Mayo and LaFrance remains important today as it highlights issues of cultural change and acculturation. They commented in 1973 (p. 389),

> In interaction among Whites, a clear indication of attention by the listener is that he looks at the speaker. . . . Blacks do not look while listening, presumably using other cues to communicate attention. When Blacks and Whites interact, therefore, these differences may give rise to communication breakdowns. The White may feel he is not being listened to, while the Black may feel unduly scrutinized. Further, exchanges of listener-speaker roles may become disjunctive, leading to generalized discomfort with the encounter.

Many African Americans have acculturated to White standards and may not show the same differences as in the past. At the same time, it is well known that minorities in the United States and Canada often have two communication styles: one for the "outgroup" (White culture) and one within their own family and cultural group. These observations on eye contact remain relevant today to Native American and Latin cultures. Cross-cultural communication occurs most often in a situation in which the majority has power and many minorities have learned to mask their emotions and nonverbal expressions.

LaFrance has turned to research on gender differences and power and here notes major differences in communication style. For example, LaFrance and Woodzicka (1998) found women's nonverbal reactions to sexist humor are quite marked and different from those of men.

Reading about nonverbal communication is helpful, but we suggest that you visit an Internet search site (e.g., www.google.com) and enter "nonverbal communication" as the key search term. You will often find helpful visuals to support what you have read here. Add the word "research" to your search, and you will find several summaries of databased work.

The web site *Nonverbal Behavior/Nonverbal Communication Links* (www3. usal.es/~nonverbal/introduction.htm) is especially useful and well worth your time.

Nonverbal communication and neuropsychology

Observation of nonverbal communication is obviously based in physiology and the ability of the brain to recognize what is occurring. As just one example in this area, there are fascinating multicultural findings that have immediate relevance to counseling and therapy process. Among them are the following reported by Eberhardt (2005):

▲ Japanese have been found to be more holistic thinkers than Westerners (Masuda & Nisbett, 2001). Expect the possibility of different cognitive/emotional styles when you work with people who are culturally different from you—but *never stereotype!*

(continued)

Box 5-2   (continued)

▲ Brain systems can be modified by life experience (Draganski et al., 2004). As you learn to observe your client more effectively in the interview, your brain is likely developing new connections. Expect your multicultural learning to become one of those new connections.

▲ Blacks and Whites both exhibit greater brain activation when they view same race faces and less when race is different (Golby et al., 2001). Note here that this could affect your work with a client whose race is different from yours, thus suggesting that discussing racial and other cultural differences early in the session can be a helpful way to build trust.

Vital to our work as counselors and therapists is awareness that brain functioning is not fixed, but also relates to cultural factors and environmental issues (e.g., recurring racism and sexism), significant interpersonal events (e.g., experience of truama, divorce, serious illness), and our own learning of new skills such as those described in this book.

Selective Attention

What are your patterns of selective attention? Clients tend to talk about what we are interested in and willing to hear. For example, behavioral counselors tend to have clients who talk about specific concrete situations, existential therapists have clients who talk about the meaning of life, and psychoanalytic therapists hear a lot about dreams. Your life experience and your "theory of choice" are very real determining factors in how you listen to others.

A famous training film has three eminent counselors (Albert Ellis, Fritz Perls, and Carl Rogers) presenting their interviewing styles by all counseling the same client, Gloria (Shostrum, 1966). Gloria changes the way she talks about and thinks about her relationship issues, responding very differently as she works with each counselor. Further, research on verbal behavior in the film revealed that Gloria tended to match the language of the varying counselors (Meara et al., 1979, 1981).

Should clients match the language of the interviewer or should you, the interviewer, learn to match your language and style with that of the client? Most likely, both approaches are relevant, but certainly at the beginning, we want to draw out clients' stories and issues from their own language perspective, not from ours. And we always need to consider the perspective of how the client talks and makes meaning.

Relating this to your own experience, think about what you consider most important in the interview. Are there topics you are less comfortable with? Some interviewers are excellent in helping clients talk about vocational issues but shy away from personal matters such as interpersonal conflict and sexuality. Others may find their clients constantly talking about interpersonal issues, leaving little time to deal with other critical practical issues such as getting a job or choosing a course of study.

Your personal style and theory will deeply affect how others respond in the session and talk about issues. What is to be the direction of your influence?

Key Words

If you listen carefully to clients, you will find that certain words appear again and again in their descriptions of situations. Noting their key words and helping them explore the facts, feelings, and meanings underlying those words may be useful. Key descriptive words are often the constructs by which a client organizes the world; these words may reveal underlying meanings. *Verbal underlining* through vocal emphasis is another helpful clue in determining what is most important to a client. Through intonation and volume, clients tend to stress the single words or phrases that are most important to them.

You will find that joining clients by using their key words facilitates your understanding and communication with them. If their words are negative and self-demeaning, reflect those perceptions early in the interview but later help them use more positive descriptions of the same situations or events. Often you will aim to help the client change from "I can't" to "I can."

Many clients will demonstrate problems of verbal tracking and selective attention. They may either stay on a single topic to the exclusion of other important issues or change the topic, either subtly or abruptly, when they want to avoid talking about a difficult issue. Perhaps the most difficult task of the beginning counselor or interviewer is to help the client stay on the topic without being overly controlling. Observing client topic changes is particularly important. At times it may be helpful to comment—for instance, "A few minutes ago we were talking about X." Another possibility is to follow up that observation by asking how the client might explain the shift in topic.

Concreteness Versus Abstraction

Where is the client on the "abstraction ladder"? Two major client styles of communication are important to observe and you will find it helpful to be prepared for both types of conversation (Ivey, Ivey, Myers, & Sweeney, 2005). Clients who talk with a *concrete/situational* style are skilled at providing specifics and examples of

Box 5-3 The Abstraction Ladder

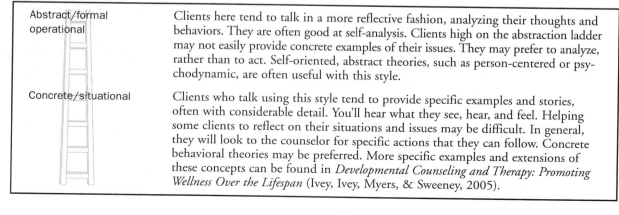

Abstract/formal operational	Clients here tend to talk in a more reflective fashion, analyzing their thoughts and behaviors. They are often good at self-analysis. Clients high on the abstraction ladder may not easily provide concrete examples of their issues. They may prefer to analyze, rather than to act. Self-oriented, abstract theories, such as person-centered or psychodynamic, are often useful with this style.
Concrete/situational	Clients who talk using this style tend to provide specific examples and stories, often with considerable detail. You'll hear what they see, hear, and feel. Helping some clients to reflect on their situations and issues may be difficult. In general, they will look to the counselor for specific actions that they can follow. Concrete behavioral theories may be preferred. More specific examples and extensions of these concepts can be found in *Developmental Counseling and Therapy: Promoting Wellness Over the Lifespan* (Ivey, Ivey, Myers, & Sweeney, 2005).

their concerns and problems. Clients who are more abstract and formal operational, on the other hand, have strengths in self-analysis and are often skilled at reflecting on their issues. Of course, most adult and many adolescent clients will talk at both levels. Children, however, can be expected to be primarily concrete in their talk— and so are many adolescents and adults.

Clients with a concrete/situational style will provide you with plenty of specifics. The strength and value of these details is that you know relatively precisely what happened, at least from their point of view. However, they often will have difficulties in seeing the point of view of others. Some with a concrete style may tell you, for example, what happened to them when they went to the hospital from start to finish with every detail of the operation and how the hospital functions. They may have a difficult interpersonal relationship and then discuss the situation through a series of endless stories full of specific facts and "He said . . . and then I said. . . ." If asked to reflect on the meaning of their story or what they have said, they may appear puzzled.

Here are some examples of concrete/situational statements:

Child, age 5: Jonnie hit me in the arm—right here!

Child, age 10: He hit me when we were playing soccer. I had just scored a goal and it made him mad. He snuck up behind me, grabbed my leg and then punched me when I fell down! Do you know what else he did, well he . . .

Man, age 45: I was down to Myrtle Beach—we drove there on 95 and was the traffic ever terrible. Well, we drove into town and the first thing we did was to find a motel to stay in, you know. But we found one for only $60 and it had a swimming pool. Well, we signed in and then . . .

Woman, age 27: You asked for an example of how my ex-husband interferes with my life? Well, a friend and I were sitting quietly in the cafeteria, just drinking coffee. Suddenly, he came up behind me, he grabbed my arm (but didn't hurt me this time), then he smiled and walked out. I was scared to death. If he had said something, it might not have been so frightening.

The details are important, but clients who use a primarily concrete style in their conversation and thinking may have real difficulty in reflecting on themselves and seeing patterns in situations.

Abstract/formal operational clients are good at making sense of the world and reflecting on themselves. But some clients will talk in such abstract broad generalities that it is hard to understand what they are really saying. They may be able to see patterns in their lives and be good at discussing and analyzing themselves, but you may have difficulty finding out specifically what is going on in their lives.

Example abstract/formal operational statements include these:

Child, 12, beginning awareness of pattern thought: He does it to me all the time. It never stops. It's what he does to everyone all the time.

Man, age 20: As I think about myself, I see a person who responds to others and cares deeply, but somehow I feel that they don't respond to me.

Woman age 68: As I reflect back on my life, I see a pattern of selfishness that makes me uncomfortable. I think a lot about myself.

Many interviewers tend to be more abstract/formal operational themselves and may be drawn to the analytical and self-reflective style. They may conduct entire interviews focusing totally on analysis, and the observer might wonder what the client and counselor are talking about.

In each of the conversational styles the strength is also possibly a weakness. You will want to help abstract/formal clients to become more concrete ("Could you give me an example?"). If you persist, most of these clients will be able to provide the needed specifics.

You will also want to help concrete clients become more abstract and pattern oriented. This is best effected by a conscious effort to listen to their sometimes lengthy stories very carefully. Paraphrasing and summarizing what they have said can be helpful (see Chapter 6). Just asking them to reflect on their story may not work ("Could you tell me what the story means to you?" "Can you reflect on that story and what it says about you as a person?"). More direct questions may be needed to help concrete clients step back and reflect on their stories. A series of questions such as these might help: "What one thing do you remember most about this story?" "What did you like best about what happened?" "What least?" "What could you have done differently to change the ending of the story?" Questions like these that narrow the focus can help children and clients with a concrete orientation move from self-report to self-examination.

Essentially, however, we all need to match our own style and language to the uniqueness of the client. It will help if you think about your own language style. If you have a concrete style yourself, abstract clients may challenge you. If you are more abstract, you may not reach those with a more concrete orientation (and they are the majority of our clients). If the client tends to be concrete, listen to the specifics and enter that client's world as he or she presents it. If the client is abstract, once again, listen and join that client where he or she is. After you have listened carefully, then consider the possibility of helping the client look at the concern from the other perspective.

"I" Statements and "Other" Statements

Clients' ownership and responsibility for issues will often be shown in their "I" and "other" statements. Consider the following:

"I'm working hard to get along with my partner. I've tried to change and meet her/him halfway."
versus
"It's her/his fault. No change is happening."

"I'm not studying enough. I should work harder."
versus
"The racist insults we get on this campus make it nearly impossible to study."

"I feel terrible. If only I could do more to help. I try so hard."
versus
"Dad's an alcoholic. Everyone suffers."

"I'm at fault. I shouldn't have worn that dress. It may have been too sexy."
versus
"No, women should be free to wear whatever they wish."

"I believe in a personal God. God is central to my life."

versus

"Our church provides a lot of support and helps us understand spirituality more deeply."

We need to be aware of what is happening in and for our clients as well as learn what occurs in their relationships with others and in their families and communities. There is a need to balance internal and external responsibility for issues.

Review the five pairs of statements above. Some of them represent positive "I" and "other" statements; some are negative. Some clients attribute their difficulties solely to themselves; others see the outside world as the issue. A woman may be sexually harassed and see clearly that others and the environment are at fault; another woman will feel that somehow she provoked the incident. Counselors need to help individuals look at their issues but also to help them consider how these concerns relate to others and the surrounding environment.

The alcohol-related statements above may serve as an example. Some children of alcoholics see themselves as somehow responsible for a family member's drinking. Their "I" statements may be unrealistic and ultimately "enable" the alcoholic to drink even more. In such cases, the task of the interviewer is to help the client learn to attribute family difficulties to alcohol and the alcoholic. In work with alcoholics themselves who may deny the problem, one goal is often to help move them to that critical "I" statement, *"I am an alcoholic."* Part of recovery from alcoholism, of course, is recognizing others and showing esteem for others. Thus, a balance of "I" and "other" statements is a useful goal.

You can also observe "I" statements as a person progresses through a series of interviews. For example, the client at the beginning of counseling may give many negative self-statements—"It's my fault." "I did a bad thing, I'm a bad person." "I don't respect myself." "I don't like myself."

If your sessions are effective, expect such statements to change to "I'm still responsible, but I now know that it wasn't all my fault." "Calling myself 'bad' is self-defeating. I now realize that I did my best." "I can respect myself more." "I'm beginning to like myself."

Similarly, other statements can become more positive. "They did it to me" may become "They didn't intend to harm me and I can see them as trying now to make up for it." "My parents made me the mess that I am" changes to "They did their best and I need to forgive them." *"Indians* get more financial help than they deserve on this campus" moves to "Now I understand that *Native People* don't get as much university aid as I thought—their scholarship funds come from their own tribes and nations. The university actually does very little to support their issues."

"I" statements and "other" statements can be useful ways for you to track client change in the session, both short and long term. Consider the following:

In the first session: I'm flunking out. This place is hopeless.
Later: These study skills are really helping me. I'm doing better and the classes make a lot more sense.

In the first session: That old man. He just sits there. I can't stand working in the nursing home anymore. It's too much for me. I am beginning to hate old people. But I need the money.

Later: I'm learning that I'm afraid of aging. I guess that's why I've been so unsympa-
thetic at the home. And I'm so tired of working and studying both. I now know
that I've got to control my frustration. Part of that is taking time for myself and
having fun. I'm starting to learn how to talk to the patients at the home and
things are working out better.

Multicultural differences in the use of "I" are important. We should remember that
English is one of the very few languages that capitalizes the word "I." A Vietnamese
immigrant comments:

> There is no such thing as "I" in Vietnamese. . . . We define ourselves in rela-
> tionships. . . . If I talk to my mother, the "I" for me is "daughter," the "you" is
> for "mother." Our language speaks to relationships rather than to individuals.
> If I talk to my children, the mother would be me and my children refer to
> themselves as sons or daughters.

Discrepancies

The variety of discrepancies clients may manifest is perhaps best illustrated by the
following statements:

"My son is perfect, but he just doesn't respect me."
"I really love my brother." (said in a quiet tone with averted eyes)
"I can't get along with Charlie." (discrepancy between the client and another person)
"I deserve to pass the course." (from a student who has done no homework and just
    failed the final examination)
"That question doesn't bother me." (said with a flushed face and a closed fist)

Once the client is relatively comfortable and some beginning steps have been taken
toward rapport and understanding, a major task of the counselor or interviewer is
to identify basic discrepancies, mixed messages, conflicts, or incongruities in the
client's behavior and life. A common goal in most interviews, counseling, and therapy is to assist clients in working through discrepancies and conflict, but first these
have to be identified clearly.

Examples of Discrepancies Internal to the Client

Discrepancies in nonverbal behaviors. A client may be talking smoothly on a topic,
    but careful observation may reveal that the client's smile is coupled with a
    tightly closed fist. Mixed messages are often conveyed when parts of the body
    lack congruence.
Discrepancies in verbal statements. In a single sentence a client may express two com-
    pletely contradictory ideas ("My son is perfect, but he just doesn't respect me"
    or "This is a lovely office you have; it's too bad that it's in this neighborhood").
    Over a longer time, a positive statement about one's job may be qualified by an
    extensive discussion of problems. Most of us have mixed feelings toward our
    loved ones, our work, and other situations. It is helpful to aid others in under-
    standing their ambivalences.
Discrepancies between what one says and what one does. A parent may talk of love for
    a child but be guilty of child abuse. A student may say that he or she deserved
    a higher grade than the actual time spent studying suggests. The client may

verbalize a support for multicultural, women's, or ecology causes, but fail to "walk the talk."

Discrepancies between statements and nonverbal behavior. "That question doesn't bother me" (said with a flushed face and a closed fist). Watch for the timing of "lint picking." The client may be talking of a desire to repair a troubled relationship while simultaneously picking at his or her clothes. Many clients make small or large physical movements away from the interviewer when they are confronted with a troubling issue and feel inadequately supported by the interviewer.

Examples of Discrepancies Between the Client and the External World

Discrepancies between people. Conflict can be described as a discrepancy between people. Noting interpersonal conflict is a key task of the interviewer, counselor, or therapist. You will find that one of the predominant issues you face in interviewing and counseling is discord and arguments among people. Mediation, in particular, focuses on this particular type of discrepancy.

Discrepancies between a client and a situation. "I want to be admitted to medical school, but I didn't make it." "I just found out I have early symptoms of arthritis." "I can't find a job." In such situations the client's ideal world is often incongruent with what really *is*. The counselor's task is to work through these issues in terms of behaviors or attitudes. Many People of Color, gays, women, or the disabled find themselves in a contextual situation that makes life difficult for them. Discrimination, heterosexism, sexism, and ableism represent situational discrepancies.

Examples of Discrepancies Between You and the Client

One of the more challenging issues occurs when you and the client are not in synchrony. And this can occur on any of the dimensions above. Your nonverbal behavior may be markedly different from the client's. You may be saying one thing, the client another. A conflict in values or goals in counseling may be directly apparent or a quiet, unsaid thing—and either can destroy your relationship.

At times, you may carefully introduce discussion of this discrepancy into the session. In such cases, you may interpret or reframe the situation differently from how the client has been presenting it. Or you may face differences squarely and openly discuss your mutual lack of communication at that moment.

Chapter 9, dealing with confrontation and conflict management, builds on this discussion of the observation of discrepancies and suggests that in these dimensions and their resolution lies much of what we term counseling and therapy.

SUMMARY: OBSERVATION SKILLS

The interviewer seeks to observe client verbal and nonverbal behavior with an eye to identifying discrepancies, mixed messages, incongruity, and conflict. Counseling and therapy in particular, but also interviewing, frequently focus on problems and their resolution. A discrepancy is often a problem. At the same time, discrepancies in many forms are part of life and may even be enjoyed. Humor, for example, is based on conflict and discrepancies. It is necessary for counselors and interviewers to work on client problems, but that emphasis often results in a tendency to view life as a problem to be solved rather than as an opportunity to be enjoyed. Even while you are working with the most complex case, it is wise to focus on client strengths and assets from time to time. The positive asset search is ideally part of every interview. The focus on positive assets is necessary to combat the tendency to search constantly for problems and difficulties. People solve problems with their strengths, not with their weaknesses!

Box 5-4   Key Points

Why?	The self-aware interviewer is constantly aware of the client and of the here-and-now interaction in the session. Clients tell us about their world by nonverbal and verbal means. Observation skills are a critical tool in determining how the client interprets the world.
What?	Observation skills focus on three areas: 1. *Nonverbal behavior.* Your own and client eye-contact patterns, body language, and vocal qualities are, of course, important. Shifts and changes in these may be indicative of client interest or discomfort. A client may lean forward, indicating excitement about an idea, or cross his or her arms to close it off. Facial clues (brow furrowing, lip tightening or loosening, flushing, pulse rate visible at temples) are especially important. Larger scale body movements may indicate shifts in reactions, thoughts, or the topic. 2. *Verbal behavior.* Noting patterns of verbal tracking for both you and the client is particularly important. At what point does the topic change and who initiates the change? Where is the client on the abstraction ladder? If the client is concrete, are you matching her or his language? Is the client making "I" statements or "other" statements? Do the client's negative statements become more positive as counseling progresses? Clients tend to use certain key words to describe their behavior and situations; noting these descriptive words and repetitive themes is helpful. 3. *Discrepancies.* Incongruities, mixed messages, contradiction, and conflict are manifest in many and perhaps all interviews. The effective interviewer is able to identify these discrepancies, to name them appropriately, and sometimes, to feed them back to the client. These discrepancies may be between nonverbal behaviors, between two statements, between what clients say and what they do, or between statements and nonverbal behavior. They may also represent a conflict between people or between a client and a situation. And your own behaviors may be positively or negatively discrepant.
How?	Simple, careful observation of the interview is basic. What can you see, hear, and feel from the client's world? Note your impact on the client: How does what you say change or relate to the client's behavior? Use these data to adjust your microskill or interviewing technique.
With whom?	Observation skills are essential with all clients. Note individual and cultural differences in verbal and nonverbal behavior. Always remember that some individuals

(continued)

Box 5-4 (continued)

	and some cultures may have a different meaning for a movement or use of language from your own personal meaning. Use caution in your interpretation of nonverbal behavior.
And?	Movement harmonics are particularly interesting and provide a basic concept that explains much verbal and nonverbal communication. When two people are talking together and communicating well, they often exhibit *movement synchrony* or *movement complementarity* in that their bodies move in a harmonious fashion. When people are not communicating clearly, *movement dissynchrony* will appear: body shifts, jerks, and pulling away are readily apparent.

COMPETENCY PRACTICE EXERCISES AND PORTFOLIO OF COMPETENCE

Many concepts have been presented in this chapter; it will take time to master them and make them a useful part of your interviewing. Therefore, the exercises here should be considered introductory. Further, it is suggested that you continue to work on these concepts throughout the time that you read this book. For example, although the next chapter deals with paraphrasing you will want to continue observing clients' verbal behavior while you practice paraphrasing. In practicing the skill of reflection of feeling (see Chapter 7) you will want to observe nonverbal expressions of emotion. In Chapter 8 you will have the opportunity to note "I" statements and whether the nature of the statements changes in a positive direction later in the interview. If you keep practicing the concepts in this chapter throughout the book, material that might now seem confusing will gradually be clarified and become part of your natural style.

Individual Practice

Exercise 1: Observation of Nonverbal Patterns

Observe 10 minutes of a counseling interview, a television interview, or any two people talking. Videotape so that repeated viewing is possible.

Visual/eye contact patterns. Do people maintain eye contact more while talking or while listening? Does the "client" break eye contact more often while discussing certain subjects than others? Can you observe changes in pupil dilation as an expression of interest?

Vocal qualities. Note speech rate and changes in intonation or volume. Give special attention to speech "hitches" or hesitations.

Attentive body language. Note gestures, shifts of posture, leaning, patterns of breathing, and use of space. Give special attention to facial expressions such as changes in skin color, flushing, and lip movements. Note appropriate and inappropriate smiling, furrowing of the brow, and so on.

Movement harmonics. Note places where movement synchrony and echoing occurred. Did you observe examples of movement dissynchrony?

Where possible, observe your own videotape so that you can view the interview several times. One useful approach is to observe 5 to 10 minutes of interaction several times. Be sure to separate behavioral observations from impressions on the Behavioral Observation Form.

1. Present context of observation. Briefly summarize what is happening verbally at the time of the observation. Number each observation.
2. Observe the interview for the following and describe what you see as precisely and concretely as possible: visual/eye contact patterns, vocal qualities, attentive body language, and movement harmonics.
3. Record your impressions. What interpretations of the observation do you make? How do you make sense of each observation unit? And—most important—are you cautious in drawing conclusions from what you have seen and noticed?

Exercise 2: Observation of Verbal Behavior and Discrepancies

Observe the same interview again, but this time pay special attention to verbal dimensions and discrepancies (which, of course, will include nonverbal dimensions). Consider the following issues and provide concrete evidence for each of your decisions as to what is occurring in the session. Again, separate observations from your interpretation and impressions.

Verbal tracking and selective attention. Pay special attention to topic jumps or shifts. Who initiates them? Do you see any pattern of special topic interest and/or avoidance? What does the listener seem to *want* to hear?

Key words. What are the key words of each person in the communication?

Abstract or concrete conversation. Is this conversation about patterns or about specifics? Are the people involved approaching this issue in a similar fashion?

"I" statements. Consider each person. What is he or she trying to say from an "I" statement framework? Are any "I" statements present?

"Other" statements. How aware is this individual of connections with others? How accurate are those perceptions?

Discrepancies. What incongruities do you note in the behavior in either person you observe? Do you locate any discrepancies between the two? What issues of conflict might be important?

Exercise 3: Examining Your Own Verbal and Nonverbal Styles

Videotape yourself with another person in a real interview or conversation for at least 20 minutes. Do not make this a role-play. Then view your own verbal and nonverbal behavior and that of the person you are talking with in the same detail as in Exercises 1 and 2. What do you learn about yourself?

Exercise 4: Classifying Statements as Concrete or Abstract

Following are examples of client statements. Classify each statement as primarily concrete or primarily abstract. You will gain considerably more practice and thus have more suggestions for interventions in later chapters. (Answers to this exercise may be found at the conclusion of this chapter.) Circle C (concrete) or A (abstract) below:

C	A	1. I cry all day long. I didn't sleep last night. I can't eat.
C	A	2. I feel rotten about myself lately.
C	A	3. I feel very guilty.
C	A	4. Sorry I'm late for the session. Traffic was very heavy.
C	A	5. I feel really awkward on dates. I'm a social dud.
C	A	6. Last night my date said that I wasn't much fun. Then I started to cry.
C	A	7. My father is tall, has red hair, and yells a lot.
C	A	8. My father is very hard to get along with. He's difficult.
C	A	9. My family is very loving. We have a pattern of sharing.
C	A	10. My mom just sent me a box of cookies.

Systematic Group Practice

Many observation concepts have been discussed in this chapter. It is obviously not possible to observe all these in one single role-play interview. However, practice can serve as a foundation for elaboration at a later time. This exercise has been selected to summarize the central ideas of the chapter.

Step 1: Divide into practice groups.   Triads or groups of four are most appropriate.

Step 2: Select a group leader.

Step 3: Assign roles for the first practice session.

❑ Client, who responds naturally and is talkative.

❑ Interviewer, who will seek to demonstrate a natural, authentic style.

❑ Observer 1, who observes client communication, using the Feedback Form: Observation, Box 5-5.

❑ Observer 2, who observes counselor communication, using the Feedback Form. Here the consultative microsupervision process usually focuses on helping the interviewer understand and utilize nonverbal communication more effectively. Ideally you have a videotape available for precise feedback.

Step 4: Plan.   State the goals of the session. As the central task is observation, the interviewer should give primary attention to attending and open questions. Use other skills as you wish. After the role-play is over, the interviewer should report personal observation of the client made during that time and demonstrate basic or active mastery skills. The client will report on observations of the counselor.

The suggested topic for the practice role-play is "Something or someone with whom I have a present conflict or have had a past conflict." Alternative topics include the following:

My positive and negative feelings toward my parents or other significant persons
The mixed blessings of my work, home community, or present living situation

The two observers may use this session as an opportunity both for providing feedback to the interviewer and for sharpening their own observation skills.

Step 5: Conduct a 6-minute practice session. As much as possible, both the interviewer and the client will behave as naturally as possible discussing a real situation.

Step 6: Review the practice session and provide feedback for 14 minutes. Remember to stop the audiotape or videotape periodically and listen to or view key items several times for increased clarity. Observers should give special attention to careful completion of the feedback sheet throughout the session and the client can give important feedback via the Client Feedback Form of Chapter 1.

Step 7: Rotate roles.

Box 5-5   Feedback Form: Observation

_____ (Date)

_____   _____
(Name of Interviewer)   (Name of Person Completing Form)

Instructions: Observe the client or counselor carefully during the role-play session and immediately afterward complete the nonverbal feedback portion of the form. As you view the videotape or listen to the audiotape, give special attention to verbal behavior and note discrepancies. If no recording equipment is available, one observer should note nonverbal behavior and the other verbal behavior.

Nonverbal behavior checklist

1. **Visuals.**   At what points did eye contact breaks occur? Staring? Did the individual maintain eye contact more when talking or when listening? Changes in pupil dilation?

2. **Vocals.**   When did speech hesitations occur? Changes in tone and volume? What single words or short phrases were emphasized?

3. **Body language.**   General style and changes in position of hands and arms, trunk, legs? Open or closed gestures? Tight fist? Playing with hands or objects? Physical tension: relaxed or tight? Body oriented toward or away from the other? Sudden body shifts? Twitching? Distance? Breathing changes? At what points did changes in facial expression occur? Changes in skin color, flushing, swelling or contracting of lips? Appropriate or inappropriate smiling? Head nods? Brow furrowing?

4. **Movement harmonics.**   Examples of movement complementarity, synchrony, or dissynchrony? Echoing? At what times did these occur?

(continued)

Box 5-5   (continued)

5. **Nonverbal discrepancies.**   Did one part of the body say something different from another? With what topics did this occur?

Verbal behavior checklist

1. **Verbal tracking and selective attention.**   At what points did the client or counselor fail to stay on the topic? To what topics did each give most attention? List here the most important key words used by the client; these are important for deeper analysis.

2. **Abstract or concrete?**   Which word represents the client? How did the counselor work with this dimension? Was the counselor abstract or concrete?

3. **"I" statements and "other" statements.**   While discussing the conflict or confusing situation, what self- and other statements did the client make? Which might be desirable to change, given a longer interview?

4. **Verbal discrepancies.**   Write here observations of verbal discrepancies in either client or counselor.

Portfolio of Competence

Determining your own style and theory can be best accomplished on a base of competence. Each chapter closes with a reflective exercise asking your thoughts and feelings about what has been discussed. By the time you finish this book, you will have a substantial record of your competencies and a good written record as you move toward determining your own style and theory.

Use the following as a checklist to evaluate your present level or mastery. Check those dimensions that you currently feel able to do. Those that remain unchecked can serve as future goals. *Do not expect to attain intentional competence on every dimension as you work through this book.* You will find, however, that you will improve your competencies with repetition and practice.

Check below the competencies that you have met to date.

Level 1: Identification and classification.
❑ Ability to note attending nonverbal behaviors, particularly changes in behavior in visuals—eye contact, vocal tone, and body language.
❑ Ability to note movement harmonics and echoing.
❑ Ability to note verbal tracking and selective attention.
❑ Ability to note key words used by the client and yourself.
❑ Ability to note distinctions between concrete/situational and abstract/formal operational conversation.
❑ Ability to note discrepancies in verbal and nonverbal behavior.
❑ Ability to note discrepancies in the client.
❑ Ability to note discrepancies in yourself.
❑ Ability to note discrepancies between yourself and the client.

Level 2: Basic competence.   Nonverbal and verbal observation skills are things that you can work on and improve over a lifetime. Therefore, use the intentional competence list below for your self-assessments.

Level 3: Intentional competence.   You will be able to note client verbal and nonverbal behaviors in the interview and use these observations at times to facilitate interview conversation. You will be able to match your behavior to the client's. When necessary, you will be able to mismatch behaviors to promote client movement. For example, if you first join the negative body language of a depressed client and then take a more positive position, the client may follow and adopt a more assertive posture. You will be able to note your own verbal and nonverbal responses to the client. You will be able to note discrepancies between yourself and the client and work to resolve those discrepancies.

Developing mastery of the following areas will take time. Come back to this list later as you have practiced other skills in this book. For the first stages of basic and intentional mastery, the following competencies are suggested as most important:

❑ Ability to mirror nonverbal patterns of the client. The interviewer mirrors body position, eye contact patterns, facial expression, and vocal qualities.

❑ Ability to identify client patterns of selective attention and use those patterns either to bring talk back to the original topic or to move knowingly to the new topic provided by the client.

❑ Ability to match clients' concrete/situation or abstract/formal operational language and help them to expand their stories in their own style.

❑ Ability to identify key client "I" and "other" statements and feed them back to the client accurately, thus enabling the client to describe and define what is meant more fully.

❑ Ability to note discrepancies and feed them back to the client accurately. (Note that this is an important part of the skill of confrontation and conflict management, discussed in detail in Chapter 9.) The client in turn will be able to accept and use the feedback for further effective self-exploration.

❑ Ability to note discrepancies in yourself and act to change them appropriately.

Level 4: Teaching competence. You will demonstrate your ability to teach others observation skills. Your achievement of this level can be determined by how well your students can be rated on the basic competencies of this self-assessment form. Certain of your clients in counseling may be quite insensitive to obvious patterns of nonverbal and verbal communication. Teaching them beginning methods of observing others can be most helpful to them. Do not introduce more than one or two concepts to a client per interview, however!

❑ Ability to teach clients in a helping session the social skills of nonverbal and verbal observation and the ability to note discrepancies.

❑ Ability to teach small groups the above skills.

DETERMINING YOUR OWN STYLE AND THEORY: CRITICAL SELF-REFLECTION ON OBSERVATION SKILLS

This chapter has focused on the importance of verbal and nonverbal observation skills and you have experienced a variety of exercises designed to enhance your awareness in this area.

Again, what single idea stood out for you among all those presented in this chapter, in class, or through informal learning? What stands out for you is likely to be important as a guide toward your next steps. What are your thoughts on multicultural differences? What other points in this chapter struck you as important? How might you use ideas in this chapter to begin the process of establishing your own style and theory?

_____

_____

_____

_____

_____

REFERENCES

Asbell, B., & Wynn, K. (1991). *Touching.* New York: Random House.

Draganski, B., Gaser, C., Busch, V., Schuierer, G., Bogdahn, U., & May, A. (2004). Neuroplasticity: Changes in grey matter induced by training. *Nature, 427,* 311–312.

Eberhardt, J. (2005). Imaging race. *American Psychologist, 60,* 181–190.

Ekman, P. (1999). Basic emotions. In T. Dalgleish & M. Power (Eds.), *Handbook of cognition and emotion.* Sussex, UK: Wiley.

Ekman, P. (2004). *Emotions revealed.* Woodacre, CA: Owl.

Golby, A., Gabrelli, J., Chiao, J., & Eberhardt, J. (2001). Differential responses in the fusiform region to same-race and other-race faces. *Nature Neuropsychology, 4,* 845–850.

Hall, E. (1959). *The silent language.* New York: Doubleday.

Hargie, O, Dickson, D, & Tourish, D. (2004). *Communication skills for effective management.* New York: Palgrave Macmillan.

Hill, C., & O'Brien, K. (1999). *Helping skills.* Washington, DC: American Psychological Association.

Ivey, A., Ivey, M., Myers, J., & Sweeney, T. (2005). *Developmental counseling and therapy: Promoting wellness over the lifespan.* Boston: Lahaska/Houghton Mifflin.

Knapp, M., & Hall, J. (2001). *Nonverbal communication in human interaction* (5th ed.). Belmont, CA: Wadsworth.

LaFrance, M., & Woodzicka, J. (1998). No laughing matter: Women's verbal and nonverbal reactions to sexist humor. In J. Swim & C. Stangor (Eds.), *Prejudice: The target's perspective.* San Diego, CA: Academic Press.

Masuda, T., & Nisbett, R. (2001). Attending holistically versus analytically: Comparing the context sensitivity of Japanese and Americans. *Journal of Personality and Social Psychology, 81,* 922–934.

Mayo, C., & LaFrance, M. (1973). *Gaze direction in interracial dyadic communication.* Paper presented at the Eastern Psychological Association meeting, Washington, DC.

Meara, N., Pepinsky, H., Shannon, J., & Murray, W. (1981). Semantic communication and expectation for counseling across three theoretical orientations. *Journal of Counseling Psychology, 28,* 110–118.

Meara, N., Shannon, J., & Pepinsky, H. (1979). Comparisons of stylistic complexity of the language of counselor and client across three theoretical orientations. *Journal of Counseling Psychology, 26,* 181–189.

Sharpley, C., & Guidara, D. (1993). Counselor verbal response mode usage and client-perceived rapport. *Counseling Psychology Quarterly, 6,* 131–142.

Sharpley, C., & Sagris, I. (1995). Does eye contact increase counselor-client rapport? *Counseling Psychology Quarterly, 8,* 145–155.

Shostrum, E. (1966). *Three approaches to psychotherapy* [Film]. Santa Ana, CA: Psychological Films.

Willis, C. (1989). *The measurement of mutual nonverbal coordination in the psychotherapeutic process.* Unpublished doctoral dissertation, University of Massachusetts, Amherst.

HOW ALLEN RESPONDED TO THE COURTROOM SITUATION

As part of his testimony on whether Horace could take on some form of employment, Allen pointed out to the judge that many traditional African Americans from the South had a different pattern of eye contact from standard middle-class White standards. Looking away in some African Americans is actually a sign of respect. It also goes back to a history when Blacks had learned it was unsafe to have direct eye contact with White people in power.

Interpreting what you see needs to be modified by constant awareness of individual and cultural differences. In addition, nonverbal communication style can overrule what is said verbally. Counseling and therapy tend to be word centered and we too often forget that eye contact, body language, and the way we listen can take precedence over what is said. As you work with this chapter, spend time observing nonverbals and note how they modify the meaning of what is said. At the same time, please do not over-generalize and stereotype what you see—"judging" is dangerous. Each client is unique.

CORRECT RESPONSES FOR EXERCISE 4

1 — C	5 — A	8 — A
2 — A	6 — C	9 — A
3 — A	7 — C	10 — C
4 — C		

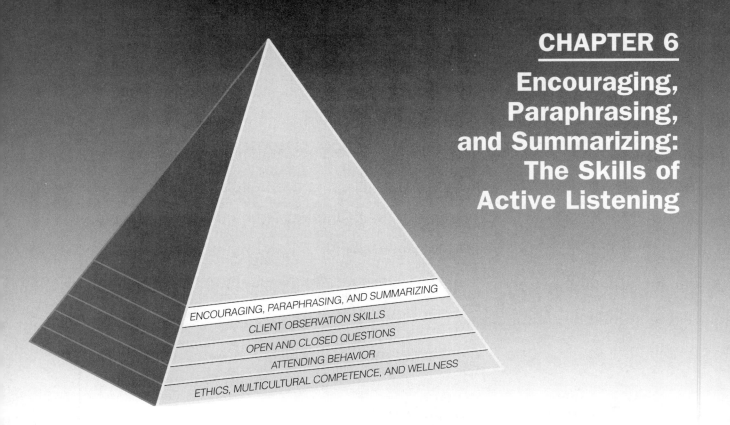

CHAPTER 6

Encouraging, Paraphrasing, and Summarizing: The Skills of Active Listening

ENCOURAGING, PARAPHRASING, AND SUMMARIZING

CLIENT OBSERVATION SKILLS

OPEN AND CLOSED QUESTIONS

ATTENDING BEHAVIOR

ETHICS, MULTICULTURAL COMPETENCE, AND WELLNESS

How can these three skills help you and your clients?

Major function

Clients need to know that the interviewer has *heard* what they have been saying, *seen* their point of view, and *felt* their world as they experience it. Encouragers and restatements, paraphrases, and summarizations are basic to helping a client feel understood. Once clients' stories have been truly heard, the clients can be much more open to and ready for change.

Secondary functions

Knowledge and skill in these dimensions result in the following:

▲ Clarifying for the client what he or she has said.
▲ Clarifying for the interviewer what the client has said. By feeding back what you have heard, you can check on the accuracy of your listening.
▲ Helping clients to talk in more detail about issues of concern to them.
▲ Helping an overly talkative client stop repeating the same facts or story, thus speeding up and clarifying the interview process.

Jennifer: (enters the room and starts talking immediately) I really need to talk to you. I don't know where to start. I just got my last exam back and it was a disaster, maybe because I haven't studied much lately. I was up late drinking at a party last night and I almost passed out. I've been sort going out with a guy for the last month, but that's over as of last night. . . . (pause) But what really bothers me is that my Mom and Dad called last Monday and they are going to separate. I know that they have fought a lot, but I never thought it would come to this. I'm thinking of going home, but I'm afraid to . . .

Jennifer continues for another three minutes in much the same vein, repeating herself somewhat, and seems close to tears. At times the data are coming so fast that it is hard to follow her. Finally she stops and looks at you expectantly.

What might be going through your mind about Jennifer at this moment? Using the ideas of active listening, what would you say and do to help her feel that you understand her and empathize?

_____

_____

_____

On page 176, you'll find what occurs for us.

INTRODUCTION: ACTIVE LISTENING

Listening is not a passive process. Whether using attending skills, encouraging, paraphrasing, or summarizing, you are actively involved in the interview. You do not just sit and listen to a story. Active listening demands that you participate fully by helping the client clarify, enlarge, and enrich the story. It requires that you be able to hear small changes in thoughts, feelings, and behaviors. It asks that you walk in the other person's shoes. Active listening demands serious attention to empathy.

Encouraging, paraphrasing, and summarizing are basic to empathic understanding and enable you to communicate to clients that they have been heard. In these accurate listening skills you do not mix your own ideas with what the client has been saying. You say back to the clients what you have heard, using their key words. You help clients by distilling, shortening, and clarifying what has been said.

Accurate empathic listening is not as common as we might wish nor is it as easy as it might sound, but its effects are profound. Ask your friend or family member to tell you a story (e.g., a conflict, a positive experience, a current challenge). Simply sit and listen to what is said, perhaps asking a few questions to enrich and enlarge the story. Then, say back to the volunteer client, as accurately as possible, what you have heard. Ask your friend how accurate your summary was and how it felt to be listened to. Use this space for your observations and the other person's reactions.

_____

_____

_____

Box 6-1 Listening Skills and Children

	The listening and observing skills are just as important to use with children. Children too often go through life being told what to do. If we listen to them and their singular constructions of the world, we can reinforce their unique qualities and help them develop belief in themselves and in their own value. Here are a few key comments on the listening skills and children:
Attending	Avoid looking down at children; whenever possible, talk to them at their level. This may mean sitting on the floor or in small chairs. Their energy is such that it helps if they have something to do with their hands; perhaps you can allow them to draw or play with clay as they talk to you. Be prepared for more topic jumps than with adults, but use attending skills later to bring them back to critical issues that need to be discussed.
Questions and concreteness	Use short sentences, simple words, and a concrete style of language, avoiding abstractions. Children may have difficulty with a broad open question such as "Could you tell me generally what happened?" Break down such abstract questions into concrete and situational language using a mix of closed and open questions such as, "Where were you when the fight occurred?" "What was going on just before the fight?" "Then what happened?" "How did he feel?" "Was she angry?" "What happened next?" "What happened afterward?" In questioning children on touchy issues, be especially careful of closed, leading questions. Seek to get their perspective, not yours.
Encouraging, paraphrasing, and summarizing	Effective elementary teachers use these skills constantly, especially paraphrasing and encouraging. Seek out a competent teacher and observe for yourself. These skills, coupled with good attending and questioning, are very important in helping children get out their stories.
Other issues	Provide an atmosphere that is suitable for children by using small chairs and interesting objects. Warmth and an actual liking for children are essential. Use names rather than pronouns, as children often get confused when under stress (as do many adults). Smiling, humor, and an active style will help.

Paraphrasing and summarizing communicate to clients that they have been listened to. Encouraging helps clients explore their feelings and thoughts more completely. A brief definition of these skills follows:

Encouragers are a variety of verbal and nonverbal means that the counselor or therapist can use to prompt clients to continue talking. They include head nods, openhanded gestures, phrases such as "Uh-huh," and the simple repetition of key words the client has uttered. Restatements are extended encouragers, the repetition of two or more words *exactly* as used by the client. In addition, appropriate smiling and interpersonal warmth are major encouragers that help clients feel comfortable and keep talking in the interview.

Paraphrasing feeds back to the client the essence of what has just been said. The listener shortens and clarifies the client's comments. Paraphrasing is not parroting; it is using some of your own words plus the important main words of the client.

Summarizations are similar to paraphrases but are used to clarify and distill what the client has said over a longer time span. Summarizations may be used to begin or end an interview, to move to a new topic, or to clarify complex issues. Most important, summarization helps both the client and you organize thinking about what is happening in the interview.

EXAMPLE INTERVIEW: THEY ARE TEASING ME ABOUT MY SHOES

Counseling children, in reality, is much like counseling adolescents and adults. You will use the same microskills, but encouraging, paraphrasing, and summarization skills are perhaps even more important. They help the child to continue talking from the child's frame of reference. And of course, adolescents and adults have an equal need to know that they have been heard. Just telling the story to a person who hears you accurately is clarifying, comforting, and reassuring. The skills of this chapter are truly important in all interviewing, counseling, and psychotherapy.

The session below presents a child's problem, but all of us, regardless of age, experience nasty teasing and put-downs, often in our closest relationships. Below you will find that Mary first draws out the child's story about teasing and then her thoughts and feelings about the story. This is followed by a focus on client strengths. The client, Damaris, and the counselor are then ready to address possibilities to restory and act differently; story—positive asset—restory—action.

The following interview is an edited version of a videotaped interview conducted by Mary Ivey with Damaris, a child actor, role-playing the problem. Damaris is an 11-year-old sixth grader. The case below illustrates socioeconomic issues that occur in many school settings; but adolescents and adults also deal with financial challenge. If you have ever wondered how to make car payments just as your car's transmission fails, the following story may make special sense to you.

In the condensed version of the interview below, Mary used many encouragers and restatements. A review of the precise video transcript reveals nine minimal encouragers ("oh . . . ," "uh, huh," and single word utterances), three positive encouragers ("that's great," "nice"), and four additional brief restatements. A viewing of the videotape would reveal smiles and head nods. In short, active listening with children demands constant involvement, showing interest, and a good humor. Children tend to respond more briefly than adults and interviews will not be as long as those with adults.

Interviewer and Client Conversation	Process Comments
1. *Mary:* Damaris, how're you doing?	The relationship between Mary and Damaris is already established as they know each other through school activities.
2. *Damaris:* Good.	She smiles and sits down.
3. *Mary:* I'm glad you could come down. You can use these magic markers if you want to doodle or	Mary first welcomes the child and offers her something to do with her hands. Many children get restless just

(continued)

Interviewer and Client Conversation	Process Comments
draw something while we're talking. I know— you sort of indicated that you wanted to talk to me a little bit.	talking to the counselor or interviewer. They may need something else to do. Damaris starts to draw almost immediately.
4. *Damaris:* In school, in my class, there's this group of girls that keep making fun of my shoes, just 'cause I don't have Nikes™.	Damaris looks down and appears a bit sad. She stops drawing. Children are well aware of their economic circumstances, particularly if they are one of the "have-nots." Some children may have only used sneakers to wear. Damaris, at least, has newer sneakers.
5. *Mary:* They keep making fun of your shoes?	Restatement using Damaris's exact key words.
6. *Damaris:* Well, they're not the best but, I mean—they're not Nikes, like everyone else has.	Damaris has a slight tone of anger mixed with her sadness. She starts to draw again.
7. *Mary:* Yeah, they're nice shoes, though. You know?	Mary offers reassurance. This is early and she's not really in touch with what is going on. A simple "uh-huh" would have been more effective. Later in this interview, the same comment might have worked. It is sometimes tempting to comfort clients rather than to listen to them. This can minimize the concerns of both child and adult clients.
8. *Damaris:* Yeah. But my family's not that rich, you know. Those girls are rich.	Clients, especially children, hesitate to contradict the counselor. Notice that Damaris says, "But . . ." When clients say, "Yes, but, . . ." interviewers are clearly off track and need to change their style.
9. *Mary:* I see. And they can afford Nike shoes, and you have these shoes, which are nice, but they're just not like they have, and they tease you about it?	Mary recognizes that her reassurance was too early. She backs off and paraphrases what has been said so far. The paraphrase uses Damaris's key words. Mary chose not to follow up on the issue of *rich* at this point. Do you agree? What if the client were an adolescent or adult?
10. *Damaris:* Yeah. . . . Well, sometimes they make fun of me and call me names, and I feel sad or— I try to ignore them, but still, the feeling inside me just hurts.	Often a client will respond with *yeah* or *yes* if you paraphrase or summarize accurately. They usually continue or elaborate their story.
11. *Mary:* It makes you feel hurt inside that they should tease you about shoes.	Mary reflects Damaris's feelings. The reflection of feeling is close to a paraphrase and is elaborated in the following chapter.
12. *Damaris:* Mmm-hmm. (pause) It's not fair.	Damaris thinks about Mary's statement and looks up expectantly as if to see what happens next. She thinks back on the basic unfairness of the whole situation.

13. *Mary:* So far, Damaris, I've heard how the kids are teasing you about not having Nikes and that it really hurts your feelings. It really is not fair. You know, I think of you, though, and I think of all the things that you do well, I get—you know, it makes me sad to hear this part because I think of all the talents you have and all the things that you like to do and—and the strengths that you have.	In a summary, we seek to bring together the main facts of the story as well as how the client thinks and feels about the story. This brief summary covers most of the important points covered in the session thus far. Mary discloses some of her own feelings about the situation. This self-disclosure used carefully and sparingly can be helpful, but if used too often, it will interfere with the helping process. Mary then turns to the positive asset search.
14. *Damaris:* Right. Yeah.	Damaris smiles slightly and relaxes a bit.
15. *Mary:* What comes to mind when you think about all the positive things you are and have to offer? Wow, I just think that they don't know the real you, with all the things that you have to offer!	An open question is followed by a summary of Mary's thoughts about Damaris from earlier contacts with her in the school.
16. *Damaris:* Well, in school, I like to read and write, and I want to be a journalist when I grow up. The teacher says I'm a good writer. The last story I wrote the teacher liked so much that she thinks it should be in the school paper.	Damaris talks a bit more rapidly and smiles.
17. *Mary:* You read and write and you want to be a journalist, 'cause you can write well? Wow!	Mary enthusiastically paraphrases Damaris's positive comments using Damaris's own key words.
18. *Damaris:* Mmm-hmm. And I play soccer on our team. It's kinda like at recess, like I'm one of the people that plays a lot, so I'm like the leader, almost, but—	Damaris is now smiling fully. Obviously, she has many things to feel good about.
19. *Mary:* So, not only are you a scholar, you are also a leader and an athlete. Other people look up to you. Is that right?	In this paraphrase, Mary has added the word *scholar* and *athlete*. She knows from observation on the playground that other children do look up to Damaris, even though her shoes are not like the rich kids' shoes. For clarification and broader understanding, it can be helpful if the counselor adds related words to expand the meaning. But as the new words are Mary's, she wisely uses the check-out "Is that right?" to see if she is on track, rather than leading the client too much.
20. *Damaris:* (small giggle) Yeah.	Damaris also looks down briefly. Some term this spontaneous movement the "recognition response." It happens most often when clients recognize something new about themselves and it makes sense to them. Looking down is not always sadness!
21. *Mary:* Uh-huh. So how do you feel when you're a leader in soccer and when you're, you know —	Mary picks up the feelings and asks an open question related to emotions.

(continued)

Interviewer and Client Conversation	Process Comments
22. *Damaris:* (interrupting Mary) It feels good! (small laugh)	Damaris has internalized the good feelings.
23. *Mary:* It feels good to be a leader in soccer! So you're a good student and you are good at soccer and a leader, and—and that makes—it makes you feel good inside.	Mary summarized the positive asset search using both facts and feelings. The summary of feeling *good inside* contrasts with the feelings of *hurt inside* discussed earlier.
24. *Damaris:* Yeah, it makes me feel good inside. I do my homework and everything, (pause) but then when I come to school, they just have to spoil it for me.	Again, Damaris agrees with the paraphrase. She feels support from Mary and is now prepared to deal from a stronger position with the teasing. She looks sad again.
25. *Mary:* They just spoil it. So you've got these good feelings inside, good that you're strong in academics, good that you're, you know, good at soccer and a leader. Now, I'm just wondering how we can take those good feelings that you feel as a student who's going to be a journalist someday and a soccer player who's a leader. Now the big question is how you can take the good, strong feelings and deal with the kids who are teasing. Let's look at ways to solve your problem now.	Mary restates Damaris's last words and again summarizes the many good things that Damaris does well. Mary changes pace and is ready to move to the problem-solving portion of the interview.

Mary had a good relationship with Damaris as the interview began. She was able to draw out the story fairly quickly. She then moved to positive assets and strengths. With this information it is then easier to sort out and deal with client problems and challenges. Later portions of the demonstration interview move to goal setting and working with Damaris's concerns.

If you were using these same skills with an adult, expect to follow a similar interviewing structure except that most adults will provide longer verbal responses. And you would usually not be expected to use as many verbal and nonverbal encouragers as Mary did.

Informed Consent and Working With Children

When you work with children, the ethical issues around informed consent become especially important. Depending on state laws and practices, it is often necessary to obtain written parental permission before interviewing a child. And you must obtain permission when you are sharing information about the interview with others. The child and family should know exactly how the information is to be shared, and interviewing records should be available to them for their comments and evaluation. An important part of informed consent is stating that they have the right to withdraw their permission at any point.

INSTRUCTIONAL READING: THE ACTIVE LISTENING SKILLS OF ENCOURAGING, PARAPHRASING, AND SUMMARIZING

Encouraging, paraphrasing, and summarizing help the client clarify issues and move into deeper exploration of concerns. They also help you make sure that you hear what the client is saying accurately. These are not passive skills; they require you to be very active in the session, listening intently and clarifying what clients have to say.

Important in this section and in all interviewing, counseling, and psychotherapy is a *nonjudgmental attitude* in which you listen to clients without evaluating them and what they say as "good" or "bad." You simply try to hear and accept what they are saying in their stories. You convey your ability to be neutral and supportive by your 3 V's + B—visuals, vocals, verbal following, and body language. All interviewing behavior can possibly convey judgmental and negative attitudes. A real challenge for most of us is to be nonjudgmental and accepting as we listen to clients.

Encouraging

Encouragers have been defined as head nods, open gestures, and positive facial expressions that encourage the client to keep talking. Minimal verbal utterances such as "Ummm" and "Uh-huh" have the same effect. Silence, accompanied by appropriate nonverbal communication, can be another type of encourager. All these encouragers affect the direction of client talk only minimally; clients are simply encouraged to keep talking.

Repetition of key words and direct restatement, on the other hand, have more influence on what clients talk about. Consider the following client statement. Let's imagine that Jennifer, the client with multiple issues presented at the beginning of this chapter, focuses on her parents' separating as the major immediate issue.

> I feel like my life is falling apart. I've always been close to both my folks and since they told me that they were breaking up, nothing has been right. When I sit down to study, I can't concentrate. And my roommate says that I get angry too easily. I guess everything upsets me. I was doing OK in my classes until this came along. I'm hurting so much for my Mom.

Jennifer still has a lot going on in her life. Eventually, we will want to focus on some of her strengths, but at the moment, she clearly needs to vent and explore her thoughts and feelings around her parents in more depth. There are several key words and ideas in this statement and the repetition of any of them is likely to lead Jennifer to expand on current issues. As a counselor, we'd tend to recommend repeating the exact key words, "You're hurting." This provides an opening for her to discuss her feelings or thoughts about Mom, herself, and, if she chooses, even other issues. "You're hurting for your Mom" would focus more narrowly, but likely would be another good choice. "Falling apart," "close to your folks," "can't concentrate," and "you get angry easily" are other possibilities. All of these will help Jennifer continue to talk, but do lead in varying directions. Taking a wellness approach, you could zero in on "doing OK in classes" as an alternative. The positive asset search and an emphasis on Jennifer's wellness strengths will be helpful in the long run, but

Box 6-2   Accumulative Stress: When Do "Small" Events Become Traumatic?

At one level, being teased about the shoes one wears doesn't sound all that serious—children will be children! However, some poor children go through their entire school life wearing clothes that others tease them about and laugh at, either directly or indirectly. At a high school reunion, Allen talked with a classmate who clearly was disturbed emotionally. During the talk, it became clear that teasing and bullying during schools days were still immediate and painful memories for him.

Small slights become big hurts if repeated again and again over the years. Athletes and "popular" students may talk arrogantly and dismissively about the "nerds," "townies," "rurals," or other outgroup. Teachers, coaches, and even counselors sometimes join in the laughter. Over time, these slights mount inside the child or adolescent. Some people internalize their issues in psychological distress; others may act them out in a dramatic fashion—witness the episodes of shootings throughout the country, including the many dead at Columbine High School in Colorado.

Discrimination and prejudice are other examples of accumulative stress and trauma. One of Mary's interns, a young African American woman, spoke with her of a recent racial insult. She was sitting in a restaurant and overheard two White people talking loudly about the "good old days" of segregation. Perhaps the remark was not directed at her, but still, it hurt. The young woman went on to speak of how common racial insults were in her life, directly or indirectly. She could tell how bad things were racially by the size of her phone bill. When an incident occurred that troubled her, she needed to talk her sister or parents and seek support. Out of such continuing indignities come feelings of underlying insecurity about one's place in the world (internalized oppression and self-blame) and/or tension and rage about unfairness (externalized awareness of oppression). Either way, the person who is ignored or insulted feels tension in the body, the pulse and heart rate increase, and—over time—hypertension and high blood pressure may result. The psychological becomes physical and accumulative stress becomes traumatic.

Soldiers at war, women who suffer sexual harassment, those who are overweight or short in height, the physically disfigured through birth or accident, gays and lesbians, and many others are all at risk for accumulative stress building to real trauma. They all suffer the dangers of posttraumatic stress.

As an interviewer, you will want to be alert for signs of accumulative stress in your clients. Are they internalizing the stressors by blaming themselves? Or are they externalizing and building a pattern of rage and anger inside that may explode? All these people have important stories to tell and at first, these stories may sound routine. Posttraumatic stress responses in later life may be alleviated or prevented by your careful listening and support.

Finally, social work's position on social justice is that the interviewer has the responsibility to act and intervene, where possible, to combat oppression and injustice. The counseling and psychology position on social action is not as clear. Where do you stand?

at the moment, emphasis on this key word should be brought up later, after Jennifer has been able to ventilate about her issues more thoroughly.

Key word encouragers contain one, two, or three words while *restatements* are longer. Both focus on staying very close to the client's language, most typically changing only "I" to "you." (*Jennifer:* "I'm hurting so much for my Mom." *Counselor:* "You're hurting.")

It may be helpful if you reread the paragraphs above, saying aloud the suggested encouragers and restatements. Use different vocal tones and note how your verbal style can facilitate others' talking or stop them cold.

All types of encouragers facilitate client talk unless they are overused or used badly. Excessive head nodding or gestures and too much parroting can be annoying and frustrating to the client. From the observation of many interviewers, we know that use of too many encouragers can seem wooden and unexpressive. However, too few encouragers may suggest to clients that you are not interested or involved. Well-placed encouragers help to maintain flow and continually communicate that the client is being listened to. Single-word encouragers often facilitate client talk toward deeper meanings.

Paraphrasing

At first glance, paraphrasing appears to be a simple skill, only slightly more complex than encouraging. In restatement and encouraging, exact words and phrases are fed back to the client, but in a shortened and clarified form. If you are able to give an accurate paraphrase to a client, you are likely to be rewarded with a "That's right" or "Yes . . . ," and the client will go on to explore the issue in more depth. Furthermore, accurate paraphrasing will help the client stop repeating a story. Some clients have complex problems that no one has ever bothered to hear accurately, and they literally need to tell their story over and over until someone indicates they have been heard clearly. Once clients know they have been heard, they are often able to move on to new topics. The goal of paraphrasing is the facilitation of client exploration and the clarification of issues. The tone of your voice and your body language accompanying the paraphrasing also indicate to the client whether you are interested in listening in more depth or wish for the client to move on.

How do you paraphrase? Client observation skills are important in accurate paraphrasing. You need to hear the client's important words and use them in your paraphrase much as the client does. Other aspects of the paraphrase may be in your own words, but the main ideas and concepts should reflect the client's view of the world, not yours!

An accurate paraphrase, then, usually consists of four dimensions:

1. A *sentence stem* sometimes using the client's name. Names help personalize the session. Examples would be, "Damaris, I hear you saying . . . ," "Luciano, sounds like . . . ," and "Looks like the situation is . . ." A stem is not always necessary and, if overused, can make your comments seem like parroting. Clients have been known to say in frustration, "That's what I just said; why do you ask?"
2. The *key words* used by the client to describe the situation or person. Again, drawing on client observation skills, the effort is to include main ideas that come from clients and their exact words. This aspect of the paraphrase is sometimes confused with the encouraging restatement. A restatement, however, is almost entirely in the client's own words and covers only limited amounts of material.

Box 6-3 Research Evidence That You Can Use: Active Listening

Microskills training enables counselors to respond in a more culturally appropriate fashion (Nwachuku & Ivey, 1991). Research on encouraging, paraphrasing, and summarizing are often treated as part of a larger whole—empathic listening. Some of the most carefully designed work in this area has been done by Bensing (1999) on physician-patient relationships. She found that physicians who established a solid relationship and *listened* carefully tended to be rated more highly and, actually, to be more likely to have patients follow their suggestions and directives. She also found that talk-time of patients markedly increased with physicians who listened to them. Nine studies on microcounseling with nurses found that they were rated more highly on empathy, focused more on the client, and made fewer therapeutic errors. Similar findings exist among counselors and therapists (Daniels & Ivey, 2006).

Smiling is one of the most encouraging things you can do. Many researchers have found that smiling "works" and is a primary way to communicate warmth and openness (Restak, 2003). Although the following is not the result of formal research, it is based on observation: Allen Ivey is a person who does not physically show a lot of emotion—he can smile, but is a bit shy and reserved. On the other hand, Mary Bradford Ivey is known for her smile and sunny disposition. Guess which person other people talk to when Allen and Mary are together? If you are a naturally warm person, this characteristic communicates itself to your client. If you are more like Allen, it may take you a bit longer, but you can still be a very effective listener!

Active listening and neuropsychology

Carter (1999, p. 87) reports:

> Expressions can . . . transmit emotions to others—the sight of a person showing intense disgust turns on in the observer's brain areas that are associated with the feeling of disgust. Similarly, if you smile, the world does indeed smile with you (up to a point). Experiments in which tiny sensors were attached to the "smile" muscles of people looking at faces show the sight of another person smiling triggers automatic mimicry—albeit so slight that it may not be visible. . . . the brain concludes that something good is happening out there and creates a feeling of pleasure.

This is a variant of movement synchrony as described in the chapter on nonverbal communication. It is also possible for you to pick up the depressed mood and style of your client and recommunicate the sadness back to the client, thus reinforcing a cycle of negativity. And, if you listen with energy and interest, and this is communicated effectively, expect your client to receive that affect as a positive resource in itself.

3. The *essence of what the client has said* in summarized form. Here the interviewer's skill in transforming the client's sometimes confused statements into succinct, meaningful, and clarifying statements is most manifest. The counselor has the difficult task of keeping true to the client's ideas but not repeating them exactly.
4. A *check-out* for accuracy. The check-out is a brief question at the end of the paraphrase, asking the client for feedback on whether the paraphrase (or summary

or other microskill) was relatively correct and useful. Some example check-outs include "Am I hearing you correctly?" "Is that close?" "Have I got it right?" It is also possible to paraphrase with an implied check-out by raising your voice at the end of the sentence as if the paraphrase were a question.

Here is a client statement followed by sample key-word encouragers, restatements, and a paraphrase:

> I'm really concerned about my wife. She has this feeling that she has to get out of the house, see the world, and get a job. I'm the breadwinner and I think I have a good income. The children view Yolanda as a perfect mother, and I do too. But last night, we really saw the problem differently and had a terrible argument.

▲ Key-word encouragers: "Breadwinner?" "Terrible argument?" "Perfect mother?"
▲ Restatement encouragers: "You're really concerned about your wife." "You see yourself as the breadwinner." "You had a terrible argument."
▲ Paraphrase: "You're concerned about your picture-perfect wife who wants to work even though you have a good income, and you've had a terrible argument. Is that how you see it?"

This example shows that the key-word encourager, the restatement, and the paraphrase are all different points on a continuum. In each case the emphasis is on hearing the client and feeding back what has been said. Both short paraphrases and longer key-word encouragers will resemble restatements. A long paraphrase is close to a summary. All can be helpful in an interview; or they can be overdone. In the paraphrase, note that the client's visual system was stressed.

Summarizing

Summarizations fall along the same continuum as do the key-word encourager, restatement, and paraphrase. Summarizing, however, encompasses a longer period of conversation; at times it may cover an entire interview or even issues discussed by the client over several interviews.

In summarizing, the interviewer attends to verbal and nonverbal comments from the client over a period of time and then selectively attends to key concepts and dimensions, restating them for the client as accurately as possible. A check-out at the end for accuracy is an important part of the summarization. The following are examples of summarizations.

To begin a session: Let's see, last time we talked about your feelings toward your mother-in-law and we discussed the argument you had with her around the time the new baby arrived. You saw yourself as guilty and anxious. Since then you haven't gotten along too well. We also discussed a plan of action for the week. How did that go?

Midway in the interview: So far, I've seen that the plan didn't work too well. You felt guilty again when you saw the idea as manipulative. Yet one idea did work. You were able to talk with her about her garden, and it was the first time you had been able to talk about anything without an argument. You visualize the possibility of following up on the plan next week. Is that about it?

At the end of the session:	In this interview we've reviewed your feelings toward your mother-in-law in more detail. Some of the following things seem to stand out: First, our plan didn't work completely, but you were able to talk about one thing without yelling. As we talked, we identified some behaviors on your part that could be changed. They include better eye contact, relaxing more, and changing the topic when you start to see yourself getting angry. I also liked your idea at the end of talking with her about these issues. Does that sum it up?

Reflecting Feelings Compared With Encouraging, Paraphrasing, and Summarizing

Reflecting feelings is closely allied to the ideas expressed above. Discussed in detail in the following chapter, reflecting feelings names and feeds back the emotional tone of the conversation or story to the client. Encouragers that focus on emotional words tend to bring out affective dimensions. A reflection of feeling will look much like a paraphrase except that the focus is on emotion related to the story rather than just the content or facts of the story. Most summaries will also contain the main feelings expressed by the client. But this is not always so, as emotional dimensions are not always central to the discussion.

Client:	I can't stand being ignored. I'm going crazy. Anwar comes home, doesn't say a word, and turns on the television. If I try to talk to him, he just looks away. No matter what I do, it isn't working.
Nonverbal encourager:	(Head nod indicating that the client should continue.)
Key-word encouragers:	"Ignoring you?" or "Going crazy?" (The first focuses on content, the second on emotion)
Restatement:	If you try to talk, he just looks away.
Paraphrase:	Anwar is ignoring you and no matter how hard you try, it's getting worse.
Reflection of feeling:	You really feel frustrated, almost crazy, over the situation.

Most often, in response to this type of client statement, a combination paraphrase/reflection of feeling might be best—"Sounds like you are really frustrated with Anwar's ignoring you all the time."

Diversity and the Listening Skills

Periodic encouraging, paraphrasing, and summarizing are basic skills that seem to have wide cross-cultural acceptance. Virtually all your clients like to be listened to accurately. It may take more time to establish a relationship with a client who is culturally different from you. But again, never generalize or stereotype.

North American and European counseling theory, by and large, expects the client to get at the problem immediately and may not allow enough emphasis on relationship building. Some traditional Native American Indians, Pacific Islanders, Aboriginal Australians, and New Zealand Maori may want to spend a full interview getting to know you and trust you before beginning counseling. If you are seen in the village or community or at pow-wows or other cultural celebrations enjoying yourself in a natural way, this is likely to be helpful in trust building. Actively involving yourself in positive community activities will be noticed.

Box 6-4   National and International Perspectives on Counseling Skills: Developing Skills to Help the Bilingual Client

Azara Santiago-Rivera, President, National Latina/o Psychological Association, and University of Wisconsin, Milwaukee

It wasn't that long ago that counselors considered bilingualism a "disadvantage." We now know that a new perspective is needed. Let's start with two important assumptions: *The person who speaks two languages is able to work and communicate in two cultures and, actually, is advantaged. The monolingual person is the one at a disadvantage!* Research actually shows that bilingual children have more fully developed capacities and a broader intelligence (Power & Lopez, 1985).

If your client was raised in a Spanish-speaking home, for example, he or she is likely to think in Spanish at times, even though having considerable English skills. We tend to experience the world nonverbally before we add words to describe what we see, feel, or hear. For example, Salvadorans who experienced war or other forms of oppression *felt* that situation in their own language.

You are very likely to work with clients in your community who come from one or more language backgrounds. Your first task is to understand some of the history and experience of these immigrant groups. Then, we suggest that you learn some key words and phrases in their original language. Why? Experiences that occur in a particular language are typically encoded in memory in that language. So certain memories containing powerful emotions may not be accessible in a person's second language (English) because they were originally encoded in the first language (for example, Spanish). And if the client is talking about something that was experienced in Spanish, Khmer, or Russian, the *key words* are not English; they are in the original language.

Here is an example of how you might use these ideas in the session:

Social worker:	Could you tell me what happened for you when you lost your job?
Maria (Spanish-speaking client):	It was hard; I really don't know what to say.
Social worker:	It might help us if you would say what happened in Spanish and then you could translate it for me.
Maria:	*Es tan injusto! Yo pensé que perdi el trabajo porque no hablo el ingles muy bien. Me da mucho coraje cuando me hacen esto. Me siento herido.*
Social worker:	Thanks; I can see that it really affected you. Could you tell me what you said now in English?
Maria:	(More emotionally) I said, "It all seemed so unfair. I thought I lost my job because I couldn't speak English well enough for them. It makes me really angry when they do that to me. It hurts."
Social worker:	I understand better now. Thanks for sharing that in your own language. I hear you saying that *injusto* hurts and you are very angry. Let's continue to work on this and, from time to time, let's have you talk about the really important things in Spanish, OK?

The above brief example provides a start. The next step is to develop a vocabulary of key words in the language of your client. This cannot happen all at once,

(continued)

Box 6-4 (continued)

but you can gradually increase your skills. Here are some Spanish key words that might be useful with many clients:

Respecto: Was the client treated with respect? For example, the social worker might say, "Your employer failed to give you *respecto.*"

Familismo: Family is very important to many Spanish-speaking people. You might say, "How are things with your *familia?*"

Emotions (see next chapter) are often experienced in the original language. When reflecting feeling, you could learn and use these key words with clients:

Aguantar: endure   *Miedo:* fear
Amor: love   *Orgullo:* proud
Carino: like   *Sentir:* feel
Coraje: anger

We also recommend learning key sayings, metaphors, and proverbs in the language(s) of your community. *Dichos* are Spanish proverbs, like the following examples:

Al que mucho se le da, mucho se le demanda.   The more people give you, the greater the expectations of you.

Vale mas tarde que nunca.   Better late than never.

No hay peor sordo que el no quiere oir.   There is no worse deaf person than someone who doesn't want to listen.

En la unión está la fuerza.   Strength is found in unity.

Consider developing a list like this, learn to pronounce them correctly, and you will find them useful in counseling Spanish-speaking clients. Indeed, you are giving them *respecto.* You may wish to learn key words in several languages.

Mary Ivey comments: There is a large population of Cambodian students in the Amherst, Massachusetts, schools. As Azara suggests, I found it very helpful to suggest to children that they talk about difficult issues in Khmer. They then translated what they had said to me. It clearly helped our relationship. I also found it important to get out into the Cambodian community. I visited homes, sat on the floor, attended weddings and funerals, and enjoyed community celebrations. If you are to work with people who are culturally different from you, you have the pleasure and responsibility of participating with them and learning from their background.

In the interview itself, if you are very different in cultural background from your client and use questioning and listening skills exclusively, the client may become suspicious and consider your careful listening untrustworthy. In such cases, self-disclosure and explanation of what you are trying to do may be helpful. The client may want directions and suggestions for action. A young adolescent may expect reaction from you before he or she has even finished the story.

A general recommendation for working cross-culturally is to discuss differences early in the interview. For example, "I'm a European American and if that is an issue, we may need to discuss it. And if I miss something, please let me know." "I know that some gay people may distrust heterosexuals. Please let me know if something bothers you." "Some White people have issues in talking with an African American counselor. If that's a concern, let's talk about it up front." "You are 57 and I (the counselor) am 26. How comfortable are you working with someone my age?" There are no absolute rules here for what is right. A highly acculturated Jamaican, Native American Indian, or Asian American might be offended by the same statements.

Some Asian (Cambodian, Chinese, Japanese, Indian) clients from traditional backgrounds may be seeking direction and advice. They are likely to be willing to share their stories, but you may need to tell them why you want to wait a bit before coming up with answers. At times, to establish credibility, you may have to commit yourself and provide advice earlier than you wish. In such a case be assured and confident, but let them know you want to learn more with them and that the advice may change as you get to know them better.

Women tend to use paraphrasing and related listening skills more than men, whereas men tend to use questions more frequently. You may notice in your own classes and workshops that men tend to raise their hands faster and interrupt more often. But there are so many exceptions to this "rule" that it should not be relied on. Nonetheless, differences in gender do exist and it is important to consider them. On some issues gender differences need to be addressed directly by both men and women.

SUMMARY: PRACTICE, PRACTICE, AND PRACTICE

We have stressed the importance of three major listening skills in this chapter—encouraging, paraphrasing, and summarizing. Regardless of your theory of choice and however you integrate the microskills into your own natural style, these skills will remain central if you are to be effective.

Basic competence comes when you use the skills in an interview and you can expect that to be helpful to your clients. Every client needs to be heard, and demonstrating that you are listening carefully often makes a real difference. However, intentional competence in these skills requires practice.

Amanda Russo was a student in a counseling course at Western Kentucky University taught by Dr. Neresa Minatrea. She shared with us how she practiced the skills and gave us her permission to pass this on to you. As you read her comments, ask yourself if you are willing to go as far as she did to ensure expertise.

> For my final project I decided to do a practice from the book *Theories of Counseling: A Multicultural Perspective* (Ivey, Ivey, & Simek-Morgan, 2002). The exercise is entitled The Positive Asset Search: Building Empathy on Strengths. I chose an exercise early on in the book because I do not have much experience with counseling and I wanted to try a fairly simple exercise to start out. I performed the same exercise on five different people to see if I would get the same results.
>
> The exercise consists of asking the client what some of their areas of strength are, getting them to share a story regarding that strength, and then for

the counselor to observe the client's gestures and be aware of any changes. The first person I tried this exercise on was Raphael, a dormitory proctor. Some of his strengths were family, friends, working out, and that he had a good inner circle/support group. As he talked about his support group and how they reminded him of the positives he started to sit in a less tense position. He seemed very relaxed, yet excited about his topic of discussion and I noticed a lot of hand gestures. In a matter of seconds I saw him change from tense and unsure to relaxed and enthusiastic about what he was saying.

The next person I practiced this exercise on was my roommate Karol. She was a bit nervous when we started and had a difficult time thinking of strengths. Once I asked her to share a story with me she became very animated. As she spoke, I could see a sparkle in her eyes. Her voice became stronger and her hands were moving every which way. She feels strongly about doing well at work, giving advice, working out, playing music, and finishing the song she is currently writing.

Once she gave me a couple of strengths they started her wheels turning and she was coming up with more and more. She felt very good about jazz practice earlier that day. She introduced a new song to the group and they really enjoyed it. She also shared a story with me about a huge accomplishment at work that day. Karol was definitely the person whose mood/persona changed the most in this exercise.   (Courtesy of Amanda Russo.)

Amanda went on to interview three more people to test out her skills and reported in detail on each one. If you seek to reach intentional competence, the best route toward this is systematic practice. For some of us, one practice session may be enough. For most of us, it will take more time. What commitments are you willing to make?

Box 6-5   Key Points

Why?	Clients need to know that their story has been *heard*. Attending, questioning, and other skills help the client open up, but accurate listening through the skills of encouraging, paraphrasing, and summarizing is needed to communicate that you have indeed heard the other person fully.
What?	Three skills of accurate listening help communicate your ability to listen:

1. *Encouragers* are a variety of verbal and nonverbal means the counselor or interviewer can use to encourage others to continue talking. They include head nods, an open palm, "Uh-huh," and the simple repetition of key words the client has uttered. Restatements are extended encouragers using the exact words of the client.
2. *Paraphrases* feed back to the client the essence of what has just been said by shortening and clarifying client comments. Paraphrasing is not parroting; it is using some of your own words plus the important main words of the client.
3. *Summarizations* are similar to paraphrases except that a longer time and more information are involved. Summarizations may be used to begin or end an

(continued)

Box 6-5   (continued)

	interview, for transition to a new topic, or to provide clarity in lengthy and complex client stories.
How?	Encouragers have just been described. It is important to add that the so-called simple repetition of key words is more important than appears at first glance. Key words repeated to the client usually lead the client to elaborate in greater detail on the meaning of that word to him or her. Interviewers and counselors find it interesting and sometimes challenging to note their own selective attention patterns as they use this first "simple" skill.

Paraphrasing and summarizing usually involve four dimensions:

1. *A sentence stem.* Often you will want to use the client's name—"Jamilla, I hear you saying . . . ," "Carlos, sounds like . . . ," but you may also paraphrase, focusing on what the client has just said.
2. *Key words.* The clients' exact own key words that they use to describe their situation.
3. *The essence of what the client has said in distilled form.* Here you use the client's key words in a brief summary of what the client has just said. Summaries are longer paraphrases that often include emotional dimensions as well.
4. *A check-out.* Implicitly or explicitly, check with the client to see that what you have fed back to him or her is accurate. "Have I heard you correctly?"

With whom?	These skills are useful with virtually any client. However, some find repetition tiresome and may ask, "Didn't I just say that?" Consequently, when you use the skill you should employ your client observation skills.
What else?	All of these skills involve active listening, encouraging others to talk freely. They communicate your interest and help clarify the world of the client for both you and the client. This skill is one of the most difficult in the microtraining framework for many people.
	As you listen to clients, seek to maintain a nonjudgmental, accepting attitude. Even the most accurate paraphrasing or summarization can be negated by oppositional nonverbal behaviors.

COMPETENCY PRACTICE EXERCISES AND PORTFOLIO OF COMPETENCE

The three skills of encouraging, paraphrasing, and summarizing are much less controversial than questions. Virtually all interviewing theories recommend and endorse these key skills of active listening.

Individual Practice

Exercise 1: Identifying Skills

Below is a client statement followed by several alternative interviewer responses. Identify encouragers (E), restatements (R), paraphrases (P), and summaries (S). One reflection of feeling (RF) is included below in anticipation of distinctions discussed in the next chapter.

Client: The visit went well. I've pretty much decided to go back and finish college. But how will I pay for it? I've got a good job now, but I'll have to move to part time and I'm not sure that they will keep me. Getting all this financed will be difficult. It is kind of scary.

_____ "Uh-huh."

_____ Silence with facilitative body language.

_____ "Scary?"

_____ "You're not sure that they will keep you."

_____ "Sounds like you've made up your mind to finish college, but financing it will be a major challenge."

_____ "It's scary to think of going back to college."

_____ "In the last interview, we talked about your going back to visit the school and so far it sounds as if it went well and you really want to do it. But at the same time, now, it is a little scary when you think of all the financial issues. Have I heard what's been happening correctly?"

Exercise 2: Generating Written Encouragers, Restatements, Paraphrases, and Summarizations

A. "Chen and I have separated. I couldn't take his drinking any longer. It was great when he was sober, but it wasn't that often he was. Yet that leaves me alone. I don't know what I'm going to do about money, the kids, or even where to start looking for work."

Write three different types of key-word encouragers for this client statement:

_____

_____

_____

Write a restatement:

_____

Write a paraphrase (include a check-out):

_____

_____

Write a summarization (generate data by imagining previous interviews):

_____

_____

_____

_____

B.   "And in addition to all that, we worry about having a child. We've been trying for months now, but with no luck. We're thinking about going to a doctor, but we don't have medical insurance."

Write three different types of key-word encouragers for this client statement:

_____

_____

_____

Write a restatement:

_____

Write a paraphrase (include a check-out):

_____

_____

Write a summarization (generate data by imagining previous interviews):

_____

_____

_____

_____

_____

_____

Exercise 3: Practice of Skills in Other Settings

Encouraging.   During conversations with friends or in your own interviews, deliberately use single-word encouragers and brief restatements. Note their impact on your friends' participation and interest. You may find that the flow of conversation changes in response to your brief encouragers. Summarize some of your observations here:

_____

_____

_____

_____

_____

_____

_____

Group Practice

Experience has shown that the skills of this chapter are often difficult to master. It is easy to try to feed back what another person has said, but to do it *accurately*, so that the client feels truly heard, is another matter.

Step 1: Divide into practice groups.   Include triads as a possibility.
Step 2: Select a group leader.
Step 3: Assign roles for the first practice session.
▲ Client
▲ Interviewer
▲ Observer 1 uses Feedback Form (Box 6-6)
▲ Observer 2 uses Feedback Form

Step 4: Plan.   Establish and state clear goals for the practice session. The interviewer should plan a role-play in which open questions are used to elicit the client's concern. Once this is done, use encouragers to help bring out more details and deeper meanings. Use more open and closed questions as appropriate, but give primary attention to the paraphrase and the encourager. End the interview with a summary (this is often forgotten). Check the accuracy of your summary with a check-out ("Am I hearing you correctly?").

For real mastery, seek to use only the three skills in this chapter and use questions only as a last resort.

The suggested topic for this practice session is the story of a past or present stressful experience that may in some way relate to the ideas of accumulative trauma. Examples include teasing; bullying; being made the butt of a joke; an incident when you were seriously misunderstood or misjudged; an unfair experience with a school teacher, coach, or counselor; or a time you experienced prejudice or oppression of some type.

Emotions may appear as the story unfolds. Feel free to paraphrase or summarize these emotions, but this time focus first on the story and the event or situation itself. We suggest that you repeat this story again when you practice the skill of reflection of feeling in the next chapter.

All of the above incidents or topics will provide observers with the opportunity to observe nonverbal behaviors and discrepancies, incongruity, and conflict. Does the client internalize or externalize responsibility and blame? Internal attribution occurs when clients see themselves as being at fault. External attribution occurs when "they" or external matters are seen as the cause. Most clients will demonstrate some balance of internal and external attribution—self-blame versus other blame.

Step 5: Conduct a 3-minute practice session.
Step 6: Review the practice session and provide feedback to the interviewer for 12 minutes.   Be sure to use the Feedback Form (Box 6-6) to ensure that the interviewer's statements are available for discussion. This form provides a helpful log of the session, which greatly facilitates discussion. And give special attention to feedback from the client, perhaps using the Client Feedback Form of Chapter 1. If you have an audio- or videotape, start and stop the tape periodically and rewind it to

hear and observe important points in the interview. Did the interviewer achieve his or her goals? What mastery level was demonstrated?

Step 7: Rotate roles.

Some general reminders.   It is important that clients talk freely in the role-plays. As you become more confident in the practice session, you may want your clients to become more "difficult" so you can test your skills in more stressful situations. You'll find that difficult clients are often easier to work with after they feel they have been heard.

Box 6-6   Feedback Form: Encouraging, Paraphrasing, and Summarizing

_____ (Date)

_____   _____
(Name of Interviewer)   (Name of Person Completing Form)

Instructions: Write below as much as you can of each counselor statement. Then classify the statement as a question, an encourager, a paraphrase, a summarization, or other. Rate each of the last three skills on a scale of 1 (low) to 5 (high) for its accuracy.

Counselor statement	Open question	Closed question	Encourager	Paraphrase	Summarization	Other	Accuracy rating
1. _____							
2. _____							
3. _____							
4. _____							
5. _____							
6. _____							
7. _____							
8. _____							
9. _____							
10. _____							

1. What were the key discrepancies demonstrated by the client?

2. General interview observations. Was responsibility for the concern placed internally or externally, or with some balance between the two?

Portfolio of Competence

Active listening is one of the core competencies of intentional interviewing and counseling. Please take a moment to review where you stand and where you plan to go in the future.

Use the following as a checklist to evaluate your present level of mastery. Check those dimensions that you currently feel able to do. Those that remain unchecked can serve as future goals. *Do not expect to attain intentional competence on every dimension as you work through this book.* You will find, however, that you will improve your competencies with repetition and practice.

Highlight the competencies that you have met to date:

Level 1: Identification and classification.

❑ Ability to identify and classify encouragers, paraphrases, and summaries.
❑ Ability to discuss issues in diversity that occur in relation to these skills.
❑ Ability to write encouragers, paraphrases, and summaries that might predict what a client will say next.

Level 2: Basic competence.   Aim for this level of competence before moving on to the next skill area.

❑ Ability to use encouragers, paraphrases, and summaries in a role-played interview.
❑ Ability to encourage clients to keep talking through use of nonverbals and through the use of silence, minimal encouragers ("uh-huh"), and the repetition of key words.
❑ Ability to discuss cultural differences with the client early in the interview, as appropriate to the individual.

Level 3: Intentional competence.

❑ Ability to use encouragers, paraphrases, and summaries accurately to facilitate client conversation.
❑ Ability to use encouragers, paraphrases, and summaries accurately to keep clients from repeating their stories unnecessarily.
❑ Ability to use key word encouragers to direct client conversation toward important topics and central ideas.
❑ Ability to summarize accurately longer periods of client utterances—for example, an entire interview or the main themes of several interviews.
❑ Ability to communicate with bilingual clients using some of the key words and phrases in their primary language.

Level 4: Teaching competence.   Teaching competence in these skills is best planned for a later time, but a client who has particular difficulty in listening to others may indeed benefit by careful training in paraphrasing. There are some individuals who often fail to hear accurately and distort what others have said to them.

❑ Ability to teach clients in a helping session the social skills of encouraging, paraphrasing, and summarizing.
❑ Ability to teach small groups the above skills.

DETERMINING YOUR OWN STYLE AND THEORY: CRITICAL SELF-REFLECTION ON THE ACTIVE LISTENING SKILLS

This chapter has focused on encouraging/restatement, paraphrasing, and summarization as critical to obtaining a solid understanding of what clients want and need. Active listening is central and these three skills are key.

What single idea stood out for you among all those presented in this chapter, in class, or through informal learning? What stands out for you is likely to be important as a guide toward your next steps. What are your thoughts on race/ethnicity? What other points in this chapter struck you as important? How might you use ideas in this chapter to begin the process of establishing your own style and theory?

_____

_____

_____

_____

_____

REFERENCES

Bensing, J. (1999). *Doctor-patient communication and the quality of care.* Utrecht, the Netherlands: Nivel.

Carter, R. (1999). *Mapping the mind.* Berkeley: University of California Press.

Daniels, T., & Ivey, A. (2006). *Microcounseling* (3rd ed.). Springfield, IL: Thomas.

Ivey, A., Ivey, M., & Simek-Morgan, L. (2002). *Counseling and psychotherapy: A multicultural approach.* Boston: Allyn & Bacon.

Nwachuku, U., & Ivey, A. (1991). Culture-specific counseling: An alternative approach. *Journal of Counseling and Development, 70,* 106–151.

Power, S., & Lopez, R. (1985). Perceptual, motor, and verbal skills of monolingual and bilingual Hispanic children: A discrimination analysis. *Perceptual and Motor Skills, 60,* 1001–1109.

Restak, R. (2003). *The new brain.* New York: Rodale.

ALLEN AND MARY'S THOUGHTS ABOUT JENNIFER

Active listening requires actions and decisions on our part. What we listen to (selective attention) will have a profound influence on how clients talk about their concerns. When a client comes in full of information and talks rapidly, we often find ourselves confused and, we admit, a bit overwhelmed. It takes a lot of active listening to hear this type of client accurately and fully. If our work was personal counseling, we would most likely focus on her parents' separation and use an encourager by restating some of her key words, thus helping her focus on what may be the most central issue at the moment (e.g., "Your Mom and Dad called earlier this week and are separating"). We'd likely get a more focused story and could learn more about what's happening. As we understand this issue more fully, we could later move to discussing some of the other problems.

Another possibility would be to summarize the main things that Jennifer was saying as succinctly and accurately as possible. We'd do this by catching the essence of her several points and saying them back to her. Most likely, we'd use what is called the "check-out" to see if we have been reasonably close to what she thinks and feels (e.g., "Have I heard you correctly so far?"). This would then be followed by our asking her, "You've talked about many things. Where would *you* like to start today?"

And if we were academic counselors, not engaging in personal issues, we'd likely selectively attend to the area of our expertise (study issues) and refer Jennifer to an outside source for personal counseling.

How does this compare to what you would have done?

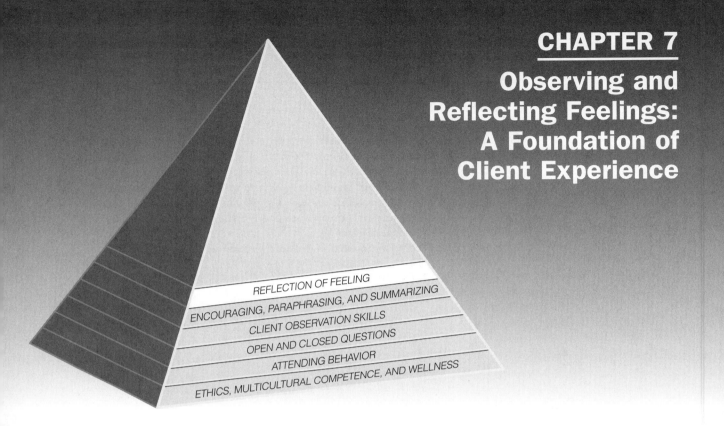

REFLECTION OF FEELING

ENCOURAGING, PARAPHRASING, AND SUMMARIZING

CLIENT OBSERVATION SKILLS

OPEN AND CLOSED QUESTIONS

ATTENDING BEHAVIOR

ETHICS, MULTICULTURAL COMPETENCE, AND WELLNESS

How can reflection of feeling help you and your clients?

Major function	Underlying clients' words, thoughts, and behaviors are feelings and emotions. The purpose of reflection of feeling is to make these implicit, sometimes hidden, emotions explicit and clear to the client. This chapter is essential to your being fully with the client, and we shall go into depth with this skill.
Secondary functions	Knowledge and skill in reflection of feeling result in the following:

▲ Bringing out the richness of the client's emotional world.
▲ Noting that most clients have mixed or ambivalent feelings toward significant others and events. You can use the skill to help clients sort out these complex feelings and thoughts.
▲ Grounding the counselor and client from time to time in basic experience. There is a tendency in much interviewing to intellectualize and move away from deeper feelings and goals.

Thomas: My Dad drank a lot when I was growing up, but it didn't bother me so much until now. (Pause) But I was just home and it *really hurts* to see what Dad's starting to do to my Mum—she's awful quiet, you know. (Looks down with brows furrowed and tense) Why she takes so much, I don't figure out. (Looks at you with a puzzled expression) But, like I was saying, Mum and I were sitting there one night drinking tea and he came in, stumbled over the doorstep, and then he got angry. He started to hit my mother and I moved in and stopped him. I almost hit him myself, I was so *angry*. (Anger flashes in his eyes) I *worry* about Mum. (A slight tinge of fear seems to mix with the anger in the eyes and you notice that his body is tensing)

Paraphrasing, as presented in Chapter 6, is concerned with feeding back the key points of what a client has said. Reflection of feeling, in contrast, involves observing emotions, naming them, and repeating them back to the client. The two are closely related and often will be found together in the same statement, but the important distinction is emphasis on content (paraphrase) and emotion (reflection of feeling).

To clarify the distinction, write in the lines following a paraphrase of the client's comments above with an emphasis on content; then write a reflection of feeling, focusing on emotion. You have not yet been asked to write a reflection of feeling, so use your intuitions and note the main feeling words of the client above. Two possible sentence stems for your consideration are presented.

Paraphrase: I hear you saying . . .

_____

_____

_____

Reflection of feeling: You feel . . .

_____

_____

_____

You may want to compare your response with the thoughts of Mary and Allen presented in the following introductory section.

INTRODUCTION: COMPARING PARAPHRASING AND REFLECTION OF FEELING

Paraphrasing client statements focuses on the content and clarifies what has been communicated. The content includes the drinking history, Mum being quiet and taking it, and of course, the actual situation when the client was last home. The paraphrase will indicate to the client that you have heard what has been said and encourage him to move further to the discussion.

Paraphrase: Thomas, your father has been drinking a long time and your Mum takes a lot. But now he's started to be violent and you've been tempted to hit him yourself. Have I heard you right?

The key content issue is escalation of violence and the need to protect Thomas's Mum. It will not help the situation if Thomas becomes part of the violence. At this point, the issue is to listen, learn more about the situation, and plan with Thomas actions for the future. In this example, we are focusing on what is happening and seeking to understand the total situation. Later in the session, we can focus on emotion in depth and help him work through issues.

The first task in eliciting and reflecting feelings is to recognize the *key emotional words* expressed by the client. In the above example, you may have recorded *really hurts*, *angry*, and *worry*. You can know with some certainty that the client has these feelings, as they have been made explicit. The most basic reflections of feeling would be "It really hurt," "You felt angry," and "You are worried." These reflections of feelings use the client's exact main words.

But there are also many *unspoken* feelings expressed in the statement—and the client may or may not be fully aware of them. These unspoken or implicit feelings are often, but not always, expressed nonverbally. For example, the client looked down with brows furrowed and body tense (an indication of likely tension and confusion), anger flashed in his eyes as he was talking about hitting, and fear in the eyes mixed with the anger.*

Feelings are also layered, like an onion. Clients may talk about emotional tones such as confused, lost, or frustrated; or they may be direct and forthright with a single clear emotion. However, further listening and reflection often reveals underlying complex and sometimes conflicting emotions. For example, clients may say that they are frustrated in their relationships with a partner. Reflecting that frustration may lead to discussions in which the client talks about anger at lack of attention, fear of being alone if the relationship breaks up, and residual deep caring for the partner. You may discover that the partner is trying to control your client. As the client becomes aware of this issue, further emotions may be deep anger at the partner's controlling behavior. Later, the client may experience feelings of relief and anticipation of better times if the relationship ends. And in the middle of all this, the hurt will likely remain important.

While we believe that focusing first on the potential violence is critical, combining the paraphrase with feelings by repeating the client's stated key feeling words is likely also appropriate. For example, "You're *really hurting* with it all right now," "You're *angry* because your Dad hit your Mum," "You're *worried* that your Dad's drinking is getting worse." Combining the feeling with the paraphrase acknowledges the client's emotions and may encourage a fuller telling of the story.

Later in the session, after the story is told more completely through your listening, you may help the client discuss and sort through the many and often conflicting emotions. There are numerous possibilities for reflecting implicit emotions that *seem to be* there. In each of the following we suggest a check-out so that the accuracy of your observations can be tested with the client. Here are three for your consideration.

---

*It also may be noted that the client says that his father's drinking didn't bother him until recently. But this seems unlikely, and it may be useful at a later point to explore his family life while he was growing up. Is the client *denying* underlying long-term deep emotions about family drinking? At this point, however, the main issue is drawing out the story and noting the client's emotions associated with the story.

Right now, you're hurting about the situation. I also see some anger. Is that part of what you feel too? (Uses a key word that reflects the implicit feeling.)

Stopping your Dad from hitting your Mum brought out a lot of emotion—I see some anger, perhaps even a little fear about what's going on. Am I close to what you're feeling? (The focus here is on unspoken feelings, seen more nonverbally than verbally. The check-out is particularly important here.)

I hear that your Dad has been drinking for many years. (Paraphrase) And I hear many different feelings—anger, sadness, confusion—and I also hear that you care a lot both for your Mum and Dad. Am I close to what you are feeling about the total situation? (This is a broader reflection of feeling that summarizes several implicit feelings and encourages the client to think more broadly.)

Which of the several *possibilities* (and more) presented above for a reflection of feeling is "right"? If you are intentionally attuned to where your client *is* at the moment and demonstrate empathy and good listening skills, any of them could be suitable. The general recommendation is to start with the actual expressed explicit feelings and use the client's actual emotional words. Later, you can move to an exploration of unspoken feelings.

THE LANGUAGE OF EMOTION

People are constantly expressing emotions verbally and nonverbally. General social conversation usually ignores feelings unless they are especially prominent. Thus, many of us are trained not to focus on the other person's emotional experience and we may even be unaware of what is happening before our eyes.

It is often helpful to start work on eliciting and reflecting feelings by establishing your own vocabulary or list of emotions.

An easy way to brainstorm about emotional words is to focus on four basic feelings—*sad, mad, glad,* and *scared.* These four words are listed with room for you to write related emotional words. Think particularly of different intensities of the same emotion. For example, *mad* might lead you to think of *annoyed, angry,* and *furious.*

Sad	*Mad*	*Glad*	*Scared*
_____	_____	_____	_____
_____	_____	_____	_____
_____	_____	_____	_____
_____	_____	_____	_____
_____	_____	_____	_____
_____	_____	_____	_____
_____	_____	_____	_____
_____	_____	_____	_____

These four emotions plus *surprise* and *disgust* are termed the primary emotions and their commonality has been validated throughout the world in all cultures (Ekman, 2004) in terms of people's facial expressions and language.

Box 7-1   National and International Perspectives on Counseling:
The Invisible Whiteness of Being

Derald Wing Sue

Multicultural issues remain a "hot button" topic and often bring out emotions that may surprise us, whether client or interviewer. I posed this open question to people on the street in San Francisco—"What does being White mean?" Here are some responses. Note the feelings that this question brought out.

Forty-two-year-old White businessman

Q: What does it mean to be White?

A: Frankly, I don't know what you're talking about.

Q: Aren't you White?

A: Yes, but I come from Italian heritage. I'm Italian, not White.

Q: Well then, what does it mean to be Italian?

A: Pasta, good food, love of wine. (obviously agitated) This is getting ridiculous.

Theme: This person denies the color of his skin and speaks superficially about his Italian heritage. He shows feelings of anger around the issue.

Twenty-six-year-old female college student

Q: What does it mean to be White?

A: Is this a trick question? . . . I've never thought about it . . . Well, I know that lots of Black people see us as prejudiced and all that stuff. I wish people would just forget about race differences and see one another as human beings. People are people and we should be proud to be Americans.

Theme: She never thinks of being White and shows feelings of defensiveness and confusion. She focuses on "people are people."

Twenty-nine-year-old Latina administrative assistant

Q: What does it mean to be White?

A: I'm not White, I'm Latina!

Q: Are you upset with me?

A: No . . . it's just that I'm light, so people always think I'm White. It's only when I speak that they realize I'm Hispanic.

Q: Well, what does it mean to be White?

(continued)

However, clients also express emotions in ways that are less clear. The client above appeared puzzled. He appeared to have mixed and conflicting emotions of caring, anger, and fear—and probably even more. A client going through a difficult time such as a divorce may express feelings of love toward the spouse at

Box 7-1   (continued)

> *A:* Do you really want to know? . . . OK, it means you're always right. It means that you never have to explain yourself or apologize. . . . You know that movie *Love is never having to say you're sorry?* Well, being White is never having to say you're sorry. It means you think you're better than us.
>
> *Theme:* Anger at being misidentified and anger and frustration with what she sees as dominant Whites who feel superior.

Thirty-nine-year-old African American salesman

> *Q:* What does it mean to be White?
> *A:* If you're White, you're right. If you're Black, step back.
> *Q:* What does that mean?
> *A:* White folks are always thinking they know all the answers. A Black man's word is worth less than a White man's. When White customers come into our dealership and see me standing next to cars, I become invisible to them. Actually, I think they see me as a well-dressed janitor. They seek out White salesmen. I talked to my boss about this, but he says I'm oversensitive. That's what being White means. It's having the authority or power to tell me what's really happening even though I know it's not. Being White means you can fool yourself into thinking that you're not prejudiced, when you are. That's what it means to be White.
>
> *Themes:* Feelings of frustration and anger around Whites who view minorities as less competent and capable—also, they have the power to define the world for others.
>
> These four examples illustrate the deep-seated emotions underlying issues of race. The Latina and African American comments illustrate feelings that many People of Color have toward White people. Whites, on the other hand, are often oblivious to how color affects their lives. Being White brings with it a privilege that European Americans are often oblivious of and unaware of having.
>
> **Allen Ivey comments:** Race can be a powerful dimension in the counseling interview, but we usually avoid the issue. As a White professional, I have been engaged in lifelong learning about racism, White privilege, and other multicultural issues. It has been both a humbling and liberating process. Both Mary and I suggest that White people spend much more time becoming aware of themselves as a cultural beings and the many emotional issues associated with race, privilege, and power.
>   Regardless of your racial and ethnic background, it is important to be aware that these issues are always in the background of every counseling and therapy session. We need to learn our own multicultural background and be ready to discuss these issues as they come up in the interview.
>
> ---
>
> This box by Derald Wing Sue is partially adapted from *Overcoming Our Racism: The Journey to Liberation,* Jossey-Bass, 2003. Reprinted by permission.

one moment and extreme anger the next. Words such as puzzled, sympathy, embarrassment, guilt, pride, jealousy, gratitude, admiration, indignation, and contempt are *social emotions* made up from primary emotions and learned in the cultural/environmental context.

You can be especially helpful to clients as they sort out social emotions. Think of how the word "guilt" combines anger toward oneself, sadness, and perhaps even some fear. The feeling of guilt has been learned through social interaction in the family and culture. The negative primary emotions are located deep in the brain whereas the social emotions and happiness are found in higher areas, combining experience with underlying primary emotional reactions (Damasio, 2003).

EXAMPLE INTERVIEW: MY MOTHER HAS CANCER, MY BROTHERS DON'T HELP

Difficult life situations bring with them many emotions. Whether you are dealing with clients who experience physical illness, interpersonal conflict, alcohol or drug abuse, or challenges in the work or school setting, learning the way they feel about the situation is vital. The intentional interviewer or counselor is always alert to emotions underlying all situations and knows how to bring them out.

The discovery of cancer, AIDS, or other major physical illness brings with it an immense emotional load. Busy physicians and nurses sometimes may fail to deal with emotions in their patients. And certainly, they will have little time to help family members of those who are ill. Illness can be a frightening experience, and family, friends, and neighbors as well as professionals may have trouble dealing with it.

The following transcript illustrates reflection of feeling in action. This is the second session and Jennifer has just welcomed the client, Stephanie, into the room. They had a brief personal exchange of greetings and it was clear nonverbally that the client was ready to start immediately.

Interviewer and Client Conversation	Process Comments
1. *Jennifer:* So, Stephanie, how are things going with your mother?	Jennifer knows what the main issue is likely to be, so she introduces it with her first open question.
2. *Stephanie:* Well, the tests came back and the last set looks pretty good. But, I'm upset. With cancer, you never can tell. It's hard . . . (pause)	Stephanie speaks quietly and as she talks, she talks in an even softer tone of voice. At the word "cancer," she looks down.
3. *Jennifer:* You're really upset and worried right now.	Jennifer uses the emotional word ("upset") used by the client, but adds the unspoken emotion of worry. With "right now," she brings the feelings to here-and-now immediacy. She did not use a check-out. Was that wise?
4. *Stephanie:* That's right. Since she had her first bout with cancer . . . (pause), I've been really concerned and worried. She just doesn't look as well as she used to, she needs a lot more rest. Colon cancer is so scary.	Often if you help clients name their unspoken feelings, they will say, "That's right" or something similar or they may nod their head. Naming and acknowledging emotions helps clarify them.
5. *Jennifer:* Scary?	Repeating the key emotional words used by clients may help them elaborate on issues in more depth.

6. *Stephanie:* Yes, I'm scared for her and I'm scared for me. They are saying that it can be genetic. She had Stage 2 cancer and we have really got to watch things carefully.	The intentional prediction comes true. Stephanie elaborates on the scary feelings and where they come from. She has a frightened look on her face and she looks physically exhausted.
7. *Jennifer:* So, we've got two things here. You've just gone through your mother's operation and that was scary. You said earlier that they got the entire tumor, but your Mom really had trouble with the anesthesia and that was frightening for a while. You had to do all the caregiving because the rest of the family is far away and you felt pretty lonely. That is scary enough. And the possibility of your inheriting the genes is pretty terrifying. Putting it all together, you feel overwhelmed. Is that the right word to use—overwhelmed?	At this point, Jennifer decides to summarize what has been said. She repeats some key feelings identified in the first interview as well as in this session. She uses a new word, *overwhelmed,* which comes from her observations of the total situation and how very tired Stephanie appears. The interviewer took a chance with the word *overwhelmed.* It might produce too much emotion in the client at the moment.
8. *Stephanie:* (immediately) Yes, I'm overwhelmed, I'm so tired, I'm scared, and I'm furious with myself. (Pause) But I can't be angry; my mother needs me. It makes me feel guilty that I can't do more. (Starts to sob)	This reflection of feeling seems to have brought out more emotion than the interviewer expected this early in the session. Stephanie is now talking about her issues using a sensorimotor style (see page 196). This is probably OK as the relationship is solid, Stephanie has not cried in the interview before, and she likely needs to allow herself to cry and let the emotions out. Caregivers such as Stephanie often burn out and need care themselves. The primary focus is on the person with illness and often little attention is given to the person who suddenly finds herself with all the responsibilities.
9. *Jennifer:* (sits silently for a moment) Stephanie, you've faced a lot and you've done it alone. Allow yourself to pay attention to you for a moment and experience the hurt. (As Stephanie cries, Jennifer comments.) Let it out . . . that's OK.	Stephanie has held it all in and needs to experience what she is feeling. If you personally are comfortable with emotional experience, this ventilation of feelings can be helpful. At the same time, there is a need at some point to return to discussion of Stephanie's situation from a less emotional frame of reference.
10. *Stephanie:* (continues to cry, but the sobbing lessens)	See Box 7-2 for ideas in helping clients deal with emotional experience.
11. *Jennifer:* Stephanie, I really sense your hurt and aloneness. I admire your ability to feel—it shows that you care. Could you sit up now and take a breath?	The client sits up, the crying almost stops, and she looks at the interviewer a bit cautiously. She wipes her nose with a tissue and takes a deep breath. Jennifer did three things here: (a) she reflected Stephanie's here-and-now emotions; (b) she identified a positive asset and strength in those emotions; and (c) she suggested that Stephanie take a breath. Conscious breathing often helps clients bring themselves together.

(continued)

Interviewer and Client Conversation	Process Comments
12. *Stephanie:* I'm OK. (Pause)	She wipes her eyes and continues to breathe. She seems more relaxed now that she has let out some of her emotions.
13. *Jennifer:* You've been holding that inside for a long time. That's the first time I've seen you cry. You had to have a lot of strength and power to do what you did. Your caring and strength really show. It's also strong to cry.	Jennifer provides feedback to Stephanie on her observations and outlines some positive strengths that she has seen. She provides a reframe (see Chapter 12), pointing out the positives inherent in Stephanie's caring attitude.
14. *Stephanie:* Thanks, but I still feel so guilty.	Stephanie is now back in control of herself.
15. *Jennifer:* You feel guilty?	This is the most basic reflection of feeling and it appears in the form of a restatement. The prediction is that Stephanie will elaborate on the meaning of the feeling.
16. *Stephanie:* Who am I to cry? My mother is the one with the pain and Stage 2 cancer. I just wish I had been able to talk my brothers into coming home to see her at least.	The prediction is confirmed and we see Stephanie elaborating on her guilt. We are beginning to see indications of Stephanie being an "overfunctioning" individual who takes on more responsibility than she needs to.
17. *Jennifer:* Your mother has gone through cancer, but you also have pain and fear, although in a different way. Do I hear you saying that you feel guilty because your brothers didn't come home?	First, Jennifer reflects Stephanie's feelings of pain and fear. She separates out the guilt, however, with a reflection of feeling in the form of a closed question. In this case, the question serves as a check-out.
18. *Stephanie:* Well, I called them daily and told them what was happening. They were fine on the phone, but they simply wouldn't come.	Stephanie talks a little faster and her fists tighten.
19. *Jennifer:* I sense a little anger at them. Is that close?	Jennifer draws on the nonverbal observations for this reflection of unspoken feelings. Wisely, she includes a check-out as it is possible that Stephanie will deny the anger.
20. *Stephanie:* I feel so guilty that I couldn't talk them into coming. (Thoughtful pause) No, that's right. They should have come. (Angrily) I know they have jobs and it's hard to get away, but THIS is their mother. (Pause) And, it's not just this time. They hardly ever call. They just seem to be in their own world. I wonder if they'll even show up THIS year for the holidays. They didn't last year.	Could it be that Stephanie has taken the caregiver role for the entire family? As she explores her feelings, she is beginning to make new discoveries.

21. *Jennifer:* Stephanie, right now I hear that you're really angry with them because they don't help and aren't involved.	A classic reflection of feeling involves 1. A sentence stem (*I hear you*), usually using the client's name or the pronoun *you.* 2. The naming of the feeling *(angry)* or feelings. 3. The underlying facts or reasons behind the feeling. 4. Bringing the emotion to the here and now of the immediate moment (*right now*).
22. *Stephanie:* And you know what else they did? (Continues with another story)	The reflection unleashes Stephanie to share some long held-back stories of frustration and anger with her brothers.
22–30. Omitted.	In this exchange, Stephanie explores her feelings toward her brothers.
31. *Jennifer:* So, Stephanie, we've talked about your anger and disappointment with your brothers. You seem very much in touch with something you weren't really aware of before. You seem to be saying, also, that feeling guilty about not getting them to shape up and take their share doesn't make sense anymore. At the same time, I sense that you still have hopes and want to involve them more. Before we go further, I wonder if we can change focus. So far, we've been talking about your concerns and difficulties. At the same time, you've managed to do a lot. You care a lot and you've managed this past month. Could you share with me some of the things and the strengths that have enabled you to manage during this past month?	The intervening discussion is summarized. Stephanie's feelings of guilt and anger are better understood. This should free her for more open discussion in the future. Note how summaries help punctuate the interview and aid the transition to other stages and issues. Jennifer has listened to problems and challenges for the entire first session and all of this interview thus far. She decides it is time for the positive asset search. If Jennifer can help the client identify some positive feelings, thoughts, and behaviors associated with this situation, then Stephanie will be better prepared and stronger to deal with the many issues, challenges, and problems that she faces.

You may have noted that the major skill used by Jennifer through this session was reflection of feelings, accompanied by a few questions to help bring out emotions. This skill is important in all theories of counseling and therapy as human change and development is often rooted in emotional experience. There are some humanistic interviewers and counselors who consider eliciting and reflecting feelings the central skill and strategy of many sessions. Encouraging, paraphrasing, summarizing, and questioning are used to facilitate emotional expression and the discovery of underlying feelings.

You are most likely beginning your work and starting to discover the importance of reflecting feelings. It will probably take you some time before you are fully comfortable using this skill. This is so because it is less a part of daily communication than the other skills of this book, but reflecting feelings is central to the helping professions.

We would suggest that you start your work in this area by first simply noting emotions and then reflecting them back through short acknowledging reflections. As you gain confidence and skill, you will eventually decide the extent and place of this skill area in your helping repertoire.

Box 7-2 Helping Clients Increase or Decrease Emotional Expressiveness

Observe nonverbals	Breath directly reflects emotional content. Rapid or frozen breath signals contact with intense emotion. Also note facial flushing, pupil contraction/dilation, body tension, and changes in vocal tone; note especially speech hesitations. You may also find apparent absence of emotion when discussing a difficult issue. This might be a clue that the client is avoiding dealing with feelings or that the expression of emotion is culturally inappropriate for this client.
Pace clients	You can pace clients and then lead them to more expression and awareness of affect. Many people get right to the edge of a feeling, and then back away with a joke, change of subject, or intellectual analysis. Some of the things you can do are the following: ▲ Say to the client that she looked as though she was close to something important. "Would you like to go back and try again?" ▲ Discuss some positive aspect of the situation. This can free the client up to face the negative. You as counselor also represent a positive asset yourself. ▲ Consider asking questions. Used carefully, questions may help some clients explore emotions. ▲ Use here-and-now sensorimotor techniques, especially in the present tense: "What are you feeling right now—at this moment?" "What's occurring in your body as you talk about this?" Use Gestalt exercises or anything to enable a client to become more aware of body feeling. Use the word *do* if you find yourself uncomfortable with emotion: "What do you feel?" or "What did you feel then?" starts to move the client away from here-and-now experiencing.
When tears, rage, despair, joy, or exhilaration come up	Your comfort level with your own emotions and feelings will affect how your client faces emotion. If you aren't comfortable with a particular emotion, your client will likely avoid it also and you may not be effective in helping the client work through feeling issues. A balance between, on the one hand, being very present with your

(continued)

Before moving further, it is important for you to reflect on yourself and your own personal style. How comfortable are you with emotional expression? If discussing feelings was not common in your experience, this skill may be difficult for you. Thus as you work through the ideas presented here, reflect now and then on your own personal history and ability to deal with emotion. If you find this area uncomfortable, you may have difficulty in helping clients explore their issues in depth. The exercises here and throughout the book may help you gain greater access to your own experiential and emotional world.

In your own practice with reflection of feeling, attempt to use the skill as frequently as possible. In the early stages of mastery it is wise to combine the skill with questioning, encouraging, and paraphrasing; most people find it awkward to use a single skill at a time. Full mastery of a skill will appear when a person can conduct a long interview segment effectively using a single skill almost constantly. Over the years the most effective interviewer, counselor, or therapist often develops— consciously or subconsciously—a level of skill such that the specific skill being used makes relatively little difference. Each is used so well that positive client benefits can be produced. Effectiveness and competence, then, do not necessarily depend on the

Box 7-2 (continued)

own breathing and showing culturally appropriate and supportive eye contact and on the other, still allowing room to sob, yell, or shake, is important.

You can also use phrases such as these:

▲ I'm here. I've been there too.
▲ I'm standing right with you in this. Let it out . . . that's OK.
▲ These feelings are just right. I hear you.
▲ I see you. Breathe with it.

Sometimes it is helpful to keep emotion expression within a fixed time; 2 minutes is a long time when you are crying. Afterward, helping the person reorient is important.

Tools for reorienting the interview include these:

▲ Slowed, rhythmic breathing
▲ Counselor and client discussion of positive strengths inherent in the client and situation
▲ Discussion of direct, empowering, self-protective steps that the client can take in response to the feelings expressed
▲ Standing and walking or centering the pelvis and torso in a seated position
▲ Positive reframing of the emotional experience
▲ Commenting that the story needs to be told many times and each time helps

A caution

As you work with emotion, there is always the possibility of reawakening issues in a client who has a history of painful trauma. This is an area where the beginning interviewer often needs to refer to a more experienced professional. Even the 2-minute expression of emotion suggested here may be too long. In such situations, seek supervision and consultation.

This box is adapted from a presentation by Leslie Brain at the University of Massachusetts.

particular skill but on the art of using it effectively. Nonetheless, being aware of and competent in each skill will facilitate your general personal and professional development as an interviewer.

INSTRUCTIONAL READING: BECOMING AWARE OF AND SKILLED WITH EMOTIONAL EXPERIENCE

Many authorities argue that our thoughts and actions are only extensions of our basic feelings and emotional experience. The skill of reflecting feeling is aimed at assisting others to sense and experience the most basic part of themselves—how they really feel about another person or life event.

At the most elementary level, the brief encounters we have with people throughout the day involve our emotions. Some are pleasant; others can be fraught with tension and conflict even though the interaction may be only with a telemarketer desperate to make a sale, with a hurried clerk in a store, or with the police as they stop you for speeding. Feelings undergird these situations just as the more complex feelings we have toward significant others underlie more intimate relationships.

Box 7-3   Research Evidence That You Can Use: Reflection of Feeling

Emotional processing—the working through of emotional states and the ability to examine feelings and body states—has been found fundamental in effective experiential counseling and therapy. For example, gains in treatment of depressed clients was found to be highly related to emotional processing skills (Pos, Greenberg, Goldman, & Korman, 2003). As you work with all clients, your skill in reflecting feelings can be a basic factor in helping them take more control of their lives.

Dealing with emotion is not only a central aspect of interviewing, counseling, and psychotherapy; it is also key to high-quality interviewing (Bensing, 1999b; Daniels & Ivey, 2006). Working with emotion requires attention to nonverbal dimensions. Head nodding and eye contact, and especially *smiling* are important. Clearly, warmth, interest, and caring are communicated nonverbally as much as or more than through the use of the skill. Moreover, Hill and O'Brien (1999) and Tamase and Kato (1990) found that using questions oriented toward affect increased client expression of emotion. Reflection of feeling, then, provides a chance for you to check out the accuracy of your hearing. Nonetheless, once a client has expressed emotion, continued use of questions may be too intrusive and the more reflective approach should be more useful.

"Several studies have shown that between 30 and 60 percent of patients in general practice present health problems for which no firm diagnosis can be made" (Bensing, 1999a). Recognizing emotional complexity can be important. Older persons tend to manifest more mixed feelings than others (Carstensen et al., 2000). Perhaps this is because life experience has taught them that things are more multi-faceted than they once thought. Helping younger clients become aware of emotional complexity may also be a goal of some interviewing sessions. Tamase, Otsuka, and Otani (1990), through their work in Japan, have provided clear indication that the

(continued)

Becoming aware of others' feelings can help you move through the tensions of the day gracefully and allow you to be helpful to other individuals in many small ways. Rather than reflecting clients' feelings about these situations, you may find that a brief, simple acknowledgment of feeling is helpful. The same structure is used in an acknowledgment as in a full reflection, but much less emphasis is given to feeling, and interaction moves on more quickly.

At another level our work and social relationships and the decisions we make are often based on emotional experience. In an interviewing situation it is often helpful to assist the client in identifying feelings clearly. For example, an employee making a move to a new location may have positive feelings of satisfaction, joy, and accomplishment about the opportunity but simultaneously feel worried, anxious, and hesitant about new possibilities. The effective interviewer notes both dimensions and recognizes them as a valid part of life experience.

A basic feeling we have toward our parents, family, and best friends is love and caring. This is a deep-seated emotion in most individuals. At the same time, over years of intimate contact, negative feelings about the same people may also appear, possibly overwhelming and hiding positive feelings; or negative feelings may be buried. A common task of counselors is to help clients sort out mixed

Box 7-3 (continued)

	reflection of feelings is useful cross-culturally. Hill and O'Brien (1999) reported on a series of studies in this area and noted the facilitating impact of reflective responses: "Clients often do not even notice when helpers are using good restatement and reflections because they are simply facilitated in their exploration" (p. 126).
Reflection of feeling and neuropsychology	Neuropsychology has found that counseling's traditional categories of emotion as presented in this chapter also appear in brain imaging (Kolb & Wishaw, 2003). The limbic system organizes bodily emotions and includes the amygdala, hypothalamus, thalamus, hippocampal formation, and cortex. Much of the research in this area has focused on brain damage. Feelings of fear and anger are located in the amygdala, disgust in the anterior insular cortex; feelings of depression appear to come from a less active brain area, particularly the frontal cortex. More positive feelings are located in the prefrontal cortex in the ventral media area. Selective attention to events and thoughts that bring about positive emotions can help calm the more turbulent parts of the limbic system.
	More complex and mixed emotions are also defined and organized in the prefrontal cortex (Kolb & Wishaw, 2003). For example, you may receive a gift from a person who is important to you, but the gift isn't to your liking. Very possibly, you may feel irritation about the gift (mild anger directed outward), positive appreciation for the thoughtfulness it represents, and guilt about not responding more positively to the giver (anger directed inward). You may even feel some bodily fear from the amygdala that your lack of enthusiasm shows too prominently. Moreover, at any point a more powerful emotion residing in the amygdala may override all these feelings and lead to an inappropriate emotional outburst of anger. This type of outburst may appear in some counseling sessions when the client appears to overreact to some event—it is almost as if certain events lead a person to "blow a fuse."

feelings toward significant people in their lives. Many people want a simple resolution and want to run away from complex mixed emotions. However, ideally the counselor should help the client discover and sort out both positive and negative feelings.

Trust between counselor and client is necessary for full emotional exploration, but in some cultures expression of feelings is discouraged (see Box 7-4). You will find that a brief acknowledgment of feeling is at times more appropriate than a deep exploration of feelings. In acknowledging feelings, you state the feeling briefly ("You seem to be sad about that," or "It makes you happy") and then move on with the interview. However, do not use individual or cultural uniqueness as an excuse to avoid talking about emotion in the interview. All clients have vital emotional lives, whether they are aware of them or not.

With children, you will find acknowledgment of feelings may be especially helpful, particularly when they themselves are unaware of what they are feeling. At the same time, many children respond well to the classic reflection of feeling, "You feel (sad, mad, glad, scared) because . . ."

When addressing client feelings and emotions in an interview, the following specifics are important to keep in mind.

Box 7-4 National and International Perspectives on Counseling Skills: Does He Have Any Feelings?

Weijun Zhang

Illustration: A student from China comes in for counseling, referred by his American roommate. According to the roommate, the client quite often calls his wife's name out loud while dreaming, which usually wakes the others in the apartment, and he was seen several times doing nothing but gazing at his wife's picture. Throughout the session the client is quite cooperative in letting the counselor know all the facts concerning his marriage and why his wife is not able to join him. But each time the counselor tries to identify or elicit his feelings toward his wife, the client diverts these efforts by talking about something else. He remains perfectly polite and expressionless until the end of the session.

No sooner had the practicing counselor in my practicum class stopped the videotape machine than I heard comments such as "inscrutable," and "He has no feelings!" escape from the mouths of my European American classmates. I do not blame them, for the Chinese student did behave strangely, judged from their frame of reference.

"How do you feel about this?" "What feelings are you experiencing when you think of this?" How many times have we heard, or used, questions such as these? The problem with these questions is that they stem from a European American counseling tradition, which is not always appropriate.

For example, in much of Asia, the cultural rationale is that the social order doesn't need extensive consideration of personal, inner feelings. We make sense of ourselves in terms of our society and the roles we are given within the society. In this light, in China, individual feelings are ordinarily seen as lacking social significance. For thousands of years, our ancestors have stressed how one behaves in public, not how one feels inside. We do not believe that feelings have to be consistent with actions. Against such a cultural background, one might understand why the Chinese student was resistant when the counselor showed interest in his feelings and addressed that issue directly.

(continued)

Observing Client Verbal and Nonverbal Feelings

When a client says "I feel sad"—or "glad" or "frightened"—and supports this statement with appropriate nonverbal behavior, identifying emotions is easy. However, many clients present subtle or discrepant messages, for often they are not sure how they feel about a person or situation. In such cases the counselor may have to identify and label the implicit feelings. However, with skilled listening you can often help such clients label their own emotions.

The most obvious technique for identifying client feelings is simply to ask the client an open question ("How do you feel about that?" "Could you explore any emotions that come to mind about your parents?" "What feelings come to mind when you talk about the loss?"). With some quieter clients, a closed question in which the counselor supplies the missing feeling word may be helpful ("Does that feel hurtful to you?" "Could it be that you feel angry at them?" "Are you glad?").

At other times the counselor will want to infer, or even guess, the client's feelings through observation of verbal and nonverbal cues such as discrepancies between

Box 7-4   (continued)

> But I am not suggesting here that Asians are devoid of feelings or strong emotions. We are just not supposed to telegraph them as do people from the West. Indeed, if feelings are seen as an insignificant part of an individual and regarded as irrelevant in terms of social importance, why should one send out emotional messages to casual acquaintances or outsiders (the counselor being one of them)?
>
> What is more, most Asian men still have traditional beliefs that showing affection toward one's wife while others are around, even verbally, is a sign of being a sissy, being unmanly, or weak. I can still vividly remember when my child was four years old, my wife and I once received some serious lecturing on parental influence and social morality from both our parents and grandparents, simply because our son reported to them that he saw "Dad give Mom a kiss." You can imagine how shocking it must be for most Chinese husbands, who do not dare even touch their wives' hands in public, to see on television that American presidential candidates display such intimacy with their spouses on the stage! But the other side of the coin is that not many Chinese husbands watch television sports programs while their wives are busy with household chores after a full day's work. They show their affection by sharing the housework!
>
> **Allen Ivey comments:** Not only is the expression of feelings as demanded in counseling inappropriate in many Asian cultures and among many Asian Americans, but you will also find that Native Americans, Latino men, African American men, and many European American men will resist or have difficulty in exploring emotions. Furthermore, some women may also resist sharing emotions, particularly if the helper is a man. Many cultures and family traditions require people to mask the expression of feelings.
>
> With clients who do not easily express emotion, you will need to have patience and first spend time developing trust and rapport. This may take more than one session, and in the meantime using brief acknowledgments of possible emotions may be helpful. And when you sense the time is ripe to tackle the client's feelings, proceed little by little and with cultural and gender sensitivity.

what the client says about a person and his or her actions, or a slight body movement contradicting the client's words. As many clients have mixed feelings about the most significant events and people in their lives, inference of unstated feelings becomes one of the important observational skills of the counselor. A client may be talking about caring for and loving parents while holding his or her fist closed. The mixed emotions may be obvious to the observer though not to the client.

Stephanie, the client in the example interview, used the following feeling words during the session: *upset, worried, concerned, scared, tired, furious, guilty, anger, disappointed,* and *hope.* Jennifer, the counselor, reflected those words, but also brought in some of her own observations. We saw Stephanie's reaction to the word *overwhelmed.* But the interviewer also sought to emphasize some positive emotions such as caring and the strengths that the client demonstrated. As the interview moves on, we can anticipate that Stephanie will express fewer difficult emotions and will move to more awareness of positive feelings and strengths. At the moment, however, she needs to be encouraged to talk through the worries and problems.

The intentional counselor does not necessarily respond to every emotion, congruent or discrepant, that has been noted; reflection of feeling must be timed to meet the needs of the individual client. Sometimes it is best simply to note the emotion and keep it in mind for possible comment later.

The Techniques of Reflecting Feelings

Somewhat like the paraphrase, reflection of feeling involves a typical set of verbal responses that can be used in a variety of ways. The classic reflection of feeling consists of the following dimensions:

1. A *sentence stem* using, insofar as possible, the client's mode of receiving information (auditory, visual, or kinesthetic) often begins the reflection of feeling ("I hear you are feeling . . . ," "It looks like you feel . . . ," "Sounds like . . ."). Unfortunately, these sentence stems have been used so often that they can almost sound like comical stereotypes. As you practice, you will want to vary sentence stems and sometimes omit them completely. Using the client's name and the pronoun *you* helps soften and personalize the sentence stem.
2. A *feeling label* or emotional word is added to the stem ("Jonathan, you seem to feel bad about . . ." "Looks like you're happy," "Sounds like you're discouraged today; you look like you feel really down"). For mixed feelings, more than one emotional word may be used ("Maya, you appear both glad and sad . . .").
3. A *context or brief paraphrase* may be added to broaden the reflection of feeling. (To use the examples above, "Jonathan, you seem to feel bad about *all the things that have happened in the past two weeks*"; "Maya, you appear both glad and sad *because you're leaving home.*") The words *about, when,* and *because* are only three of many that add context to a reflection of feeling.
4. The *tense* of the reflection may be important. Reflections in the present tense ("Right now, you *are* angry") tend to be more useful than those in the past ("You felt angry then"). Some clients will have difficulty with the present tense and talking in the "here and now." But occasionally, a "there and then" review of past feelings can be quite helpful. You'll find that experiencing feelings and thinking back on them in the past are often quite different.
5. A *check-out* may be used to see whether the reflection is accurate. This is especially helpful if the feeling is unspoken ("You feel angry today—am I hearing you correctly?").

Individual and cultural diversity in the way one expresses feelings needs to be respected. The student from China discussed in Box 7-4 is an example of cultural emotional control. Emotions are obviously still there, but they are expressed differently. Do not expect all Chinese or Asians to be emotionally reserved, however. Their style of emotional expression will depend on their individual upbringing, their acculturation, and other factors. Many New England Yankees may be fully as reserved in emotional expression as the Chinese student described by Weijun Zhang. But again, it would be unwise to stereotype all New Englanders in this fashion.

Emotions are very personal. Regardless of cultural background, each client will be unique. Build trust and relationship first. When you work with clients who are emotionally reticent, you might want to try the suggestions for helping them express their feelings presented in Box 7-2. But be sure the clients know what is going to happen and express a willingness to move into their feelings more deeply with your help. A brief acknowledgment of feeling may be sufficient in the early stages until adequate trust is generated.

The Place of Positive Emotions in Reflecting Feelings

Positive emotions, whether joyful or merely contented, are likely to color the ways people respond to others and their environments. Research shows that positive emotions broaden the scope of people's visual attention, expand their repertoires for action, and increase their capacities to cope in a crisis. Research also suggests that positive emotions produce patterns of thought that are flexible, creative, integrative, and open to information (Gergen & Gergen, 2005).

"Sad, mad, glad, scared." This is one way to organize the language of emotion. But perhaps we need more attention to glad words such as pleased, happy, contented, together, excited, delighted, pleasured, and the like. If you were to take just a moment now and think of specific situations when you experienced each of the positive emotions listed in the previous sentence, it is very likely that you would smile, your body tension would be reduced, and even your blood pressure might change in a more positive direction.

When you experience emotion, your brain signals bodily changes. Thus, when you feel sad or angry, a set of chemicals floods your body and usually these changes will show nonverbally. Emotions change the way your body functions and thus are a foundation for all our thinking experience (Damasio, 2003). As you help your clients experience more positive emotions, you are also facilitating wellness and a healthier body. The route toward health, of course, often entails confronting negative emotions.

Research examining the life of nuns found that those who had expressed the most positive emotions in early life lived longer than those who expressed a less positive past (Danner, Snowdon, & Friesen, 2001). Stress reaction to the 9/11 bombing disaster found that students who had access to the most positive emotions showed fewer signs of depression (Fredrickson, Tugade, Waugh, & Larkin, 2003).

It thus becomes apparent that searching for wellness strengths and positive assets will likely be helpful to you and your clients. Obviously, we need to explore negative and troubling emotions, but if your clients can start from a positive base of emotion, they may be better able to cope with the negative.

Some specific ways to bring positive emotions into the session include the following.

As part of a wellness assessment, be sure that you reflect the positive feelings associated with aspects of wellness. For example, your client may feel safety and strength in the spiritual self, pride in gender and/or cultural identity, caring and warmth from past and/or present friendships, and the intimacy and caring of a love relationship. It would be possible to anchor these emotions early in the interview and draw on these positive emotions during more stressful moments. Out of a wellness inventory can come a "backpack" of positive emotions and experiences that are always there and can be drawn on as needed.

When reflecting feelings, observe your client and search for strengths. These can become part of your reflection of feeling strategy. For example, the client may be going through the difficult part of a relationship breakup and crying and wondering what to do. We don't suggest that you should interrupt the emotional flow, but with appropriate timing, reflecting back the positive feelings that you have observed can be helpful.

Couples with relationship difficulties can be helped if they focus more on the areas where things are going well—what remains good about the relationship. Many couples focus on the 5% where they disagree and fail to note the 95% where they have been successful or enjoyed each other. Some couples respond well when asked to focus on the reasons they got together in the first place. These positive strengths can help them deal with very difficult issues.

When providing your clients with homework assignments, have them engage daily in activities associated with positive emotions. For example, it is difficult to be sad and depressed when running or walking at a brisk pace. Meditation and yoga are often useful in generating more positive emotions and calmness. Seeing a good movie when one is down can be useful, as can going out with friends for a meal. In short, help clients remember that they have access to joy, even when things are at their most difficult.

Service to others often helps people feel good about themselves. When one is discouraged and feeling inadequate, volunteering for a church work group, working on a Habitat for Humanity home, or giving time to work with animal rights can all be helpful in developing a more positive sense of self.

Important caution: But please do not use the above paragraphs as a way to tell your clients that "everything will be OK." There are some interviewers and counselors who are so afraid of negative emotions that they never allow their clients to express what they really feel. Do not minimize difficult emotions by too quickly focusing on the positive.

Noting Emotional Intensity: A Developmental Skill

Clients have varying levels of intensity with which they describe emotional experiences. You will find that some clients are overwhelmed by emotion and feelings while others are more remote and may use thinking and cognition to avoid really looking at how they feel. Ability and willingness to explore emotion varies from client to client.

In developmental counseling and therapy (DCT), key observational skills have been identified to help organize the depth of client emotional experiencing (see Ivey, Ivey, Myers, & Sweeney, 2005, for elaboration of ideas presented here). Feelings are not just "feelings." They vary in intensity and the ways they are expressed. It is important that you be able to determine how a client reacts emotionally. Once you have awareness of a client's style of emotional expressiveness, you will have a better idea of how you can help the client explore the complex world of affect and feeling.

DCT's four emotional styles are described in the following paragraphs. Spend some time noticing these important differences as *practice is critical* for developing competence. The effort will pay off.

Sensorimotor emotional style.   These clients *are* their emotions. They *experience* emotions rather than naming them or reflecting on them. The body is fully involved. They may cry, they may laugh, but emotional experience is primary and there is only limited separation of thought and feeling. The positives of this style of emotional involvement are access to the real and immediate experiences of being sad, mad, glad, or scared in the moment. On the negative side, these clients may be overwhelmed by too much emotion. The reaction of Stephanie at 8 to the word *overwhelmed* is a clear example. If you are to help clients reexperience issues, you will often want to encourage full sensorimotor experiencing of emotion at some appropriate point. A question that may enhance or open sensorimotor, deeper emotions is, "What are you feeling/experiencing at this moment?" You may also suggest that the client develop a visual, auditory, or kinesthetic image: "Can you develop an image of that experience—what are you seeing/hearing/feeling at this moment?" This imagery exercise can be powerful. Use with care and a clear sense of ethics.

Concrete emotional style. The skills of reflection of feeling as presented in this chapter are primarily focused on a concrete emotional style. See Jennifer 3, 5, 15, 21 for four examples. Feelings are named (*early* concrete) with statements such as "You seem to feel sad." Late concrete emotion is emphasized when we add causation statements to the reflection—"You feel sad because . . ." Many of your clients will come to you with only a vague sense of the emotions underlying their concerns. Concrete exploration of emotions helps clarify issues. The resulting increased awareness can form a foundation for moving to more in-depth emotional exploration within the concrete style and can also help clients understand emotions within sensorimotor and other styles. Note that as we make emotions concrete or examine their patterns, we move away from direct sensorimotor emotional expression. For clients who need to become self-reflective, this is most useful. But for clients who avoid really looking at how they feel, concrete and formal emotional reflection can be a way of avoiding feelings.

Most of this chapter is focused on concrete emotional experience. The first task is concretely *naming* or labeling the emotion—"What are you feeling?" If necessary, ask, "What name would you give to that feeling?" The simplest, most direct concrete reflection of feeling is "You are feeling (or felt) X?" For example, "You feel sad" (or glad, mad, or scared). This naming of feelings itself can be very therapeutic to some clients as they may be very out of touch with their emotions.

Another type of concrete reflection entails context and situation and may even become causal: "You feel X because Y," or "You feel X when Y happens." For example, "You feel sad because your teacher scolded you." "You feel scared when you encounter a new situation."

Abstract formal-operational emotional style. The client becomes less concrete and more abstract, *reflecting on* emotions, and thus may avoid experiencing them at the sensory level. Formal clients may be quite effective at seeing repeating patterns of emotion, but they may also have difficulty within the concrete style. Stephanie at 22 starts the process of *thinking about* her emotions. You may find that some clients are very good at reflecting abstractly on their feelings, but they never allow themselves to experience emotion with the full sensorimotor style. Reflection is very useful in helping clients understand their patterns and modes of emotionality. "As you look back on your situation/yourself, what types of feelings do you notice?" "As you *reflect* on *that feeling* you just talked about, what do you think?" "What are your patterns of emotion?" "Do you feel that way a lot?" At this point emotion is more abstract. Clients think back rather than directly experiencing their feelings. Jennifer at 31 seeks to help Stephanie develop a more reflective style.

Abstract dialectic/systemic emotional style. Clients using this style are very effective at analyzing their emotions, and their emotionality will change with the context. A client may say, "I am terribly sad to lose my lover through AIDS, but I am proud of how he/she lived effectively despite it. In a way, I am joyful at the triumph over adversity my lover has demonstrated." Note that this is a more analytic and multiperspective view of emotionality, which moves even farther away from direct, here-and-now experiencing. Jennifer does not illustrate this type of emotion as yet.

Dialectic/systemic emotions are complex, and they change in context. The style tends to be theoretical, and emotions are analyzed more than experienced. "How do your emotions change when you take another perspective on your issue(s)?" "Where do you think you learned that pattern of emotional expression—in your family or elsewhere?" These are only two of many possible questions that lead to multiperspective thought on emotional experience.

What can the DCT view of emotions do for you? If you assess where clients are experiencing emotion, you'll be better able to empathize and understand how they are experiencing a situation. For example, with clients who are predominantly abstract in style, it may be helpful to aid them in experiencing emotion more immediately within the concrete or sensorimotor styles. A person experiencing significant loss (divorce, death, job loss, illness) will usually benefit from talking about her or his issues using all four emotional styles. With concretely oriented clients, a long-term interviewing goal may be to help them get more distance on their emotions and think more abstractly. Emotion is holistic, and no one way of experiencing emotion is best.

Box 7-2 presents techniques for working with emotion in the session. Note that specific ways are suggested to help clients get in touch with their feelings. At other times, you may want to assist clients in slowing down the emotional flow. Each of the questions associated with the emotional styles can expand or contract emotional discussion depending on the timing of the skill. As you become more experienced in dealing with clients' emotions, you may find these guidelines helpful.

SUMMARY: A CAUTION ABOUT REFLECTION OF FEELINGS

Reflection has been described as a basic feature of the counseling process, yet it can be overdone. Many times a short and accurate reflection may be the most helpful. With friends, family, and fellow employees, a quick acknowledgment of feelings ("If I were you, I'd feel angry about that . . ." or "You must be tired today") followed by continued normal conversational flow may be most helpful in developing better relationships. In an interaction with a harried waiter or salesperson, an acknowledgment of feelings may change the whole tone of a meal or business interchange. Similarly, with many clients a brief reflection of feeling may be more useful than the detailed emphasis outlined here. Identifying unspoken feelings can be helpful, and as clients move toward complex issues, the sorting out of mixed feelings may be a central ingredient of successful counseling, be it vocational interviewing, personal decision making, or in-depth individual counseling and therapy.

Nevertheless, it is important to remember that not all clients will appreciate or welcome your commenting on their feelings. Clients tend to disclose feelings only after rapport and trust have been developed. Less verbal clients may find reflection puzzling at times or may say, for instance, "Of course I'm angry; why did you say that?" With some cultural groups, reflection of feeling may be inappropriate and represent cultural insensitivity. Some men, for example, may believe that expression of feelings is "unmanly," yet a brief acknowledgment may be helpful to them. Be aware that an empathic reflection can sometimes have a confrontational quality that causes clients to look at themselves from a different perspective; it may therefore seem intrusive to some clients. Though noting feelings in the interview is essential, acting on your observations may not always be in the best interests of the client. Timing is particularly important with this skill.

Box 7-5   Key Points

Why?	Emotions undergird our life experiences. Emotions are the source of many of our thoughts and actions. If we can identify and sort out clients' feelings, we have a foundation for further action.
What?	Emotions and feelings may be identified through labeling client behavior with affective words such as *sad, mad, glad,* and *scared.* The counselor will want to develop an array of ways to note and label client emotions. In labeling client feelings, it is important to note the following:
	1. Emotional words used by the client
	2. Implicit emotional words not actually spoken
	3. Nonverbally expressed emotions discovered through the observation of body movement
	4. Mixed verbal and nonverbal emotional cues, which may represent a variety of discrepancies
How?	Emotions may be observed directly, drawn out through questions ("How do you feel about that?" "Do you feel angry?"), and then reflected through the following steps:
	1. Begin with a sentence stem such as "You feel . . ." or "Sounds like you feel . . ." or "Could it be you feel . . . ?" Use the client's name.
	2. Feeling word(s) may be added (*sad, happy, glad*).
	3. The context may be added through a paraphrase or a repetition of key content ("Looks like you feel happy *about the excellent rating*").
	4. In many cases a present-tense reflection is more powerful than one in the past or future tense. "You feel happy *right now*" rather than "you felt" or "you will feel."
	5. Following identification of an unspoken feeling, the check-out may be most useful. "Am I hearing you correctly?" "Is that close?" This lets the client correct you if you are either incorrect or uncomfortably close to a truth that he or she is not yet ready to admit.
	Brief reflections of feeling may be particularly helpful with friends, family, and people met during the day. Deeper reflections and a stronger emphasis on this skill

(continued)

Box 7-5   (continued)

	may be appropriate in many counseling situations, but they require a relatively verbal client. The skill may be inappropriate with clients of certain cultural backgrounds. The acknowledgment of feeling puts less pressure on clients to examine their feelings and may be especially helpful in the early stages of interviewing a client culturally different from you. Later, as trust develops, you can explore emotion in more depth.
What else?	Developmental skills and the observation skills of Chapter 5 will prove especially helpful in improving your skill in reflecting feeling. The concept of concreteness may be useful to add to reflection of feeling. For example, "You seem to be angry with your spouse. Could you give me a specific example of a situation in which you feel this anger?" Following this, other feelings and thoughts may be identified, and the question "What does this mean to you?" may be helpful. Just because you observe a feeling does not mean it should be reflected. Too much reflection may overintensify the feeling; also, some counselors focus too much attention on negative feelings. Accentuate the positive too! Where you discover a strong negative feeling, there is frequently an unseen positive contrasting feeling as well.

COMPETENCY PRACTICE EXERCISES AND PORTFOLIO OF COMPETENCE

Feelings are basic to human experience. Although we may observe them in daily interaction, we usually ignore them. In counseling and helping situations, however, they can be central to the process of understanding another person. Further, you will find that increased attention to feelings and emotions may enrich your daily life and bring you to a closer understanding of those with whom you live and work.

Individual Practice

Exercise 1: Increasing Your Feeling Vocabulary

Return to the list of affective words you generated at the beginning of this chapter. Take some more time to add to that list. One way to lengthen the list is to consider two categories of feeling words that would provide additional ideas about how the client feels about the world.

The first category is words that represent mixed or ambivalent feelings. In such cases the feelings are often very unclear, and your task is to help the client sort out the deeper emotions underlying the surface, expressed word. List here words that represent confused or vague feelings (for instance, *confused, anxious, ambivalent, torn, ripped, mixed*).

Mixed, Vague Feeling Words

_____   _____   _____

_____   _____   _____

A common mistake is to assume that these words represent the root feelings. Most often they cover deeper feelings. The word *anxiety* is especially important to consider in this context: It is sometimes a vague indicator of mixed feelings. If you accept client anxiety as a basic feeling, counseling may proceed slowly. An important task of the interviewer when noting mixed feeling words is to use questions and reflection of feeling to help the client discover the deeper feelings underlying the surface ambivalence. Underlying anxiety or confusion, for example, you may find anger, hurt, love, or other feelings.

Now take two of the words from your mixed feelings list and lay out more basic words that might underlie affectively oriented words such as *confused* or *frustrated*. Again, the basic words *mad, sad, glad, scared* may be helpful in this process.

Word 1   *Word 2*

_____   _____

_____   _____

_____   _____

_____   _____

_____   _____

_____   _____

Feelings are often presented through metaphors, concrete examples, and similes. It is often more descriptive of your emotions to say that you feel like a limp dishrag than to say you are tired and exhausted. Other examples might include "down in the pits," "high as a kite," "crashed worse than a computer," or "proud as a peacock." Because metaphors are often masks for more complex feelings, at times it is inappropriate to accept such descriptions as presented; you may want to search for the underlying feelings. After you have developed a list of at least five metaphors here, you may wish to generate a list of basic feeling words underlying the metaphor.

Metaphor   *Basic Feeling Words*

_____   _____

_____   _____

_____   _____

_____   _____

_____   _____

_____   _____

Exercise 2: Distinguishing a Reflection of Feeling From a Paraphrase

The key feature that distinguishes a reflection of feeling from a paraphrase is the affective word. Many paraphrases contain reflection of feeling; such counselor statements are classified as both. Consider the two following examples. In the first example, you are to indicate which of the leads is an encourager (E), which a paraphrase (P), and which a reflection of feeling (RF).

> "I am really discouraged. I can't find anywhere to live. I've looked at so many apartments, but they are all so expensive. I'm tired and I don't know where to turn."

Mark the following counselor responses with an E, P, RF, or combination if more than one skill is used.

_____ "Where to turn?"

_____ "Tired . . ."

_____ "You feel very tired and discouraged."

_____ "Searching for an apartment simply hasn't been successful; they're all so expensive."

_____ "You look tired and discouraged; you've looked hard but haven't been able to find an apartment you can afford."

> For the next example, write an encourager, paraphrase, reflection of feeling, and a combination paraphrase/reflection of feeling in response to the client.

> "Right, I do feel tired and frustrated. In fact, I'm really angry. At one place they treated me like dirt!"

Encourager: _____

Paraphrase: _____

_____

Reflection of feeling: _____

Combination paraphrase and reflection of feeling: _____

_____

_____

Exercise 3: Acknowledgment of Feeling

We have seen that the brief reflection of feeling (or acknowledgment of feeling) may be useful in your interactions with busy and harried people during the day. At least once a day, deliberately tune in to a server/waitstaff person, teacher, service station attendant, telephone operator, or friend, and give a brief acknowledgment of feeling ("You seem terribly busy and pushed"). Follow this with a brief self-statement ("Can I help?" "Should I come back?" "I've been pushed today myself, as well") and note here what happens:

_____

_____

_____

_____

Exercise 4: Developmental Skills Area 1—Recognizing Varying Styles Toward Emotional Expression (see pages 196–198 for definitions)

Classify the following emotions as either sensorimotor (S), concrete (C), formal-operational (FO), or dialectic/systemic (D/S).

A client discusses arguments he (she) has with his (her) parents that occur just before leaving home.

_____ (tears) "I'm overwhelmed."

_____ "I feel really sad because of the argument I had with my parents last week."

_____ "As I think about it, I feel bad because we have so many arguments. It seems to be a pattern and we argue every time I am about to leave home for school."

_____ "I suppose we could look at it from several perspectives. First, it really hurts to have these arguments, but I know I have to find my own space and perhaps it is part of my becoming a separate person. I know my parents care for me; perhaps that's why we argue just before I leave home."

A friend discusses reactions she (he) has to anxiety about an examination.

_____ "It's maddening and it made me angry when the professor didn't bring the exam to class today."

_____ "I suppose I can see the professor's frame of reference. After all, she has 40 papers to look over and I know she looks awfully hassled. But it sure does make it difficult for me to know where I am. A lot of students are really angry."

_____ "I'm scared. I can't eat. My stomach hurts. I'm confused."

_____ "Professor Jones is often late. It's typical of me to feel angry and upset when I have to wait. It's an emotional pattern for me."

Exercise 5: Developmental Skills Area 2—Facilitating Clients' Exploration of Emotion Using Varying Styles

Assume that you are working with one of the preceding clients who is overwhelmed by emotion. How would you help this person move away from direct, here-and-now sensorimotor experiencing?

_____

_____

_____

Assume you are working with one of the preceding clients who avoids really looking at here-and-now emotional experiencing. How would you help this client increase affect and feeling?

_____

_____

_____

Systematic Group Practice

One of the most challenging skills is reflection of feeling. Mastering this skill, however, is critical to effective counseling and interviewing.

Step 1: Divide into practice groups.
Step 2: Select a group leader.
Step 3: Assign roles for the first practice session.
▲ Client
▲ Interviewer
▲ Observer 1, who gives special attention to noting client feelings, using the Feedback Form in Box 7-6. The focus of microsupervision needs to be on the ability of the interviewer to bring out and deal with emotions.
▲ Observer 2, who gives special attention to interviewer behavior and writes down each specific interviewer lead.

Step 4: Plan.   We suggest that you examine a past or present story of a stressful experience (bullying, teasing, being seriously misunderstood, an unfair situation with school personnel, going through a hurricane or flood, or a time when you experienced prejudice or oppression of some type).

Most of us experienced real stress and some form of trauma as we watched planes fly into the World Trade Center. You very likely have a story to tell about where you were when you saw or heard the news. What were you seeing, hearing, and feeling? How did it affect you both then and now?

Another possibility is to explore accumulative trauma (see page 160). How have repeating events in one's life added up to real trauma?

Give some attention to observing styles of emotional expression. With what style does the client present the story: sensorimotor, concrete, abstract formal-operational, or abstract dialectic/systemic? You may also wish to explore specific questions to facilitate emotional discussion using different styles of experience (see pages 196–198).

Establish clear goals for the session. You can use questioning, paraphrasing, and encouraging to help bring out data. Periodically, the interviewer reflects feelings. This may be facilitated by one-word encouragers that focus on feeling words and by open questions ("How did you feel when that happened?"). The practice session should end with a summarization of both the feelings and the facts of the situation. Examine the basic and active mastery goals in the "Self-Assessment and Follow-Up" section to determine your personal objectives for the interview.

The observers should use this time to examine the feedback forms and to plan their own sessions.

Step 5: Conduct a 5-minute practice session using this skill.
Step 6: Review the practice session and provide feedback to the interviewer for 10 minutes.   How well did the interviewer achieve goals and mastery objectives and what feedback does the client provide verbally and perhaps through the Client Feedback Form of Chapter 1? As skills and client role-plays become more complex, you'll find that this time is not sufficient for in-depth practice sessions and you'll

want to contract for practice time outside the session with your group. Again, it is particularly important that the observers and the interviewer note the level of mastery achieved by the interviewer. Was the interviewer able to achieve specific objectives with a specific impact on the client?

Step 7: Rotate roles.
A reminder.   The client may be "difficult" if he or she wishes, but must be talkative. Remember that this is a practice session, and unless affective issues are discussed, the interviewer will have no opportunity to practice the skill.

Box 7-6   Feedback Form: Observing and Reflecting Feelings

_____ (Date)

_____   _____
(Name of Interviewer)   (Name of Person Completing Form)

Instructions: Observer 1 will give special attention to client feelings via notations of verbal and nonverbal behavior below. On a separate sheet, Observer 2 will write down the wording of interviewer reflections of feeling as closely as possible and comment on their accuracy and value.

1. Verbal feelings expressed by the client. List here all words that relate to emotions.

2. Nonverbal indications of feeling states in the client. Facial flush? Body movements? Others?

3. Implicit feelings not actually spoken by the client. Check these out with the client later for validity.

4. Reflections of feelings used by the interviewer. As closely as possible, use the exact words of the interviewer and record them on a separate sheet of paper.

5. Orientation to emotional expression. Within what developmental style or styles was this client telling the story: here-and-now sensorimotor, detailed concrete description, abstract formal-operational reflection, or abstract dialectic/systemic with multiple perspectives?

6. Comments on the reflections of feeling. What sentence stems were typically used? Were the feeling words reflected by the interviewer implicitly or explicitly expressed by the client? Was the interviewer's use of the skill accurate and valid? Was the check-out used?

Portfolio of Competence

Skill in reflection of feeling rests in your ability to observe client verbal and non-verbal emotions. Reflections of feeling then can vary from brief acknowledgment to exploration of deeper emotions. You may find this a central skill as you determine your own style and theory.

Use the following as a checklist to evaluate your present level of mastery. Check those dimensions that you currently feel able to do. Those that remain unchecked can serve as future goals. *Do not expect to attain intentional competence on every dimension as you work through this book.* You will find, however, that you will improve your competencies with repetition and practice.

Highlight competencies that you have met to date.

Level 1: Identification and classification.

❑ Ability to generate an extensive list of affective words.
❑ Ability to distinguish a reflection of feeling from a paraphrase.
❑ Ability to identify and classify reflections of feeling.
❑ Ability to discuss, in a preliminary fashion, issues in diversity that occur in relation to this skill.
❑ Ability to write reflections of feeling that might encourage clients to explore their emotions.
❑ Ability to recognize the developmental styles of emotion: sensorimotor, concrete, formal operational, and dialectic/systemic.

Level 2: Basic competence.   Aim for this level of competence before moving on to the next skill area.

❑ Ability to acknowledge feelings briefly in daily interactions with people outside of interviewing situations (restaurants, grocery stores, with friends, and the like).
❑ Ability to use reflection of feeling in a role-played interview.
❑ Ability to use the skill in a real interview.

Level 3: Intentional competence.   Ask yourself the following questions, all related to predictability and evaluation of the effectiveness of your abilities in working with emotion. These are skill levels that may take some time to achieve. Be patient with yourself as you gain mastery and understanding.

❑ Ability to facilitate client exploration of emotions. When you observe clients' emotions and reflect them, do clients increase their exploration of feeling states?
❑ Ability to reflect feelings so that clients feel their emotions are clarified. They may often say, "That's right . . . and . . ." They then continue to explore their emotions.
❑ Ability to help clients move out of overly emotional states to a period of calm.
❑ Ability to facilitate client exploration of multiple emotions one might have toward an important relationship (confused, mixed, positive, and negative feelings).
❑ Ability to recognize and facilitate client exploration within the four styles of emotional expression—sensorimotor, concrete, formal operational, and dialectic/systemic.

Level 4: Teaching competence.   Teaching competence in these skills is best planned for a later time, but a client who has particular difficulty in listening to

others may indeed benefit by training in observing emotions. Many individuals fail to see the emotions occurring all around them. Empathic understanding is rooted in awareness of the emotions of others. All of us, including clients, can benefit from bringing this skill area into use in our daily lives.

❏ Ability to teach clients in a helping session how to observe emotions in those around them.

❏ Ability to teach clients how to acknowledge emotions—and, at times, to reflect the feelings of those around them.

❏ Ability to teach small groups the skills of observing and reflecting feelings.

DETERMINING YOUR OWN STYLE AND THEORY: CRITICAL SELF-REFLECTION ON REFLECTION OF FEELING

This chapter has focused on emotion and the importance of grounding both yourself and your client. Special attention was given to identifying four varying styles of emotional expression as well as how you might help clients express more or less emotion, as appropriate to their situations.

What single idea stood out for you among all those presented in this chapter, in class, or through informal learning? What stands out to you is likely to be important as a guide toward your next steps? What are your thoughts on diversity? What other points in this chapter struck you as important? How might you use ideas in this chapter to begin the process of establishing your own style and theory?

_____

_____

_____

_____

_____

_____

_____

_____

REFERENCES

Bensing, J. (1999a). *Doctor-patient communication and the quality of care.* Utrecht, the Netherlands: Nivel.

Bensing, J. (1999b). The role of affective behavior. *Communication,* 1188–1199.

Carstensen, L., Pasupathi, M., Mayr, U., & Nesselroade, J. (2000). Emotional experience in everyday life across the life span. *Journal of Personality and Social Psychology, 79,* 644–655.

Damasio, A. (2003). *Looking for Spinoza: Joy, sorrow, and the feeling brain.* New York: Harvest.

Daniels, T., & Ivey, A. (2006). *Microcounseling* (3rd ed.). Springfield, IL: Thomas.

Danner, D., Snowdon, D., & Friesen, W. (2001). Positive emotion in early life and longevity. *Journal of Personality and Social Psychology, 80,* 804–813.

Ekman, P. (2004) *Emotions revealed.* Woodacre, CA: Owl.

Fredrickson, B., Tugade, M., Waugh, C., & Larkin, G. (2003). A prospective study of resilience and emotion following the terrorist attacks on the United States on September 11, 2001. *Journal of Personality and Social Psychology, 84,* 365–376.

Gergen, K., & Gergen, M. (2005). The power of positive emotions. *The Postive Aging Newsletter,* www.healthandage.com, February.

Hill, C., & O'Brien, K. (1999). *Helping skills.* Washington, DC: American Psychological Association.

Ivey, A., Ivey, M., Myers, J., & Sweeney, T. (2005). *Developmental counseling and therapy: Promoting wellness over the lifespan.* Boston: Lahaska/Houghton Mifflin.

Kolb, B., & Wishaw, I. (2003). *Fundamentals of neuropsychology* (5th ed.). New York: Worth.

Pos, A., Greenberg, L., Goldman, R., & Korman, L. (2003). Emotional processing during experiential treatment of depression. *Journal of Clinical and Consulting Psychology, 73,* 1007–1016.

Tamase, K., & Kato, M. (1990). Effect of questions about factual and affective aspects of life events on an introspective interview. *Bulletin of Institute for Educational Research* (Nara University of Education), *39,* 151–163.

Tamase, K., Otsuka, Y., & Otani, T. (1990). Reflection of feeling in microcounseling. *Bulletin of Institute for Educational Research* (Nara University of Education), *26,* 55–66.

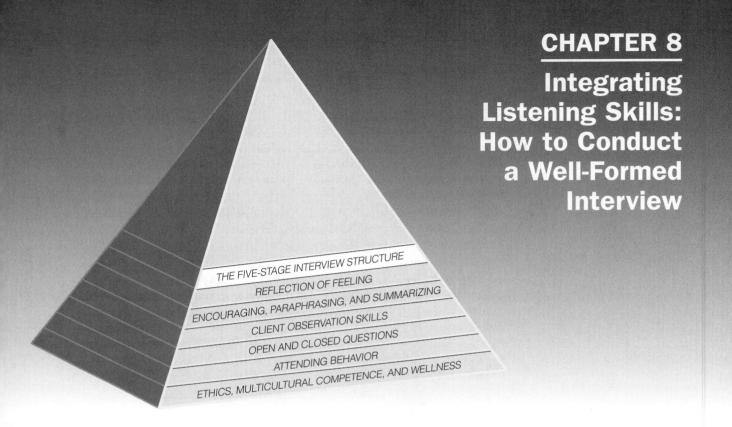

THE FIVE-STAGE INTERVIEW STRUCTURE

REFLECTION OF FEELING

ENCOURAGING, PARAPHRASING, AND SUMMARIZING

CLIENT OBSERVATION SKILLS

OPEN AND CLOSED QUESTIONS

ATTENDING BEHAVIOR

ETHICS, MULTICULTURAL COMPETENCE, AND WELLNESS

CHAPTER 8

Integrating Listening Skills: How to Conduct a Well-Formed Interview

How can the skills and concepts of this chapter be used to help you and your clients?

Here you are at a transition point as you move toward intentional competence. It is a time to look back at the skills and strategies covered thus far. The listening skills and the five-stage interviewing structure provide a springboard for the influencing skills. But as you explore the later chapters of this book, please recall that the true foundation of effective interviewing is listening to the client.

When you listen to your clients carefully, you have given them something important—the gift of being heard by another person. One of the best indicators that you have developed well your ability to hear and listen is that you can complete a full, well-formed interview using only listening skills. This foundation of effective listening and structuring the interview can be used with multiple theories of counseling.

To help you as you prepare to conduct a complete five-stage interview and plan for the future, this chapter has three central themes that are defined and described here:

Theme and Definition	*Function*
Ivey Taxonomy. The microskills hierarchy is presented in summary form—the listening skills, structuring the interview, confrontation, focusing, reflecting meaning, and the influencing skills.	The use of each skill, strategy, or concept has anticipated results in the interview. Intentional competence in the microskills provides you with predictions of how you can help clients in the

Theme and Definition	*Function*
	session. However, clients will often say or do something different from what you may expect. Intentional competence requires you to flex and generate a new alternative for helping when the first skill or strategy produces an unexpected result.
Empathic understanding. The listening skills presented thus far form the behavioral basis of empathy. The qualitative dimensions such as concreteness, immediacy, and a nonjudgmental attitude are also important.	The empathic dimensions supplement the microskills. For example, they will enable you to rate the quality and helpfulness of your interventions. It would be possible, for example, to ask an "off-track" open question that does not facilitate client exploration. A 5-point scale of empathic understanding will help you examine the usefulness of helping leads.
Five stages/dimensions of the interview. (1) Initiating the session, (2) gathering information, (3) mutual goal setting, (4) working, (5) terminating and generalizing learning to daily life. Wellness and the positive asset search are central to all stages/dimensions.	The five *stages* serve to ensure purpose and direction to the interview. They help define specific outcomes. The word *dimensions* is important as it refers to the fact that each interview and each client are unique and that the interview is a holistic experience. The five stages can appear in many ways. Different theories of interviewing and counseling give varying attention to each stage or dimension. The ability to conduct a well-formed interview using only listening skills is a prime competency of the intentional interviewer or counselor.

INTRODUCTION: A REVIEW OF CULTURAL INTENTIONALITY AND INTENTIONAL COMPETENCE

A critical issue in interviewing is that the same skills may have different effects on people with varying individual and cultural backgrounds. Diversity will always characterize the mainstream of interviewing and counseling. In each interview you have, you will encounter people with varying life experiences. To this must be added the many issues of multiculturalism (e.g., ethnicity/race, people with disabilities, sexual orientation, spirituality/religion). In effect, all interviewing is multicultural.

To this diversity must be added individual differences. One Arab American client cannot be expected to behave the same as the next. Veterans from the Iraq war are not all the same. And you will find that the individual client will behave differently from one interview to the next. Almost as soon as you think you fully understand a client, a new side of her or his personality will appear.

A special joy and opportunity we have in the helping fields is discovering the richness of individual and human diversity. Cultural and intentional competence asks you to have many responses available to help the constantly varying individuals with whom we work.

The microskills offer us a way to have some predictability in our work with individuals, but we must always be aware that anticipating specific results from our interventions is also potentially dangerous. Our wide-ranging clientele will constantly vary. What "works" as expected one time may not the next. Intentional competence requires flexibility and the ability to move and change in the moment with constantly shifting client needs.

INSTRUCTIONAL READING 1: THE IVEY TAXONOMY—ANTICIPATING THE RESULTS OF SKILL USAGE

The Anticipated Consequences of Skill Usage: An Overview of the Microskills

Box 8-1 presents the skills and strategies of the microskills hierarchy—what has been presented thus far and what will be offered in later chapters. The Ivey Taxonomy (IT) includes expected consequences in client conversation as a result of your choice of skills and strategies. Thus, if you ask an open question, anticipate that the client will offer more conversation; reflect feelings and expect the client to explore emotions. But, as noted above, be prepared for individual variation in clients. What is anticipated does not always happen.

You have already covered the first four sections of the taxonomy—ethics and multicultural competence, attending behavior, observing, and the basic listening sequence. In this chapter you will work with the five stages/dimensions of the interview and, with practice, demonstrate your ability to complete a full session using only the listening skills.

All the influencing skills presented later in this book depend on effective use of listening and observing. You may wish to skim the influencing skills in the taxonomy so that you can see the overall structure of this book and the goals that you can obtain. But do not be concerned about the influencing skills at this point. Your major task is to solidify your listening skills and apply them to a complete interview.

As preparation for the five stages, let us examine the basic listening sequence once again.

The Basic Listening Sequence: A Review

Observations of interviews in counseling and therapy as well as in management, medicine, and other settings have revealed a common thread of skill usage. Many successful interviewers begin their sessions with an open question followed by closed questions for diagnosis and clarification. The paraphrase checks out the content of what the client is saying, and the reflection of feeling (usually brief in the early stages) examines key emotions. These skills are followed by a summary of the concern expressed by the client. Encouragers may be used throughout the interview to enrich it and help evoke details.

Though these skills can be used in many different situations, they need not be used in any rigid sequence. Each counselor or interviewer adapts these skills to meet the needs of the client and the situation. The effective interviewer uses client observation skills to note client reactions and intentionally *flexes* at that moment to provide the support the client needs. As you learn other skills and observe individual

Box 8-1   The Ivey Taxonomy: An Expansion of the Microskills Hierarchy

Skill and Strategy Area With Brief Definition	Anticipated Consequence of Skill or Strategy Usage
Ethics and Multicultural Competence Ethical and multicultural issues are present in all interviewing, counseling, and psychotherapy sessions.	*Expected consequence:* Interviewers who intentionally base their behavior on an ethical approach with an awareness of the many issues of diversity have a solid foundation for a lifetime of personal and professional growth. Anticipate that your clients will appreciate, respect, and learn from your increasing knowledge in ethics and multicultural competence.
Wellness Helping clients become aware of and use positive strengths in themselves and their environment to help resolve their issues.	*Expected consequence:* Through an emphasis on wellness and the positive asset search, clients will develop increased self-esteem and self-confidence. Their inner and outer strengths and supports may enable them to work through and solve even the most challenging issues and problems more effectively.
Attending Behavior Individually and culturally appropriate visuals, vocals, verbals, and body language.	*Expected consequence:* Clients will talk more freely and respond openly, particularly around topics to which attention is given. Depending on the individual client and culture, we can anticipate fewer eye contact breaks, a smoother vocal tone, a more complete story (with fewer topic jumps), and a more comfortable body language.
Basic Listening Sequence Open and closed questions; client observation skills; encouraging, paraphrasing, reflecting feeling, and summarizing.	*General expectation:* The client's story, problem, or concern will be presented with the basics of the client's thoughts and feelings, and the facts and behaviors related to the issues presented.
Open and Closed Questions Open questions often begin with *who, what, when, where,* and *why.* Closed may start with *is* or *are.*	*Expected consequences:* Clients will answer open questions in more detail. Clients will give shorter answers to closed questions, providing specific information. Effective questions—open or closed—may encourage more talk regardless.
Client Observation Skills Observing one's own and the client's verbal and nonverbal behavior as well as discrepancies and incongruities that may occur in the session.	*Expected consequence:* The interviewer is able to observe verbal and nonverbal behaviors in self and in clients and use these observations as a foundation for use of the various microskill skills and strategies. The client, in turn, will respond to the interviewer. The smoothly flowing interview will often demonstrate movement symmetry and complimentarity. Through the observation of discrepancies and conflict, the interviewer will be able to provide appropriate helping leads and interventions.
Encouraging Means to help a client keep talking—verbal (repeating key words and short	*Expected consequence:* Clients will elaborate on the topic, particularly when encouragers and restatements are used in a questioning tone of voice.

(continued)

Box 8-1 (continued)

Skill and Strategy Area With Brief Definition	Anticipated Consequence of Skill or Strategy Usage
statements) and nonverbal (head nods and smiling).	
Paraphrasing Feeds back to the client the essence of what has just been said, thus shortening and clarifying client comments.	*Expected consequence:* Clients will feel heard. They will tend to go on further and not repeat the same story again. If a paraphrase is inaccurate, it provides the client with an opportunity to correct the interviewer. If a questioning tone of voice is used, the client may elaborate further.
Summarizing Similar to paraphrase, but used over a longer time span. Attention to feelings is often part of an effective summarization.	*Expected consequence:* Clients will feel heard and often learn how their stories are integrated. Particularly useful is to summarize and integrate the thoughts, emotions, and behaviors of the client and relevant others. This tends to facilitate a more centered discussion. The summary also provides a more coherent transition from one topic to the next or as a way to begin and end a full session.
Reflection of Feeling Identifying the key emotions of a client and feeding back and clarifying affective experience.	*Expected consequence:* Clients will go more deeply into their emotional experience. They may correct the reflection with another word. With some clients, the brief acknowledgment of feeling may be more appropriate.
The Five-Stage Interview Structure The five interview aspects provide a checklist to ensure a well-formed interview.	*General expectation:* The client will establish a positive relationship with the interviewer, will tell her or his story, will set realistic goals, develop a new story or way of viewing issues, and transfer new learning to daily life.

Important in all of the following is the *6th dimension, wellness and the positive asset search (PAS).* The PAS is the hub of a holistic interview and can be expected to be part of each of the five stages/dimensions of the session. If strengths and positives are found, anticipate that clients can more easily effect change and achieve their goals.

Initiating the session—Rapport and structuring ("Hello"). *Expected consequence:* The client will feel at ease and know what to expect.

Gathering data—Drawing out stories, concerns, problems, or issues ("What's your concern?" "What are your strengths and resources?"). *Expected consequence:* The client will share thoughts, feelings, and behaviors. With the positive asset search, the client will also be expected to share strengths and resources that may be available for problem resolution.

Mutual goal setting ("What do you want to happen?"). *Expected consequence:* The client will discuss directions in which he or she might want to go, new ways of thinking, desired feeling states, and, behaviors that might be changed. The client might also seek to learn how to live |

(continued)

Box 8-1 (continued)

Skill and Strategy Area With Brief Definition	Anticipated Consequence of Skill or Strategy Usage
	more effectively with situations or events that cannot be changed at this point (rape, death, an accident, an illness). A more ideal story ending might be defined.
	Working—Exploring alternatives, confronting client incongruities and conflict, restorying ("What are we going to do about it?"). *Expected consequence:* The client may reexamine individual goals in new ways and start the move toward new stories and action through confronting discrepancies, facing new challenges, and experiencing a variety of influencing skills from the interviewer. Creative problem solving may be an important part of this stage.
	Terminating—Generalizing and acting on new stories ("Will you do it?"). *Expected consequence:* If all stages are completed successfully, expect the client to demonstrate change in behavior, thoughts, and feelings in daily life outside of the interview. If this stage is ignored, the chances for change may be reduced.
Confrontation A supportive challenge in which incongruities and discrepancies are noted and then fed back to the client.	*Expected consequence:* The client will respond to the confrontation of discrepancies and conflict with new ideas, thoughts, feelings, and behaviors and these will be measurable on the Confrontation Impact Scale (see Chapter 9).
Focusing The interviewer's response emphasizes conversation around the client, main theme/problem, others, family, mutuality, interviewer, cultural/environmental/ context. This skill may appear within any other microskill and thus the client statement has a double (or triple) classification.	*Expected consequence:* Clients will tend to focus their conversation or story on the focus dimensions used by the interviewer. With multiple focus, the story or concern may be elaborated from multiple perspectives.
Reflection of Meaning The stress is on client's deeper, often unsaid, thoughts relating to underlying values and attitudes. Closely allied to paraphrasing, reflection of feeling, and the interpretation/ reframe.	*Expected consequence:* The client will discuss stories, issues, and concerns in more depth with a special emphasis on deeper meanings, values, and understandings. Meanings are considered close to core experiencing.

(continued)

Box 8-1   (continued)

Skill and Strategy Area With Brief Definition	Anticipated Consequence of Skill or Strategy Usage
Influencing Skills and Strategies All of the skills and strategies of this book may lead to reframes, new ways of thinking, change, and restorying. However, there are some skills that are primarily identified with interpersonal influence.	*General expectation:* Intentional competence in influencing skills leads clients to change behavior, thoughts, and feelings and may also enable them to find new stories or ways of thinking. The Confrontation Impact Scale, discussed in the next chapter, may be used to provide a way of measuring the degree of change or growth.
Interpretation/Reframe Provides a new way of thinking about a concern or problem. May link or join previously separated or different thoughts or feelings together.	*Expected consequence:* The client may discover another perspective on a story, issue, or problem. The perspective could have been generated by a theory or by simply looking at the situation afresh.
Logical Consequences Helps client anticipate the consequences of actions or of thoughts, feelings, and behaviors. "If you do this . . . , then. . . ."	*Expected consequence:* Clients will better anticipate consequences of actions and change thoughts, feelings, and behaviors—or, at least, indicate that they are considering the possibility of the impact of consequences on them.
Self-Disclosure Counselor's sharing of himself or herself through personal life experience. Often starts with an "I" statement.	*Expected consequence:* Used briefly by the interviewer, self-disclosure may encourage the client to self-disclose in more depth. Effective self-disclosures often enable the client to feel more comfortable in the relationship.
Feedback Provides accurate data on how the interviewer or	*Expected consequence:* The client obtains the perspectives of the interviewer and/or others on their thoughts, feelings, and behaviors and this may lead to change.

(continued)

and cultural differences, you may find it appropriate to begin some interviews with a self-disclosure or even a directive instead of the usual open question.

In using the basic listening sequence (BLS), you have a three-part goal. Whenever you are working with a client on any topic, you will want to elicit the following:

1. *An overall summary of the issue.* This is done initially through the open question "Could you tell me your story?" or through simple attending. At the close of a section of the interview, you may want to summarize the client's main facts and feelings.

Box 8-1   (continued)

Skill and Strategy Area With Brief Definition	Anticipated Consequence of Skill or Strategy Usage
others view or may view the client.	
Information/Advice Sharing of specific information with the client—e.g., career information, choice of major, where to go for information on AIDS or community services, social skill ideas.	*Expected consequence:* If given sparingly and effectively, the client will use this new information and ideas to act in new ways.
Directives Giving a client specific actions to take in the interview, even to the point of telling her or him what is to happen and what to expect as a result.	*Expected consequence:* The client will listen to and follow the directives, opinions, and structures suggested by the leader. Directives are often given when clients engage in assertiveness or social skills training or when the counselor suggests career information or specific exercises in the session such as the Gestalt "hot seat" or use of imagery.
Skill Integration Integrating the microskills into a well-formed interview and generalizing the skills to situations beyond the training session or classroom.	*Expected consequence:* Each of us will integrate skills in our own way and add them to our natural style. The consequences for each of us will vary, but we all can anticipate that increasingly we will know what to do and what to expect in the richness of interviewing, counseling, and psychotherapy. Interview planning and treatment planning will be an important part of this process.
Determining Personal Style and Theory	*General expectation:* This is a lifelong process in which you will constantly evaluate and examine your behavior, thoughts, feelings, and deeply held meanings as you work with clients and think about yourself. In the early stages, it is often helpful to examine your own preferred skill usage and what you do in the session. This will help you find your natural style. But you will also want to constantly build on your natural style as you learn more and more about theory and practice in interviewing, counseling, and psychotherapy.

2. *The key facts of a situation.* These are obtained through *what* questions, encouragers, and paraphrases. They include client thoughts about what happened.
3. *The central emotions and feelings.* You reach these through questions (such as "Could you share your feelings about that issue?"), reflection of feeling, and encouragers that focus on emotional words.

The basic listening sequence provides you with specific skills to ensure that you understand the basic structure of the client's story. Later, depending on your theoretical orientation and natural style, you may want to bring out other thoughts, behaviors, and meanings.

Table 8-1   Three examples of the basic listening sequence

Skill	Counseling	Management	Medicine
Open question	"Could you tell me what you'd like to talk to me about . . ."		
Closed question	"Did you graduate from high school?" "What specific careers have you looked at?"	"Who was involved with the production line problem?" "Did you check the main belt?"	"Is the headache on the left side or on the right? How long have you had it?"
Encouragers	Repetition of key words and restatement of longer phrases.		
Paraphrases	"So you're considering returning to college."	"Sounds like you've consulted with almost everyone."	"It looks like you feel it's on the left side and may be a result of the car accident."
Reflection of feeling	"You feel confident of your ability but worry about getting in."	"I sense you're upset and troubled by the supervisor's reaction."	"It appears you've been feeling very anxious and tense lately."
Summarization	In each case the effective counselor, manager, or physician summarizes the story from the client's point of view *before* bringing in the interviewer's point of view.		

For the beginning counselor or interviewer, mastery of the basic listening sequence can be most beneficial. Table 8-1 gives examples of the basic listening sequence in counseling, management, and medical interviewing. A special advantage of mastering this sequence is that once a basic set of skills is defined, it can be used in many different situations. It is not unusual for a person skilled in the concepts of intentional interviewing to be conducting career counseling at a college in the morning, training parents in communication skills in the afternoon, and working as a management consultant in the evening. The microskills approach can be applied in many settings. In each case the BLS has the objective of bringing out client data—facts, thoughts, and feelings—for later interviewer and client action.

Counseling, interviewing, and psychotherapy can be difficult experiences for some clients. They have come to discuss their problems and resolve conflicts, so the session can rapidly become a depressing litany of failures and fears.

People grow from their strengths. Taking a wellness approach and using the positive asset search is a useful method to ensure a more optimistic and directed interview. Rather than just ask about problems, the effective interviewer seeks constantly to find positive wellness strengths upon which the client can focus. Even in very complex issues, it is possible to find good things about the client and things that he or she does right. Emphasizing positive assets also gives the client a sense of personal power in the interview.

To conduct a positive asset search the interviewer simply uses the basic listening sequence to draw out the client's strengths and resources and then reflects them back. This may be done systematically, as a separate part of the interview, or used constantly throughout the session.

Box 8-2   Research Evidence That You Can Use: Overview of Microskill Research

Daniels (2007) has assembled and reviewed 450 databased studies on microskills. This review is available in the accompanying CD-ROM. Among his central observations and conclusions that have direct relevance to interviewing practice are the following:

▲ The microskills listed in the taxonomy do exist and can be classified with excellent reliability. However, the basic listening sequence skills have more research validation than the influencing skills discussed later in this text. Microcounseling skill training appears to be as effective or more effective than alternative systems.

▲ Maintenance of skills requires practice and use. Some research has found that if the skills are not used in actual practice, they may disappear over time. Continued practice is necessary if these skills are to be transferred to your own interviewing sessions. Understanding does not represent ability to perform.

▲ Research in the Netherlands and Japan clearly indicates the cross-cultural validity of the skills. Studies with medical practitioners in the Netherlands reveal the importance of establishing solid rapport with patients and show that it can lead to better results. The skill-training program has been translated into at least 17 languages. For example, it has been used in Northern Canada to train Inuit social workers, in Africa by UNESCO to train AIDS peer counselors, in Sweden and Indonesia with top-level managers, and in Japan with counselors and refugees, social workers, psychologists, nurses, businesspeople, and many others.

▲ A number of studies have shown that intentional skill usage in counseling results in predictable client impact and satisfaction with the session. Hill and O'Brien's (1999) review of parallel research on counseling skills produced similar findings on this important issue. What you say and do in the interview affects what clients say and do.

▲ The basic interviewing skills of this book all can be used effectively with children, particularly if you use concrete language and match their cognitive/emotional style (Ivey, Ivey, Myers, & Sweeney, 2005; Van Velsor, 2004).

▲ Teaching clients skills of communication can be effective. Research has shown that clinical practice with clients who may demonstrate avoidant personality style (shyness), depression, schizophrenia, and other diagnoses has been particularly successful. Individual, group, and family communication skills can be taught. Clearly, more research needs to be conducted in this important area.

Theoretically, the positive asset search may be described as a psychoeducational intervention that emphasizes human development rather than remediation of problems. The concept appears under many different guises in the various forms of interviewing, counseling, and therapy. At times, the positive asset search for wellness assets can replace or supplement traditional problem solving in the session, as client strengths naturally overcome their weaknesses.

INSTRUCTIONAL READING 2: EMPATHY AND MICROSKILLS

Carl Rogers (1957, 1961) brought the importance of empathy to our attention. He made it clear that it is vital to listen carefully, enter the world of the client, and communicate that we understand the client's world *as the client sees and experiences it.* The client's frame of reference is central to empathic understanding. Empathy is often defined as experiencing the world as if *you* were the client, but with awareness that the client remains separate from you. Attending behavior and the microskills of the basic listening sequence of the previous chapters are deeply involved with empathy.

Many others have followed and elaborated on Rogers's influential definition of empathy (cf. Carkhuff, 2000; Egan, 2002; Ivey, D'Andrea, Ivey, & Simek-Morgan, 2002). A common practice is to describe three types of empathic understanding:

▲ *Basic empathy:* Interviewer responses are roughly interchangeable with those of the client. The interviewer is able to say back accurately what the client has said. Accurate use of the basic listening sequence is a way to demonstrate basic empathy.

▲ *Additive empathy:* Interviewer responses add something beyond what the client has said. This may be adding a link to something the client has said earlier or it may even be a congruent idea or frame of reference that helps the client see a new perspective. Skilled use of listening skills and/or influencing skills (see the later chapters of this book) enable an interviewer to become additive.

▲ *Subtractive empathy:* Interviewer responses sometimes give back to the client less than what the client says and perhaps even distort what has been said. In this case, the listening or influencing skills are used inappropriately.

You will find it possible to rate interviewing skills to establish the quality of empathic understanding. A 5-point scale for examining empathy and its related constructs is presented in the following example.

Client:	I don't know what to do. I've gone over this problem again and again. My husband just doesn't seem to understand that I don't really care any longer. He just keeps trying in the same boring way—but it doesn't seem worth bothering with him anymore.
Level-1 Counselor:	(subtractive) That's not a very good way to talk. I think you ought to consider his feelings, too.
Level-2 Counselor:	(slightly subtractive) Seems like you've just about given up on him. You don't want to try anymore.
Level-3 Counselor:	(basic empathy or interchangeable response) You're discouraged and confused. You've worked over the issues with your husband, but he just doesn't seem to understand. At the moment, you feel he's not worth bothering with. You don't really *care*.
Level-4 Counselor:	(slightly additive) You've gone over the problem with him again and again to the point that you don't really *care* right now. You've tried hard. What does this mean to you?
Level-5 Counselor:	(additive) I sense your hurt and confusion and that right now you really don't care anymore. Given what you've told me, your thoughts and feelings make a lot of sense to me. At the same time, you've had a reason for trying so hard. You've talked about some deep feelings of caring for him in the past. How do you put that together right now with what you are feeling? (Or, if you truly sense that the relationship is hopeless, it may be wise to help the client acknowledge that fact.)

To be empathic means to take risks, and higher-level responses are not always well received by clients. As the counselor attempts to move toward additive responses, the risk goes up. Risk, in this case, means risk of error. You may have excellent listening skills, but when you seek to add your own perceptions you may be out of sync with client needs.

Box 8-3 National and International Perspectives on Counseling Skills: Is Empathy Always Possible?

Kathryn Quirk, Student in Counseling Program at Cambridge College

As beginning students in counseling, one of the first concepts we run into is empathy—experiencing and understanding the world of the client. Certainly this is core to the helping interview.

But, as I read my text (not this one), I felt increasingly uncomfortable. Was this almost magical concept really possible? I'll tell you why. First the happy ending. I am now the delighted mother of a lovely child, the darling of my life. But, Ryan did not come easily and my husband and I needed the help of two fertility clinics.

The "simpler" strategies of getting pregnant failed for three years. Those years were agonizing, but only a sample of the trauma we were to face (yes, dear reader, going through fertility procedures meets the full definition of trauma). We then moved to complicated in vitro procedures involving Petri dishes and surgery. The first three procedures failed and the fourth resulted in a pregnancy that ended when twins died after three months. I don't like the word "fetus"—and grieving for lost babies was horrible. We moved to a new clinic and our fifth try was fantastically successful.

How does all this relate to empathy? I recall a pleasant and expert nurse who counseled a group of us experiencing primary infertility. She was helpful and had good suggestions, but when things got emotional and we cried, she simply didn't get it. She would say that she understood and knew what we were going through. But, let's face it; *she hadn't been there herself.* How could she truly understand the physical pain or the feelings of failure, shame, and hopelessness? She didn't understand our loneliness and, perhaps worst of all, the crushed hopes. How can she understand what we were really feeling? I resented it when she said she understood when she clearly did not and could not. Fortunately, those in the group who had "been there" supplied the needed empathy and support.

Does this mean that if you haven't experienced the inner world and actual experience of the client that you can't be empathic? At first I thought that understanding of my experience was impossible except for those who had experienced what I had gone through. However, I've softened my thoughts somewhat as I learn about and think about good counseling. I still feel that *being there* is what serves as a foundation for the deepest empathy. But the nurse could have provided a deeper empathy than she did if she had admitted openly that she understood our feelings and experience only partially. She did, after all, have more experience listening to people with pregnancy challenges than we did. She did have something to offer.

By failing to discuss and admit that she was different from us suggested to me and to others in the group that she did not understand. What could she have done? First, I think she should have said early in her work with us that she herself had not experienced the difficulties that we went through and, as such, she could have admitted that her understanding and empathy were only partial. But she could have pointed out that she understood pain and loss and perhaps even shared some of her own difficult experiences. Saying this and also outlining her expertise and knowledge would have developed more trust and given us all a deeper feeling that she was an empathic person.

(continued)

Box 8-3   (continued)

As I've gone through my counseling program, I increasingly become aware that I too will have problems with being truly empathic and communicating the understanding I do have. When I meet clients who are different from me multiculturally (e.g., race, sexual orientation, religious commitment), I now know that I need to discuss these issues upfront. And I have the obligation to learn as much about the cultural background of these clients as I can. To maximize my empathic potential, I need to read, get out in the community where these people live, and participate with them when I can.

This also holds true for me when I work with alcoholics, cancer survivors, and those who have been raped. I haven't been there, but I have a responsibility to learn more about those whose life experience is different from mine.

Empathy is clearly important, but it is not learned just from classes and books. We all need to examine the human experience and become more fully aware of the life of those around us.

Lisa Gebo comments: Kathryn captures a very important paradox. That is, while I personally believe that there are common human experiences that connect us all at some very foundational level and that those connections can transcend race, class, gender, and culture, I will never truly know what it is like to walk in another person's shoes, what it is like to be a Person of Color in this society, or what it is like to go through the trauma of infertility. Those who have experienced infertility, cancer, or heart attacks all represent unique cultural groups that may require you to have "been there" to truly understand at the deepest level. At the same time, counselors through careful listening and empathic understanding can get close to experiences that they have not had and their special expertise and objectivity can be immensely helpful. But do be careful of saying "I understand" as it may not always be received well. Some variation of "I can imagine how difficult it is for you" may be more acceptable to those whose life experience different from you.

Right now, I'm trying to help my aging father, but I may never even know what it is like to experience advanced old age like my dad is now. If you haven't been there, you may not fully understand. Like the nurse described above, I try to empathize and be accepting of his anger over loss of strength, friends, independence. I think I can understand how that feels and it makes me profoundly sad. But, also like the nurse, I have not had my dad's individual life experience or his experience as a man, nor have I had the experience of being a man of his particular generation, and so on. As connected as I feel to his humanness, and of course, to him as my loving dad, these things separate us, too, and I can see that it is just as important to acknowledge this "disconnect" as it is to stress the points of commonality in an effort to be with him at this point in his life. Such is the paradox of life and I imagine, the paradox of counseling. I so admire people who can counsel, who can hold these opposing realities in their minds and hearts.

Source: Kathryn Quark, Counseling graduate student at Cambridge College.
Lisa Gebo is Mary and Allen's editor at Thomson Brooks/Cole. We thought her reactions to Kathryn's comments on empathy provided an unusually clear picture of some of the challenges we face as counselors and therapists.

A Level-3 interchangeable response is fairly safe and direct. When counselors strive for higher level, additive responses, they may sometimes be completely off the mark, and the client will respond negatively. This does not necessarily mean the counselor response was bad; it may be simply that the client wasn't ready for it at that moment.

In any case, how the client responds to your interviewing lead is more important than any external rating of what you say. No interviewer can always predict client responses to leads. It is here that client observation skills and the ability of the interviewer to flex and change the next lead to fit the needs of the client are most important. Flexing requires the interviewer to note the client's response to interventions and intentionally provide another intervention more in sync with the client's needs at that moment. The following 1-2-3 pattern illustrates the flow necessary for flexing and true empathic responding.

1. The interviewer observes the client's verbal and nonverbal behavior and consciously or unconsciously selects a verbal lead (skill) with the potential for facilitating client development. (The counselor in the previous example might select the Level-5 response as likely to facilitate talk and deeper exploration.)
2. The client reacts to the counselor's statement with verbal and nonverbal behavior. (In this case, the client might say angrily, "I don't put that together. What I want at the moment is to get away. Are you pushing me back toward him?")
3. The interviewer again observes the client's verbal and nonverbal behavior and selects another verbal lead (skill) with the possibility of facilitating client development. (After the client's angry response the counselor might try a Level-3 response: "What I just said made you feel angry. You want very much to get away from him right now.")

Number 1 in the pattern is the behavior of the counselor, number 2 is the reaction of the client, and number 3 is the consequent behavior of the counselor. True empathy requires the interviewer constantly to flex and be ready to change and adapt to each unique client. Although many counselors are anxious to move to the deeper Level-4 and -5 responses, if these are used out of context and without regard to client readiness, failure of even the "best" responses is inevitable. True empathy requires you to be where the client is able to hear you.

Empathy has been further refined through specific ways to enhance the quality of the interviewing relationship—positive regard, respect and warmth, concreteness, nonjudgmental attitude, and authenticity or congruence. All of the following *empathic dimensions* can also be rated on a 5-point scale.

Positive Regard

Positive regard, closely related to wellness and the positive asset search, is responding to the client as a worthy human being. More concretely, it may be defined as selecting positive aspects of clients' stories and selectively attending to positive aspects of client statements. A 5-point scale may be developed to measure positive regard (as well as other qualities discussed in this section), with the same points of reference as for empathy.

Level	1	2	3	4	5
	Subtractive		Interchangeable		Additive

In a subtractive response, the counselor finds something wrong with the client. In an interchangeable response, the counselor notes or reflects accurately what the client has talked about. In an additive response the counselor points out how, even in the most difficult situation, the client is doing something positive. For example, the counselor may say the following to a client suffering from depression and talking about many problems: "John, I respect your ability to talk about and analyze your issues and problems so well. I can understand at least some of your feelings as you have described them. Clearly, you've got good insight into what is happening. Let's change the focus just a bit. When isn't the problem occurring? Tell me about a time in the past when you were able to feel better about yourself and others." All too often counselors fall into the trap of listening only to the negative. Search constantly for positives and strengths in your clients.

Respect and Warmth

Respect and warmth may be most easily rated from a kinesthetic and nonverbal point of view. You show respect and warmth by your open posture, your smile, and your vocal qualities. Your ability to keep your comments congruent with your body language is an indicator of respect and warmth. Capture yourself on videotape and obtain feedback on how much respect and warmth you show the client. Use the 5-point scale above.

Concreteness

Concreteness has been stressed throughout this book. It is important to seek specifics rather than vague generalities. As interviewers, we are most often interested in specific feelings, specific thoughts, and specific examples of actions. As has been stressed many times, one of the most useful of all open questions is "Could you give me a specific example of . . . ?" Concreteness makes the interview live and real. Likewise, communication *from* the interviewer—the directive, the feedback skill, and interpretation—needs to be highly specific or it may become lost in the busy world of the client.

There are times, nevertheless, when concreteness is not the most appropriate response. Some problems are best discussed in more general and abstract terms, and some cultural groups tend to be more subtle. Individual differences of expression in empathy, respect and warmth, and concreteness must always be kept in mind.

Immediacy

Immediacy is often described as being in the moment with the client. What are you and the client experiencing here and now? Immediacy is also a useful concept to give the interview timeliness. It is most easily described in terms of language. You may respond in three tenses to a client who is angry: "You *were* angry . . . you *are* angry . . . you *will be* angry." We tend to respond to others in the same tense in which they are speaking. You will find that some clients always talk in the past tense; they may profit from present-tense discussion. Other clients are always thinking of the future; still others are constantly in the present. The most useful response is generally made in the present tense. A change of tense may be used to speed up or slow down the interview. It seems, however, that for counseling styles of many types, responses that include all three tenses tend to be the most useful.

Another way to view immediacy is in terms of the relationship between the counselor and the client (Egan, 2002). The more personal the relationship is, the more immediate it is. As issues of closeness between interviewer and client arise, this type of immediacy may become very powerful. A relationship is made immediate in this sense by a focus on the counselor and client ("I–you" talk) and by staying in the present tense.

You will find that as interviews move more to present tense immediacy, your presence in the interview will become more powerful and important. You may even find that clients start responding to you in the here and now similarly to ways they responded to significant people from their past. It is in such instances that the immediacy of "I–you" talk between counselor and client becomes especially important. The skill of artful self-disclosure (see Chapter 12) will be useful here.

Nonjudgmental Attitude

A nonjudgmental attitude is difficult to describe. Closely related to positive regard and respect, a nonjudgmental attitude requires that you suspend your own opinions and attitudes and assume value neutrality in relation to your client. Many clients have attitudes toward their issues and concerns that may be counter to your own beliefs and values. If you listen to your clients carefully, you will come to an understanding of why they might have taken that position or action. People who are working through difficulties and issues do not need to be judged or evaluated.

A nonjudgmental attitude is expressed through vocal qualities and body language and by statements that indicate neither approval nor disapproval. However, as with all qualities and skills, there are times when your judgment may facilitate client exploration. There are no absolutes in counseling and interviewing.

For a moment, stop and think of a client whose behavior troubles you personally. It may be someone whom you regard as dishonest, a perpetrator of violence, or one who shows clear sexism and racism in the interview. These are challenging moments for the nonjudgmental attitude. You do not have to give up your personal beliefs to present a nonjudgmental attitude; rather you need to suspend your private thoughts and feelings. If you are to help these people change and become more intentional, presenting yourself as nonjudgmental is critical. You do not have to agree with or approve of the thoughts and behaviors of the client to be nonjudgmental. At times, you may express your disagreement and your own value stance, but still be nonjudgmental to the person before you. You regard the total person nonjudgmentally, as change most often comes from a basis of trust and honesty.

Authenticity and Congruence

Are you personally *real?* Authenticity and congruence are the reverse of discrepancies and mixed messages. The hope is that the counselor or interviewer can be congruent and genuine and not display many discrepancies. Needless to say, however, life is full of discrepancies and paradoxes, and your ability to be flexible in response to the client may be the most basic demonstration of your authenticity.

Box 8-4   Research Evidence That You Can Use: Empathy and Neuropsychology

Empathy is clearly central to all interviewing, counseling, and psychotherapy, regardless of the theory used. Empathy, however, is not just an abstract idea—empathy is identifiable and measurable in the physical brain. Fascinating research on brain activity validates what the helping field has been saying for years. "The basic building blocks (of empathy) are hardwired into the brain and await development through interaction with others. . . . empathy (is) an intentional capacity" (Decety & Jackson, 2004, pp. 71, 93).

When we see something happening to others, our brains react, even if we are not directly experiencing the same event. Consider the following pain experiment with volunteer couples. A painful electric shock was administered to one member of each couple and brain scans revealed brain activity throughout the full pain-processing network of the shocked individual. Important for our understanding of empathy, when the partner merely watched the shock given to a loved one, *the affective section of the observer's brain (somatosensory cortex) was activated, but not the sensory component where physical pain is felt (somatosensory cortex)* (Singer et al., 2004). In effect, we see here that brain patterns related to pain appear in the person who does not directly experience pain.

We should also mention that research reveals that the antisocial, criminal personality has a reduced ability to appreciate the emotions of others. This deficit appears to be a dysfunction of the amygdala (Blair, 2001).

What we learn here is that the empathic person's brain responds to another person's experience, even though he or she does not actually experience the other person's world. Many studies over the years back up this central point. For example, children around their second year indicate concern for others cognitively, emotionally, and behaviorally by comprehending others' difficulties and trying to help (Zhan-Waxler et al., 1992). You also may have seen two young children playing together. One falls and starts crying. Even though the second child has not been hurt, he or she also cries. This ability to observe the feelings of others could be considered the developmental roots of empathic understanding.

(continued)

INSTRUCTIONAL READING 3: THE FIVE STAGES/DIMENSIONS OF THE WELL-FORMED INTERVIEW

Virtually anything we and our clients do can be viewed as the solving of a problem or making a decision. This section summarizes a basic decisional or problem-solving structure that you can use in virtually all types of interviewing, counseling, and therapy. This framework is applicable in individual, family, and group counseling (Ivey & Matthews, 1984; Ivey, Pedersen, & Ivey, 2001). You will find it helpful later as you work with varying theories of counseling and psychotherapy. Equally important, this section also shows how listening skills can be used effectively in each segment of the interview. At this point, with listening skills, you have the basics for a full interview using only listening skills.

Chapter 13 outlines decisional counseling in more detail. The structure is based on a classic problem-solving model that focuses on

1. Defining the problem.
2. Defining goals.

Box 8-4   (continued)

Awareness of self, awareness of others, and the ability to differentiate yourself from the client are essential for empathic understanding. This area of research has been summarized by Decety and Jackson (2004):

> Whenever an action is taking place, it activates an intentional schema, a structure internal to every person involved in that action. The intentional schema has the capacity of coordinating first- and third-person information; according to the input schemas available, the action is attributed to the self or the other person. (p. 83)

Here we clearly see the connections among brain research, empathy, and intentionality as described in this book. It is fascinating to see counseling's ideas validated this way. Our task as counselors and therapists is to resonate or empathize with the client's story, but we also need to separate ourselves from that client. In addition, we need to work intentionally with an array of skills and strategies that can help the client to grow further. And that client growth will be shown not only in behavior, but also in measurable aspects of brain functioning.

At the same time, if you have not "been there" in terms of your own life experience, you may not be able to have full empathic understanding. For example, recovered alcoholics or substance abusers who enter the helping fields have a head start when it comes to empathizing with clients who have similar difficulties.

In addition, even if you have "been there" as an alcoholic or faced another life challenge, your personal experience is not the same as that of the other person. Always be careful in projecting your experiences on another.

But it is also obvious that no person will have "been there" for all human experience. None of us can be expected to live through and understand all the challenges of the world. And this is why effective and empathic listening becomes particularly important. There is a common element to human pain, joy, and other emotions. Draw on your parallel life experiences, but never assume that you can understand every client fully.

3. Generating alternative solutions and selecting a new, potentially, more effective approach.

To this basic triad, we add developing a relationship with the client as you start the interview. And, after a solution has been selected, ensuring that the client actually does something after the interview to ensure change.

Much of interviewing, counseling, and therapy are about problem solving and decision making. Alternative theories, of course, approach decisions very differently. Thus, this is not the only interviewing structure, but you will find that it provides an organization that can produce results. Later in this book, you will see how this structure can be used in many approaches to counseling and therapy, even though the theories appear to be very different from one another.

Each interview will be unique and different from all others. Nonetheless, there are many issues in common. All interviewers use microskills and strategies and they often follow a sequence of stages from the beginning to the end:

1. Initiating the session—Rapport and structuring
2. Gathering data—Drawing out stories, concerns, problems, or issues

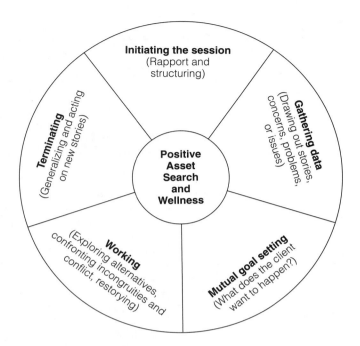

Figure 8-1   The circle of interviewing stages

3. Mutual goal setting—What does the client want to happen?
4. Working—Exploring alternatives, confronting client incongruities and conflict, restorying
5. Terminating—Generalizing and acting on new stories

The positive asset search and wellness are the core or hub of the five stages/dimensions. Finding strengths as you initiate the session can make an important difference; helping clients do a positive asset search as you gather data may provide you with many strengths for later problem resolution. When you turn to goal setting, some clients may benefit from your helping them reframe their objectives in more positive ways. During the working and termination phases, frequent reference to positives and a "can do" attitude may make an important difference.

The five-stage model of the interview was developed out of a linear decision-making model. Not all clients will feel comfortable with the stages as presented, regardless of cultural background. For example, it should be clear that issues of rapport and structuring will be important throughout the interview. At times, new information later in the session will result in your returning to Stage 2 and redefining client concerns in a new way.

Although all five dimensions need to be considered in the interview and treatment plan, it is not always necessary to follow them in a specific order, and consequently we have what is called the circle of interviewing stages (Figure 8-1). After you have mastered the five stages in a linear fashion, consider them an important checklist to ensure that you have covered all the issues.

A circle has no beginning or end and can sometimes be used as the symbol of an egalitarian relationship in which interviewer and client work together on concerns. You will find that some clients prefer that you take charge, particularly in the early stages as you gain credibility as a helper. However, eventually you will want to work with clients on a mutual basis in helping them reach their goals.

The importance of Stage 1 structuring and Stage 3 mutual goal setting between you and your client cannot be overstressed. If you and your client are clear about what is to happen in counseling and interviewing and agree to mutual goals for the process (structuring of how the two of you will work together) and goals (Stage 3), you have a solid basis for trust and effective helping.

Failure to treat is one of the most common causes of malpractice suits. Failure-to-treat issues most often appear when counselors and clients have only vague ideas about what is to happen in counseling and what counseling can actually accomplish. If you know what you are doing and you and the client are clear as to goals and structure, you have a good foundation for your series of interviews. Clients and counselors who agree on goals in a clear fashion can work toward them and revise the goals as the sessions move forward. Clients who participate in goal setting and understand the reasons for your helping interventions may be more likely to participate in the process and more open to change.

The circle of interviewing may remind us that helping is a mutual endeavor between client and counselor and that we need to be flexible in our use of skills and strategies.

1. Initiating the Session—Rapport and Structuring ("Hello")

Building Rapport

"Hello, _____." A prime rule for establishing rapport is to use the client's name and repeat it periodically through the session. Some interviewers give extensive attention to the rapport stage, whereas others simply assume rapport and start immediately. Introducing the interview and building rapport obviously are most important in the first interview with a client. In some cases rapport building may be quite lengthy and blend into treatment; an example is in reality therapy with a delinquent youth, in which playing Ping-Pong and getting to know the client on a personal basis may be part of the treatment. It may take several sessions before some who are culturally different from you develop real trust. In most Western counseling and interviewing, however, this stage is quite short. After a brief "Hello" the interviewer my immediately move to a discussion of what the client wants.

The most important microskills for building rapport are attending behavior and client observation skills. Basic attending is used to demonstrate that you understand the client and are interested. Client observation is critical at this stage of the interview. Is the client comfortable and relaxed? All observations are helpful in the process of rapport development. Self-disclosure on your part may be helpful with some clients as well. In a continuing series of sessions, summarization may be used so that past interviews are integrated with the current session to maintain rapport.

Wellness and the positive asset search may be an important part of rapport building. With a nervous or insecure client, taking time to outline specific and concrete client assets provides the client with a secure base from which to confront difficult problems. What is most important is that the interviewer is open, authentic, and congruent with the client, and flexibly meets the needs expressed by that client.

Nevertheless, you should remember that many—perhaps most—interviews begin with some variation on "Could you tell me how I might be of help?" In much of interviewing rapport can be assumed, but it is often the most important issue throughout the entire interview. Unless the client has some liking for and trust in you, you won't get far. Again, your ability to observe clients will tell you when it is appropriate to move to Stage 2. One of the best clues as to when to start is the point at which the client begins talking spontaneously about concerns and/or you note that body language is mirrored between you and the client. Movement symmetry and movement complementarity are excellent ways to judge the level of empathy between you and the client.

Structuring

Structuring is the second part of this first stage. This is where informed consent and ethical issues may be discussed with the client. When you conduct an interview using only attending and listening skills, it may be helpful to inform your client that you are going to listen to him or her carefully. Later, as you make a more detailed theoretical commitment, you will want to inform your client of the purposes of your interview, your general methods, and other structural issues important to the relationship.

Some clients need to have the interview explained for them. This may be their first interview and they may not know how to behave. Setting the stage for this type of client can be extremely important. In such cases the interviewer explains the purpose of the interview and what he or she can or cannot do. Interviewers conducting a welfare intake interview, for example, find that they can better assist clients if they indicate very early in the session what their powers are. If the client has a different need, immediate referral is possible, with minimal frustration for both client and interviewer.

Multicultural Considerations

As we consider this stage in depth, it is important to recall some key multicultural issues. First, you constantly need to be aware that different cultural groups develop relationships in varying ways. In traditional Native American or Dene culture, for example, almost the entire first helping session may need to be devoted to relationship building. In Latin cultures, the concepts of respect and dignity are particularly important and may require a more formal approach. The more experienced your client is in English-speaking cultures, the more likely he or she will be to understand and accept traditional counseling and interviewing theory and approaches.

What about cross-cultural counseling where your race and ethnicity differ significantly from the client's? Authorities are in increasing agreement that cultural and ethnic differences need to be addressed straightforwardly relatively early in counseling, often in the first interview (for example, see Cheatham & Stewart, 1990; Kim, Hill, Gelso, Goates, Asay, & Harbin, 2003). Your ability to recognize and respect differences may be essential for the success of the interview or counseling series.

2. Gathering Data—Drawing Out Stories, Concerns, Problems, or Issues ("What is your concern?" "What are your strengths and resources?")

Listening to the Client's Story

"What's your concern?" The first task of the interviewer is to find out why the client is there and what the problem is. Coupled with that is gathering necessary information about the client. Clients often confuse interviewers with a long list of issues and concerns. One simple rule is that the *last* item a client presents in a "laundry list" of problems is often the one of central concern; watch for it, but be prepared to redefine the problem as you listen further. Never forget, however, that the entire list is important.

The basic listening sequence is crucial to bringing out the story. Open and closed questions will help define the issue as the client views it. Encouragers and paraphrases will provide additional clarity and an opportunity for you to check out that you have heard correctly. Acknowledgment of emotions through reflection of feeling will provide an important beginning understanding of the emotional underpinnings. And finally, summarization provides a good way to put the several ideas the client has presented into an orderly format.

The central task at this point is defining the issues as the client describes them. This can be supplemented by gathering information and data about clients and their perceptions. The basic *who, what, where, when, how,* and *why* series of questions provides one short and often useful framework to make sure you have covered the most important aspects of information gathering and problem definition.

In your attempts to define the central client concerns, always ask yourself, what is the real world of the client? What problem seeks resolution, or what opportunity needs to be actualized? Failure to clearly answer these questions often results in an interview that wanders and lacks purpose.

Positive Asset Search and Wellness

At the same time, it is important to give attention to client strengths. The positive asset search should be part of this stage of the interview or systematically included later in the session. Always remember one of the central points of this text: *clients grow from strength.* The dimensions of wellness presented in Chapter 2 can serve as one guideline as you search for resources.

Multicultural Considerations

Recall that the word *problem* can be a problem for some clients. Consider using *story, issue,* or *challenge* as alternatives. You may wish to keep in mind that the problem-focused language of much of counseling and therapy may be inappropriate for some clients. Furthermore, the second stage of the interview, for many clients from all cultural backgrounds, may be better placed in the third stage, which focuses on client goals. If you establish joint goals with your client, you empower that client and the two of you may then return to concerns oriented toward helping that client reach her or his goals. In fact, in brief counseling the interview often starts with a focus on client goals (see Chapter 14).

3. Mutual Goal Setting ("What do you want to happen?")

The third stage focuses on client goals. Where does the client want to go? Many counselors and interviewers will summarize the problem or challenge of the client and then ask a question such as "What would you imagine the ideal solution to be?" "Where do you want to go with this?" "Could you take a moment to sit back and develop a fantasy of what you would like to happen?" After the client proposes some solutions, you may use the basic listening sequence, and more details of the client's thoughts and feelings about ideal situations will be generated. Often clients can solve their own issues in this stage of the interview.

The word *mutual* is included as part of the goal-setting process. Your active involvement in client goal setting is important. You will have clients who constantly talk about their problems, but when asked, "What might you want to do about it? What is your goal?" they ignore you and return to their issues. *If the client and you don't know where the interview is going, you may end up somewhere else!*

Too often the client and counselor assume they are working toward the same outcome when actually each of them wants to head in a different direction. A client may be satisfied with sleeping better at night, but the counselor wants complete personality reconstruction. The client may want brief advice about how to find a new job, whereas the counselor wants to give extensive vocational testing and suggest a new career. On the other hand, clients often expect and want more than the interviewer can deliver. Clarifying this issue early in the session through questioning can save a great deal of time and effort.

Some clients prefer to have this stage of the interview early in the session, even before gathering data about the issue. For example, with high school discipline problems, less verbal clients, and those of some cultural groups, moving quickly to define a clear outcome may help rapport development. Some clients dislike lengthy analysis of their concerns and want action *now.* If you adapt your interviewing style to meet this simple but obvious need, you will be able to achieve counseling success with many clients that more traditional problem-focused methods cannot reach.

Clients who are confused about their careers profit from exploration in this phase of the interview, particularly if their concerns are vague and ambiguous. Sorting out a clear goal early may later result in better data gathering and asset identification.

A maxim for the confused interview—be it discipline, career, or even marital counseling—is "Define a goal, make the goal explicit, search for assets to help facilitate goal attainment, and only then examine the nature of the concern." At times, clear goal definition and a solid asset search can make problem identification unnecessary. Brief, solution-oriented interviewing and counseling (Chapter 14) indeed makes goal attainment the issue and at times even ignores Stage 2, gathering data.

The question of determining outcomes is of interest from a theoretical as well as a practical perspective. Rogers (1957, 1961) talks about many clients who have incongruencies between the real self and the ideal self. Behavioral psychologists often talk about present behavior compared to desired behavioral goals. Reality therapists talk about fulfilling unmet needs, and brief counselors speak of facilitating a client's decision to act. Many theoretical orientations ask, "Where is the client, and where does she or he want to go?" "What is the difference between the real world and the desired world?" Blanchard and Johnson (1981), in their continuing best-seller *The One-Minute Manager,* summarize the issue succinctly:

If you can't tell me what you'd like to be happening . . . you don't have a problem yet. You're just complaining. A problem only exists if there is a difference between what is actually happening and what you desire to be happening. (p. 3)

Once the discrepancies—between the real self and the ideal self, between the real situation and the desired situation, between the nature of the relationship now and the desired relationship, and so on—are clear, it is possible to confront issues clearly and precisely. A wide variety of skills, techniques, and theories are available to explore and confront the conflict faced by the client.

For example, consider the following five model sentences:

▲ *Decision making:* "On the one hand your concern may be summarized as . . . and on the other hand your desired outcome is . . . and you have the following assets and strengths to help you reach your goals. . . ."
▲ *Rogerian client-centered:* "Your real self, as you describe yourself, is. . . . Yet you see your ideal self as . . . and you have several positive qualities, such as. . . ."
▲ *Cognitive-behavioral:* "Your present behavior is . . . but you would like to behave differently. For example, you would really like to . . . and we note the following positive behaviors and actions from the past. . . ."
▲ *Marital counseling:* "Your present relationship is described as . . . but you would like to see it change as follows. . . . As a couple you seem to have several strengths, such as . . . that will be helpful in resolving the conflict."
▲ *Career counseling (confused client):* "You are searching for a college major (or life career) and aren't really sure of your possibilities. Yet you've described your short- and long-term goals rather clearly. . . . You've had several positive work experiences in the past. . . . How do you put that together?"

Note that all these model confrontation sentences bring together and point out the discrepancy between the problem definition and the desired outcome. The positive asset summary is used to help the client realize that he or she is personally capable of problem resolution.

Many verbal clients can use such a summary as a springboard to action. They will resolve the discrepancy on their own with your support. Clients who are extremely emotionally involved with their problems may require more influencing skills and active direction on your part. Observing your client's verbal and nonverbal reaction to this summary will help you determine your style of interview in Stage 4. Clear definition of goals can make an immense difference.

4. Working—Exploring Alternatives, Confronting Client Incongruities and Conflict, Restorying ("What are we going to do about it?")

"What are we going to do about it?" The purpose of this stage of the interview is resolving concerns and finding relief for the client. The problem may lie in deciding between two positive alternatives, in making a vocational choice, or in any of the issues described earlier. The client at this stage is stuck and unable to come up with productive alternatives. The task of the counselor or interviewer is to explore possibilities and to assist the client in finding new ways to act more intentionally in the world.

How does the interviewer confront and explore incongruity? Two major routes appear to be available. First, one could summarize the client conflict and frame of reference and use the basic listening sequence to facilitate the client's resolution of the issue(s). Second, one could summarize the client conflict and frame of reference, then *add* one's own frame of reference through influencing skills (feedback, self-disclosure, instruction, directives, interpretation) and/or by applying alternative helping theories (Gestalt, psychodynamic, behavioral). It is preferable for most interviewers to concentrate on the first alternative, that of facilitating clients through the use of listening skills only.

Let us consider a school counselor talking with an acting-out teen who had just had a major showdown with the principal. With the five-stage interview structure here, the first task of the counselor is to establish rapport. The teen may be expected to challenge you, particularly as he or she expects you to support the school administration. Your task is not to judge but to gather data from the teen's point of view. If this is done effectively, the teen will feel that he or she has been heard. You can follow this by finding what the teen would like to have happen. Often, if you hear the client's story first, her or his goals often become more workable and realistic. It does little good, of course, to work with unrealistic goals. Work with the teen to find a way to "save face" and move on. Recall that at times you may have to become an ally of teens if you are to be effective in important conflicts in the future. If things are going reasonably well, teens in such situations may be able to describe the conflict from the principal's point of view.

To bring the teen to Stage 4, the model summarizations described in detail in the discussion of Stage 3 can be helpful. "On one hand, you see the situation as . . . and your goal is. . . . But, on the other hand, the principal tells a different story and, as you say, the principal's goal is likely to be. . . ." Given that you, the counselor, have some specific goals, what do you think you can do to reach this client and find a solution that works? If you have listened well and developed rapport, many teens at this point are able to generate ideas to help resolve the situation. Your first goal in Stage 4 is to encourage your clients to find their own resolution.

In addition, you will find that the basic listening sequence and skilled questioning are useful in facilitating client exploration of answers and solutions. Here are some useful questions to assist client problem solving:

"What other alternatives can you think of?"
"Can you brainstorm ideas—just anything that occurs to you?"
"What has worked for you before?"
"What part of the problem is workable if you can't solve it all right now?"
"Which of the ideas that you have generated appeals to you most?"
"What would be the consequence of your taking that alternative?"

In effect, all of these are oriented toward opening client thought leading to new solutions. You will also find that encouraging skills are useful in helping clients stop and explore possibilities. Repeat key words that might lead the client to new alternatives for action.

Counseling and long-term therapy both try to resolve issues in clients' lives in a similar fashion. The counselor needs to establish rapport, define the problem, and establish certain desired client outcomes. The distinction between the problem and the desired outcome is the major incongruity the therapist seeks to resolve. This incongruity or discrepancy may be resolved in three basic ways. The counselor can use attending skills to clarify the client's frame of reference and then feed back a summary of client concerns and the goal. Often clients will generate their own synthesis and resolve their challenges. If clients do not generate their own answers, then the therapist can add interpretation, self-disclosure, and other influencing skills in an attempt to resolve the discrepancy. In that case, the counselor would be working from a personal frame of reference or theory. Finally, in systematic problem solving the counselor and client might together generate or brainstorm alternatives for action and set priorities for the most effective possibilities.

Decisional counseling requires alternatives. It is good to keep the basic problem-solving model in mind throughout the interview, particularly in this stage:

1. Define the concern, keeping in mind the goal or desired outcome.
2. Generate alternatives.
3. Decide on action.

During this stage it is particularly important to keep the issue, or challenge, in view while generating alternatives for a solution and for an eventual decision for action. However, a decision for action is not enough. You also need to plan to make sure that feelings, thoughts, and behaviors generalize beyond the interview itself. Stage 5 of the interview speaks to this task.

5. Terminating—Generalizing and Acting on New Stories ("Will you do it?")

"Will you do it?" The information conveyed, the concepts learned in the interview, the new behaviors suggested might all be for naught if systematic thought is not given to the transfer and generalization of the interview to daily life. The complexities of the world are such that taking a new behavior back to the home setting is difficult.

You will find that some therapies work on the assumption that behavior and attitude change will come out of new unconscious learning; they "trust" that clients will change spontaneously. This indeed can happen, but there is increasing evidence that planning for change greatly increases the likelihood of its actually happening.

Consider the situation again with the teen in conflict with the principal. You may have generated some good ideas together, but unless the teen follows up on them, nothing is likely to happen and soon he or she will be back in your office again and the principal may be wondering why you aren't more effective. (And if the problem is not with the teen but with an insensitive teacher and/or principal, then your challenge is that much greater.) Your task is to find something "that works" and changes the repeating story. Think through the following paragraph and the list of generalization suggestions as you read them. What might you try to help this teen act and find a new way of restorying her or his life?

Change does not come easily, and maintaining any change in thoughts, feelings, or behavior is even more difficult. Behavioral psychology has given considerable thought to the transfer of training and has developed an array of techniques for transfer; even so, clients still revert to earlier, less intentional behaviors. At this point, we suggest that you turn to page 451 where you will find the "Maintaining Change and Relapse Prevention" form. More and more interviewers, therapists, and counselors are using some variation of this form to help ensure that the hard work done in the session has relevance and impact in the "outside world."

Here are examples that different theoretical schools have used to facilitate the transfer of learning from the interview. Many of these examples are elaborated on later in this book.

Role-playing.   Just as in the practice sessions of this book, the client can practice the new behavior in a role-play with the counselor or interviewer. This emphasizes the specifics of learning and increases the likelihood that the client will recognize the need for the new behavior after the session is over.

Imagery.   Ask the client to imagine the future event and also imagine what he or she will specifically need to do to manage the situation more effectively.

Behavioral charting and progress notes.   The client may keep a record of the number of times certain behaviors occur and report back to the counselor. With other clients an informal diary of personal subjective reactions may be more helpful.

Homework.   The interviewer may suggest specific tasks for the client to try during the week as a follow-up on the interview. This is an increasingly common practice in counseling.

Family or group counseling.   Sometimes individual problems are deeply merged within difficult marriage, family, or work-group arrangements. Therefore, an increasing number of counselors now seek to involve spouses and families in the counseling process. In work settings, too, managers and personnel people increasingly see organizational development and team building as critical to improving individual skills and transfer of behavior.

Follow-up and support.   It may be helpful to ask the client to return periodically to check on the maintenance of behavior. The telephone or e-mail may be used. At this time the counselor can also provide social and emotional support through difficult periods.

As you can see, the use of influencing skills is particularly important in the final stages of the interview. When conducting an interview using only listening skills, the follow-up and support element is perhaps the most relevant. Here are some questions you can use to help clients plan their own generalization from the interview:

"What one thing from the interview stands out for you right now that you might take home?"
"You've generated several ideas and selected one to try. How are we going to know if you actually do it?"
"What comes to your mind to try as homework for next week that we can look at when we get together?"

Each of these can be coupled with the basic listening sequence to draw out the generalization plan in more detail. You may want to ask your client at the close of the

Table 8-2 The five stages/dimensions of the interview

Definition of Stage/Dimension	Function and Purpose	Commonly Used Skills
1. Initiating the session— Rapport and structuring ("Hello")	To build a working alliance with the client and to enable the client to feel comfortable with the interviewer. Structuring may be needed to explain the purpose of the interview. Structuring helps keep the session on task and helps inform the client of what the counselor can and cannot do.	Attending behavior to establish contact with the client and client-observation skills to determine appropriate method to build rapport. Structuring often involves the influencing skill of information giving and instructions. Discuss informed consent and related ethical issues.
2. Gathering data—Drawing out stories, concerns, problems, or issues ("What's your concern?" "What are your strengths and resources?")	To find out why the client has come to the interview and listen to the story. Skillful problem definition will help avoid aimless topic jumping and give the interview purpose and direction. It also helps to identify clearly positive strengths of the client.	Most common are the attending skills, especially the basic listening sequence. Other skills may be used as necessary. If problems aren't clear, you may need more influencing skills. The positive asset search often reveals capabilities in the client that are useful in problem resolution.
3. Mutual goal setting ("What do you want to happen?")	To find out the ideal world of the client. How would the client like to be? How would things be if the problem were solved? This stage is important in that it enables the interviewer to know what the client wants. The desired direction of the client and counselor should be reasonably harmonious.	Most common are the attending skills, especially the basic listening sequence. Other skills are used as necessary. If outcome is still unclear, more influencing skills may be helpful. With some clients, this phase should often precede Stage 2. Solution-oriented interviewing and counseling (Chapter 14) gives central attention to this dimension.
4. Working—Exploring alternatives, confronting client incongruities and conflict, restorying ("What are we going to do about it?")	To work toward resolution of the client's issue. This may involve the creative problem-solving model of generating alternatives and deciding among those alternatives. It also may involve lengthy exploration of personal dynamics. This stage of the interview may be the longest.	May begin with a summary of the major discrepancies. Depending on the issue and theory of the interviewer, more influencing skills (discussed in Chapter 12) may be expected. Listening skills provide balance.
5. Terminating—Generalizing and acting on new stories ("Will you do it?")	To facilitate changes in thoughts, feelings, and behaviors in the client's daily life. Many clients go through an interview and then do nothing to change their behavior, remaining the same as when they came in.	Influencing skills, such as directives and information/explanation, are particularly important. Attending skills used to check out client's understanding of importance of generalizing interview learning to daily life.

interview "Will you do it?" as a form of contract between the two of you for the future.

These are but a few of the many possibilities to help develop and maintain client change. Each individual will respond differently to these techniques, and client observation skills are called for to determine which technique or set of techniques is most likely to be helpful to a particular person. For maximal impact and behavior transfer, a combination of several techniques is suggested. Evidence makes clear that behavior and attitudes learned in the interview do not necessarily transfer to daily life without careful planning. The five stages of the interview are summarized in Table 8-2.

EXAMPLE INTERVIEW: I CAN'T GET ALONG WITH MY BOSS

This interview illustrates how listening skills can be used to help the client understand and cope with interpersonal conflict. The interview has been edited to show portions that demonstrate skill usage and levels of empathy as clearly as possible. When you conduct your own interview and develop a transcript indicating your own ability to use listening skills, you may want to arrange your transcript in a similar fashion.

The client in this case is a 20-year-old part-time student who is in conflict with his boss at work. You will find him relatively verbal. For the most part, a verbal, cooperative client is required for a counselor to work through a complete interview using only listening skills.

Stage 1: Rapport/Structuring

Counselor/Client Statement	Process Comments
Machiko: Robert, do you mind if we tape this interview? It's for a class exercise in interviewing. I'll be making a transcript of the session, which the professor will read. Okay? And if you wish, we can turn the recorder off at any time. I'll show you the transcript if you are interested. If you decide later you don't want me to use this material, I won't and we'll start again.	Closed question followed by structuring information. It is critical that you obtain client permission before taping, and offer personal control over the material. As a beginner or student you cannot legally control confidentiality, but it is nonetheless your responsibility to protect your client.
Robert: Sounds fine; I do have something to talk about.	Robert seems at ease and relaxed. As the taping was presented casually, he is not concerned about the use of the recorder. Rapport was easily established.
Machiko: What would you like to share?	Open question, almost social in nature, designed to give maximum personal space to the client.
Robert: My boss. He's pretty awful.	Robert indicates clearly through his nonverbal behavior that he is ready to go. Already, Machiko and he have some body mirroring. Therefore, Machiko decides to move immediately to Stage 2. With some clients, several interviews may be required to reach this level of rapport.

Stage 2: Gathering Information, Defining the Problem, and Identifying Assets

Counselor/Client Statement	Process Comments
Machiko: Could you tell me about it?	Open question, oriented toward obtaining a general outline of the problem the client brings to the session.
Robert: Well, he's impossible.	Instead of the expected general outline of the concern, Robert gives a brief answer. The predicted consequence didn't happen.
Machiko: Impossible?	Encourager. Intentional competence requires you to be ready with another response.
Robert: Yeah, really impossible. It seems that no matter what I do he is on me, always looking over my shoulder. I don't think he trusts me.	Clients often elaborate on the specific meaning of a problem if you use the encourager. The prediction holds true this time.
Machiko: Could you give me a more specific example of what he is doing to indicate he doesn't trust you?	Open question eliciting concreteness. Robert is a bit vague in his discussion.
Robert: Well, maybe it isn't trust. Like last week, I had this difficult customer who was lipping off to me. He had a complaint about a shirt he had just bought. I don't like that type of thing, so I just started talking back. No one can do *that* to me! And of course the boss didn't like it and chewed me out. It wasn't fair.	As we make events concrete through specific examples, we can understand more fully what is going on in the client's life and mind. The underlying meaning of trust and of the relationship with the boss is changing.
Machiko: As I hear it, Robert, it sounds as though this guy gave you a bad time and it made you angry, and then the boss came in.	Paraphrase and reflection of feeling. Represents Level-3 empathy, as Machiko's response is relatively similar to what Robert said. As she does not include the important dimension of fairness, some might call it a Level-2 response.
Robert: Exactly! It really made me angry. I have never liked anyone telling me what to do. I left my last job because the boss was doing the same thing.	Accurate listening often results in the client's saying "exactly" or something similar.
Machiko: So your last boss wasn't fair either?	Paraphrase with a questioning voice. Machiko has brought back Robert's key word *fair.* This would be a Level-3 empathic response. In addition, Machiko's vocal tone and body language communicate warmth and respect. She is nonjudgmental as she listens.

(The interview continues to explore Robert's conflict with customers, his boss, and past supervisors. There appears to be a pattern of conflict with authority figures over the past several years. This is a common pattern among young males in their early careers. After a detailed discussion of the specific conflict situation and several other examples of the pattern, Machiko decides to conduct a positive asset search.)

Machiko: Robert, we've been talking for a while about difficulties at work. I'd like to know some things that have gone well for you there. Could you tell me about something you feel good about at work?	Paraphrase, structuring, open question.
Robert: Yeah; I work hard. They always say I'm a good worker. I feel good about that.	Robert's increasingly tense body language starts to relax with the introduction of the positive asset search. He talks more slowly.
Machiko: Sounds like it makes you feel good about yourself to work hard.	Reflection of feeling, emphasis on positive regard, Level-3 empathy.
Robert: Yeah. For example, . . .	

(Robert continues to talk about his accomplishments. In this way Machiko learns some of the positives Robert has in his past and not just his problems. She has used the basic listening sequence to help Robert feel better about himself. Machiko also learns that Robert has several important assets to help him resolve his own problems—among them, determination and willingness to work hard. However, with many clients, the positive asset search might begin earlier in the session.)

Stage 3: Determining Outcomes

Counselor/Client Statement	Process Comments
Machiko: Robert, given all the things you've talked about, what would an ideal solution for you be? How would you like things to be?	Open question. The addition of a new possibility for the client represents additive empathy, a Level-4 response, as it enables Robert to think of something new.
Robert: Gee, I guess I'd like things to be smoother, easier, and less full of conflict. I come home so tired and angry.	
Machiko: I hear that. It's taking a lot out of you. Tell me more specifically how things might be better.	Paraphrase, open question oriented toward concreteness.
Robert: I'd just like less hassle. I know what I'm doing, but somehow that isn't helping. I'd just like to be able to resolve these conflicts without always having to give in.	Robert is not as concrete and specific as anticipated. But he brings in a new aspect of the conflict—giving in.
Machiko: Give in?	Encourager.

(In the goal-setting process, you will often find yourself changing the problem definition. Here Machiko learns another dimension of Robert's conflict with others. Subsequent use of the basic listening sequence brings out this pattern with several customers and employees. [This is a shortened version of the discussion.])

Machiko: *So,* Robert, I hear two things in terms of goals. One that you'd like less hassle, but another, equally important, is that you like not to have to give in. Have I heard you correctly?	Summary. Most likely this is a Level-4 summary, as Machiko has helped Robert clarify his problem even though no resolution is yet in sight. Note the care and time Machiko has given to the problem-definition and goal-setting stages.
Robert: You're right on, but what am I going to do about it?	

Stage 4: Exploring Alternatives and Confronting Client Incongruity

Counselor/Client Statement	Process Comments
Machiko: So, Robert, on the one hand I've heard you have a long-term pattern of conflict not only with supervisors and bosses, but also with customers who give you a bad time. Clearly, you are a good worker and like to do a good job. On the other hand, I heard just as loud and clear your desire to have less hassle and not to have to give in to others all the time. Given all this, what occurs to you that you can do about it?	Major summary of the entire interview to this point. Machiko, in this Level-4 response, has distilled and clarified what the client has said. While her words are interchangeable with those of Robert and *appear* to represent Level-3 empathy, the clarity of the interview was brought about largely by her own skills in listening. She remains nonjudgmental and appears to be very congruent with the client in terms of both words and body language.
Robert: Well, the first thing that occurs to me is that if I am a good worker—and I'm not dumb—perhaps I can take these conflicts as another chance to do well. This would be a place where I need to work harder.	Many times clients can use their already existing positive assets to solve their own problems. Robert's self-image of being hard-working is an asset that can be used for problem resolution.
Machiko: Uh-hummm . . .	Encourager. Machiko's body language and vocal tone are supportive as well. She leans forward.
Robert: I could see this as a problem to be solved. I think I've been fighting it too much. I think I've let the boss and the customers control me too much. I think what I'd like to do the next time a customer comes in and complains is not say a word and simply fill out the refund certificate. Why should I take on the world?	Robert talks more rapidly. He, too, leans forward. However, his brow is furrowed indicating some tension. He is now "working hard."
Machiko: So one thing you can do is keep quiet. Sounds as if you are thinking you could maintain control in your own way by doing so, and you would not be giving in.	Paraphrase. Level-4 empathy, as Machiko is using Robert's key words and feelings from earlier in the interview to reinforce his present thinking.
Robert: Yeah, that's what I'll do, keep quiet.	He sits back, his arms folded.
Machiko: Sounds like a good beginning, but I'm sure you can think of other things as well. What are some other things you could do, especially at those times when you simply can't be quiet? Can you brainstorm more ideas?	Machiko gives Robert brief feedback, but realizes that there is more work to do. Her open question is a Level-4 response adding to the interview. She is aware that his closed nonverbals suggest more is needed.

(Clients are often too willing to seize the first idea that comes up. It may not be the best thing for them. Here it is helpful to use a variety of questions and listening skills to draw out the client further. With Robert, some ideas came easily but others more slowly. Eventually, he was able to generate two other useful suggestions: (a) to talk frankly with his boss about the continuing problem and seek his advice; and (b) to plan an exercise program after work to help blow off steam and energy. In addition, Robert began to realize that his problem with his boss was but one example of a continuing problem. He and Machiko discussed the possibility of talking more or for him to visit a professional therapist. Robert decided he'd like to talk with Machiko a bit more. A contract was made: if the situation did not improve within 2 weeks, Robert would seek professional help.)

Stage 5: Generalization and Transfer of Learning

Counselor/Client Statement	Process Comments
Machiko: So we've talked a great deal about the issue and you've decided that the most useful idea of all is to talk with your boss. But the big question is "Will you do it?"	Paraphrase, open question.
Robert: Sure, I'll do it. The first time the boss seems relaxed.	
Machiko: As you've described him, Robert, that may be a long wait. Could you set up a specific plan so we can talk about it the next time we meet?	Paraphrase, open question. To generalize from the interview, it is important to encourage specific and concrete action in your client.
Robert: I suppose you're right. Okay, occasionally he and I drink coffee in the late afternoon at Rooster's. I'll bring it up with him tomorrow.	
Machiko: What, specifically, are you going to say?	Open question, again eliciting concreteness.
Robert: I think it would be smart to tell him that I like working there, but am concerned about how to handle difficult customers. I think I'll ask his advice and how he does it. In some ways, it worries me; I don't want to give in to the boss.	Robert is able to plan something that might work. With other clients, you may find it helpful to plan role-plays, give advice, and lead the generalization plan more. You will also note that Robert is still concerned about "giving in."
Machiko: Would you like to talk more about giving in the next time we meet? Maybe through your talk with him we can figure out how to deal with that. Sounds like a good contract. Robert, you'll talk with your boss and we'll meet later this week or next week.	Open question, structuring.

It would have been wise to specify the follow-up contract even more precisely, but this would most likely entail the use of influencing skills, prescribing homework, and so forth. Machiko presented an especially important response during

Stage 5 when she asked what Robert was going to do specifically. Again, you'll find that concreteness is very important in assisting clients to make and act on decisions.

Note Taking

A frequently asked question is whether one should take notes during an interview. As might be expected, the answer is dependent on you, the client, and the situation. If you personally are relaxed about note taking, it will seldom become an issue in the interview. If you are worried about taking notes, it likely will be a problem. When working with a new client, obtain permission early in the session. We find it useful to say something like the following, according to your own natural style:

> I'd like to take a few notes while you talk. Would that be OK? It often helps us refer back to important thoughts. I also like to write down your exact words for the important points. I'll make a copy of the notes before you leave so you have the same record I do. As you know, all notes in your file are open to you at any time.

Clearly, detailed note taking where the writing takes precedence over listening is to be avoided. Beyond that one major warning, you and your client can usually work out an arrangement suitable for both of you. We also suggest that any case notes be made available to clients as well, and we recommend that you share with your volunteer clients any notes or transcripts of interviews you complete with practice sessions in this book. Their feedback on the interview can be most helpful in thinking about your own style of helping.

There is nothing wrong with *not* taking notes. Probably most counselors write their notes later, and most interviewers may take notes during the session. In-session note taking is often most helpful in the initial portions of interviewing and counseling and will become less important as you get to know the client better.

Audiotaping and videotaping the session follow the same guidelines. If you are relaxed and share on an equal basis with your client, making this type of record of the interview generally goes smoothly. Some clients find it helpful to take audio recordings of the session home and listen to them, thus enhancing learning from the interview.

SUMMARY

The five-stage structure of the microskills decisional interview has been demonstrated in the preceding example, showing that it is indeed possible to integrate all the skills and concepts of this book presented thus far into a meaningful, well-formed session. You may find it challenging to work through the systematic five-stage interview and not use advice and influencing skills, yet it can be done. It is a useful format to use with individuals who are verbal and anxious to resolve their own issues. You will also find this decisional structure useful with resistant clients who want to make their own decisions. By acting as a mirror and asking questions, we can encourage many of our clients to find their own direction.

Theoretically and philosophically, the decisional style using only listening skills is related to Carl Rogers's person-centered therapy (Rogers, 1957). Rogers developed guidelines for the "necessary and sufficient conditions of therapeutic personality change," and the empathic constructs described in this chapter are derived from his thinking. Rogers originally was opposed to the use of questions but in later life modified his position. Implicit in your ability to conduct an interview without using information, advice, and influencing skills is a respect for the person's ability to find her or his own unique direction. In conducting an interview using only attending, observation, and the basic listening sequence, you are using a very person-centered approach to counseling and interviewing.

Box 8-5   Key Points

Ivey Taxonomy (IT) basic listening sequence (BLS)	The IT elaborates on the microskills hierarchy. Skills and strategies are listed with brief definitions. *Intentional competence is reached when you can anticipate the consequences of skill or strategy usage and, if the prediction fails, you can flex intentionally and use another skill or strategy.* The BLS is vital in this process.
Empathy	Empathy means experiencing the client's world as if *you* were the client. It requires attending skills and using the important key words of the client, but distilling and shortening the main ideas.
Additive empathy	The interviewer may *add* meaning and feelings beyond those originally expressed by the client. If done ineffectively, it may *subtract* from the client's experience. Empathy is best assessed by the client's reaction to a statement, not by a simple rating of the interviewer's comments.
Positive regard	This means selecting positive aspects of client experience and selectively attending to positive aspects of client statements.
Respect and warmth	Respect and warmth are attitudinal dimensions usually shown through nonverbal means—smiling, touching, and a respectful tone of voice—even when differences in values are apparent between interviewer and client.
Concreteness	Being specific rather than vague in interviewing statements constitutes concreteness.
Immediacy	An interviewer statement may be in the present, past, or future tense. Present tense statements tend to be the most powerful. Immediacy is also viewed as the immediate "I–you" talk between interviewer and client.
Nonjudgmental attitude	Suspend your own opinions and attitudes and assume a value neutrality with regard to your clients.
Authenticity and congruence	These are the opposite of incongruity and discrepancy. The interviewer is congruent with the client and is authentic in their relationship.
Five stages of the interview	*Stage 1:* Rapport and structuring ("Hello.") *Stage 2:* Gathering information and defining issues ("What's your concern?" "What are your strengths?") *Stage 3:* Determining outcomes ("What do you want to happen?") *Stage 4:* Exploring alternatives and client incongruities ("What are we going to do about it?") *Stage 5:* Generalization and transfer of learning ("Will you do it?")

(continued)

Box 8-5 (continued)

Circle of decision making	The five stages of the interview need not always follow the five steps in order. Think of the stages as dimensions that need to be considered in each session. Also, give special attention to mutuality in the session, clear structuring of goals for the interview or interview series, and joint goal setting between client and counselor. Wellness and the positive asset search were presented as the hub of the circle.

COMPETENCY PRACTICE EXERCISES AND PORTFOLIO OF COMPETENCE

Mastery of the skills of this chapter is a complex process that you will want to work on over an extended period of time. Some basic exercises for the individual and systematic group practice in each skill area follow.

Basic Listening Sequence

Exercise 1: Illustrating How the BLS Functions in Different Settings

Write counseling leads as they might be used to help a client solve the problem "I don't have a job for the summer." In this case you will have to imagine that the client has responded. Write responses that represent the BLS.

Open question: _____

_____

Closed question: _____

_____

Encourager: _____

Paraphrase: _____

_____

Reflection of feeling: _____

_____

Summary: _____

_____

_____

Now imagine you are talking with a client who has just been told that her or his parents are going to seek a divorce after more than 25 years of marriage. Your task is to use the BLS to find out how the client is thinking, feeling, and behaving in reaction to this news.

Open question: _____

_____

Closed question: _____

_____

Encourager: _____

Paraphrase: _____

_____

Reflection of feeling: _____

_____

Summary: _____

_____

_____

Finally, how would you use these skills in talking with an elementary school student who has come to you crying because no one will play with her or him?

Open question: _____

_____

Closed question: _____

_____

Encourager: _____

Paraphrase: _____

_____

Reflection of feeling: _____

_____

Summary: _____

_____

_____

Exercise 2: The Positive Asset Search and the Basic Listening Sequence (BLS)

Imagine that you are role-playing a counseling interview. Write counseling leads, using the BLS to draw out the client's positive assets and strengths.

In a career interview, the client says, "Yes, I am really confused about my future. One side of me wants to continue a major in psychology, while the other—thinking about the future—wants to change to business." Use the BLS to draw out this client's positive assets. In some cases, you will have to imagine client responses to your first question.

Open question: _____

_____

Closed question: _____

_____

Encourager: _____

Paraphrase: _____

_____

Reflection of feeling: _____

_____

Summary: _____

_____

_____

You are counseling a couple considering divorce. The husband says, "Somehow the magic seems to be lost. I still care for Chantell, but we argue and argue—even over small things." Use the positive asset search to bring out strengths and resources in the couple on which they may draw to find a positive resolution to their problems. In marriage counseling in particular, many counselors err by failing to note the strengths and positives that originally brought the couple together.

Open question: _____

_____

Closed question: _____

_____

Encourager: _____

Paraphrase: _____

_____

Reflection of feeling: _____

_____

Summary: _____

_____

_____

Empathy

Exercise 1: Writing Helping Statements Representing the Five Levels of Empathy

I'm having trouble here at the community college. I'm the first one in my family who has ever even attempted college. The work doesn't seem all that hard, but when I turn papers in, my grades seem so low. It's hard to make friends. I have to work and I don't have as much money as the other kids seem to.

Write statements here that represent the five levels of empathic response to this client's concern. These statements can employ any skill, but paraphrasing and reflection of feeling are perhaps the clearest and easiest statements to write.

Level 1 (subtractive): _____

_____

Level 2 (slightly subtractive): _____

_____

Level 3 (interchangeable response): _____

_____

Level 4 (slightly additive): _____

_____

Level 5 (additive): _____

_____

Carlena says (near tears), "Alexander and I just broke up. I don't know what to do. We've been living together for almost a year. I don't have any place to live and I'm so confused."

Again, write statements representing the five levels of empathic responding.

Level 1 (subtractive): _____

_____

Level 2 (slightly subtractive): _____

_____

Level 3 (interchangeable response): _____

_____

Level 4 (slightly additive): _____

_____

Level 5 (additive): _____

_____

Exercise 2: Rating Interview Behavior Using Empathic Dimensions

Use any systematic group practice exercise from this chapter or the whole book and rate the interviewer's empathic response. Alternatively, you may wish to use an interview on audiotape, videotape, in transcript form, or a live interview. Provide specific and behavioral evidence for your conclusions using the Feedback Form (Box 8-6).

Box 8-6   Feedback Form: Empathy

_____ (Date)

_____
(Name of Interviewer)

_____
(Name of Person Completing Form)

Instructions: Observers are to (1) view an interview or segment of an interview, rating the empathic responding on a 5-point scale, and (2) provide specific behavioral evidence for their decisions.

	Level 1 (subtractive)	Level 2	Level 3 (interchangeable)	Level 4	Level 5 (additive)
1. Overall empathy rating					
2. Positive regard					
3. Respect and warmth					
4. Concreteness					
5. Immediacy					
6. Nonjudgmental attitude					
7. Authenticity and congruence					
8. Other observations					

(continued)

Box 8-6   (continued)

Provide specific behavioral evidence in the space provided to justify our rating of empathic behaviors on the chart.

1. _____

_____

2. _____

_____

3. _____

_____

4. _____

_____

5. _____

_____

6. _____

_____

7. _____

_____

8. _____

_____

A Practice Interview Using Only Attending and Listening Skills

The goal here is a challenging one—can you go through a full interview using only attending behavior and the microskills of the basic listening sequence? You may even wish to try to complete a full session *without* any questions or with minimal use of questions. You are not fully prepared to encounter the influencing skills of later chapters until you demonstrate that you can hear the client's story through listening carefully—and this is best shown through your ability to conduct a full session using only listening skills.

We suggest that you find a volunteer client who is relatively verbal and willing to talk about something of real interest. Negotiate the topic before you start. Let the client know what you plan to do by sharing the interview stages with him or her beforehand or early in the session. It is OK openly to enlist client cooperation both in practice and real situations. This can result in a more egalitarian interviewing or counseling relationship.

After the session is over, ask the client to complete the Client Feedback Form of Chapter 1. You may also ask the client to give you immediate feedback on what was helpful and things that might have been missed. Ask the key questions, "What did we miss discussing today?" and "What else?" You may also want to review the audio- or videotape with your client and start and start at various points to obtain her or his impressions.

Under ideal circumstances, you would conduct this session in a room equipped with a one-way mirror through which an observer could see the interview directly. This, of course, will not ordinarily be possible. We suggest that you ask a classmate to serve as observer and he or she can complete the observation form by listening to your audiotape or observing your videotape. See the Feedback Form in Box 8-7. Needless to say the Client Feedback Form of Chapter 1 is important to help you obtain needed feedback, particularly if you do not have an external observer.

One possible topic is career issues. Many of us have problems in our present job, we may want to find a new position, or we may be undecided as to what career path to follow. Career counseling is an important part of helping our clients grow, and we suggest that you and your client consider it as a possibility for this important practice session. Following is an outline of things to consider as you work through the five stages/dimensions of the well-formed interview.

1. *Initiating the session (rapport and structuring).* Develop rapport with your client in your own style. Structure the interview by informing the client about video or audio recording and how the tape will be used. If using a career interview, you may wish to say something like "What we are going to do today is first discuss the issues around your career or work situation. A choice of specific career is a good topic (or we agreed to talk about . . .). Then we'll explore that and search out some of your strengths, examine goals, and look at possibilities. Finally, we will talk about how you might want to take home some of the things we've talked about today. Is that OK?" Include informed consent and ethical issues.

2. *Gathering data (drawing out stories, concerns, problems, or issues).* Use the basic listening sequence to draw out the client's career story (or other topic that was agreed to as the session started). Be sure to obtain at least one positive asset that may later be helpful in facilitating decision making.

3. *Mutual goal setting (what does the client want to happen?).* Using the basic listening sequence, work with the client to find objectives and what would represent a satisfactory solution. With some clients, goal setting may precede Stage 2—in fact, brief counseling and solution-oriented interviewing often use goal setting as one of the first agenda items.

4. *Working (exploring alternatives, confronting incongruities and conflict, restorying).* Summarize the real and ideal world from 2 and 3 above—this may appear as a summary confrontation ("On one hand, your story/concern may be summarized as . . . On the other hand, your goals seem to be . . . How can we put these together in some new ways?"). Work with the client to brainstorm and discover new ways of thinking, feeling, and behaving. Discuss and examine discrepancies and incongruities. Work with the client to help rewrite old stories into new narratives.

5. *Terminating (generalizing and acting on new stories).* Review and summarize the session and work specifically on things that the client can do concretely to follow up on what happened in this session.

Box 8-7   Feedback Form: Practice Interview Using Only the Basic Listening Sequence

_____ (Date)

_____   _____
(Name of Interviewer)   (Name of Person Completing Form)

Instructions: *Interviewer:* Conduct a brief five-stage/dimension session using only the skills of the basic listening sequence. We suggest sharing the steps beforehand with your volunteer client and working through each stage/dimension together. The suggested topic is career decision or a work issue. Or you and your client could take any current life issue that is not too complex so that you can focus on practicing skills.

Observer: Please provide feedback and commentary to the interviewer. Was this interviewer able to conduct a session using only listening skills?

1. **Stage/Dimension 1: Initiating the session.** Nature of rapport? Enough established before interview continued to next stage? Did interviewer provide structuring?

2. **Stage/Dimension 2: Gathering data, defining concerns, and identifying assets.** Is the story told using only listening skills? Was at least one positive supportive asset of the client examined?

3. **Stage/Dimension 3: Mutual goal setting.** Was specific outcome or goal outlined for the client through use of listening skills?

(continued)

Box 8-7 (continued)

4. **Stage/Dimension 4: Working.** Was the interviewer able to assist client in generating new ideas through use of listening skills only?

5. **Stage/Dimension 5: Terminating and generalizing.** Were specific plans for taking ideas home made and contracted for? Systematic plan of action for follow-up?

6. **Intentional competence: Did specific skill usage result in predicted outcomes?** When the expected result did not occur, was the interviewer able to flex intentionally and use a different skill?

Portfolio of Competence

You will find that a lifetime can be spent increasing one's understanding and competence in the ideas and skills from this chapter. You are asked here to learn and perhaps even master the basic ideas of predictability from skill usage in the session, several empathic concepts, and the five stages of the well-formed interview. We have learned student mastery of these concepts is indeed possible, but for most of us (including Allen and Mary) we find that reaching beginning competence levels makes us aware that we face a lifetime of practice and learning.

You should feel good if you can conduct an interview using only listening skills. Focus on that accomplishment and use it as a building block toward the future. As you do, you are even better prepared for developing your own style and theory.

Use the following as a checklist to evaluate your present level of mastery. Check those dimensions that you currently feel able to do. Those that remain unchecked can serve as future goals. *Do not expect to attain intentional competence on every dimension as you work through this book.* You will find, however, that you will improve your competencies with repetition and practice.

Highlight competencies that you have met to date:

Level 1: Identification and classification.
❏ Ability to identify and classify the microskills of listening.
❏ Ability to identify and define empathy and its accompanying dimensions.
❏ Ability to identify and classify the five stages of the structure of the interview.
❏ Ability to discuss, in a preliminary fashion, issues in diversity that occur in relation to these ideas.

Level 2: Basic competence.
Aim for this level of competence before moving on to the next skill area.

❏ Ability to use the microskills of listening in a real or role-played interview.
❏ Ability to demonstrate the empathic dimensions in a real or role-played interview.
❏ Ability to demonstrate five dimensions of a well-formed interview in a real or role-played session.

Level 3: Intentional competence.
Ask yourself the following questions, all related to predictability and evaluation of the effectiveness of your abilities in working with emotion. These are skill levels that may take some time to achieve. Be patient with yourself as you gain mastery and understanding.

❏ Ability to produce anticipated results in clients as a result of the use of the listening microskills, according to predicted results of the Ivey Taxonomy.
❏ Ability to facilitate client comfort, ease, and emotional expression as a result of your ability to be empathic.
❏ Ability to enable clients to reach the objectives of the five-stage interview process (specifically—(1) develop rapport and feel that the interview is structured; (2) share data about the concern and also positive strengths that might be used to facilitate problem resolution; (3) identify and perhaps even change the goals

of the interview; (4) work toward problem resolution; and (5) actually generalize ideas from the interview to their daily lives).

Level 4: Teaching competence.   Teaching competence in these skills is best planned for a later time, but those who might run meetings or need to do systematic planning can profit from learning the five-stage interview process. It can serve as a checklist to ensure that all-important points are covered in a meeting or planning session.

❑   Ability to teach clients the five stages of the interview.
❑   Ability to teach small groups this skill.

DETERMINING YOUR OWN STYLE AND THEORY: CRITICAL SELF-REFLECTION ON INTEGRATING LISTENING SKILLS

As indicated earlier, we are now at the stage when you can truly start initiating your own construction of the interviewing process. You certainly cannot be expected to agree with everything we say. You likely have found that some skills work better for you than others. You are finding it very likely that your values and history deeply affect the way that you conduct an interview. Some of this past you'd like to keep and, most likely, some you might like to change.

We encourage you to look back on these first eight chapters as you consider the following basic questions leading toward your own style and theory.

What single idea stood out for you among all those presented in these first eight chapters, in class, or through informal learning? Allow yourself time to really think through the one key idea or concept—and it well may be something that you discovered yourself. What stands out for you is likely to be important as a guide toward your next steps.

Also, what specific points in this chapter struck you as important? How might you use ideas in this chapter to begin the process of establishing your own style and theory? What are your present thoughts about diversity as part of the interviewing process?

Continue your development of your own style and theory below.

_____

_____

_____

_____

_____

_____

REFERENCES

Blair, R. (2003). Neurocognitive models of aggression, the antisocial personality disorders, and psychopathy. *Journal of Neurology, Neurosurgery, and Psychiatry, 71,* 727–731.

Blanchard K., & Johnson, S. (1981). *The one-minute manager.* San Diego, CA: Blanchard-Johnson.

Carkhuff, R. (2000). *The art of helping in the 21st century.* Amherst, MA: HRD Press.

Cheatham, H., & Stewart, J. (1990). *Black families: Interdisciplinary perspectives.* New Brunswick, NJ: Transaction.

Daniels, T. (2007). A review of research on microcounseling: 1967–present. In A. Ivey & M. Ivey, *Intentional interviewing and counseling: Your interactive resource* (CD-ROM) (3rd ed.). Pacific Grove: CA: Brooks/Cole.

Decety, J., & Jackson, P. (2004) The functional architecture of human empathy. *Behavioral and Cognitive Neuroscience Reviews, 3,* 71–100.

Egan, G. (2002). *The skilled helper* (6th ed.). Pacific Grove, CA: Brooks/Cole.

Hill, C., & O'Brien, K. (1999). *Helping skills.* Washington, DC: American Psychological Association.

Kim, B., Hill, C., Gelso, C., Goates, M., Asay, P., & Harbin, J. (2003). Counselor self-disclosure: East Asian American client adherence to Asian cultural values, and counseling process. *Journal of Counseling Psychology, 50,* 324–332.

Ivey, A., D'Andrea, M., Ivey, M., & Simek-Morgan, L. (2002). *Theories of counseling and psychotherapy: A multicultural perspective.* Boston: Allyn & Bacon.

Ivey, A., Ivey, M., Myers, J., & Sweeney, T. (2005). *Developmental counseling and therapy: Promoting wellness over the lifespan.* Boston: Lahaska/Houghton-Mifflin.

Ivey, A., Pedersen, P., & Ivey, M. (2001). *Intentional group counseling: A microskills approach.* Pacific Grove, CA: Brooks/Cole.

Ivey, A., & Matthews, W. (1984). A meta-model for structuring the clinical interview. *Journal of Counseling and Development, 63,* 237–243.

Rogers, C. (1957). The necessary and sufficient conditions of therapeutic personality change. *Journal of Consulting Psychology, 21,* 95–103.

Rogers, C. (1961). *On becoming a person.* Boston: Houghton Mifflin.

Singer, T., Seymour, B., O'Dougherty, J., Kaube, H., Dolan, R., & Frith, C. (2004). Empathy for pain involves the affective but not sensory components of pain. *Science, 303,* 1157–1161.

Van Velsor, P. (2004). Revisiting basic counseling skills with children. *Journal of Counseling and Development, 82,* 313–318.

Zhan-Waxler, C., Radke-Yarrow, M., Wagner, E., & Chapman, J. (1992). Development of concern for others. *Developmental Psychology, 28,* 128–136.

Helping Clients Generate New Stories That Lead to Action: Influencing Skills and Strategies

Story—positive asset—restory—action is the model of interviewing and counseling suggested in this book. The listening skills and the five-stage interview structure are oriented to a helping style that allows clients to share their thoughts, feelings, and behaviors. Our first task is to hear the story and our second is to find clients' positive strengths to help these individuals generate new stories. You may have noticed that the five-stage interview structure provides a system that for many clients will be complete in itself—simply telling the story in a positive, supportive atmosphere is often sufficient for developmental change to occur.

There are many ways to view a client's story, and the influencing skills and strategies of Chapters 9–12 will suggest ways in which you can actively intervene to facilitate the process of change. The ideas presented in Section III are most useful for discovering new strategies that lead to restorying, the generation of a more workable story. In turn, the new insights of the new story or stories often lead more effectively to individual change and personal growth.

This section begins, in Chapter 9, with an examination of the skill of confrontation, which some consider the most important agent of change in the interview. Confrontation builds on your present ability to observe discrepancies in the client, but adds an attempt to facilitate your client's resolving those discrepancies with new thoughts and behaviors. This resolution of discrepancies results in client developmental change, which can be described as the aim of effective interviewing, counseling, and psychotherapy. Following an effective confrontation, you will find it possible to assess your client's developmental change and progress.

Chapter 10, on focusing, extends the concept of confrontation and illustrates how to ensure that you have made a comprehensive examination of your client's story. Too many counselors and interviewers focus narrowly on their clients' problems and miss contextual issues such as family, work, and cultural challenges.

Reflection of meaning is presented in Chapter 11. There you will examine the relationship between and among behaviors, thoughts, feelings, and their underlying meaning structure. You will find this skill complex and rich; it will provide you with a deeper understanding of each client's issues and history.

Six skills and strategies of interpersonal influence are discussed in Chapter 12. Skills such as directives, logical consequences, feedback, and reframe/interpretation are explored with specific suggestions for facilitating client restorying and action.

The sum and substance of this section is to suggest ways you can take a more active stance with clients, helping them to move on by adding your own knowledge and input. The listening and observing skills of the first section provide the foundation for these more action-oriented skills and strategies.

As you develop competence in the skills of this section, you may accomplish the following:

1. Master the art of confrontation and the ability to assess your client's developmental change in response to your interventions.
2. Demonstrate the ability to change focus in the interview and to facilitate client exploration of the full complexities of the story.
3. Use the skill of reflection of meaning to help clients move to deeper levels of self-exploration and self-understanding.
4. Use an array of influencing skills and strategies to assist client developmental progress, particularly when the more reflective listening skills fail to produce change and understanding.

If you are a beginning counseling and interviewing student, do not expect to achieve intentional competence in all the concepts of this section in your first course. Confrontation, for example, is a complex skill—one on which even the most experienced counselor or therapist can always improve. Furthermore, assessing developmental-level responses to your confrontation is a concept new to the helping field and is likely to be challenging even for experienced interviewers.

Thus, it is suggested that you stress the first two levels of practice—identification and basic competence. With this solid base, you can gradually move to the predictability associated with intentional competence. The effective interviewer is always in process—growing and changing in response to new challenges.

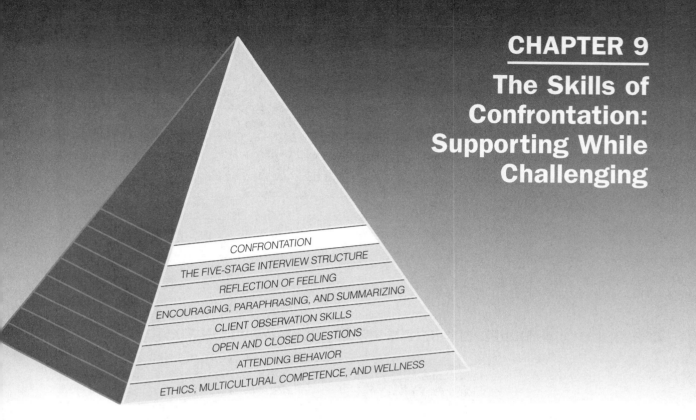

CHAPTER 9
The Skills of Confrontation: Supporting While Challenging

The pyramid from bottom to top:

ETHICS, MULTICULTURAL COMPETENCE, AND WELLNESS
ATTENDING BEHAVIOR
OPEN AND CLOSED QUESTIONS
CLIENT OBSERVATION SKILLS
ENCOURAGING, PARAPHRASING, AND SUMMARIZING
REFLECTION OF FEELING
THE FIVE-STAGE INTERVIEW STRUCTURE
CONFRONTATION

How can confrontation help you and your clients?

Major function	Although all counseling skills are concerned with facilitating change, it is the confrontation of discrepancies that acts as a lever for *the activation of human potential.* Most clients come to an interview seeking some sort of movement or change in their lives. Yet, at times, they may resist your efforts to help bring about the very transformation they seek. Your task is to help them move beyond their issues and problems to realize their full potential as human beings. An understanding of confrontation is basic to helping clients restory their lives.
Secondary functions	Knowledge and skill in confrontation result in the following: ▲ Increased ability to identify incongruity, discrepancies, or mixed messages in behavior, thought, feelings, or meanings. ▲ Ability to increase client talk with a view toward explanation and/or resolution of conflict and discrepancies. ▲ Ability to identify client change processes occurring during the interview and throughout the treatment period, using confrontation or other skills. ▲ Ability to utilize confrontation skills and the five-stage interview as part of mediation and conflict resolution.

Chris: I'm really between a rock and hard place. Here I am, ready to go to grad school, but my dad just had a heart attack and won't be back at work any time soon. He says that it doesn't make any difference and I should go. My mom insists that I stay home and take over Dad's garage. I've been planning and saving for school for two years and now everything might totally fall apart. But I'd feel so guilty if I didn't stay here. What should I do? I've been sitting on this for 2 weeks.

Chris is obviously stuck over this decision. How would you respond?

_____

_____

_____

_____

You can compare your responses with ours on page 291 at the end of this chapter.

INTRODUCTION: HELPING CLIENTS MOVE FROM INACTION TO ACTION

Cultural intentionality is not just a goal for interviewers; it is also a goal for clients. A client comes to an interview "stuck"—having either no alternatives for solving a problem or a limited range of possibilities. The task of the interviewer is to eliminate stuckness and substitute intentionality. *Stuckness* is an inelegant but highly descriptive term coined by Fritz Perls of Gestalt therapy in 1969 to describe the opposite of intentionality. Other words that represent the same condition include *immobility, blocks, repetition compulsion, inability to achieve goals, lack of understanding, limited behavioral repertoire, limited life script, impasse,* and *lack of motivation.* Stuckness may also be defined as an inability to reconcile discrepancies and incongruity. In short, clients often come to the interview because they are stuck for a variety of reasons and seek intentionality—they need a new story.

Our clients often become stuck because of internal and/or external conflict. Internal conflict or incongruity occurs when they have difficulty making important decisions, they feel confusion or sadness, or they have mixed feelings and thoughts about themselves. External conflict could be with others (friends, family, co-workers, or employers) or with a difficult situation such as coping with failure, living effectively with success, or having difficulty in achieving an important goal financially or socially. External conflict inevitably leads to internal feelings of incongruity. And internal conflict can lead to conflict with others—for example, the inability to make a decision may also result in conflict with a significant other who needs to know what is going to happen.

Development as manifested in interviewing and counseling may be described as the resolution of conflict and incongruity, the working through of an impasse or developmental delay, and the process of transformation and change by which clients learn to manage their own lives. A new way of looking at issues often involves a new story.

Later in this chapter, the process of change is discussed, as we draw from observations of clients working through loss, working their way out of alcoholism, and working through the change process in general. But first we consider the skills of confrontation.

Confrontation may be defined briefly as noting conflicts and discrepancies in clients and feeding these back by use of the attending skills (or magnifying and directing them via the influencing skills). You will read about a specific sequence for noting and working with stuckness, discrepancies, and incongruity.

Mediation and conflict resolution principles are presented in a box at the close of the chapter. In this box, listening skills, the five-stage structure of the interview, and confrontation skills are presented as key ingredients in helping people work through conflict.

INSTRUCTIONAL READING: CHALLENGING CLIENTS IN A SUPPORTIVE FASHION

Confrontation is one of the most powerful of the microskills, but it rests solidly on effective listening and observing. It is actually a complex of skills that often results in clients' examination of core issues. When client discrepancies, mixed messages, and conflicts are confronted skillfully and nonjudgmentally, clients are encouraged to talk in more detail and to resolve their problems and issues. Confrontation can be defined in this way:

> Confrontation is *not* a direct, harsh challenge. Think of it, rather, as a more gentle skill that involves listening to the client carefully and respectfully and then seeking to help the client examine self or situation more fully. Confrontation is not "going against" the client; it is "going with" the client, seeking clarification and the possibility of a new resolution of difficulties. Think of confrontation as a *supportive challenge*.

Confrontation involves three major steps. The first—identifying conflict via mixed messages, discrepancies, and incongruity—has been discussed in Chapter 5 on observation, but is repeated here for emphasis.

The second step is pointing out these issues clearly to clients and helping them work through the conflict to resolution. Therefore, the skills of questioning, observation, reflective listening, and feedback loom large in effective confrontation. It is here that you demonstrate your support.

The final step of confrontation is evaluating the effectiveness of your intervention on client change and growth. Confrontational effectiveness can be measured by placing the client's response (denial, acceptance, and so on) along the continuum of the Confrontation Impact Scale. If your confrontation is not accepted by the client, you will need to move to other skills such as focusing, reflection of meaning, and the influencing skills and strategies discussed in later chapters. All of these influencing skills, however, also have the underlying goal of facilitating client development through the resolution of incongruities.

The following interview conversation has been abbreviated and edited. The brief session is designed to illustrate the fundamentals of confrontation—supporting while challenging.

Box 9-1   Research Evidence That You Can Use: Confront, but Also Support

Attending and listening skills are used frequently in the session, but you'll find that skills such as confrontation and influencing are used only occasionally. A review of research found that confrontations account for only 1% to 5% of interviewer statements (Hill & O'Brien, 1999). The reviewers noted that confrontations are useful, but that they also often make clients uncomfortable. Especially important, they noted that clients frequently become defensive and may not deal fully with feelings and issues following a confrontation. They also note that confrontations that are intentionally modified to meet individual needs were more effective. Our point of *supportive challenge* becomes all the more important.

Counselor eye contact affected client perception of rapport in the session. Specifically less direct eye contact early in the session when discussing sensitive matters was helpful, and at this point the clients appreciated a nonconfrontational approach. As the interview progressed, more eye contact and more confrontation were acceptable (Sharpley & Sagris, 1995).

The original microskills work that led to the specific delineation of confrontation as the identification and feedback of discrepancies goes back to Ivey (1973), when he videotaped hospitalized psychiatric patients. Much of this work involved helping patients see their verbal and nonverbal discrepancies and was a very powerful confrontation strategy. But this confrontation had to be within a supportive, empathic, and trusting atmosphere, or it simply made things worse. The first task in confrontation was and remains to establish a solid base of relationship, best demonstrated by your ability to listen.

Thus, when you challenge clients and point out their discrepancies, your ability to support them, listen to them, and be empathic are central. In effect, listen before you leap! If you have a good relationship and have established trust, you may expect your confrontations to be useful. After the confrontation (or supportive challenge) is offered, it is vital that you observe client reactions, then encourage clients to examine their feelings about the challenge, always within an atmosphere of listening and support.

Confrontation and neuropsychology

Interviewing and counseling are very much concerned with helping clients create new ways to think and behave, as well as developing new perspectives on emotions. What some call "creativity" appears to be located in the connections between the holistic right brain and the linear left brain as well as the participation of the mainly unconscious limbic system (Carter, 1999). Research suggests that new learning (neuroplasticity) occurs when the two hemispheres synchronize their activity (Goodwin & Sherrard, 2005). A confrontation that points out incongruities in a person's life is, by definition, an effort to open the person for a new way of thinking. Gentle and supportive confrontations often can reach underlying emotional structures as the empathic atmosphere provides the setting for creative new learning.

Step 1: Identify the conflict; note mixed messages and incongruity.

Client: (the body language shows excitement) I've found this great friend on the Internet. He sounds wonderful and we're doing e-mail at least four times a day. It feels great. I think I'd like to meet him.

Instances of incongruity or discrepancies include desire and excitement about meeting the Internet friend, but with internal anxiety and hesitation: the inevitable external conflict of being involved with

(Her body language becomes more hesitant and she breaks eye contact.) But it means I may have to go out of town. I wonder what my partner would think if he found out. It makes me a bit anxious, but I really want to meet this guy.	two people at once. Which discrepancy might you discuss first? In addition, clients who discuss these mixed feelings and conflicts usually also show them in their body.
Counselor: You really want to meet him. On the other hand, you're a bit anxious.	This paraphrase and reflection of feeling confront the mixed feelings in the client. This catches both verbal and nonverbal observations.
Client: Yes, but what would happen if my partner found out? It scares me. I've got so much involved with him over the past 2 years. But, wow, this guy on the Internet . . .	The client responds, but turns her focus to the discrepancy between her and the partner. There are clearly two issues at least in this situation.

Step 2: Point out issues of incongruity and work to resolve them.

Counselor: Could I review where we've been this far? I know you have been having some difficulties with your partner and you've detailed them over the last two sessions. I also hear that you want to work things out despite your present anger at him. You have a lot of positive history together that you'd hate to give up. But on the other hand, you've found this man on the Internet and he doesn't live that far away. You seem really excited about the possibility. In the middle of all this, I sense you feel pretty conflicted. Have I got the issues right?	This summary indicates the counselor has been listening. Both verbally and nonverbally, the counselor communicates respect and a nonjudgmental attitude. We support by listening and searching for strengths (in this case, the positive history). The counselor summarizes the major discrepancies that led to internal and external conflict and checks out with the client to see if the listening has been accurate.
Client: Yes, I think you've got it. As I hear you, it makes me think that I've got to work a bit harder on the present relationship, but—wow—I sure would like to meet that guy.	Through having her thoughts and feelings said back to her, the client starts some movement. Resolution of conflict and discrepancy best occurs after the situation is understood fully.

The conversation continues over the next 10 minutes and the client's thoughts and feelings evolve to a new perspective.

Client: (said with conviction) It's beginning to make sense to me. I've got so much time invested in my partner and I've really got to try harder. (Her nonverbals again show hesitancy.) But, how am I going to work this out with my Internet friend?	The conflict is moving and the client is starting to show evidence of new ways of thinking that weren't there in the first two sessions. Nonetheless, conflict remains.
Counselor: (solid supportive body language and vocal tone) It looks like you really want to work it out with your partner. You sounded and looked very sure of yourself. But let us explore a bit more what the possibilities are with your Internet friend.	You can confront and help clients face discrepancies, incongruity, and conflict if you are able to listen and be fully supportive.

Step 3: Evaluate the change.
The effectiveness of a confrontation is measured by how the client responds. In the example above, the client kept moving toward a new resolution. However, there are many clients who stay stuck in the same place and repeat the same discussion of their concerns over and over again. Later in this chapter, the Confrontation Impact Scale (CIS) is presented. This is a systematic way to evaluate the effectiveness of confrontations and whether clients are moving to new ways of thinking, feeling, and behaving.

With this brief introduction, let us turn to a more detailed examination of supporting while challenging.

Step 1: Identify Conflict via Mixed Messages, Discrepancies, and Incongruity

Chapter 5, on observation skills, discussed observing discrepancies. Your ability to observe incongruities and mixed messages in the interview is fundamental to effective confrontation. (The words *discrepancy, incongruity,* and *mixed message* are used interchangeably in this chapter. In addition, conflict is often central to the issue.) The major types of discrepancies are summarized briefly below and you are encouraged to return to Chapter 5 to review the observation concepts. Also, recall the importance of listening to support clients.

What Are the Internal Conflicts?

Discrepancies internal to the client include mixed messages in nonverbal behavior, incongruities in verbal statements, discrepancies between what the person says and what he or she does, and discrepancies between statements and nonverbal behavior. The client discussing relationship difficulties with the partner in the preceding section might also present a mixed message in nonverbal behavior, such as smiling inappropriately while talking about her mixed feelings. The client's statements balancing her present partner against the excitement of the new Internet friend represent an internal conflict. The client has discrepancies within herself that need to be resolved so that she can make a decision.

What Are the External Conflicts?

Discrepancies between the client and the external world highlight conflict between people and between clients and the situation in which they find themselves. Much of your counseling and interviewing work will focus on discrepancies that clients have with their external world. In the example of the client with the conflict regarding the Internet decision, she likely has other conflicts with her partner. The external conflict with the partner, of course, leads to internal conflict. Thus, we need to examine discrepancies between the client and the partner as part of the counseling process. The basic listening sequence will be important in drawing out the underlying nature of the relationship conflict. Beyond that, the client may have difficulties at work, problems with her parents, or other situational issues that relate to external discrepancies.

Discrepancies between you and the client can be challenging. Interviewers and counselors often like to avoid disagreement and may hide differences they have with clients through reflective listening skills. In fact, if you listen long and carefully, you will find most discrepancies between you and clients disappearing as you understand how they came to think and behave as they do. So if you sense difference between you and your client, first support the client by listening. The first task is simply to notice your own or the client's discomfort with possible differences and work to understand them by internally questioning yourself silently and then carefully drawing out from the client how he or she experiences the situation.

Resistance can be an important dimension of incongruity between the interviewer and the client. Some theorists talk of "breaking down" client resistance and see client hesitancy as inappropriate. Our view is that many clients need to defend themselves from what they see as external threat, and simply sharing important information can be frightening to some clients. In this sense, so-called resistance is actually a good thing for the client. It protects the internal self from dealing with obvious contradiction. See resistance as an opportunity, not as a problem.

However, the protection offered by resistance is often faulty. Your task is to respect client resistance, listen to it, and learn how it helps the client. Once you have observed resistance and listened for the reasons it is there, you are better prepared for confronting the client in a respectful, supportive fashion.

Step 2: Point Out Issues of Incongruity and Work to Resolve Them

As noted earlier, simply labeling the incongruity through a nonjudgmental confrontation may be enough to resolve a situation. More likely, however, incongruity will remain a problem to be resolved. Remember the importance of focusing on the elements of incongruity rather than on the person. Confrontation is too often thought of as blaming a person for her or his faults; rather, the issue is facing the incongruity squarely through such measures as the following:

1. Identify the incongruity or conflict in the story or comment clearly. Using reflective listening skills, summarize it for the client. Often the simple question "How do you put these two together?" will lead a client to self-confrontation and resolution.

2. Through the use of questioning and other listening skills, draw out the specifics of the conflict or mixed messages. One at a time, give attention to each part of the mixed message, contradiction, or conflict. If two people are involved, attempt to have the client examine both points of view. It is important at this stage to be nonjudgmental and nonevaluative—aim for facts. Allow your nonjudgmental stance to be reflected in your tone of voice and body language.

3. Periodically summarize the several dimensions of the incongruity. The model confrontation statement "On the one hand . . . , but on the other hand . . ." appears to be particularly useful in summarizing incongruity. Variations include "You say . . . but you do . . . ," "I see . . . at one time, and at another time I see . . . ," and "Your words say . . . , but your actions say. . . ." Follow this with a check-out (for example, "How does that sound to you?"). When you point out incongruities as in these examples, the client is confronted with facts. As a result, the importance of being nonjudgmental and including the check-out as a final part of the confrontation cannot be overstressed.

4. The positive asset search with special attention to wellness issues can be a very helpful means to facilitate client change and development. When we challenge

clients, we often place them off balance. One of the values of confronting is indeed disturbing client complacency and comfort. We all seek our "center," and effective confrontation builds new strengths in the client. You may simply comment that the client is describing the problem clearly—for example, "I like the way you are dealing with some very real challenges." Your personal support can help clients build new behaviors, thoughts, and meanings. You will find that many clients are not comfortable with confrontive and challenging approaches. Then it becomes important to facilitate their development through helping them identify strengths and wellness qualities in themselves and their relationships. It will often be wise to take time out from confrontation to develop strengths. Then when clients are aware of their wellness strengths and positive qualities, facing difficult confrontations becomes possible.

5. If necessary, provide feedback giving your opinions and observations about the discrepancies. You may wish to use influencing skills such as directives, logical consequences, and others (see Chapter 12) to facilitate resolution.

If the incongruity is not resolved by this process, it may be necessary to say, "You see it that way, the other person sees it another way. We'll have to go at it again." Don't give up on positions you believe are correct, but allow your point of view to be modified by input from the client. Many clients are unaware of their mixed messages and discrepancies; pointing these out gently but firmly can be very beneficial to them. Finally, a wide variety of attending and influencing skills may be used to follow up and elaborate on confrontations.

Individual and Multicultural Cautions

Confrontation of discrepancies can be highly challenging to any client, but particularly to one who is culturally different from you. The confrontation process may be made acceptable if the helper takes time to establish a solid relationship of trust and rapport before engaging in confrontation. If you have a fragile client or if the relationship is not solid, confrontation skills need to be used with sensitivity, ethics, and care.

You will find that direct, aggressive confrontations are not necessary if the client contradiction is stated kindly and with a sense of warmth and caring. Direct, blunt confrontations are likely to be especially culturally inappropriate for Asian, Latina/Latino, and Native American clients. But even here, if good rapport and understanding exist between you and the client, confrontations can be most helpful.

There are several ways to present a confrontation that will help you meet varying individual and cultural needs. The presentation in this chapter focuses on *supportive, empathic confrontation.* Here you clearly try to listen to the client's story and gently and respectfully summarize the differences, encouraging the client to generate her or his resolutions.

McMinn (1996) has outlined several approaches that may be useful in broadening your thinking beyond the basic empathic, supportive confrontation. *Silence* is a second approach to confrontation. Here you simply listen to client discussion and sit quietly while the client struggles with internal or external contradiction. This type of confrontation is especially challenging to clients who may ask your

Box 9-2 National and International Perspectives on Counseling Skills: Confrontation in the Real World—Children in Northern Ireland

Owen Hargie, Professor, University of Ulster, Northern Ireland

Conflict looms large in Northern Ireland, as it does in other international "hot spots" such as between Israel and the Middle East and between India and Pakistan. Conflicts around religion and race/ethnicity may appear in counseling sessions no matter where you work. I'd like to share an example from my country and discuss implications for counseling.

The context of Catholic-Protestant conflict

After centuries of often problematic relations between Great Britain and Ireland, the predominantly Catholic Republic (south) of Ireland was established as an independent nation in 1922. Simultaneously, Northern Ireland continued as part of Great Britain. Originally, the south was 90% Catholic and 10% Protestant, but over the years the Protestant minority has dropped to some 2%. Northern Ireland currently is deeply divided between 40% Catholics and 60% Protestants and in recent years was one of the most violent places in the world. The ongoing conflict, or "Troubles," has culminated in a mortality toll of over 3,700 people—the pro rata equivalent of 600,000 deaths in the U.S.A. Not surprisingly, this violence has impacted upon almost every aspect of the lives of the population. The two groups are almost totally segregated, and despite a current downturn in violence, indications are that they are separating further.

Not least of the problems in Northern Ireland has been that of diametrically opposite political aspirations. The Protestant/Unionist community wishes to remain part of Great Britain, while the Catholic/Nationalist community seeks unification with the Republic of Ireland. The entrenched divisions, emanating from these politico-religious differences, have led to overt physical combat.

What happens to children in this context?

The tension and violence is likely to continue, and even young children are indoctrinated into an ideology of hate and mistrust. Here 10-year-old David and Liam (both Catholic) report on a visit to a Protestant school. The visit was designed to promote understanding.

Interviewer:	Why did you not like the Protestant school?
David:	'Cos they're Protestants [they giggle and look at the interviewer in amazement as if it is obvious].
Liam:	'Cos they're slabbering [calling names] to us and all, and we were like messing about.
David:	They were slabbering to me so I . . .
Interviewer:	Why did they slabber at you?
Liam:	'Cos they don't like us.
David:	See the wee Protestant, ahm, I was beside, he said to me [whispers], "I'm going to block you out," and I said [leans forward as though he didn't hear him], "Wha?" and he was going to dig [hit] me.

Research has shown that children even as young as three have these attitudes. Hate clearly is engendered at an early age.

(continued)

Box 9-2 (continued)

Working toward understanding	The skills of constructive challenge/confrontation and mediation are obviously important parts of building toward better understanding and community. My experience and research suggest the following:

▲ There is a need for actual contact and communication between combatants. You can't establish a relationship through a closed door.

▲ This contact needs to be positive and rewarding to both sides of the conflict. A win-win attitude is needed as both parties expect a win-lose result.

▲ Intimacy and getting to know the other person as a person are essential. There is need to search for common goals and areas where they might agree.

▲ Power needs to be equal between the two individuals or groups.

▲ Conflict cannot easily be resolved alone. Two individuals or two groups need support from the larger community. Key people beyond the counselor, educator, or mediator need to support resolution. This is a situation where the person is truly a person-in-community.

With severe conflict such as this, I question the need or wisdom of drawing out the detailed story underlying the conflict. While I think in-depth sharing is important and vital in counseling, when we move to severe conflict we are going to have to focus on positive assets and strengths and use a wellness approach. Without extensive community and governmental support and increased contact among groups, we are going to see ever more "recreational rioting" in which young people riot because it becomes the stimulating and fun thing to do.

As counselors and therapists, we can become part of the solution to these complex issues by first learning how to listen effectively and then working to resolve contradictions and conflict in our individual client, family, and group work. With this experience, we can then move to the real challenges of broadly and deeply based misunderstandings. It is a large task, but it is one we must undertake.

Allen Ivey comments: We face many conflicts and contradictions in this world and religious violence is one of the most challenging. Here we see the intergenerational legacy of distrust and hate. If we were to meet the individuals behind the attitudes discussed above, most of them would come across as kind and generous, similar to people on our own street. But, when it comes to certain topics, these same people become very different. I like to think of counseling as a mini–peace process. Through listening and finding wellness strengths and positive assets, we can help people find new stories and ways of being that are positive for themselves and others. Carl Rogers used his person-centered theory and skills in an effort to facilitate communication in Northern Ireland. Let us use his example to remind us that we have a larger responsibility than just individual clients.

opinion—"Don't you agree with me that my partner is wrong?" If you say nothing, the client will have to encounter your silence and clearly become aware that the answer lies within. But be aware that some could interpret your silence as disapproval. At the same time we should also be aware that silence is a value within much of the Native American, Dene, or Inuit tradition. You may sit with these clients at times for several minutes as they sort out issues. How comfortable are you with silence? Each mode of confrontation must be personally authentic and meaningful for the client or it is likely to fail.

Questioning and elaboration can be a useful approach to confrontation. Rather than challenging the client immediately, select various dimensions of the discrepancy or conflict and sort out the stories carefully. For example, "Tell me more about your feelings when you come home from work so tired. What's going on for you when you walk in the door?" After having heard that story and the accompanying emotions, ask about the partner: "What is your partner's typical story of one day like? What thoughts and feelings do you notice in her or him?" In these situations, you are helping the client to explore the discrepancy, and new alternatives for resolution may result.

It may be helpful to take "time out" from confronting discrepancies and focus on positive stories and wellness assets. For example, if clients are encouraged to talk about past successes, they may then generate the emotional courage to look at more difficult issues. Or, in working with relationship issues, a particularly useful method is to have the individual or couple go back to when their connection began—ask them to tell you a story of what brought them together, or to tell you about the last time things were going well between them. Working with the two of them in a wellness assessment can be a useful strategy.

Direct challenge can be a powerful and helpful confrontive dimension. Many of the influencing skills of Chapter 12 confront the client with a new frame of reference in which issues can be considered. The direct challenge may take the form of sharing your own thoughts with the client ("It might be helpful for your relationship if you considered how your partner experiences those first few moments when you both arrive home from work. Homecoming, or the end of the day, is often difficult for tired, busy couples."). Skills such as logical consequences (what happens if you don't change), the interpretation/reframe (telling a new story about the situation), and specific directives (you may want to consider couple counseling) are all direct challenges to the status quo.

Not confronting is one final issue for you to consider. With a forward-moving client, simply listening to the story may be sufficient. Pushing for change at times can get in the client's way. And sometimes life is such that we must accept and learn to live with discrepancies that simply cannot be resolved.

Step 3: Evaluate the Change

In North American culture, change is considered in a positive light, and somehow we expect clients to actively seek and move toward change. Yet we often find clients who resist and sabotage change. In this section we review the change process as originally defined by Kübler-Ross (1969) in her well-known, classic work on death and dying. Then we expand her model and use it as a general framework for examining change in our clients. Finally, we look at a scale for measuring change in the interview—the Confrontation Impact Scale.

In short, Step 3 of the confrontation process asks you to observe client thinking and behaving. You will find that you can (1) determine where your client is functioning in terms of change at any time in the interview and (2) discover how effective your interventions have been, thus providing you with important feedback for helping think through further actions on your part.

Death and Dying Theory and Concepts of Change

A general model of change processes can be important when thinking about the interview. Early work in the area of death and dying provides us with a background before we move on to practical implications for the here and now of the interview. Five stages of change in thinking patterns are important.

When individuals learn that they have a terminal disease, they seem to go through a natural process of change in attitudes and emotions about death and dying. Kübler-Ross (1969) found that at first many people often deny the reality of approaching death. Furthermore, she observed a general developmental process in individuals and families as they came to terms with dying.

Individuals (and their families) differ markedly in their reactions to the inevitability of death. Some cancer patients, for example, may deny that they are going to die even until the final moments. Other patients may first start with denial and then move gradually to acceptance and finally to a peaceful and transcendent death. Death and dying theory emphasizes respecting each person's approach to the issue. There is no one "correct" way.

You will find that many of your clients who are confronting issues of change will work through levels of denial, acceptance, and positive change in a fashion similar to that first described by Kübler-Ross. For example, an alcoholic may first react with denial to any helper questions aimed at acknowledging a drinking problem. Then through effective confrontation, that person may move to greater openness and awareness. A person who abuses others may first deny that possibility and then later move to awareness and change.

The following five levels of attitudes, beliefs, and behaviors related to death and dying are an adaptation of Kübler-Ross's observations (Ivey, 2000/1986). Her ideas may be related to a general model of developmental change, which will be presented later in this chapter. Five stages are listed below, but each person will have a unique approach to grieving. For example, some may start with acceptance, then move to bargaining or anger.

1. *Denial.* The patient cannot accept that he or she will die and denies this reality. "It won't happen—not to me!" "The lab tests must be wrong." With some patients, this form of denial can continue until the final moments. Family members may respond in similar ways and talk in detail about a trip planned for the next year or about what they are going to do for the holidays. The inevitable is ignored.

2. *Partial acceptance of reality: bargaining and anger.* Kübler-Ross talks about "bargaining," in which the patient, family, or both engage in what is almost magical thinking. They may say, "If I lead a better life, then God will let me live." Prayers may involve promises and pleas. Family members may reconcile old differences in the hope that the individual may live.

   Another type of partial acceptance is the anger stage, in which the affected person may become angry at God ("Why me? I've lived a good life") or at the overall unfairness of it all ("It isn't right"). Anger or other less effective emotions may be expressed in the family. This is a marked move from denial, but the anger actually masks deeper emotions of sadness and fear. You will often find that the anger emotion is a cover-up for more basic issues, not only in death and dying, but in other situations as well.

3. *Acceptance and recognition.* The fact of dying is acknowledged by the individual and the family and along with this comes a marked shift of emotion. The underlying emotions of sadness and fear surface, and grief reactions are manifest. The fact of loss is faced rationally but with appropriate emotions. There may be considerable depression at this stage, which itself can be a major counseling problem. But depression and sadness are also natural responses to loss and may be part of healthy grieving.

You have likely talked with a number of friends and family members who have experienced major losses, and perhaps you can think of your own reactions to the death of a loved one. Kübler-Ross points out that some people go to their death in a constant state of denial, bargaining, or anger, and never reach the stage of acceptance. Each of us has our own way of dealing with difficult issues, and it is important to understand and acknowledge that each person you deal with will not necessarily work through the stages. Your task as an interviewer or counselor is to be with the client wherever he or she is.

It is also possible to move beyond acceptance to other levels of cognition and emotion, both involving change and moving beyond the fact of death.

4. *Generation of a new solution—early transcendence.* Here death may be reframed as a new developmental challenge to be surmounted, as an opportunity to forgive and forget old family arguments, or as an opportunity to meet God and loved ones in heaven. The patient may decide to donate organs to someone else. A new and potent meaning has been given to death that will allow both acceptance and some degree of transcendence.

5. *Development of new, larger, and more inclusive constructs, patterns, or behaviors—transcendence.* In some cases, transcendence becomes the predominant mode of being. Sadness around loss of the joy of living continues to be part of the person's being, but the individual becomes peaceful and serene. Several of the new solutions mentioned in number 4 may be implemented, and the person may appear to have made a major change in consciousness.

There is value in all five levels, but most feel that Levels 3, 4, and 5 are preferable, as they provide the individual with an opportunity to review his or her life, say good-bye to friends and family, and achieve peace. Again, recall that people do not necessarily move through the five stages in any special order. Even denial has value—at times the best way to deal with an impossible situation is to ignore it.

Now let us review Kübler-Ross's five stages from the storytelling framework. At the first stage of denial, we often hear the story, but it may be one that fails to deal with reality. Soon the individual is forced to confront the reality of the death, and bargaining and partial acceptance occur—the story is changing. At Level 3, acceptance, the reality of the story is acknowledged and here storytelling is often most accurate and complete. However, it is possible to move to Levels 4 and 5, which involve new solutions and transcendence. Now the client is active in restorying. Moreover, if changes in behaviors and cognitions are sufficient, we see the new story moved into action. For example, a family who loses a child starts a support group for other parents who have experienced a similar event.

The following section takes these ideas of developmental change and applies them in a more general fashion to the helping interview. You will find that your client stories often change dramatically as clients confront the reality of their situations. It is an honor, a privilege, and a major responsibility to work with clients as they accept the necessity of confronting discrepancies and move to restory their situations and/or their lives.

The movement through denial to bargaining and finally to change seems to be a general model of change in consciousness and thinking. You will find that those who experience divorce, who must face alcohol and drug problems, or who may confront a major change such as a move, a new job, or any of a host of issues will have reactions to the change process much like those described by Kübler-Ross.

Five Levels of Change You May Encounter in the Interview

Clients often are highly resistant to change. They may come seeking a difference in their lives, but when confronted with the incongruity of their behavior, thoughts, or emotions in their stories, they may deny your confrontation. Nowhere does this denial show more prominently than in work with alcoholics. It is well known that simply breaking through denial is the foundation of effective treatment, and acknowledging alcoholism is the first of Alcoholics Anonymous's 12 steps.

Having broken through denial, many clients may bargain with you; they may become angry and defensive or present some other ineffective or inappropriate emotion. They may attack you personally and show anger. Clients at the partial acceptance stage may have moved beyond denial, but they are not always easy to work with.

Acceptance and acknowledgment of a problem, the third level, is where the change process can begin. For some clients, simply accepting reality is enough, but with alcoholics and many other clients it is possible to acknowledge and accept a problem *and to do nothing about it.* Any number of alcoholics know that they have a drinking problem, but they continue to drink. Similarly, you will find that many clients acknowledge that they need to start looking for a job, spend more quality time with the family, or stop destructive eating patterns, but they continue to do nothing about their problems.

Real change occurs at Level 4, when the client actually stops drinking, looks for a job, takes time for children, or starts eating properly. A new story leading to a solution has been generated. The solution is often related to the specific issue that brought the client into counseling.

Level 5 change, the development of new, larger, and more inclusive constructs, patterns, or behaviors, occurs when the alcoholic not only stops drinking but takes on a new way of life, well described at the later levels of 12-step programs. The individual who once would not look for a job not only finds a job but develops new skills and ways of relating. The person with a family problem or eating problem not only solves the problem but is doing better on the job and with friends as well. People may comment on a major personality change. Restorying and action are present.

You will find that virtually any problem a client presents to you may be assessed at one of the five levels. If your client starts with you at Level 1 or Level 2 and then moves with your help to Level 3 or Level 4, you have clear evidence of the effectiveness of your interviewing process. The five levels may be seen as a general way to view the change process in interviewing, counseling, and therapy.

Using These Change Concepts in the Interview

You will find that the way clients respond to you in the interview itself can be assessed using the same concepts. When you confront clients, ask them a key question, or provide them with a powerful intervention, you will find they may have a variety of responses. Ideally, they will respond actively to your interventions, generate new ideas, and move forward. On the other end of the spectrum, they may ignore or deny the fact that you have challenged or confronted them. Most often, however, your intervention will be acknowledged and absorbed as part of a larger process of change.

It is possible to assess the direct impact of your confrontation or other microskill by using the Confrontation Impact Scale (CIS) presented in Box 9-3. This scale will give you a frame of reference for determining how well clients have responded to your confrontation, question, interpretation, or other skill.

Implications of the Confrontation Impact Scale

Virtually every helping lead you use, whether it's a question, reflection of feeling, or directive, may lead to client reactions that can be placed along this 5-point scale. You will find clients denying your question or feedback at times, whereas at other times they will recognize it and use it to move through thoughts, feelings, or behavior. When things are going well, you will find clients transforming their concepts into new ideas, thoughts, and plans for action in their daily lives: It is these Level 4 and Level 5 responses that you seek for clients. Although change at this level is often slower and more difficult than we would like, confrontation can speed and facilitate such progression.

During the process of counseling, you will find the Confrontation Impact Scale a helpful informal measuring tool to keep in mind, allowing you to track your client's reactions to you and your interventions. When you are off track, you will find clients responding at Level 2. When your clients seek to avoid your confrontation, you will likewise find them subtly avoiding the full issue. Your task is to recognize what is happening and to seek to keep the client at the minimum level of responding that allows acknowledgment of conflict and incongruity. Over time, clients will move toward Levels 4 and 5, but, depending on the issue, that level of change may be slow. For example, do not expect immediate change in a person who is an alcoholic or has experienced abuse. A movement to Level 2 or Level 3 for some clients is a real triumph. *Movement on the scale can occur in one interview or it may take a year or more to help a person move out of denial or the partial acceptance of bargaining and anger.*

Chapter 13 provides more examples of scoring with the CIS. A full interview there exemplifies what often happens: The client starts at Level 2 at the beginning

Box 9-3   The Confrontation Impact Scale*

When you work with clients individually, as couples, or in family situations, you will encounter conflict, which may be internal to one person or between two or more individuals. The general concepts of change presented here can be organized into a working Confrontation Impact Scale (CIS). You will find that your clients talk and think about change in various ways. Identifying where they are in the change process can help you identify the effectiveness of your interventions and help you plan with the clients for further exploration. The way a client deals with concerns such as examination anxiety, facing cancer or death, reviewing a serious life trauma, coping with racism or sexism, or making a critical vocational decision can all be considered using the CIS.

Example of the Confrontation Impact Scale

The effectiveness of a confrontation with a client about mixed messages concerning an impending divorce may be measured on a 5-point scale. The example here is of reactions to a divorce, but any time clients are working through change, you will find them talking about their issues with varying levels of awareness. Using this scale during the interview can be helpful in following client movement both during a single session and over a series of interviews.

1. *Denial.* The individual may deny that an incongruity or mixed message exists or fail to hear that it is there. ("I'm not angry about the divorce. These things happen. I do feel sad and hurt, but definitely not angry.")
2. *Partial examination.* The individual may work on a part of the discrepancy but fail to consider the other dimensions of the mixed message. ("Yes, I hurt and perhaps I should be angry, but I can't really feel it.")
3. *Acceptance and recognition, but no change.* The client may engage the confrontation fairly completely but make no resolution. Much of counseling operates at this level or at Level 2. Until the client can examine incongruity, stuckness, and mixed messages accurately, developmental change later will be most difficult. ("I guess I do have mixed feelings about it. I certainly hurt about the marriage. I hurt, but I'm really angry at what's been done to me.")
4. *Generation of a new solution.* The client moves beyond recognition of the incongruity and puts things together in a new and productive way. ("Yes, you've got it. I've been avoiding my deep feelings of anger and I think it's getting in my way. I sure hurt, but if I'm going to move on, I'll have to allow myself to feel that angry part of myself too.")
5. *Development of new, larger, and more inclusive constructs, patterns, or behaviors—transcendence.* A confrontation is most successful when the client recognizes the discrepancy, works on it, and generates new thought patterns or behaviors to cope with and perhaps resolve the incongruity. ("I like the plan we've worked out. You've helped me see that mixed feelings and thoughts are part of every relationship. I've been expecting too much, even before. If I expressed both my hurt and anger more effectively, perhaps I wouldn't be facing a divorce. I'm going to call my [spouse] and see if we can develop a new way of thinking about the meaning of the relationship.")

Ivey, A., Ivey, M., Myers, J., & Sweeney, T. (2005). *Developmental counseling and therapy: Promoting wellness over the lifespan.* Boston: Lahaska/Houghton Mifflin.
*A paper-and-pencil measure of the Confrontation Impact Scale was developed by Heesacker and Pritchard and was later replicated by Rigazio-Digilio (cited in Ivey et al., 2005). Factor analytic study of over 500 students and a second study of 1,200 revealed that the five CIS levels are identifiable and measurable.

of the session ("I'm not sure what I want to do for a career"); this is followed by extensive Level 3 examination when she faces her issues more directly; finally at the end she approaches Level 4 where she intends to take some beginning action.

The following Example Interview illustrates many of the points just covered.

EXAMPLE INTERVIEW: BALANCING FAMILY RESPONSIBILITIES

The excerpt that follows is designed to show you how confrontation can be used in the interview. This excerpt has been edited down from a longer interview and the conversation simplified to clarify the concepts described in this chapter. You may want to read the segment several times and study it carefully. At this point, aim for understanding. With experience and practice, these concepts will be useful and important to you in working with influencing skills and in your interviewing practice.

A lot happens in any counseling session. You'll find that's the case here, as well. Three main points are reviewed simultaneously in this interview analysis: (1) listening skills are used to obtain client data, (2) confrontations of client discrepancies are noted, and (3) the effectiveness of the confrontations is considered using the Confrontation Impact Scale.

The following interview presents a conflict that is common to many working couples—balancing home tasks. Male attitudes and behavior are changing, but we find that many working women are still burdened with the responsibility for most tasks at home. In this example we have both internal and external incongruity. Dominic is struggling internally with the discovery that things have changed and externally with the realization that his wife is behaving differently. Arguments in such situations may be particularly intense as two tired people come home from a hard day's work to face needy children and undone housework. Couples may blame each other rather than seeing their issues as caused externally.

Interviewer and Client Conversation	Process Comments
1. *Dominic:* I'm having a terrible time with my wife right now. She's working for the first time and we've been having lots of arguments. She isn't fixing meals like she used to and isn't watching over the kids. I don't know what to do.	On the five levels of the CIS, this client would be rated as Level 2 because he is partially aware of the problem. At the same time, he is denying his role in the issue. At Level 1, he might say, "Everything's fine" and be unaware of family issues.
2. *Ryan:* So, on one hand she's working now and has lots of things to do, but on the other hand, you expect her to continue with all the old things she did. You feel pretty confused right now about things.	Confrontation presented as paraphrase; reflection of feeling.
3. *Dominic:* You damn betcha she's expected to do what she's always done—I'm not confused about that.	CIS Level 1—denial.

(continued)

Interviewer and Client Conversation	Process Comments
4. *Ryan:* I see. You're not confused; you really don't like what's going on. Could you give me a specific example of what's happening—something that goes on between you when she gets home?	Paraphrase, open question oriented to concreteness.
5. *Dominic:* (sighs, pauses) Yeah, that's right, I don't like what's going on. Like last night, Sara was so tired that she just lay on the sofa—she didn't get around to fixing dinner 'til half an hour late. I was hungry and tired myself. We had a big argument. This has been going on for 3 weeks now.	CIS Level 2, as Dominic is able to talk about part of the contradiction originally posed by Ryan's confrontation. For example, he does not seem aware of how his wife feels; he seems insensitive to the fact that his wife is working and he expects her to do everything she did in the past for him.
6. *Ryan:* I hear you, Dominic. You're pretty angry about the situation. Let's change focus for a minute. As I recall, you and Sara have some pretty special memories. I recall your talking about being present at the birth of your child and that it was a special moment for the two of you. Sometimes when dealing with conflict, it helps to concentrate on positive things in a relationship. Out of those positives we can often find ideas that help move toward resolution.	Here we have an incongruity between the difficult present situation and a positive past presented via a summary and the positive asset search.
7. *Dominic:* Yes, I am angry and discouraged. Sara and I were doing pretty well until the baby came. Somehow things just got off kilter. (5-second pause and silence) . . . Well, let me try it your way. One of the things that always worked well for us was our fun times together. We both like outdoor activity and doing things together. We never seem to have time for that now.	It is often useful in couples work to remind them of positive stories from the past. It would be useful to take time out and explore positives in the relationship in more depth than presented here. Couples all too often forget why they got together in the first place.
8. *Ryan:* So, again, you have a good history and have enjoyed each other. I'm wondering if part of the solution isn't finding time just to be together doing fun things. Let's make that part of our discussion later. But for the moment, let's go back to the idea that you'd like her to continue doing things around the house. Could you give me a specific example of what's happening—something that goes on between you when she gets home?	Paraphrase, suggestion, open question oriented to concreteness. Getting specifics helps clarify the situation.
9. *Dominic:* Well, I've had a lot of pressure lately on the job. They've been downsizing and morale there is really bad. I worry that I won't be able to continue, so I try really hard and when I get home, I haven't much left. I just want to sit. But Sara's got the same thing. She's got a new boss	It is very common for partners to get angry with each other when external stressors hit one or both individuals in the relationship. One can't argue with the boss or colleagues easily, so the partner is scapegoated.

(continued)

who just seems to want more all the time. I guess when she comes home, she's about as exhausted and confused as I am.	
10. *Ryan:* I see. That gives me a clearer picture. She's working all day, you both come home exhausted, and, Dominic, you would like to be taken care of like you used to be. Do you think Sara physically can work and take care of you like she used to?	Summary, closed question. Ryan's implicit confrontation is now more concrete: "On one hand Sara is working and is exhausted; on the other hand you expect her to continue to take care of you."
11. *Dominic:* I guess I hadn't thought of it that way before. If Sara is working, she isn't going to be physically able to do what she did. But where does that leave me?	Developmental assessment: CIS Level 3, as Dominic is able for the first time to see that Sara can't continue as she has in the past. You may note that he is still thinking primarily of himself. To move to higher levels on the CIS, Dominic would have to be able to take Sara's perspective—for example, "I can see how Sara would feel—she must be too tired to do anything." Ideally, the counselor would like to see the client move to this level of thinking, but it isn't always possible. Consequently, practical, workable compromises in counseling style and goals may have to be made.
12. *Ryan:* Yes, where does that leave you? Dominic, let me tell you about my experience. My wife started working and I, too, expected her to continue to do the housework, take care of the kids, do the shopping, and fix the meals. She went to work because we needed the money to make a down payment on a small house. Well, what I found was that my wife couldn't work unless I helped around the house. I had to decide which was more important—getting the house for all of us or maintaining our traditional roles at home. What I've done is share some of the household work with her. I don't like it, but it seems like it's got to be done. I've taken over the shopping and now I pick the kids up at day care, too. How does that sound to you? Do you think you want to continue to expect Sara to do it all?	Self-disclosure followed by a check-out. Ryan is operating like a coach. He is speaking up directly with his ideas, but he is also allowing the client to react to them. Clearly, Ryan is trying to get Dominic's thinking and behavior to move. His last statement contains an important implicit confrontation: "On the one hand, your wife is working and if she continues, she'll need some help; on the other hand, perhaps you don't want her to work—what is your reaction to this?"
13. *Dominic:* Uhhh . . . we need the money. We've missed the last car payment. It was my idea that Sara go to work. But isn't housework "women's work"?	CIS Level 3, in that Dominic is now facing up to the contradiction he is posing. He has not yet synthesized his desire for his wife to work with the need for his sharing the workload at home, but at least he is moving toward a more open attitude. As he acknowledges his part in the situation and the need for more money, he is beginning to come to a new understanding, or synthesis, of the problem—he is less incongruent and is taking beginning steps toward resolving his discrepancies.

Box 9-4   Conflict Resolution and Mediation

You will find that confrontation skills are important in the mediation process. In conflict resolution, whether between children or between adolescents who have been fighting or in mediating arrangements with a couple initiating divorce, you will find the following steps useful.

1. *Develop rapport and outline the structure of your session.* Pay equal attention to each participant in a neutral fashion. A clear understanding of the stages of the negotiation process may help make the process less emotional. Four useful rules for children are "(a) agree to solve the problem, (b) no name calling or put-downs, (c) be as honest as you can, and (d) do not interrupt" (Lane & McWhirter, 1992). Agreeing to some variation of these with adults can be helpful as well in obtaining commitment to the process of mediation.

2. *Define the problem (concern).* Use the basic listening sequence to draw out clearly and *concretely* the point of view of each person involved in the dispute. Acknowledgment of feeling rather than reflection of feeling is recommended so as to avoid emotional outbursts. Summarize clearly each person's frame of reference toward the issue and check out the accuracy of your summarization carefully with each. In addition, you may wish to ask each disputant to state the opponent's point of view clearly and accurately.

   After you have summarized each person's frame of reference clearly, outline and summarize the points of agreement and the points of disagreement, perhaps in written form. You may use the model confrontation summary "On one hand, person 1 sees the problem as . . . while on the other hand, person 2 sees the problem as . . . Points of agreement are . . . while points of disagreement are . . ." The points of agreement serve much the same function as the positive asset search.

3. *Set goals.* Use the basic listening sequence to draw out each person's wants and desires for satisfactory problem solution, again focusing primarily on concrete

(continued)

For each developmental task completed in the interviewing process, it often seems that a new problem arises. Just as the counselor is beginning to facilitate client movement, a new obstacle ("women's work") arises. To make the progress shown in the transcript thus far took the counselor half the session. Working through the concept of "women's work" and changing it to "work in the home that must be shared if the car payments are to be met" took the rest of the session. This last statement can be located at Level 4 on the CIS—generating a new solution.

But as Dominic begins this developmental movement, note that a new problem appears: How can he help his wife when he considers housework women's work? A new, broader problem has been generated that may require a major change in consciousness for Dominic. Achieving a new level of awareness of male–female culture would require a Level 5 change on the Confrontation Impact Scale—a major transformation in thinking and behavior. Needless to say, this type of change may require several interviews, group sessions, and time for Dominic to internalize these new ideas.

Box 9-4   (continued)

facts rather than emotions and abstract intangibles. This is the beginning of the negotiation process, and as a result of this discussion the problems and concerns may be redefined and clarified. Again, summarize the goals for each person, using the model confrontation summary just mentioned.

In some situations, it may be wise to follow rapport building and structuring with some attention to broad goals, which can then be addressed again later in the session with more concreteness.

4. *Generate solutions.* It is here that negotiation begins in earnest. For the most part, rely on your listening skills to see whether the parties can generate their own satisfactory solutions. At times, when a level of concreteness and clarity of the issues has been achieved (Steps 2 and 3), the parties involved may be led to a stage close to agreement. If the parties are very conflicted, you may need to meet each one separately as you brainstorm alternative solutions. With touchy issues, it may be important to summarize them in writing. You will find many of the influencing skills covered later in this book useful in the process of negotiation.

5. *Contract and generalize.* Again, using the basic listening sequence, summarize the agreed-upon solution (or parts of the solution if negotiations are still in progress). Make the solution as concrete as possible and write down touchy main issues to make sure each party understands the agreement. Obtain agreement about subsequent steps. With children, congratulate them on their hard work and ask each child to tell a friend of the resolution.

A poster on display in the Martin Luther King Jr. Center (1989) summarizes six steps for nonviolent change that are closely related to the mediation model above: (1) information gathering; (2) education; (3) personal commitment; (4) negotiations; (5) direct action; (6) reconciliation. When you work on complex issues of institutional or community change, a review of Dr. King's model may be helpful in thinking through your approach to major challenges.

SUMMARY

We have covered three steps of confrontation and change:

1. Identify conflict via observing incongruities, discrepancies, and mixed messages.
2. Point out issues of incongruity and work toward resolution.
3. Evaluate the change process via the Confrontation Impact Scale.

Confrontation itself is a not a distinct skill; it is a set of skills that may be used in different ways. The most common confrontation uses the paraphrase, reflection of feeling, and summarization of discrepancies observed in the client or between the client and her or his situation. However, questions and influencing skills and strategies can also lead to client change. Silence can be a powerful and useful confrontation in itself.

Remember that each individual works through change in her or his own way. Some clients will move rapidly through all five levels in one session. Other clients may move more slowly. If you work with a major grief reaction around divorce or a highly significant change such as stopping drinking, do not expect clients to respond to your confrontations very rapidly. Developing and acting on a new story take time.

Box 9-5   Key Points

Why?	Clients come to us stuck and immobilized in their developmental processes. Through the use of microskills—confrontation in particular—we facilitate change, movement, and transformation—restorying and action.
What?	Confrontation has been defined as a supportive challenge in which you note incongruities and discrepancies and then feed back or paraphrase those discrepancies to the client. Our task is then to work through the resolution of the discrepancy. We seek to help our clients *change* their ways of thinking and behaving. We have examined the ideas of Kübler-Ross and the notion that clients often work through five identifiable stages as they change their thoughts, feelings, and behaviors: (1) denial, (2) partial acceptance of reality/bargaining/anger, (3) acceptance and recognition, (4) generation of a new solution, and (5) development of new, larger, and more inclusive constructs, patterns, or behaviors. The Confrontation Impact Scale is a tool to examine the effect microskills and confrontation have on client verbalizations immediately in the interview. At the lowest level clients may deny their incongruities; at middle levels they may acknowledge them; at higher levels they may transform or integrate incongruity into new stories and action. Mediation and conflict resolution can be facilitated by using listening skills, confrontation, and the five-stage interview model.
How?	An explicit confrontation can be recognized by the model sentence "On one hand . . . but on the other . . . How do you put those two together?" In addition, many interviewer statements contain implicit confrontations that can be helpful in promoting client growth and developmental movement. For example, you may summarize client conversation, pointing out discrepancies, or use an influencing skill such as the interpretation/reframe (see Chapter 12) to confront a client.
With whom?	Confrontation is believed to be relevant to all clients, but it must be worded to meet individual and cultural needs. A narcissistic or self-centered client may resist confrontation, and the microskills of interpretation and feedback may be more helpful. Clients from more direct and outspoken cultures such as European Americans and African Americans may respond well. Cultures that place more emphasis on subtlety and an indirect approach such as Asian groups may prefer gentler, more polite confrontations. Modification in style to accommodate various individuals will be necessary. Do not expect individuals in any cultural group always to follow one pattern—avoid stereotyping.
What else?	If your confrontation doesn't work, shift your style and use either a more direct or a gentler approach, depending on your client. You may also wish to share with clients how they did or did not respond to your confrontation. Give special attention to the positive asset search to help clients deal with the challenges of confronting their issues.

COMPETENCY PRACTICE EXERCISES AND PORTFOLIO OF COMPETENCE

This chapter is designed to help you construct a view of helping oriented toward change. If you master the cognitive concepts of the reading material and the exercises that follow, you will be able to promote client change and assess the effectiveness of your interventions. Again, this is an area that takes practice and experience. Apply the ideas here throughout the rest of your work with this book.

Individual Practice

Exercise 1: Identifying Discrepancies, Incongruity, and Mixed Messages and Strengths Leading Toward Resolution

Exercises 1–3 of Chapter 5 on observation skills are basic. Viewing videotapes of interviews, especially your own, is the best way to practice identification skills. The following discussion will also be useful as it encourages self-examination. Unless you can identify incongruity in your own self, seeing it in others may be difficult or even inappropriate.

Discrepancies internal to the self. Can you identify specific times in which your non-verbal behavior contradicted your verbal statements and gave you away? Are there times when you say two things at once and your verbal statements are incongruous? Have you done one thing while saying another?

Discrepancies between you and the external world. Part of life is living with contradictions. Many of these are unresolvable, but they can give considerable pain. What are some of the discrepancies between you and other individuals? What are some of the mixed messages, contradictions, and incongruity you face in your world of schooling or work?

Discrepancies between you and the client. You may have already experienced this and can easily summarize times when you felt out of tune and discrepant from the client. Or if you have not interviewed extensively, it may be helpful to think of situations where you had major differences with someone else. Often we have typical situations that "push our buttons" and move us toward actions that are too quick. Self-awareness in this area can be most helpful.

Specific strengths. Resolution of conflict and discrepancy is often made from a positive frame of reference. Can you identify personal strengths and wellness assets that can help you resolve internal and external differences? What strengths do you admire in others that you might like to add to your repertoire?

Exercise 2: Practicing Confrontation of Incongruity

Write confrontation statements for the following situations. Using the model sentence "On the one hand . . . , but on the other hand . . ." provides a standard and useful format for the actual confrontation. Of course, you may also use variations such as "You say . . . but you do . . ." and remember to follow up the confrontation with a check-out.

A client breaks eye contact, speaks slowly, and slumps in the chair while saying, "Yes, I really like the idea of getting to the library and getting the career information you suggest. Ah . . . I know it would be helpful for me."

_____

_____

_____

_____

"Yes, my family is really important to me. I like to spend a lot of time with them. When I get this big project done, I'll stop working so much and start doing what I should. Not to worry."

_____

_____

_____

_____

"My partner is good to me most of the time—this is only the second time he's hit me. I don't think we should make a big thing out of it."

_____

_____

_____

_____

"My daughter and I don't get along well. I feel that I am really trying, but she doesn't respond. Only last week I bought her a present, but she just ignored it."

_____

_____

_____

_____

Exercise 3: Practicing With the Confrontation Impact Scale

Here are some statements made by clients. Identify which of the five levels each client statement represents.

1. Denial
2. Partial examination
3. Acceptance and recognition
4. Generation of a new solution
5. Development of new, larger, and more inclusive constructs, patterns, behaviors— transcendence

Health issues. Look for movement from denial to new ways of taking care of one's body.

_____ I can't have a heart attack. It will never happen to me. I need to eat real food.

_____ Oh, I suppose I am overweight, but if I cut down a bit on butter and perhaps no more milk shakes, I'll be okay.

_____ I guess I can see that I need to balance my diet, but the busy life I lead won't really allow that to happen.

_____ I'm now able to cut out fats. At least that's taken care of.

_____ I've completely changed my way of doing things. I eat right—no fat at all—I exercise, and I'm even getting to like relaxation and stress management.

Career planning. Look for a movement from inaction or randomness to action.

_____ Okay, I guess I see your point. I've been released from two work–study programs because I didn't show up on time. But those were the bosses' fault. They should have made what they wanted clearer.

_____ The teacher referred me to you. Everyone has to have a job plan, but I see no need to worry about it so much. I'll be OK.

_____ Yes, I need a job plan. I can see now that is necessary. I'll write one and bring it to you tomorrow.

_____ I've got a job! The plan worked and I interviewed well and now I'm on my way.

_____ The plan has been helpful. I think I see now how to interview more effectively and present myself better.

Awareness of racism, sexism. Look for movement from denial that these issues exist to awareness and action.

_____ I feel committed. I've started action at home and at work, and I'm really going to concentrate on a more active approach to this issue. It's so important.

_____ Well, some people do discriminate, but I think that many people are just exaggerating.

_____ I don't really believe there is such a thing as racism or sexism. It's just people complaining.

_____ I've starting working with my family and children on being more tolerant, fair, and understanding of people different from us.

_____ There is a fair amount of prejudice, racism, and sexism everywhere.

Exercise 4: Writing Model Confrontation Statements

Review the Confrontation Impact Scale described on page 576. Then read the following confrontations.

Dominic: How can she expect me to work around the house? That's women's work!

Ryan: On one hand, I hear you wanting that second income she brings in. On the other hand, you seem to want her to keep up her housework as she did in the past without any help. How do you put that together?

Dominic could respond to that confrontation level by using denial, or he could work toward new ways of thinking. Can you write below example statements for Dominic representing the five levels of the Confrontation Impact Scale?

Level 1 (Denial): _____

_____

_____

Level 2 (Partial examination): _____

_____

_____

Level 3 (Acceptance and recognition): _____

_____

_____

Level 4 (Generation of a new solution): _____

_____

_____

Level 5 (Development of new, larger, and more inclusive constructs, patterns, or behaviors):

_____

_____

Client: I'm getting tired of talking with you. You always seem to think I'm taking the easy way out.

Counselor: Sounds as if you're telling me that on the one hand you want to change—that's why you started counseling—but on the other, now that it's getting close, you want to leave. That seems similar to the way you handle your relationships with the opposite sex: When someone gets close, you leave. How do you respond to that?

Level 1 (Denial): _____

_____

_____

Level 2 (Partial examination): _____

_____

_____

Level 3 (Acceptance and recognition): _____

_____

_____

Level 4 (Generation of a new solution): _____

_____

_____

Level 5 (Development of new, larger, and more inclusive constructs, patterns, or behaviors): _____

_____

Group Practice

Step 1: Divide into groups.
Step 2: Select a group leader.
Step 3: Assign roles for the first practice session.

❑ Client

❑ Interviewer

❑ Observer 1, who will rate each client statement using the Feedback Form (Box 9-6) and, during a replay of an audio- or videorecording, will stop the tape after each client statement and rate it carefully, together with the others.

❑ Observer 2, who will record the key words of each interviewer statement on a separate sheet of paper, thus making it possible to construct a picture of the interviewer as well. Pay special attention to the microskill leads of the interviewer.

Step 4: Plan.   State the goal of the session. The interviewer's task is to use the basic listening sequence to draw out a conflict in the client and then to confront this conflict or incongruity. Your ability to observe and note discrepancies on the spot during the session and to feed them back to the client will be important.

A useful topic for this practice session is any issue on which the volunteer client feels conflicted within or without. Internal conflict often shows around a difficult decision, past or present. External conflict most often appears when one has difficulty in dealing with a family member, a friend, or someone at work. Usually you will find both internal and external conflict in the client. Potentially useful topics include these:

❑ An important purchase

❑ A career decision involving a choice between a larger income and work that would be enjoyed more fully

❑ Virtually any type of interpersonal conflict

❑ Moral decisions ranging from telling the truth when one has held it back, to differences of opinion on abortion, divorce, or making a commitment, to issues of diversity or the role of spirituality in one's family

Step 5: Conduct a 5-minute practice session using confrontation skills as part of your listening and observation demonstration.
Step 6: Review the practice session using confrontation skills.
Step 7: Rotate roles.
Some general reminders.   The volunteer client may be asked to complete the Client Feedback Form of Chapter 1. This exercise is an attempt to integrate many of the skills and concepts used thus far in this book. Allow sufficient time for thinking through and planning this practice session, and recall the potential value of the positive asset search, coupled with full awareness of wellness potential.

Box 9-6   Feedback Form: Confrontation

_____ (Date)

_____   _____
(Name of Interviewer)   (Name of Person Completing Form)

Instructions: Summarize the main words of the client insofar as possible so that the response of the client to each counselor lead may be noted. Give special attention to client responses to explicit and implicit confrontations by the interviewer.

Confrontation Impact Scale rating form			Level		
	1	2	3	4	5
	Denial	Partial acceptance of reality: bargaining and anger	Acceptance and recognition	Generation of a new solution	Development of new, larger, and more inclusive constructs, patterns, or behaviors
1. _____					
2. _____					
3. _____					
4. _____					
5. _____					
6. _____					
7. _____					
8. _____					
9. _____					
10. _____					
11. _____					
12. _____					
13. _____					
14. _____					

Portfolio of Competence

Skill in confrontation depends on your ability to listen first and then to take an active role in the helping process. This needs to be done in a nonjudgmental fashion with respect for differences. As you work through this list of competencies, think ahead to how you would include confrontation skills in your own Portfolio of Competence.

Use the following as a checklist to evaluate your present level of mastery. Check those dimensions that you currently feel able to do. Those that remain unchecked can serve as future goals. *Do not expect to attain intentional competence on every dimension as you work through this book.* You will find, however, that you will improve your competencies with repetition and practice.

Level 1: Identification and classification.

❑ Ability to identify discrepencies and incongruities manifested by a client in the interview.

❑ Ability to classify and write counselor statements indicating the presence or absence of elements of confrontation.

❑ Ability to identify client change processes through observation on the Confrontation Impact Scale.

Level 2: Basic competence.

❑ Ability to demonstrate confrontation skills in a real or role-played interview.

❑ Ability, in the here and now of the interview, to observe and identify client responses on the five levels of the Confrontation Impact Scale.

❑ Ability to utilize wellness and the positive asset search to help clients find strengths that might help them move forward toward positive change when confronted.

Level 3: Intentional competence.
You will be able to use confrontational skills in such a manner that clients improve their thinking and behaving on the CIS.

❑ Ability to help clients change their manner of talking about a problem as a result of confrontation. This may be measured formally by the CIS or by others' observations.

❑ Ability to move clients from a discussion of issues at the lower levels of the CIS, when beginning discussion of a problem, to discussion at higher developmental levels at the end of the interview, or when the topic has been fully explored.

❑ Ability to identify client responses inferred from the CIS on the spot in the interview and change counseling interventions to meet those responses.

Level 4: Teaching competence.
Are you able to teach change and confrontation concepts to clients and to others? These concepts are generally intended more for counselors and interviewers than for clients. However, you will find change concepts useful in assisting self-directed thinkers in examining their situations. For example, those going through the stages of grief associated with death may find it helpful to have the change stages identified for them, thus enabling them to understand their feelings and thoughts more fully. The same stages may be expected among clients as they move through alcoholism, rape, or other difficult life issues.

DETERMINING YOUR OWN STYLE AND THEORY: CRITICAL SELF-REFLECTION ON CONFRONTATION

Confrontation is based primarily in listening skills, but it does require you to move more actively in the session through highlighting discrepancies and conflict. The Confrontation Impact Scale (CIS) was presented to show that you can assess the influence of your interventions in the here and now of the session.

What single idea stood out for you among all those presented in this chapter, in class, or through informal learning? What stands out that is likely to be important as a guide toward your next steps? How might confrontation relate to diversity issues? What other points in this chapter struck you as important? How might you use ideas in this chapter to begin the process of establishing your own style and theory?

_____

_____

_____

_____

_____

_____

REFERENCES

Carter, R. (1999). *Mapping the brain.* Berkeley: University of California Press.

Goodwin, L., & Sherrard, P. (2005). *Guided imagery: A therapeutic intervention to facilitate breakthrough insight.* Unpublished paper, University of Florida, Gainesville.

Hill, C., & O'Brien, K. (1999). *Helping skills.* Washington, DC: American Psychological Association.

Ivey, A. (1973). Media therapy: Educational change planning for psychiatric patients. *Journal of Counseling Psychology, 20,* 338–343.

Ivey, A. (2000/1986). *Developmental therapy: Theory into practice.* San Francisco: Jossey-Bass.

Ivey, A., Ivey, M., Myers, J., & Sweeney, T. (2005). *Developmental counseling and therapy: Promoting wellness over the lifespan.* Boston: Lahaska/Houghton-Mifflin.

Kübler-Ross, E. (1969). *On death and dying.* New York: Macmillan.

Lane, P., & McWhirter, J. (1992). A peer mediation model: Conflict resolution for elementary and middle school children. *Elementary School Guidance and Counseling, 27,* 15–23.

McMinn, M. (1996). *Psychology, theology, and spirituality in Christian counseling.* Wheaton, IL: Tyndale.

Sharpley, C., & Sagris, I. (1995). Does eye contact increase counsellor-client rapport? *Counselling Psychology Quarterly, 8,* 144–145.

ALLEN AND MARY'S THOUGHTS ABOUT CHRIS

Every day in the counseling process we encounter clients facing difficult decisions. Our first task is to identify the key discrepancies, conflicts, or contradictions faced by the client. In this case, Chris has several obvious conflicting issues that he needs to explore before making a decision:

▲ The central issue, at the moment, is that of graduate school versus staying at home and caring for his parents.

▲ He is dealing with mixed family messages, one that he continue on to college, the other that he stay home and help.

▲ We have incomplete data on his finances. If he has been saving up for 2 years for school, why is he worried that grad school will disappear if he doesn't go now?

▲ There is also the important issue of mixed, conflicting emotions. He has a desire to continue his education but guilt about the possibility of leaving home.

▲ Multicultural issues come in as well. Within certain cultures, the idea of leaving home after a family crisis may not even occur to the client. In a very individualistic culture, there may not even be a question—Chris should go on to graduate school. Thus, staying or leaving can also represent a cultural conflict to the client.

Our first goal would be to hear Chris's story more fully and at first we'd likely focus on the central issue, recognizing that later what appears to be the central dimension of decision may change as we learn more.

"Chris, I really hear the pressure you feel you are under. On one hand, you do want to care for your parents, but your dad suggests that they can take care of themselves. On the other hand, it leaves you feeling guilty . . . and your mom adds to that conflict. Tell me more."

We think this lead catches the essence of the conflict, although we have left out (for the moment) his desire for grad school. We call this type of response a confrontation because it focuses on conflicting, contradictory possibilities.

As our work with Chris evolves and we hear his story more completely, we can explore the other issues listed above—and likely, there will be new issues to encounter. One may be a cultural tradition of responsibility to family—or the cultural traditional may be that the individual needs to make the decision totally on his own. Resolution of the contradictions occurs when Chris is able to find a solution that satisfies him and important others.

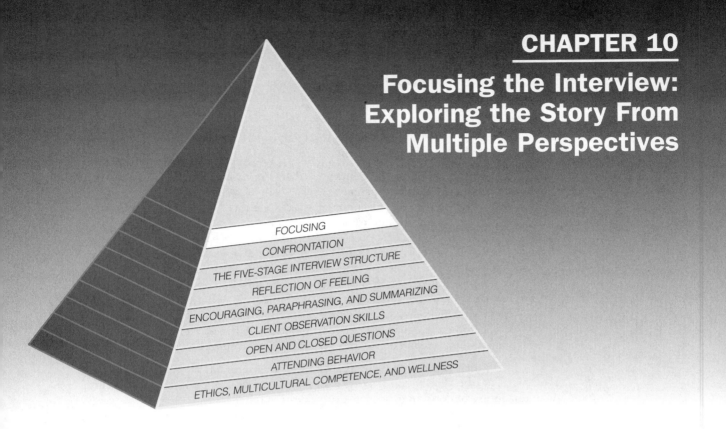

FOCUSING

CONFRONTATION

THE FIVE-STAGE INTERVIEW STRUCTURE

REFLECTION OF FEELING

ENCOURAGING, PARAPHRASING, AND SUMMARIZING

CLIENT OBSERVATION SKILLS

OPEN AND CLOSED QUESTIONS

ATTENDING BEHAVIOR

ETHICS, MULTICULTURAL COMPETENCE, AND WELLNESS

How can focusing help you and your clients?

Major function	Focusing is a skill that enables multiple telling of the story and will help you and clients think of new possibilities for restorying. Client issues are often complex, and the systematic framework of focusing can help in reframing and reconstructing problems, concerns, issues, and challenges.
Secondary functions	Knowledge and skill in focusing result in the following: ▲ Increased client cognitive and emotional complexity. Too often issues are considered from only one frame of reference. ▲ Better understanding of the viewpoints of others with an accompanying increase in empathic understanding and cultural intentionality. ▲ More complete involvement of family and cultural issues, particularly through family and community genograms. ▲ Awareness of the role of advocacy and social change as part of the focus of your interviewing practice.

Vanessa: (Walks swiftly into the office and starts talking even before she sits down) I'm really glad to see you. I need help. My sister and I just had an argument. She won't come home for the holidays and help me with Mom's illness. My last set of exams was a mess and I can't study. I just broke up with the guy I was going with 3 years. And now I'm not even sure where I'm going to live next term. And my car wouldn't start this morning until I got a jump start . . . (She continues with her list of issues and begins repeating stories almost randomly, but always with energy and considerable emotion.)

Most of us face times like this when everything seems to fall apart. Some clients come to the session with many issues and problems to solve. You may even feel overwhelmed at the barrage of information they throw at you while you wonder which direction to take.

What would you do with Vanessa or a client like her? How could you help her focus? What do you see as her number one issue? Put your thoughts and a possible response here.

_____

_____

_____

You can compare your thoughts with ours on page 323.

INTRODUCTION: PUTTING STORIES IN CONTEXT

As we have said many times, the first task of an interviewer is to hear the story as the client tells it and to identify key client thoughts, feelings, and behaviors. Focusing is a skill area that seeks to enlarge client storytelling and facilitate the discovery of new narratives and new ways of thinking about an issue. In this chapter we will also present the family and community genograms as strategies to help you and the client see how individuals exist in relationship to others.

The concept of self-in-relation may be helpful here. Increasingly, theorists are challenging the idea of a totally autonomous self. The term *being-in-relation* has been suggested by Baker Miller, Stiver, and Hooks (1998). The person-as-community is stressed by Obonnaya (1994), who points out that our family and community history live within each of us (see also Rigazio-DiGilio, Ivey, Grady, & Kunkler-Peck, 2005). The family and community genograms of this chapter may be useful in helping clients to gain new perspectives on themselves and their relationships to others.

Selective attention (see Chapter 3) is basic to focusing—clients tend to talk about topics to which you give your primary attention. Through your attending skills (visuals, vocal tone, verbal following, and body language), you indicate to your client what is of most importance. Clients sometimes will follow your lead rather than talk about what they really want to say, so it is important to be aware of both your conscious and unconscious patterns of selective attention. If you focus solely on individual issues, clients will talk about themselves and their frame of reference. This, of course, is important and central to the interviewing process. On the other hand, if you help clients to see themselves in relation to others and their social context, they can learn new ways of being in the world. It is not the individual versus social context; rather, social context enriches the uniqueness of each person and client.

Selective attention also shows in other ways. For example, if you look away when clients talk about sexual issues, conversation on this topic will most likely be limited. On the other hand, if you constantly lean forward and attend to sexuality, this may be the main topic of your sessions in many or even most of your interviews. Your style of listening deeply affects what clients say.

Your task in this chapter is to become aware of focusing and how you can broaden the session so that clients are aware of themselves more fully in social context. You can also become more aware of your own style of choosing which of many possibilities you take for a focus. In the long run, this awareness of complexity will make you more intentional in your work with clients.

INSTRUCTIONAL READING 1: MULTIPLE CONTEXTUAL PERSPECTIVES ON CLIENT CONCERNS

Individual counseling is obviously for the person sitting in front of you. Our major focus needs to be on the client and her or his unique concerns and issues. But if we are to discover fully the client's uniqueness, we also need to understand the social context of that client. If we know the social setting of the client (friends, family, community), we are better prepared to understand and to help. Multiple focusing and the family and/or community genograms (see the next section) provide a framework for understanding and action.

There are many ways to focus attention in the interview. Focusing on various aspects of client issues helps bring broader perspectives and potential new solutions. Below are brief definitions of each focus area followed by an example.

1. *Individual focus:* Beginning interviewers need to start by focusing on the individual client. Interviewing, counseling, and psychotherapy are for and about the client. Personal pronouns and the person's name help us to focus on the individual. ("Jillian, you're feeling sad and discouraged right now about having to place your mother in a nursing home. You seem to be really hurting.")

2. *Main theme or problem focus:* We also need to spend time listening to client concerns and issues. Drawing out the client's story also represents this type of focus. In much of the interview, you will find yourself focusing on both the individual clients and their issues; thus, a dual focus may appear frequently. ("I hear you saying that this home may not be the best place for her, but you had no practical alternative as she now needs constant supervision.")

3. *Other focus:* The client will often talk about significant others including friends, family members, or supervisors. When we focus on others, there is danger of failing to consider the individual who sits before us. Beware of sessions in which all the talk is about people not present. The interview is for the client, not for the absent person. However, as part of data gathering and problem solution, it is obviously important to focus briefly on significant others from time to time. ("You say that her friends are angry with you for putting her in the home. What have they done to help?" "Tell me more about the nurse you met there and your worries about her.")

4. *Family focus:* The word *family* is clear here. We are all products of our family history. Family can be defined in many ways—nuclear, extended, single-parent, or even as a very close living or friendship group. Gay or lesbian clients may see the gay community as their family. The family genogram strategy helps clients see themselves in relation to their present and past families. Be sure to be flexible in your definition of family. You will find that helping individuals see themselves in relation to family history can be an invaluable experience, but if you fail to bring the individual focus back in, the unique person may be lost. ("What help are you getting from your family?" "What's going on with your children?") In addition, many clients find it helpful to look at their family from a cultural frame of reference. ("You've commented that you are . . . Italian/Polish/Vietnamese . . . Could you tell me a bit more about some of your family traditions and how they affect you?")

5. *Mutuality focus:* This focus can be controversial as it brings immediacy to examination of the interviewer/client relationship. It should be used relatively infrequently. It can be quite powerful as it uses the interviewer as an instrument—how the client reacts to the interviewer can be an indication of how the client develops in relation to other people. At another level, mutuality focus puts the client and interviewer on an equal level. ("How can *we* work *together* on this issue?" "How would you like me to help?" "How do you feel toward me right now as we are talking?")

6. *Interviewer focus:* The interviewer may provide feedback, opinions, or advice from her or his perspective. This focus should be used only occasionally and the focus should be brought back immediately to the client. ("I had to put my father in a home 2 years ago myself. It's very difficult and hurt a lot. Is that close to what you are feeling?")

7. *Cultural/environmental/context focus:* The person-as-community summarizes this focus. If you and your client complete community genograms, you will discover that who you are as an individual was deeply affected by your family and how you grew up in your community(ies) of origin. The cultural/environmental/context focus can be expanded to include issues of gender, race/ethnicity, sexual orientation, spirituality/religion, socioeconomic status, and multiple contextual issues. ("How has your spirituality helped you get through all this?" "How are the finances going to be handled?" "You say it is always the women who have to take charge and your brothers won't help.")

You have had a brief introduction to focus analysis. Now let us turn to a different type of approach for this instructional reading. The following material, oriented to specific, written exercises, will help you better define the value and merits of each focus possibility. Please write a response that you might make to each of the client statements below.

As an interviewer, counselor, or psychotherapist, you will encounter cases that are controversial and you will work with clients who have made different decisions from those that perhaps you would make. One challenging issue, on which there are several points of view, is abortion. Before you start working through this section, take some time to think through your own thoughts on this divisive issue, which is part of what is sometimes called the "culture wars." There are deeply felt beliefs on

Box 10-1 National and International Perspectives on Counseling: Where to Focus—
Individual, Family, or Culture?

Weijun Zhang

Case study: Carlos Reyes, a Latino student majoring in computer science, was referred to counseling by his adviser because of his recent academic difficulties and psychosomatic symptoms. The counselor was able to discern that Carlos's major concern was his increasing dislike of computer science and growing interest in literature. While he was intrigued about changing his major, he felt overwhelmed by the potential consequences for his family, in which he is the oldest of four siblings. He is also the first in his family ever to attend college. Carlos has received some limited financial support from his parents and one of his younger siblings, and the family income is barely above the poverty line. The counseling was at an impasse, for Carlos was reluctant to take any action and instead kept saying, "I don't know how to tell this to my folks. I'm sure they'll be mad at me."

During class discussion of this case, almost everyone argued that Carlos's problem is that he does not give priority to his personal career interests, that he should learn to think about what is good for his own mental health, and that he needs assertiveness training. I did not quite agree with my fellow students, who are all European Americans. I thought they were failing to see a decisive factor in the case: Carlos is Latino!

In traditional Hispanic culture, the extended family, rather than the individual, is the psychosocial unit of cooperation. The family is valued over the individual, and subordination of individual wants to the family needs is assumed. Also, traditional Hispanic families are hierarchical in form; parents are authority figures and children are supposed to be obedient. Given this cultural background, to encourage Carlos to make a major career decision totally on his own was impossible. Any counseling effort that does not focus on the whole family is doomed to fail.

Because it is the financial support from the family that made his college education possible, Carlos may be expected to contribute to the family when he graduates. This reciprocal relationship is a lifelong expectation in Hispanic culture, and the oldest son is especially responsible in this regard. Changing his major in his junior year does not only mean he will be postponing the date when he will be able to help his family financially, but it also means he may not be able to do so at all, for we all understand how hard it is to find a well-paying job in the field of literature. When interdependence is the norm among Hispanic Americans, how can we expect Carlos to focus entirely on his personal interests without giving more weight to his family's pressing economic needs?

(continued)

each side of this issue. The emotions around this issue can run high and even the language of "pro-choice" and "pro-life" positions can be upsetting to some.

What is your personal position around this challenging issue?

_____

_____

As a counselor, it is vital that you understand the situations, thoughts, and feelings of those who take varying positions around abortion, whether you agree with them or not. Now, can you state some of the thoughts and feelings of those who

Box 10-1   **(continued)**

If I were Carlos's counselor, rather than focusing immediately on his needs, I would first support him with his family loyalty and then help him understand that there are not just two solutions: either . . . or. . . . Together, we might brainstorm to generate some alternatives, such as having literature as his minor now and as his pastime after he graduates, changing his career when his younger siblings are off on their own, or exploring possibilities that might combine the two. He could, for example, design computer programs to help schoolchildren learn literature. Each of these takes into account family needs as well as those of Carlos.

The professor praised me highly for my "different and sensitive perspective," but I shrugged it off; this is just common sense to most Third World minority people and, probably, many Italian and Jewish Americans as well. (I remember years ago, when I was trying to make major career decisions with my parents; at least 10 of my relatives were involved. And these days, I am still obligated to help anyone in my extended family who is in financial need.) It took me almost a year to come to realize that when I am asked, here in the United States, "How's your family?" I usually need tell how only my wife and child are faring, not my parents, grandparents, and siblings.

If the meaning of family in Hispanic culture is confusing to many counselors, the traditional extended family clan system of Native American Indians, Canadian Dene, or New Zealand Maori can be even more difficult for them to grasp. This family extension can include at times several households and even a whole village. Unless majority group counselors are aware of these differences in family structure, they may cause serious harm through their own ignorance.

Mary Bradford Ivey comments: Working with a Puerto Rican high school student, the counselor invited the family to a session. The parents, grandparents, godparents, and a neighbor all appeared. It is incumbent on all of us to be aware that what is defined as family varies with individual and cultural backgrounds. Feminist theory and multicultural counseling and therapy have challenged the "I" focus of much of counseling. Decisions often need to be made in terms of relationships and keeping as many people as possible comfortable and part of the process. We, of course, can't stereotype women any more than we can Hispanics or Italians. Focus analysis as presented in this chapter is an attempt to bring broader perspectives to the interview and consider the self-in-relation, the self-in-context. You will find that developing family and community genograms as described later in this chapter is useful in thinking about the self-in-relation.

have a different position from your own? How might their situation be different from your own?

_____

_____

We all work with clients who have different beliefs from us on powerful issues. Whether it is politics, the role of women, affirmative action, or abortion, you are going to counsel clients who think differently from you. Sharing your honest thoughts with a client on very sensitive issues is often dangerous and potentially destructive to the client. Equally problematic are the situations when you unconsciously direct the client in a direction that you favor without awareness of what you are doing. Seek supervision and consultation if you find yourself in a challenging situation in which your own thoughts about client issues can get in the way of a successful counseling relationship.

Counseling is not teaching clients how to live or what to believe. It for is helping clients make their own decisions. At the same time, you may need at times to help clients learn to recognize their own sexism, racism, and other -isms, as well as to carefully think through unexplored biases. Moreover, whether you are opposed to abortion or are a pro-choice advocate, you may find yourself using the interview to further your own position. Most would agree that counselors should carefully work against bias in counseling. It may be useful to your clients to help them understand more than one position on abortion.

As you can see, when it comes to controversial, highly charged beliefs and issues, there is a danger that counseling could become political and biased in one direction or another. The art and mastery of being an effective counselor involves awareness and respect for beliefs merged with unbiased probing in the interest of client self-discovery, autonomy, and growth. The following exercises are offered with the intention of helping you hone these skills in yourself and in clients.

Some school systems have written policies forbidding any discussion of abortion. Further, if you are working within certain agencies (e.g., a faith-based agency, a women's counseling program, or a gay support therapy group), the agency may have their own specific policies around these issues. Ethically, we believe that clients should be aware of such agency beliefs and policies before counseling in an agency is undertaken. Work in these settings may start from a different value base from that taken by others in the counseling profession.

There are no final answers to these difficult issues. However, we wish to bring them to your attention. Again, counseling is for the client, not for a personal political or moral agenda. When clients ask you for your opinion, your first task is to throw that question back at the clients and encourage them to make their own decisions.

Imagine that a client comes to you after just having gone through the termination of a fetus. How would you help this client, who clearly needs to tell her story?

For each statement below, write a client-centered focus and also a main theme/problem-centered focus:

"I just had an abortion."

Client focus: _____

Problem/main-theme focus: _____

Choosing where to focus as you listen to clients can be most challenging. Too many beginners focus only on the main theme or problem ("Tell me more about the abortion"), which may result in a voyeuristic interview in which many facts are obtained but little is learned about the client. An extremely important task is drawing out the client's story ("I'd like to hear *your* story. What do *you* want to tell me?" "How was this for *you*?"). While there are no final rules on where to focus, generally, we want to hear the unique client's experience. Thus focusing on the individual is usually where to start—note the three times the word *you* was used in the example in parentheses.

Through each of the following examples think about how you can use the concept of confrontation to facilitate focusing and the development of awareness.

Teresa: I just had an abortion and I feel pretty awful. The staff counselor and medical staff were great and the operation went smoothly. But Cordell won't have anything to do with me and I can't talk with my parents.

Client focus: _____

Problem/main-theme focus: _____

Focusing on Teresa's feelings about the abortion would be an individual focus. The focus on the main theme or problem has several possibilities, including more details about the facts of the abortion and the communication problems with Cordell and her parents.

At this point it becomes apparent that other key figures (Cordell, friends, the physician and nurse, and perhaps a minister, priest, imam, or rabbi) are part of a larger picture. What are their stories? How do they relate to Teresa as a "self-in-relation"? You, in turn, can understand her situation more fully with these other perspectives, which might bring out other stories or viewpoints.

Focus on others (Cordell): _____

_____

Focus on others (friends): _____

_____

_____

Focus on others (defined by your own questions and interests): _____

_____

_____

If you focus solely on Cordell, issues involving the parents or Teresa's feelings toward the supportive physician and nurses are temporarily lost. It is important to keep all significant others in mind in the process of problem resolution and problem examination, perhaps saving questions or other listening skills until later. For a full understanding of the abortion experience, all of these relationships (and probably others) need to be explored. In turn, the client will likely have personal feelings and thoughts toward each of these people, so it may be necessary to focus again on the client.

The client may continue her discussion as follows:

"I feel everyone is just judging me. They all seem to be condemning me. I even feel a little frightened of you."

A mutual focus often emphasizes the "we" in a relationship. Use the following space for a mutual-focus statement:

Focus on mutual issues: _____

_____

One possibility here is "Right now, *we* have an issue. Can *we* work together to help you? What are some of your thoughts and feelings about how *we* are doing?" The emphasis is on the relationship between counselor and client. Two people are working on an issue in this example, and the counselor accepts partial ownership of the problem. Some counseling theories would be against this relational approach. Among most people in Western cultures, emphasizing the distinction between "you" (client focus) and "me" (interviewer focus) would perhaps be more common. Among some Asian and Southern European peoples, the "we" focus may be especially appropriate: *"We* are going to solve this problem." The "we" focus provides a sharing of responsibility, which is often reassuring to the client regardless of his or her background. Many feminist counselors emphasize "we."

Families and family relationships are vital to each human being. In a sometimes overly individualistic society, the focus can be too much on individuals and problems and miss important family dimensions. For example, if you work with the child of an alcoholic, it is important to explore what that child or adult child (adult child of an alcoholic, ACOA) experienced growing up in the family environment. If an individual is to make a decision, the total family context needs to be involved.

For example, the client may say, "My family is really strongly against abortion; it makes me feel all the more guilty. I could never tell them." How might you focus on the family in response to her statement?

Focus on family: _____

_____

_____

If you have developed a family genogram, you may be able to locate where Teresa has resources and models who might be helpful to her. If her parents are not emotionally available, perhaps an aunt or grandmother may be helpful. The family is, of course, where personal values and ethics are first learned.

How does Teresa define "family"? There are many styles of family beyond the nuclear. African American and Hispanic clients may think of the extended family; a gay male may see his supportive family as the gay community. Issues of single parenthood and alternative family styles continue to make the picture of the family more complex.

Another type of focus is for the interviewer to refer to herself or himself. Look back on the case so far. For our example, what would a focus on the interviewer be? Below, write a statement that you might share with your client that focuses on yourself.

Interviewer focus: _____

_____

_____

An interviewer focus could be a self-disclosure of feelings and thoughts about the client or situation—for example, "*I* feel concerned and sad over what happened; *I* want to help" or "*I*, too, had an abortion . . . *my* experience was . . ." or perhaps even some personal advice. Opinions vary on the appropriateness of interviewer or counselor involvement, but the value and power of such statements are increasingly

Box 10-2 Research Evidence That You Can Use: Focusing

Zalaquett (2004) found that training students to focus on cultural/environmental/contextual issues resulted in greater awareness and willingness to discuss racial and gender differences early in the session and to make these issues a consistent part of the interview. Moos (2001) has reviewed much of the contextual literature and points out that the way we appraise a situation can be self-centered or environmental/contextual. Clients often come to the interview with a focus that may work against their own best interest. Too much of an "I" focus may result in self-blame and lack of awareness of context. On the other hand, too much of a contextual focus may lead to lack of attention to internal responsibility. The extensive research on attribution theory further highlights this point (cf. Harvey & Manusov, 2001). By focusing on multiple dimensions, you can help the client find a suitable balance.

Recent work on training students in multicultural counseling provides an excellent research model for the future (Torres-Rivera et al., 2001). The researchers used a set of multicultural skill training videos prepared by Arredondo et al. (2000). They showed these videotapes of culture-specific counseling to their students and found that the students' multicultural effectiveness and understanding increased. Moos (2001) noted that teaching clients the context of their issues helps them understand themselves in new ways and "makes possible a transformative experience."

Focus and neuropsychology

The skill of focusing is closely related to selective attention (see Chapter 3). Client selective attention is guided by existing patterns in the mind (Allport, 1998) and focusing is an intentional interviewing skill that can open up more possibilities for client thoughts, feelings, and actions. There are a number of regions of the prefrontal cortex that "are activated selectively during different aspects of attentional task preparation and execution" (Allport, 1998). Self-regulation and understanding of others (empathy) is also deeply affected by attentional systems in the brain (Decety & Jackson, 2004).

being recognized. They must not be overused, however, or the client may end up listening at length to the counselor's concerns. Keep self-disclosures brief.

Perhaps the most complex focus dimension is the cultural/environmental/context. Some topics within these broad areas are listed here, along with possible responses to the client.

▲ *Moral/religious issues:* "How can your spiritual background be helpful here?"
▲ *Legal issues:* "The topic of abortion brings up some legal issues in this state. How have you dealt with them?"
▲ *Women's issues:* "A support group for women is just starting. Would you like to attend?"
▲ *Economic issues:* "You were saying that you didn't know how to pay for the operation . . ."
▲ *Health issues:* "How have you been eating and sleeping lately? Do you feel after-effects?"
▲ *Educational/career issues:* "How long were you out of school/work?"
▲ *Ethnic/cultural issues:* "What is the meaning of abortion among people in your family/church/neighborhood?"

Any one of these issues, as well as many others, could be important to a client. With some clients all of these areas might need to be explored for satisfactory problem resolution. The counselor or interviewer who is able to conceptualize client issues broadly can introduce many valuable aspects of the problem or situation. Note that much of cultural/environmental/contextual focusing depends on the interviewer's bringing in concepts from her or his own knowledge and not simply following ideas presented by the client.

Advocacy and Social Justice

You are going to face situations when your best counseling efforts are insufficient to help your clients resolve their issues and move on with their lives. The social context of poverty, racism, sexism, and many other forms of unfairness may leave your client in an impossible situation. Or the problem may be bullying on the playground, an unfair teacher, or an employer who refuses to follow fair employment practices. It is safer when we sit in the counseling office and work for personal growth and individual problem solving, but much more challenging when we examine the societal stressors that our clients may face and consider advocacy issues.

Advocacy is speaking out for your clients; working in the school, community, or larger setting to help clients; and also working for social change. What are you going to do on a daily basis to help improve the systems that affect your clients? Here are some examples that may require you to take an advocacy position:

▲ As an elementary counselor, you counsel an elementary child who is being bullied on the playground.

▲ You are a high school counselor and work with a 10th grader who is teased and harassed by his classmates about being gay. The classroom teacher quietly watches and says nothing.

▲ You work in the human relations division as a personnel officer. Through contact with several distressed clients, you discover systematic bias against promotion for women and minorities.

▲ Working in a community agency, you have a client who speaks of abuse in the home and a fear of leaving because she sees no financial support for her in the future.

Counselors who care about their clients also become advocates for them when necessary. These counselors are willing to move out of the counseling office and seek social change. The elementary counselor can work with school officials to set up policies about bullying and harassment, actively changing the environment that allowed bullying to occur. The high school counselor faces an especially challenging issue as interview confidentiality may preclude immediate classroom action. If addressing the individual situation is not possible, then the counselor can initiate school policies and awareness programs that mitigate against oppression in the classroom. It is also possible that the passive teacher might become more active and involved through training offered to all teachers in the school.

"Whistle-blowers," those who identify and name problems that others like to avoid, can face real difficulty. The company may not want to have their systematic bias exposed. On the other hand, through careful consultation and data gathering, the human relations staff may be able to help managers develop promotion programs that are more fair. Again, the issue of policy becomes important. Counselors can advocate policy changes in their work settings.

The counselor in the community agency knows that advocacy is the only possibility when abuse is apparent. For these clients, advocacy in terms of support in getting out of the home, finding new housing, and learning how to set up a restraining order may be far more important than self-examination and understanding.

Working for social change is also an important role for the interviewer and counselor. This will often involve you beyond regular office hours. You may work with others on a specific cause or issue to facilitate general human development and wellness (e.g., pre-term pregnancy care, child care, fair housing, the homeless, athletic fields for low-income areas). This will require you to speak out, to develop skills with the media, and to learn about legal issues.

Such activities are termed *ethical witnessing,* moving beyond working with victims of injustice to the deepest level of advocacy (Ishiyama, 2005). Counseling, social work, and human relations are inherently social justice professions, but speaking out for social concerns needs more of our time and attention.

Moving to a New Story

There are many perspectives and stories about any issues. Though it may not be possible to take them all into account because of time considerations, going over the story from different frames of reference can be helpful to the client in generating a new story.

INSTRUCTIONAL READING 2: FAMILY AND COMMUNITY GENOGRAMS

Clients often come to the interview thinking that they alone are the issue and central focus when it comes to resolving their problems. Again, it is important to place the main focus on the client sitting before you. But to help you truly understand the client, you need a broader perspective, and the family genogram and the community genogram can be invaluable in making graphic the client's total context. In fact, we often post both family and community genograms on the interviewing room wall during our sessions. They help remind us and the client that we all exist in a web of interpersonal relationships. We often turn to the family and/or community genograms during the session to help the clients and ourselves see the clients' issues in a broader social and cultural context.

Wellness and the positive asset search need to be central issues in your mind as you help clients construct both family and community genograms. We can learn a lot about problems and hear many negative stories about either family or community background, but spend at least half of the time on strengths and positives. You will find that there will almost inevitably be memories of positive experiences, heroes, and specific things the client did that are strengths. Rather than considering these strategies diagnostic instruments only, think of them also as *strength inventories.* What can you draw from them to facilitate positive problem solving and restorying?

The Family Genogram

The individual develops in a family within a culture. You and your clients will more easily understand the self-in-relation concept if you help them draw a family genogram. We suggest that you consider developing family and community genograms

Box 10-3   The Family Genogram

This brief overview will not make you an expert in developing or working with genograms, but it will provide a useful beginning with a helpful assessment and treatment technique. First go through this exercise using your own family; then you may want to interview another individual for practice.

1. List the names of family members for at least three generations (four is preferred) with ages and dates of birth and death. List occupations, significant illnesses, and cause of death, as appropriate. Note any issues with alcoholism or drugs.
2. List important cultural/environmental/contextual issues. These may include ethnic identity, religion, economic, and social class considerations. In addition, pay special attention to significant life events such as trauma or environmental issues (e.g., divorce, economic depression, major illness).
3. Basic relationship symbols for a genogram are shown on the left, and an example of a genogram is shown below.
4. As you develop the genogram with a client, use the basic listening sequence to draw out information, thoughts, and feelings. You will find that considerable insight into one's personal life issues may be generated in this way.

Close	═══
Enmeshed	≡≡≡
Estranged	─//─
Distant	------
Conflictual	∧∧∧∧
Separated	─/─

Figure 10-1
Basic relationship symbols

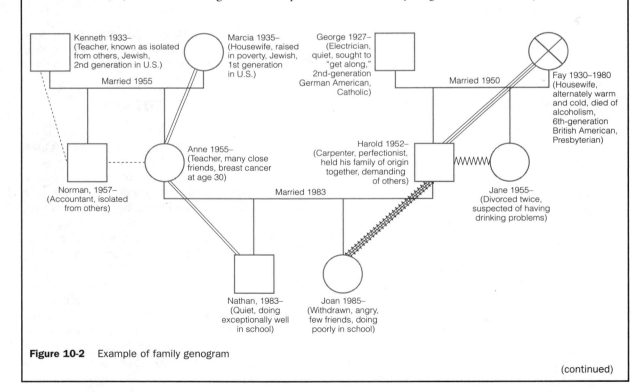

Figure 10-2   Example of family genogram

(continued)

(see the following section) with many of your clients. Moreover, some clients find them comforting as the genograms bring family history with the clients to the interview. In a sense, we are never alone; our family and community histories are always with us.

Box 10-3 presents the major "how's" of developing a family genogram. Specific symbols and conventions have been developed that are widely accepted and help

Box 10-3   (continued)

> Developing a genogram with your clients and learning some of the main facts of family developmental history will often help you understand the context of individual issues. For example, what might be going on at home that results in Joan's problems at school? Why is Nathan doing so well? How might intergenerational alcoholism problems play themselves out in this family tree? What other patterns do you observe? What are the implications of the ethnic background of this family? The Jewish and Anglo backgrounds represent a bicultural history. Change the ethnic background and consider how this would impact counseling. Four-generation genograms can complicate and enrich your observations. (Note: The clients here have defined their ethnic identities as shown. Different clients will use different wording to define their ethnic identities. It is important to use the client's definitions rather than your own.)

professionals communicate information to each other. The family genogram is one of the most fascinating exercises that you can undertake. You and your client can learn much about how family history affects the way individuals behave in the here and now. The classic source for family genogram information is McGoldrick and Gerson (1985).

The interviewer who uses the family genogram typically fits the client into an existing framework. We have found family genograms helpful and use them frequently; however, there are situations in which some clients find them less satisfying than the community genogram that follows. There is a Western, linear perspective to the family genogram that does not fit all individuals and cultural backgrounds. Obviously, it is important to adapt the family genogram flexibly to meet individual and cultural differences. You will find *Ethnicity and Family Therapy* a most valuable and enjoyable tool to expand your awareness of racial/ethnic issues (McGoldrick, Giordano, & Garcia-Preto, 2005).

Much important information can be collected in a family genogram. Many of us have important family stories that are passed down through the generations. These can be sources of strength (such as a story of a favorite grandparent or ancestor who endured hardship successfully). Family stories of this kind are real sources of pride and can be central in the positive asset search. There can be a tendency to look for problems in the family history and, of course, this is appropriate. But use this important strategy positively as well. Be sure to search for positive family stories as well as problems. How can family strengths help your client?

Children often enjoy the family genogram and a simple adaptation called the "family tree" makes it work for them. The children are encouraged to draw a tree and put their family members on the branches, wherever they wish. This strategy has the advantage of allowing children to present the family as they see it, thus permitting easy placement of extended family and important support figures as well as immediate family members. Many adolescents and adults may also respond better to this more individualized and less formal approach to the family.

The Community Genogram: Identifying Cultural Strengths

We developed the community genogram as some of our clients were uncomfortable with the family genogram. Specifically, the family genogram is most effective with a client who has a nuclear family and actually can trace the family over time. Clients who have been adopted often find the genogram uncomfortable. Single-parent families may also feel "different," particularly when important caregivers such as extended family and close community friends are not included. We have talked with gay and lesbian clients who have very differing views of the nature of their family. There is a convention of placing an "X" over departed family members. Once we were demonstrating the family genogram strategy with a client and she commented, "I don't want to cross out my family members—they are still here inside me all the time."

Much of our individuality comes from our distinctive relationship to our environment. For example, the decade of your birth is highly influential in the type of community in which you lived as well as your worldview. Clients who grew to maturity in the 1960s and 1970s lived in a very different world from those born in more recent decades. Your older clients, born before World War II, did not have television or jet planes. Their economic attitudes and expectations are often quite different because of the environment they experienced. A client who grows up in Mt. Vernon, Washington, or Bemidji, Minnesota, has a very different life experience than if he or she were raised in Los Angeles, Miami, or New York City.

The community genogram will help you understand the cultural background of your clients more completely for it is in community that we learn the culture (Ivey, 1995; Rigazio-DiGilio, Ivey, Grady, & Kunkler-Peck, 2005). The major goal of the community genogram is to help the client generate a sense of connection and how we all develop in a community/cultural context. We work with clients to help them develop community narratives as sources of strength. At the same time, many of us had problems in growing up in our community, some of which continue today. But all of us are survivors—what has occurred that has strengthened us to meet the challenges we all face? Search for both the positive narratives and personal resources that reside within the individual and the supports that exist in the past and present communities.

The strategy works equally well with children and can be an invaluable way to understand the child's total situation. Symon (2005) suggests that pets may be an important part of a child's community support and this is true of adults as well. Develop the community genogram relating to the special needs and interests of the client.

The specific steps for developing a community genogram follow. Note especially that this is a "free-form" activity in which we encourage people to develop their own style of presenting community. We encourage you first to do your own community genogram so you have a better sense of yourself as a person-in-community.

Step 1: Develop a Visual Representation of the Community

1. Consider a large piece of paper as representing your broad culture and community. It is recommended that you select the community in which you were primarily raised, for the community of origin is where you tended to learn the most about culture. But any other community, past or present, may be used.
2. Place yourself or the client in that community, either at the center or at another appropriate place. Represent yourself or the client by a circle, a star, or any other significant symbol. Two visual examples of community genograms are shown in Box 10-4.

Box 10-4 The Community Genogram: Two Visual Examples

We encourage clients to generate their own visual representations of their "community of origin" and/or their current community support network. The two examples below are only samples of the possibilities. Some clients like to draw concrete pictures while others prefer abstract symbols.

1. *The map:* The client draws a literal or metaphoric map of the community, in this case a rural setting. Note how this view of the client's background reveals a close extended family and a relatively small experiential world. The absence of friends in the map is interesting. Church is the only outside factor noted.

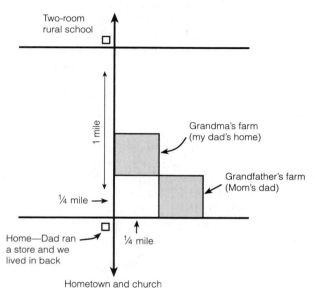

2. *The star:* Janet's world during elementary school tells us a good bit about a difficult time in her life. Nonetheless, note the important support systems.

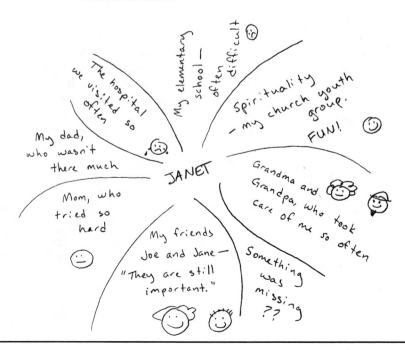

3. Place your own or the client's family or families on the paper, again represented by the symbol that is most relevant for you. The family can be nuclear or extended or both.

4. Place important and most influential groups on the community genogram, again representing them by circles or other visual symbols. School, family, neighborhood, and spiritual groups are most often selected. For teens, the peer group is often particularly important. For adults, work groups and other special groups tend to become more central.

5. Connect the groups to the focus individual, perhaps drawing heavier lines to indicate the most influential groups.

Step 2: Search for Images and Narratives of Strengths

Though many individual difficulties and problems arise in a family, community, and cultural context, the community genogram focuses on strengths. The community genogram provides a frame of reference to help the client see self-in-relation. The client is asked to generate narratives of key stories from the community where he or she grew up. If relevant, key stories from the present living community may also be important. The emphasis is on positive stories from the community and positive images. The community genogram is kept and posted on newsprint during the entire counseling series of interviews (Rigazio-DiGilio et al., 2005).

The specifics of the community positive asset search follow:

1. Focus on one single community group or the family. You or the client may want to start with a negative story or image. Please do not work with the negative until positive strengths are solidly in mind. Start with positive stories.

2. Develop a visual, auditory, or kinesthetic sensorimotor image that represents an important positive experience. Allow the image to build in your mind and note the positive feelings that occur with the image. If you allow yourself or the client to fully experience that positive image, you may experience tears and/or feelings such as rapid breathing, increased heart rate, or perhaps a profound state of relaxation. These anchored body experiences represent positive strengths that can be drawn on to help you and your clients deal with difficult issues in counseling and in life.

3. Tell the concrete story of the image. If it is your story, you may want to write it down in journal form. If you are drawing out the story from a client, listen sensitively.

4. Develop at least two more positive images from different groups within the community. It is useful to have one positive family image, one spiritual image, and one cultural image. Again, many of us will want to focus on negative issues. Hold to the search for positive resources.

5. Summarize the positive images in your own words and *reflect* on them. Encourage your clients to summarize their learning, thoughts, and feelings in their own words. As you or your client thinks back, what occurs? Record the responses, for they can be drawn on in many settings in counseling or in the daily life of the client.

The community genogram can be a very emotional and dramatic strategy, particularly if you use the strategies of developmental counseling and therapy (see pages 196–198). At a minimum, it will help you to understand the special cultural background of your client, and it serves as a reservoir of positive experiences that can be drawn on to help you and the client throughout counseling.

EXAMPLE INTERVIEW: IT'S ALL MY FAULT—HELPING THE CLIENT UNDERSTAND SELF-IN-RELATION

An individual comes to the interview with you to discuss her or his issues, concerns, and problems. We live in a culture that focuses on individuality, individual responsibility, and individual achievement. Thus, it is natural that counseling and therapy focus on the immediate person who appears in the here and now of the session. Carl Rogers's person-centered counseling has had an immense and lasting influence on the way we conduct helping sessions. It is a clear fact that we can only work with the individual before us and thus the "I" focus on the individual remains central and perhaps this always should be so.

Over time, however, the helping field has moved to awareness that issues of gender, ethnicity/race, and social history affect the person, the way he or she deals with personal issues, and the manner in which he or she can problem solve. No individual is alone for we all live in an environmental context that affects our individuality. Many would argue that real *individual* counseling and therapy require us to consider the context in which a person lives. How are we selves-in-relation, persons-in-community, or beings-in-relation?

The following interview with Janet focuses first on her individual issue of taking virtually all the responsibility for difficulties in her relationship with Sander. This is the second interview, and during the first session Janet completed a community genogram of the home where she grew up in Eugene, Oregon (see Box 10-4). The community genogram is used to help Janet understand how her history might affect her present behavior, feelings, and thoughts.

Interviewer and Client Conversation	Process Comments
1. *Samantha:* Nice to see you, Janet. How have things been going?	A solid relationship was established during the first session. When Samantha met Janet outside her office, Janet seemed anxious to start the session.
2. *Janet:* Well, I tried your suggestion. I think I understand things a bit better.	Samantha had suggested that Janet spend the week listening carefully to Sander and noting what seemed to be going on in their relationship. She also advised continuing her behavior as in the past so that she could note patterns of behavior more easily.
3. *Samantha:* You understand better? Could you tell me more?	Encourage. Focus on problem and individual.
4. *Janet:* Well, I tried listening to Sander as you suggested, just so I could understand what he really wants from me. What I found was that	Self-disclosure on observations. Focus on self (the client), others (Sander), and the main theme of relationship.

(continued)

Interviewer and Client Conversation	Process Comments
he really wants a lot. He wants me to do a better job with my cooking, he doesn't like the way I keep house, and the more I listened and observed, the more he seemed to want. I guess I've not been attentive enough to him. I should be doing more and I should be doing it better.	
5. *Samantha:* Janet, it looks like you feel sad and a little guilty for not doing more. Am I hearing you right?	Reflection of feeling, focus on Janet with secondary focus on the problem.
6. *Janet:* Yes, I do feel it's my responsibility to keep things together at home. He works so hard. I really feel sad and guilty that I don't do more.	The key word *responsibility* appears. Janet seems to believe that she is responsible for keeping the relationship together and that what she does is vital to that process. The acceptance of individual responsibility (and blame) is often characteristic of, but certainly not exclusive to, women of Northern European American background.
7. *Samantha:* Sounds like you're punishing yourself, even though you're trying so very hard and being so responsible. Could you tell me a bit more about how Sander is reacting?	Samantha makes an interpretation ("punishing yourself"). The focus is first on Janet, and then changes to Sander.
8. *Janet:* Yes, I get angry with myself for not doing better. But Sander wants the house to run smoothly. It's hard with my working too. I try and I try, but I always seem to miss something. Then Sander blows up.	Focus on self and other in connection with the relationship problem.
9. *Samantha:* I hear you trying very hard. Could you give me a specific example when Sander blew up at you?	Paragraph and open question. The focus changes from Janet to Sander in a search for concreteness.
10. *Janet:* Last night I made a steak dinner—I don't like meat much, but I try to fix what he wants. It was all perfect and he seemed so pleased. But I forgot to buy steak sauce and then he just blew up. It really shook me up . . . then, as we went to bed, he wanted to make love, but I was so tense that I couldn't. He got angry all over again.	Examples help us understand the specifics of situations. Some clients talk in vague generalities and asking for specifics can make a real difference.
11. *Samantha:* And then what happened?	Open question/encourage, focus on problem/main theme. Search for more concreteness.
12. *Janet:* He turned over and went to sleep. I just lay there and shook. As I calmed down, I realized that	Focus on self and Sander and actions that might be taken on her part to resolve or prevent the difficulties.

I need to listen to him more and do a better job and then maybe he won't get so angry with me.	
13. *Samantha:* Let's see if I understand what you are saying. You make it sound like you are doing everything you can to make the relationship with Sander work. Then small things come up, he blows up, and then you try harder. As part of that you feel the problem is your fault. Have I heard you correctly?	Samantha summarizes the situation. She recognizes it as a pattern in Janet's relationship with her husband. No matter how hard she tries, the situation always seem to escalate and Janet takes the blame solely on herself. Focus on Janet, Sander, and the main theme/problem.
14. *Janet:* It is my fault, isn't it?	Self-focus. Again note the acceptance of individual responsibility for the difficulties.
15. *Samantha:* I'm not so sure. Sometimes women allow themselves to be taken advantage of. Let's take a look at your community genogram hanging here on the wall. We've heard about your problem with Sander. But I'd like hearing for a moment about something that went right for you in the past. As you look at the wall there, what do you see that reminds you of good times, when things were going well?	Focus on interviewer and problem followed by focus on cultural/environmental context and a search for positive assets and wellness strengths. Many times, strengths and positive behaviors from the past can be helpful in understanding present situations.
16. *Janet:* (pause) Well, I loved visiting Grandma and Grandpa. They were like Mom. They were always friendly and helped me with my problems. Mom was the same way—she kept pointing out that I could do things. It was so hard with Dad being sick all the time.	Focus on family.
17. *Samantha:* How did they help? Could you tell me a story about Grandma and Grandpa or your Mom when they made things better for you?	Open questions focused on positive stories in the family.
18. *Janet:* Well, they were always supportive to me. Mom had to work and carry the family and I had the most fun with Grandma and Grandpa. I remember one time when kids were teasing me at school about my braces. I thought I was funny-looking because I was different. They listened to me; then they told me that I was beautiful and smart. They taught me to ignore the others, just try harder, and it would all work out. And I've usually found that it does—if I try harder, it does usually work out.	We learn here about a supportive family. At the same time, we also hear about a problem-solving style that seems to focus on ignoring underlying issues and trying harder—exactly what Janet is doing with Sander.
19. *Samantha:* So you learned in your family that ignoring things and trying harder usually helped work things out.	Paraphrase with focus on family and client. Linking of past history with present situation. Again, we see the emphasis on individual responsibility.

(continued)

Interviewer and Client Conversation	Process Comments
20. *Janet:* Usually, but it doesn't work with Sander. It used to, but it seems to be getting worse. No matter what I do, I get in trouble.	Focus on problem, self, and Sander.
21. *Samantha:* So, with Sander, trying harder simply doesn't work so well. I wonder where your desire to try so hard came from?	Paraphrase, focus on Sander. Open questions looking to family history and family patterns.
22. *Janet:* (pauses and looks at community genogram hanging on the wall) But, you know, my Dad sometimes made impossible demands of Mom. She talked with the minister of our church, and after that, she started standing up for herself more. I learned in church that it is important to care for others, but that I can't care for others unless I care for myself.	Focus on community, the family and the church—the individual develops in a family within a community. Janet is beginning to draw on resources from the past that might help her cope more effectively in the present. We are seeing the beginnings of Janet's awareness of her being a person-in-community. Helping her build connections may help her find a new balance in personal attribution of responsibility.
23. *Samantha:* There is a need to care for yourself if you are to care for others. Connections are very important to you.	Restatement emphasizing a new way of thinking about the present situation.
24. *Janet:* Yes, I wonder if perhaps I should pay more attention to me and then perhaps I can help Sander more. Right now, I feel so defeated that nothing I do seems to be going right. (pause) But, how could I start caring for myself more? What should I do? I know that I should use my spiritual resources more.	Focus on self and Sander and the problem. Janet now seems ready to start problem solving.
25. *Samantha:* I like that, Janet. I really appreciate your wanting to support Sander—you are a very caring person. But you come across so caring that you seem to be totally ignoring your own needs. Too many women fall into that trap. That could be a family lesson of caring, but we also need to focus on your Mom finally starting to care for herself. That made a difference for her and your family. I wonder if it would make the same difference for you?	Feedback with focus on interviewer. This is followed by a multiple focus on Sander, Janet, and her family with an interpretation or reframe. This reframe is a slight extension of what Janet seemed to be saying. Interpretations and reframing of the problem tend to be best received by clients if not too far from their present thinking.
26. *Janet:* I think so. I so appreciate your listening to me. I wonder if I've become too much of a doormat for Sander. But I worry about what he would do if I stood up for myself. That seems to be my family history—I'm too much like my Mom. And, certainly the women in my church always focused on serving the men. Yet,	Janet focuses on herself as a person in a family within the community.

church was where I got such a feeling of security and trust.	
27. *Samantha:* Janet, at this point I'd like to try some assertiveness training so we can look at your style of relationship. We'll do some role-plays so that I can understand the situation a bit more and then perhaps we can come up with some ideas to help you deal with the situation more effectively. But before that, I'd like to understand your relationship with the church and how spirituality works in your life. Could you share with me how spirituality has been helpful to you?	Focus on client and providing information on what might be done to facilitate change. Assertiveness training (see Chapter 14) often helps women find their own position and direction, but it needs to be presented with care as too much assertiveness with some men can be dangerous. Spiritual dimensions are to be focused on first, however. What resources may be found here?

The Confrontation Impact Scale can also be used to measure change in the session, regardless of skills used. At the beginning of this session, Janet took almost total responsibility for the problems she experienced with Sander. After review and discussion of the community genogram, she seems to have moved from a "1" on the CIS to a "3." She clearly has a new way of thinking and is beginning to see herself as a being-in-relation. But real and lasting change would require new behavior that works successfully in the relationship. Whether Janet can move to action on her new cognitions and feelings remains to be seen.

Samantha, in this session, sought to help Janet examine the situation in its cultural/environmental context. This approach is typical of those who advocate a feminist or multicultural approach to helping. The family and/or community genogram can be helpful in aiding a person who places too much responsibility on herself or himself. This can help both clients and helping professionals see the client as a person-in-relation or a person-in-community.

As you can see this is a more leading approach, although there remains a consistent attempt to listen carefully to what the person has to say. As focus shifts to various dimensions of the larger client story, the client and the counselor begin to understand the multidimensionality of the issue more fully. Janet's problem is not just "Janet's problem." Rather her issues interact with many aspects of her past and present situation.

SUMMARY

Focusing is one route toward broadening the concept of the individual. We must start, of course, by focusing on the special individual before us. Interviewing and counseling are for the person. However, by focusing on various dimensions, we can help clients expand their horizons. Connection and interdependence are as important for mental health as are independence and autonomy (Jordan, Walker, & Hartling, 2004).

The community genogram places the client in connection with the family in a cultural context. Rather than focusing on the many possible negatives in our communities of origin, the genogram reveals that these are places where we also learned strengths. The imaging of stories about community and cultural strengths can be an important resource in interviewing and counseling.

We recommend that you consider developing family and community genograms with your clients and placing them on the wall throughout the counseling series. In this way, both you and the client are reminded of the self-in-relation and the need to take multiple perspectives on any issue.

It is not necessary to generate the genograms with every client. What is most important is to be aware that for any client issue, there may be multiple explanations and multiple new stories.

Box 10-5   Key Points

Why?	Client stories and issues have many dimensions. It is tempting to accept problems as presented and to oversimplify the complexity of life. Focusing helps interviewer and client to develop an awareness of the many factors related to an issue as well as to organize thinking. Focusing can help a confused client zero in on important dimensions. Thus, focusing can be used to either open or tighten discussion.
What?	There are seven types of focuses. The one you select determines what the client is likely to talk about next, but each offers considerable room for further examination of client issues. As a counselor or interviewer, you may do the following: ▲ *Focus on the client:* "*Tari, you* were saying last time that *you* are concerned about *your* future . . ." ▲ *Focus on the main theme or problem:* "Tell me more about your *getting fired.* What happened specifically?" ▲ *Focus on others:* "So you didn't get along with the *sales manager.* I'd like to know a little more about *him* . . ." ▲ *Focus on family:* "How supportive has your *family* been?" ▲ *Focus on mutual issues or group:* "*We* will work on this. How can *you and I (our group)* work together most effectively?" ▲ *Focus on interviewer:* "*My* experience with difficult supervisors was . . ." ▲ *Focus on cultural/environmental/contextual issues:* "It's a time of *high unemployment.* Given that, what issues will be important to you as a woman seeking a job?"
How?	Focusing can be consciously added to the basic microskills of attending, questioning, paraphrasing, and so on. Careful observation of clients will lead to the most appropriate focus. In assessment and problem definition it is often helpful to consciously and deliberately assist the client to explore issues by focusing on all dimensions, one at a time. Advocacy and social action may be necessary when you discover that the client's issues cannot be resolved through the interview alone. Counseling could be described as a social justice profession.
With whom?	Focusing will be useful with all clients. With most clients the goal is often to help them focus on themselves (client focus), but with many other people, particularly those of a Southern European or African American background, the family and community focuses may at times be more appropriate. The goal of much North American counseling and therapy is individual self-actualization,

(continued)

Box 10-5 (continued)

	whereas among other cultures it may be the development of harmony with others—self-in-relation.
	Deliberate focusing is especially helpful in problem definition and assessment, where the full complexity of the problem is brought to light. Moving from focus to focus can help increase your clients' cognitive complexity and their awareness of the many interconnecting issues in making important decisions. With some clients who may be scattered in their thinking, a single focus may be wise.
What else?	Recall that counseling is for the client. Though expanding awareness of context and self-in-relation and understanding alternative stories of a situation are obviously useful, ultimately the unique client before you will be making decisions and acting. The bottom line is to assist that client in writing her or his own new story and plan of action.

COMPETENCY PRACTICE EXERCISES AND PORTFOLIO OF COMPETENCE

This chapter has presented several practice exercises within the instructional reading; therefore, the number of exercises in this section will be reduced to two, followed by the usual Self-Assessment section.

Individual Practice

Exercise 1: Writing Alternative Focus Statements

A 35-year-old client comes to you to talk about an impending divorce hearing. He says the following:

> I'm really lost right now. I can't get along with Elle, and I miss the kids terribly. My lawyer is demanding an arm and a leg for his fee, and I don't feel I can trust him. I resent what has happened over the years, and my work with a men's group at the church has helped, but only a bit. How can I get through the next 2 weeks?

Fill in the client's main issue as you see it and complete the blanks for the several alternative focus categories in the spaces that follow. Be sure to brainstorm a number of cultural/environmental/contextual possibilities.

Main issue as presented: _____

_____

_____

Client focus: _____

_____

Problem/main theme focus: _____

_____

Others focus: _____

_____

Family focus: _____

_____

Mutual, group, "we" focus: _____

_____

Interviewer focus: _____

_____

_____

Cultural/environmental/contextual focus: _____

_____

Now write alternative focus statements, as indicated.

Reflection of feeling focusing on the client:

_____

_____

_____

Open question focusing on the problem/main theme:

_____

_____

Closed question focusing on others:

_____

_____

Open question focusing on the family:

_____

_____

Reassurance statement focusing on "we":

_____

_____

Self-disclosure statement focusing on yourself, the interviewer:

_____

_____

Paraphrase focusing on a cultural/environmental/contextual issue:

_____

_____

An imaginary confrontative summary (assuming a longer interview) in which you demonstrate a mixed focus, pointing out a major client discrepancy:

_____

_____

_____

_____

_____

Exercise 2: Developing a Family Genogram

Using information from the chapter and the illustrations in Box 10-3, develop a family genogram with a volunteer client or classmate. After you have created the genogram, ask the client the following questions and note the impact of each. Change the wording and the sequence to fit the needs and interests of the volunteer.

What does this genogram mean to you? (individual focus)

As you view your family genogram, what main theme, problem, or set of issues stands out? (main theme, problem focus)

Who are some significant others, such as friends, neighbors, teachers, or even enemies who may have affected your own development and your family's? (others focus)

How would other members of your family interpret this genogram? (family, others focus)

What impact do your ethnicity, race, religion, and other cultural/environmental/contextual factors have on your own development and your family's? (C/E/C focus)

What I have learned as an interviewer working with you on this genogram is _____ (state your own observations). How do you react to my observations? (interviewer focus)

Summarize here some of what you've learned from this exercise. What questions did you find most helpful?

_____

_____

_____

_____

_____

_____

_____

Exercise 3: Developing a Community Genogram

Pages 305–309 present specific step-by-step instructions for developing a community genogram. For your first effort, we suggest that you use your own personal experience in your community of origin. As an alternative, work through the process with a class member and each of you practice drawing out the data. Present the completed genogram and briefly summarize important observations below.

_____

_____

_____

_____

_____

_____

Group Practice

Step 1: Divide into groups.
Step 2: Select a group leader.
Step 3: Assign roles for the first practice session.
▲ Client, who has completed a family or community genogram
▲ Interviewer
▲ Observer 1, who will give special attention to focus of the client, using the Feedback Form, Box 10-6. The key microsupervision issue is to help the interviewer continue a central focus on the client while simultaneously developing a comprehensive picture of the client's contextual world.
▲ Observer 2, who will give special attention to focus of the interviewer, using the Feedback Form, Box 10-6

Step 4: Plan.   Establish clear goals for the session. The task of the interviewer in this case is to go through all seven types of focus, systematically outlining the client's issue. If the task is completed successfully, a broader outline of issues related to the client's concern should be available.

A useful topic for this role-play is a story from your family or community. Your goal here is to help the client see the issues in broader perspective.

Observers should take this time to examine the feedback form and plan their own interviews. The client may fill out the Client Feedback Form of Chapter 1.

Step 5: Conduct a 5-minute practice session using the focusing skill.
Step 6: Review the practice session and provide feedback for 10 minutes.   Give special attention to the interviewer's achievement of goals and determine the mastery competencies demonstrated.

Step 7: Rotate roles.
Some general reminders.   Be sure to cover all types of focus; many practice sessions explore only the first three. In some practice sessions three members of the group all talk with the same client, and each interviewer uses a different focus.

Box 10-6   Feedback Form: Focus

_____ (Date)

_____   _____
(Name of Interviewer)   (Name of Person Completing Form)

Instructions: Observer 1 will give special attention to the client and Observer 2 to the interviewer. Note the correspondence between interviewer and client statements. In the space provided, record the main words used. Classify each statement by checking a box.

Main words	Client							Interviewer						
	Client (self)	Main theme/problem	Others	Family	Mutual/group/"we"	Interviewer	Cultural/environmental/contextual	Client	Main theme/problem	Others	Family	Mutual/group/"we"	Interviewer (self)	Cultural/environmental/contextual
1. _____														
2. _____														
3. _____														
4. _____														
5. _____														
6. _____														
7. _____														
8. _____														
9. _____														
10. _____														
11. _____														
12. _____														
13. _____														
14. _____														

(continued)

Box 10-6 (continued)

Observations about client verbal and nonverbal behavior:

_____

_____

Observations about interviewer verbal and nonverbal behavior:

_____

_____

Portfolio of Competence

The history of counseling and therapy has provided the field with a primary "I" focus in which the client is considered and treated within a totally individualistic framework. The microskill of focusing is key to the future of interviewing, counseling, and psychotherapy as it broadens the way both interviewers and clients think about the world. This does not deny the importance of the "I" focus. Rather, the multiple narratives made possible by use of microskills actually strengthen the individual, for we all live as selves-in-relation. We are not alone. The collective strengthens the individual.

At the same time, the above paragraph represents a critical theoretical point. Some might disagree with the emphasis of this chapter and argue that only the individual and main/theme focus are appropriate. What are your thoughts and feelings on this important point? As you work through this list of competencies, think ahead to how you would include or adapt these ideas in your own Portfolio of Competence.

Use the following as a checklist to evaluate your present level of mastery. Check those dimensions that you currently feel able to do. Those that remain unchecked can serve as future goals.

Level 1: Identification and classification. You will be able to identify seven types of focus as interviewers and clients demonstrate them. You will note their impact on the conversational flow of the interview.

❑ Ability to identify focus statements of the interviewer.
❑ Ability to note the impact of focus statements in terms of client conversational flow.
❑ Ability to write alternative focus responses to a single client statement.

Level 2: Basic competence. You will be able to use the seven focus types in a role-play interview and in your daily life.

❑ Ability to demonstrate use of focus types in a role-play interview and draw out multiple stories.
❑ Ability to use focusing in daily life situations.

Level 3: Intentional competence. You will be able to use the seven types of focus in the interview, and clients will change the direction of their conversation as you change focus. You will also be able to maintain the same focus as your client, if you choose (that is, not topic jumping). You will be able to combine this skill with earlier skills in this program (such as reflection of feeling and questioning) and use each skill with alternative focuses. Check those skills you have mastered and provide evidence via actual interview documentation (transcripts, tapes).

❑ My clients can tell multiple stories about their issues.
❑ I can maintain the same focus as my clients.
❑ During the interview I can observe focus changes in the client's conversation and change the focus back to the original one if it is beneficial to the client.

❑   I can combine this skill with skills learned earlier. Particularly, I can use focusing together with confrontation to expand client development.

❑   I am able to use multiple-focus strategies for complex issues facing a client.

Level 4: Teaching competence.

❑   I am able to teach focusing to clients and other persons. The impact of teaching is measured by the achievement of students, given the preceding criteria.

DETERMINING YOUR OWN STYLE AND THEORY: CRITICAL SELF-REFLECTION ON FOCUSING

Theoretically, this chapter asks you to think about the interview in a new way. The ideas of self-in-relation, person-in-relation, and the family and community genograms all suggest a more contextual approach. These and other focus dimensions may not be accepted by all theoretical orientations to helping. What are your thoughts?

What single idea stood out for you among all those presented in this chapter, in class, or through informal learning? What stands out for you is likely to be important as a guide toward your next steps. What are your thoughts on multicultural issues and the use of the focusing skill? What other points in this chapter struck you as important? How might you use ideas in this chapter to begin the process of establishing your own style and theory?

_____

_____

_____

_____

_____

_____

_____

_____

REFERENCES

Allport, A. (1998). Visual attention. In M. Posner (Ed.), *Foundations of cognitive science* (pp. 631–682). Cambridge, MA: MIT Press.

Arredondo, P., Ajamu, A., D'Andrea, M., Daniels, J., Ivey, A., Iwamasa, G., Leong, F., Kim, B., Martinez, A., Martinez, M., Orjuela, E., Parham, T., Santiago-Rivera, A., Vazquez, L., & Sue, D. W. (2000). *Culturally competent counseling and therapy: Five videotapes.* North Amherst, MA: Microtraining Associates.

Baker Miller, J. (1991). The development of a woman's sense of self. In J. Jordan, A. Kaplan, J. Baker Miller, I. Stiver, & J. Surrey (Eds.), *Women's growth in connection* (pp. 11–26). New York: Guilford.

Baker Miller, J., Stiver, I., & Hooks, T. (Eds.). (1998). *The healing connection: How women form relationships in therapy and in life.* Boston: Beacon.

Decety, J., & Jackson, P. (2004). The functional architecture of human empathy. *Behavioral and Cognitive Neuroscience Reviews, 3,* 71–100.

Harvey, J., & Manusov, L. (Eds.). (2001). *Attribution, communication behavior and close relationships: Advances in personal relationships.* Cambridge, UK: Cambridge University Press.

Ishiyama, I. (2005). *Anti-discrimination response training (A.R.T.) program.* Framingham, MA: Microtraining Associates.

Ivey, A. (1995). *The community genogram: A strategy to assess culture and community resources.* Paper presented at the American Counseling Association Convention, Denver.

Jordan, J., Walker, M., & Hartling, L. (Eds.). (2004). *The complexity of connection: Writings from the Stone Center's Jean Baker Miller Training Institute.* New York: Guilford.

McGoldrick, M., & Gerson, R. (1985). *Genograms in family assessment.* New York: Norton.

McGoldrick, M., Giordano, J., & Garcia-Preto, N. (2005). *Ethnicity and family therapy* (3rd ed.). New York: Norton.

Moos, R. (2001). *The contextual framework.* Presentation at the American Psychological Association, San Francisco, August.

Obonnaya, O. (1994). Person as community: An African understanding of the person as intrapsychic community. *Journal of Black Psychology, 20,* 75–87.

Rigazio-DiGilio, S., Ivey, A., Grady, L., & Kunkler-Peck, K. (2005). *The community genogram.* New York: Teachers College Press.

Symon, R. (2005). Pets and the community genogram. Personal e-mail communication, June.

Torres-Rivera, E., Pyhan, L., Maddux, C., Wilbur, M., & Garrett, M. (2001). Process vs. content: Integrating personal awareness and counseling skills to meet the multicultural challenge of the twenty-first century. *Counselor Education and Supervision, 41,* 28–40.

Walker, M., & Rosen, W. (Eds.). *How connections heal.* New York: Guilford.

Zalloquett, C. (2004). *An analysis of the effectiveness of multicultural skill training.* Unpublished paper. Tampa: University of South Florida.

ALLEN AND MARY'S THOUGHTS ABOUT VANESSA

There are many clients like Vanessa, who have numerous issues in their lives. We, perhaps like you, often feel overwhelmed when we get 3 to 5 minutes straight of stories about different complex problems. Sometimes there is an insistence that we do "something" immediately and start solving the issues. We have noted that when we have fallen into solving problems for clients they often refuse to listen to us and generate still more problems and difficulties.

So what needs to be done here? Each client, of course, is unique and there is no magic answer. But one rule really helps us settle down and start working with the client. *Counseling is for the individual client.* Our responsibility is to focus on what we can see and work with—specifically, the unique human being in front of us. We can't see the family, we can't see the boyfriend, and we can't study for the client.

Thus, our first efforts are to focus on the individual client. Although this chapter emphasizes a broadly based contextual perspective, we always want to recognize that counseling is most effective when we focus on the client as the core of our interviewing and treatment plan.

Most likely, with Vanessa, we would listen to her for no more than 3 to 5 minutes—and there are clients who will engage in nonstop talking for that time and longer. At some point, we would likely interrupt and say something like this:

> *Vanessa,* could we stop for a moment. I really hear *you* loud and clear. There are a lot of things happening right now. One of the best ways to approach these issues is to focus on how *you* are feeling and what's happening with *you.* Once I understand *you* a bit more, we can move to working on the issues that *you* describe.
>
> I get the sense that *you, Vanessa,* are hurting a lot right now and are confused as to what to do next. Could *you* take a deep breath and tell me what *you* are feeling and thinking this very minute—right now? What are these things doing to *Vanessa*? What's happening with *you*?

In these two paragraphs, we have used the words *Vanessa* and *you* a total of 12 times. The goal here is to help Vanessa focus on herself. If you go back to the beginning of the chapter, you will see that the primary focus of conversation has been on others and problems. We have heard very little about Vanessa, the person.

Once we have a better grasp of the person before us, we can work more effectively toward problem resolution.

Another possibility is to summarize what you have heard the client say and then ask her or him to name the one thing that he or she would like to start working on first. You can comment that we can explore other issues later.

While this chapter aims to broaden the view you can take of the client through multiperspective thought and bring in many ways to enlarge the conception of an issue, counseling is still for the individual person. We need to focus our efforts on the unique individual.

CHAPTER 11

Eliciting and Reflecting Meaning: Helping Clients Explore Values and Beliefs

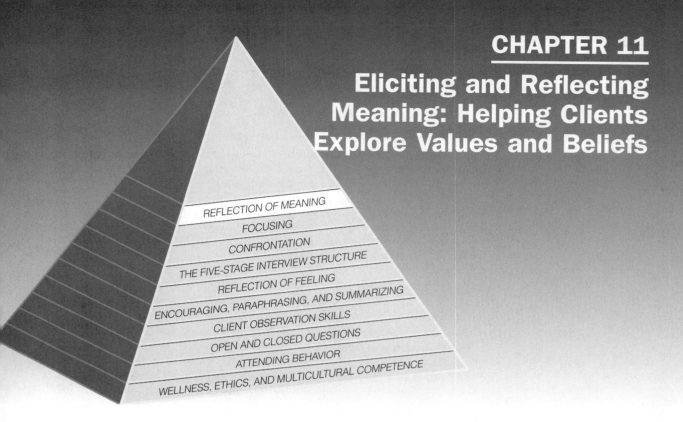

REFLECTION OF MEANING

FOCUSING

CONFRONTATION

THE FIVE-STAGE INTERVIEW STRUCTURE

REFLECTION OF FEELING

ENCOURAGING, PARAPHRASING, AND SUMMARIZING

CLIENT OBSERVATION SKILLS

OPEN AND CLOSED QUESTIONS

ATTENDING BEHAVIOR

WELLNESS, ETHICS, AND MULTICULTURAL COMPETENCE

> He who has a *why* to live can bear with almost any *how*.
>
> (Nietzsche)

How can reflection of meaning help you and your clients?

Major function	Reflection of meaning is concerned with finding the deeply held thoughts and feelings (meanings) underlying life experience. If you use reflection of meaning, you may expect clients to search into deeper aspects of their life experience. Interviewing, counseling, and therapy most often speak of thoughts, feelings, and behaviors. Underlying this basic triad is the issue of meaning—what is the purpose or significance of it all? What sense does anything make? Reflection of meaning moves beyond the surface to the underlying structure of one's life.

(continued)

This chapter is dedicated to Viktor Frankl, who inspired it. Mary and Allen had the good fortune of meeting with him in Vienna. We had just lectured at Jagiellonian University in Krakow, Poland, and followed with a visit to Auschwitz, the concentration camp where Frankl was imprisoned by the Nazis. Thus our time with him was all the more powerful. Most memorable to us was his fiery determination and his spiritual will to find meaning among life's most difficult challenges. Despite his greatness as a person and scholar, he commented, "The best of us did not survive." When asked about his relationship to Freud, who escaped Vienna to London while Frankl was unable to leave, he said: "We all build on the shoulders of giants." You will find Frankl's book *Man's Search for Meaning* (1959) very valuable to share with clients who face major tragedies and the potential loss of meaning.

For many of your clients, meaning is also closely related to spiritual issues and the role of God in their lives. You may find the concepts here useful in facilitating client awareness of spiritual and religious issues.

Secondary functions

Knowledge and skill in reflection of meaning result in the following:

▲ Facilitating clients' interpretation of their own experiences. Reflection of feeling and the skill of interpretation are closely related to this skill. However, clients can use reflection of meaning to interpret for themselves what their experience means.

▲ Understanding client deeper meanings.

▲ Assisting clients to explore their values and goals in life.

▲ Relating this skill to person-centered counseling as presented here and in Chapter 14.

▲ Facilitating client discernment of life mission.

An early heart attack downs Charlis, a workaholic, 45-year-old middle manager. After several days of intensive care, she has been moved to the floor where you are the hospital social worker, who also works with the heart attack aftercare team. It is clear that Charlis is motivated. She is following physician directives and making every effort to progress as rapidly as possible. She listens carefully to diet and exercise suggestions and, in many ways, seems the ideal patient with an excellent prognosis. One issue, however, is prominent. It is clear that she wants to return to her high-pressure job and continue her move upward through the company. At the same time, you observe some fear and puzzlement about what's happened. "Why me?" she says. As you listen to her, you sense that she feels that something is missing in her life and that she also wants to reevaluate where she is going and what she is doing.

You recognize that Charlis is reevaluating the meaning of her life. She asks questions that are hard to answer—"What is the meaning of my life? What is God saying to me? Am I on the wrong track? What should I *really* be doing?" These are issues that cannot be resolved only by problem solving or by behavior change. Something else is required.

How might you help Charlis? What thoughts occur to you? What do you see as her key issues as they might relate to the meaning of life and purpose?

_____

_____

_____

_____

You may want to compare your ideas with ours on page 354 at the end of this chapter.

INTRODUCTION: DISCOVERING AND ENCOUNTERING DEEPER ISSUES

Meaning issues often become prominent after a person has experienced a serious illness (AIDS, cancer, heart attack, loss of sight), encountered a life-changing experience (death of a significant other, divorce, loss of a job), or gone through serious trauma (war, rape, abuse, suicide of a child). You will also find these issues prominent among older clients who face major changes in their lives. These are all situations that cannot be changed, although the way one thinks about them can. They are now part of life experience.

You can help clients work through these serious issues by listening to their stories and understanding their thoughts and feelings. Effective use of basic listening skills is, of course, helpful and essential. But often clients need a deeper experience as they reflect on the meaning of their lives, which often have been totally changed by the traumatic experience.

New questions and issues arise after life-altering experiences. "How can I change my lifestyle now that I've had a heart attack?" "With AIDS, how can I most meaningfully use the rest of my life?" "Since the rape, I no longer feel safe anywhere." "I keep thinking about my wife dying; life has no meaning for me."

"What is your life about?" "What means the most to you?" "What are your basic values?" Questions such as these catch the essence of this chapter. We all have life challenges, problems and issues to resolve, and decisions to make. Underlying all these concerns are values and meaning issues. If we can help clients find what they really want and mean, their decisions and actions will flow more easily. *If you help your clients find a* why *through exploration of meaning, they will often find their own* how *or way to achieve their objectives.*

Beyond this is the area of spirituality and religion where many of your clients will find help in finding direction and purpose. While this chapter does not focus on these issues, it is important that you maintain awareness of this dimension as values are often central spiritual guidelines. Later pages on discernment of life mission also closely relate to spiritual and religious issues.

In helping our clients find their *why* and *how,* we need to be aware of levels of communication. Two levels of speech and action may be identified. Thus far, we have been working primarily with *explicit,* observable levels. Attending, questioning, encouraging, paraphrasing, and summarizing all focus on reacting specifically to what a client says and does. In using reflection of feeling it is necessary to distinguish the concepts of implicit and explicit feelings. We know that our feelings may sometimes run deeper than our own awareness of them. Meanings run as deep as, or deeper than, feelings; meanings provide basic organizing constructs for our lives. The skill of discovering and reflecting meaning often operates at an *implicit* level.

It is important, at this point, to consider the relationship of reflection of meaning to other skills. Figure 11-1 provides a pictorial representation of meaning and its relation to behaviors, thoughts, and feelings. The following points are important in relating the figure and its concepts to your own practice.

1. All four dimensions are operating simultaneously and constantly in any individual or group. We are systems, and any change in one part of the system affects the total.

2. As a general rule, paraphrases speak to thoughts, reflections of feelings to feelings, attending behavior and client observation to behaviors, and reflection of meaning to meaning. Additionally, adherents of different helping methods and theories view different areas of the model as most important.

3. In using these skills, we attempt to break down the complex behavior of the client into component parts. It is possible that attacking only one dimension (for

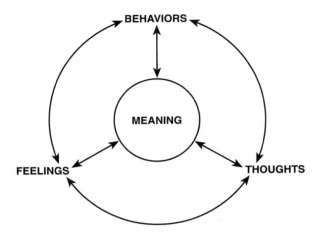

Figure 11-1 Pictorial representation of the relations among behaviors, thoughts, feelings, and meanings

instance, thoughts through paraphrasing) will lead to a change in client behavior, which in turn affects changes in feelings and meaning. *A change in any one part of the system may result in a change in other parts as well.* Even though, for practical purposes, we divide the client and the counseling interview into component parts, we cannot escape the interaction of those parts.

4. For many clients meaning is the central issue, and it is here that the most profound change may occur. For other clients, however, change in thoughts (rational-emotive therapy, cognitive-behavior modification) may be most helpful. Still others may change behavior (behavior modification, reality therapy) or work on feelings (Rogerian, encounter therapy). Meaning-oriented therapies include psychoanalysis, logotherapy, and person-centered therapy (see also Chapter 14).

5. Integrative approaches such as developmental counseling and therapy and multimodal therapy consider meaning as a central aspect (Ivey, Ivey, Myers, & Sweeney, 2005; Lazarus & Fay, 2000).

To obtain some sense of the distinctions among the concepts, let us review them in terms of the experience of divorce:

▲ *Behaviorally,* the person undergoing divorce is *doing* many things. He or she is consulting with a lawyer, perhaps arguing with the spouse on the telephone, finding new housing, and so on. You will find the words *do, doing,* and *acting* characteristic of the behavioral mode.

▲ *Feelings* are the emotions occurring in the individual while he or she is acting or thinking. For example, the client engaging in the behavior of finding new housing may feel sad, angry, or glad, perhaps all three, or even more feelings. The words *emotion, affect,* and *feelings* and the many feeling words you generated in Chapter 7 exemplify this mode. Research on the brain indicates that feelings are the most fundamental of the four concepts.

▲ *Thoughts* are mental messages that occur simultaneously with behavior and feeling. For example, the client searching for a house may have many different thoughts constantly running through his or her mind: "How can I afford decent housing?" or "I'm guilty; I shouldn't have had that affair" or "I wonder what the kids are doing now?" or even "I'm a bad person." Many such almost-random thoughts run through our minds constantly as we go through our daily routines. Sometimes a client gets stuck on one repeating type of destructive thought pattern—for example, "I'm a bad person." In such cases, cognitive techniques such as thought-stopping can be useful.

▲ *Meaning* is closely allied to feelings and thoughts and may be considered an important driving force of behavior. Meaning can often override and control emotions as values can become central to one's life goals. Moreover, the meaning level represents a clustering or grouping of thoughts and feelings into a more coherent whole. Meanings may be described as the underlying, major constructs that we use to organize our experience: our thoughts, feelings, and behavior. Meaning is manifested by more overt and directly observable thoughts, feelings, and behavior. Key words associated with meaning are *values, beliefs, unconscious motivators* (things that impel our behavior, of which we are not presently aware), and *making sense of things.*

The world around us is often complex and confusing. We process data from this world and try to make sense of it. Some of these patterns of organization come from society and culture, others come from parents and close friends, and some are totally unique to us. Some theorists suggest we use a form of *inner speech* in which we "talk" to ourselves and make sense of things. All of us seem to have some system of ordering meaning, concepts, and judgments, but with varying degrees of clarity.

Counseling can help individuals clarify underlying meanings. It is, of course, helpful to assist a client in exploring a divorce or the death of a significant person through questioning, paraphrasing, and reflecting feeling. But the thoughts and feelings are not usually worked through until they are organized into some meaningful value pattern, which often provides *reasons* for what happened. "Death, I guess, is part of life." "I can't help missing him, but at least I know he's in heaven." "I guess my extreme reaction is partly because I feel guilty and afraid of dying myself." These are just three examples of organized meaning systems' responses to the idea of death. Issues of spirituality could add further depth to the discussion.

Discovering and reflecting meaning may at times be a difficult skill; yet, used sparingly and effectively, it can help clients find themselves and their direction more clearly.

In practice, eliciting and reflecting meaning looks and sounds very much like questioning, a reflection of feeling, or a paraphrase. However, it is distinctly different in tone or purpose. You will often find a reflection of meaning following a "meaning probe" or question ("What does that mean to you?") or following an encourager focusing on an important single key word ("Divorce . . ."). The reflection of meaning is structured like the reflection of feeling with the exception that "You feel . . ." becomes "You mean. . . ." The reflection of meaning then paraphrases the important ideas of meaning expressed by the client. The skill is particularly important in cognitive-behavior theory, existential approaches to counseling, and logotherapy. The following transcript illustrates reflection of meaning in action.

EXAMPLE INTERVIEW: SURVIVING DIVORCE

In the following session, Travis is reflecting on his recent divorce. When relationships end, the thoughts, feelings, and underlying meaning of the other person and the time together often remain an unsolved mystery. Moreover, some clients are likely to repeat the same mistakes in their relationships when they meet a new person.

Terrell, the interviewer, seeks to help Travis think about the word *relationship* and its meaning. As you read this session, you'll note that Travis stresses the importance of connectedness with intimacy and caring. The issue of self-in-relation to others will play itself out very differently among individuals in varying cultural contexts. Many clients will focus on their need for independence.

Interviewer and Client Conversation	Process Comments
1. *Terrell:* So, Travis, you're thinking about the divorce again . . .	Encourager/restatement.
2. *Travis:* Yeah, that divorce has really thrown me for a loop, I tell ya. I really cared a lot about Ashley and . . . ah . . . we got along well together. But there was something missing.	
3. *Terrell:* Uh-huh . . . something missing?	Encouragers appear to be closely related to meaning, in many cases. You will find that clients often supply the meaning of their key words if you repeat them back exactly.
4. *Travis:* Uh-huh, we just never really shared something very basic. You know . . . it was like the relationship didn't have enough depth to go anywhere. We liked each other, we amused one another, but beyond that . . . I don't know . . .	Travis elaborates on the meaning and he appears to be searching for a closer, more significant relationship than he had with Ashley.
5. *Terrell:* You amused each other, but you wanted more depth. What sense do you make of it?	Paraphrase using Travis's key words followed by a question to elicit meaning.
6. *Travis:* Well, in a way, it seems like the relationship was somewhat shallow. When we got married, there just wasn't much . . . ah . . . depth there that I had hoped for in a meaningful relationship. The sex was good, but after awhile, I even got bored with that. We just didn't talk much. I needed more . . .	Note that Travis's personal constructs for discussing his past relationship center on the word *shallow* and the contrast *meaningful.* This polarity is probably one of Travis's significant meanings around which he organizes much of his experience.
7. *Terrell:* Mm-hmmm . . . you seem to be talking in terms of shallow versus meaningful relationships.	Reflection of meaning followed by a question designed to elicit further exploration of meaning.

(continued)

What does a meaningful relationship feel like to you?	
8. *Travis:* Well, I guess . . . ah . . . that's a good question. I guess for me, in order to be married, there has to be some real, you know, some real caring beyond just on a daily basis. It has to be something that goes right to the soul. You know, you're really connected to your partner in a very powerful way.	Connection appears to be a central dimension of meaning.
9. *Terrell:* So, connections, soul, deeper aspects strike you as really important.	Reflection of meaning. Note that this reflection is also very close to a paraphrase, and Terrell uses Travis's main words. The distinction centers around issues of meaning. A reflection of meaning could be described as a special type of paraphrase.
10. *Travis:* That's right. If I'm married to somebody, I have to be more than just a roommate. There has to be some reason for me to really want to stay married, and I think with her . . . ah . . . those connections and that depth were missing, and we didn't miss each other that much. We liked each other, you know, but when one of us was gone, it just didn't seem to matter whether we were here or there.	
11. *Terrell:* You'd like a relationship that is meaningful. Even when she was gone, you didn't miss her. You'd value more closeness.	Reflection of meaning plus some reflection of feeling. Note that Terrell has added the word *values* to the discussion. In reflection of meaning it is likely that the counselor or interviewer will add words such as *meaning, understanding, sense,* and *value.* Such words seem to produce a very different discussion and lead the client to interpret experience from her or his own frame of reference.
12. *Travis:* Uh-huh.	
13. *Terrell:* Ah . . . could you fantasize how you might play out those thoughts, feelings, and meanings in another relationship?	Open question oriented to meaning.
14. *Travis:* Well, I guess it's important for me to have some independence from a person, but I'd like that independence such that, when we were apart, we'd still be thinking of the other one. Depth and a soulmate is what I want.	Travis's meaning and desire for a relationship are now being more fully explored.
15. *Terrell:* Um-humm.	

(continued)

16. *Travis:* In other words, I don't want a relationship in which we are always tagging along together. You know, the opposite of that is where you don't care enough whether you are together or not. That isn't intimate enough. I guess what it boils down to is that I really want the intimacy in a marriage. My fantasy is to have a very independent partner whom I care very much about and who cares very much about me, and we can both live our lives and be individuals and have that bonding and that connectedness.	Again, we find that connectedness is an important meaning issue for Travis. With other clients, independence and autonomy may be the issue. With still others, the meaning in a relationship may be a balance of the two.
17. *Terrell:* Let's see if I can put together what you're saying. The key words seem to be independence with intimacy and caring. It's these concepts that can produce bonding and connectedness, as you say, whether you are together or not.	This reflection of meaning becomes almost a summarization of meaning. Note that the key words and constructs have come from the client in response to questions about meaning and value.

A logical place to move from this point is further exploration of meaning. Then, the next issue is moving these underlying meanings to actions in life. Counseling would seek to bring behavior into accord with thoughts. Other past or current relationships could be explored for how well they achieve the client's important meanings as well as for behaviors that illustrate or do not illustrate the meaning in action.

INSTRUCTIONAL READING: DISCERNMENT OF LIFE DIRECTION AND THEORY AND MEANING SYSTEMS

The instructional reading on the skill of reflection of meaning is divided into three sections. The first explores some alternative meaning and value systems, the second discusses practice of the skill, and the third illustrates reflection of meaning in theoretical context.

Alternative Meaning and Value Systems

Meaning does not arise from empty space. Many consider the way we make sense of things a result of our family history, our communities, our cultural and ethnic background, our spirituality, and other factors. Others would point out that these dimensions interact with one another uniquely in the individual life experience of the individual. Some might say that meaning is given to us by our background, while others locate meaning-making totally in the individual.

Given these widely varied sources of meaning, the concept of focusing can be helpful in facilitating client exploration of meanings and values. An individual will often generate a deeper sense of self as he or she discovers how deeply values have been influenced by religious background, parental models, or traumas (such as war, accident, or personal violation—rape, being held hostage, physical or emotional abuse). Just as meaning is closely tied to thoughts and feelings, and behaviors, it is also closely related to our significant life experiences.

Box 11-1 National and International Perspectives: What Can You Gain From Counseling Persons With AIDS and Serious Health Issues?

Weijun Zhang

A good friend of mine had just started counseling persons with AIDS. When I asked him, "What does this mean to you?" he started to grumble: "It's being around people with serious illness who could die at any time. It also means that no one gets cured, despite my best efforts." What a bleak picture he painted. No wonder some counselors are reluctant to work with AIDS clients or those facing truly serious health issues.

Can something as miserable and difficult as AIDS be meaningful? How can one work as a counselor in a kidney dialysis unit? What about working in a hospice where all are expected to die within a reasonably short time? Certainly, there is much to learn and profit from in this type of work. What my friend was missing is that there are precious rewards available.

Years ago, I happened upon an enlightening article by Craig Kain (1992), author of "No Longer Immune: A Counselor's Guide to AIDS." He maintains that one can gain precious rewards from doing AIDS work. I wondered what that meant.

1. Working with people who have no choice but to live in the present can change our perspective. It can help us find a deeper meaning in life and learn "to watch the world in a wondrous way, as if today were my last living day."
2. Having direct contact with clients who face unthinkable pain and suffering but who still strive to live fully can help us understand the great strength that is the human spirit. This courage in the face of adversity can be very contagious.
3. Doing AIDS work will enable us to witness a lot of truly unconditional love and help us to become bigger-hearted, more loving, and caring persons.

I have always liked doing work with the seriously ill because of the selfless giving needed on the part of the counselor. But now I understand that it can mean a tremendous taking, learning, and satisfaction for those who help others most in need. Though this taking should never be our primary motive in doing AIDS or related work, it is certainly rewarding to see the light when we deal with the darkness associated with major life challenges. This positive insight I wish to share with my good friends.

Allen Ivey comments: Not only do clients make meaning, but we counselors and helpers do as well. If we make negative meanings and interpretations of life experience, we inevitably pass on our bitterness and sadness to others. There are also some who wish to avoid dealing with disappointment and hurt who shield their clients from difficult issues; they take away from clients an important part of life. You will find that some of the most satisfying work a counselor can do is with those who face the most challenging life issues. Helping these clients make positive meanings is one of counseling's most important tasks.

We invite you to pause and think for a moment about your most significant life influences. What are those experiences in your life that were most responsible for your being who you are at this moment? It is quite appropriate to list important ordinary experiences as well—they are often the most meaningful. You may want to be silent for a moment—perhaps even moving yourself to a meditative state before making a list.

_____   _____
_____   _____
_____   _____
_____   _____
_____   _____

Out of significant life experiences and our own natural way of being we generate a sense of self and/or a sense of self-in-relation. For many people spirituality and religion will be central influences. For others, the time spent in the natural landscape gives center to their being.

Life-changing events are important in the generation of meaning systems. Some examples include college graduation, a move to a new location, meeting the right (or wrong) person, serving in the army, coping with a life-threatening illness, failing an important task, coming out as gay or lesbian, and identifying the importance of one's racial-ethnic heritage.

Both ordinary and extraordinary experiences, such as a stable family life or a constant life of physical and emotional harassment, lead to the development of meaning systems. Daily experiences in school, with co-workers, or in other situations all generate thoughts, feelings, and behaviors that ultimately reinforce or change our meaning system.

In considering the skill of reflection of meaning, we ask the following questions: What does this mean to the client? How does what the client is doing reflect a value structure? Can we find presently existing values that become the instrument of positive change and development?

Defining Skills of Eliciting and Reflection of Meaning

Clients do not usually volunteer discussion of meaning issues, even though they may be central to their concerns. As such, the first step in reflecting meaning often is to help them think in meaning terms through the use of a variety of questions. The following strategies to elicit meaning may be useful.

1. The behaviors, thoughts, and feelings need to have been made explicit and clear through attending behavior, client observation, and the basic listening sequence. A general understanding of the client is essential as a first step.
2. Consider storytelling as a useful way to discover the background of a client's meaning-making. Critical life events such as illness, loss of a parent or loved one, accident, or divorce often force people to encounter deeper meaning issues. If a major life event is critical, illustrative stories can form the basis for exploration of meaning. If spiritual issues come to the fore, draw out one or two concrete example stories of the client's religious heritage. From a narrative frame of reference, we are interested in learning how the client's story was generated in social context. The issue then becomes how values and meanings internalized from this experience serve or sometimes conflict with current life experiences. The resolution of this type of conflict may come with generating a new story that helps resolve issues in a positive fashion.

Speaking of multicultural and spiritual issues in counseling, Fukuyama (1990, p. 9) has outlined some useful questions for eliciting stories and client meaning systems. Adapted for this chapter, they include the following:

"What significant life events have shaped your beliefs about life?"
"What are your earliest childhood memories as you first identified your ethnic-cultural background? Your spirituality?"
"What are your earliest memories of church, synagogue, mosque, a higher power, or lack of religion?"
"Where are you now in your life journey? Your spiritual journey?"
"When in your life did you have existential or meaning questions? How have you resolved these issues thus far?"

3. Questions in which content is oriented toward meaning may be asked. For example:

"What does this *mean* to you?"
"What *sense* do you make of it?"
"What *values* underlie your actions?"
"*Why* is that important (or unimportant) to you?"
"Could you give me examples of some *values* that are *important in your life* decisions? How have those *values* been *implemented* in your life?"
"What are some of the *reasons* you think that happened?"
"Which of your *personal values* support/oppose that behavior/thought/feeling?"
"*Why?*" (by itself, used carefully)
"What do you want to pass on to your family?"

4. The key meaning and value words of the client are reflected. It is very important to use the exact, key words of the client for the major ideas. Your task is reflecting meaning, values, and the way a client makes sense of the world. It is his or her own unique system, not yours. A reflection of meaning is structured similarly to a paraphrase or reflection of feeling (see the examples in the next section).

Once the implicit meaning is brought forward, it is relatively easy to reflect meaning. Simply change "You feel . . ." to "You mean. . . ." Other variations could include "You value . . . ," "You care . . . ," "Your reasons are . . . ," or "Your intention was. . . ." Distinguishing among a reflection of meaning, a paraphrase, and a reflection of feeling may at times be difficult. Often the skilled counselor will blend the three skills together. For practice, however, it is useful to separate out meaning responses and develop an understanding of their import and power in the interview. Noting the key words that relate to meaning (*meaning, value, reasons, intent, cause,* and the like) will help distinguish reflection of meaning from other skills.

Reflection of meaning becomes vastly more complicated when meanings or values conflict. Just as clients may express mixed and confused feelings about an issue, so may explicitly or implicitly conflicting values underlie their statements. A client may feel forced to choose between loyalty to family and loyalty to spouse, for instance. Underlying meanings of love for both parties may be complicated by a value of dependence fostered by the family and the independence represented by the spouse. In making a

Box 11-2 Research Evidence That You Can Use: Reflection of Meaning

For information on the skill of reflection of meaning, it is best to turn to the literature on meaning-oriented counseling and psychotherapy theory—person-centered counseling, logotherapy, and Gestalt therapy. All these systems are oriented to helping clients find deeper meanings in their lives (Ivey, D'Andrea, Ivey, & Simek-Morgan, 2002). Carl Rogers brought meaning issues to center stage in counseling and therapy, whereas Viktor Frankl provided both philosophical and practical applications of meaning in counseling. A solid relationship with your client helps give meaning to your encounter.

Classic research by Fiedler (1950) and Barrett-Lennard (1962) set the stage for the present when they found that relationship variables (closely related to the listening skills) were vital to the success of all forms of interviewing, *regardless of theory.* Now, you will find the relationship issues are termed "common factors" and the idea of relationship as central has become almost universally accepted. Those working in the "Heart and Soul of Change Project" cite data suggesting that 30% or more of successful therapy is based on relationship (Miller, Duncan, & Hubble, 2005).

Research has demonstrated that "families that seek support and try to accept what happened after a traumatic injury may experience less injury-related stress and family dysfunction over time" (Wade et al., 2001, p. 412). Turning to religion was the second most used strategy among parents whose children suffered traumatic injury. Connectedness with others and the comfort of spirituality can be a most important positive asset and wellness strength for many clients. Religiously oriented clients do better in cognitive-behavioral therapy when their spirituality becomes part of the process (Probst, 1996). Recovery from heart surgery has been found more rapid among those with religious involvement, particularly among women (Contrada et al., 2004, p. 227).

Neuropsychology and meaning

At the surface, the broad idea of meaning would appear to be beyond measurement in a physical sense. Our sense of meaning brings our thoughts, feelings, and behavior into a whole, enabling us to make sense of our experience. Surprisingly, this seemingly abstract concept of meaning can be pinpointed rather precisely (Carter, 1999, p. 197):

Meaningfulness is inextricably bound up with emotion. Depression is marked by wide-ranging symptoms, but the cardinal feature of it is the draining of meaning from life. . . . By contrast, those in a state of mania see life as a gloriously ordered, integrated whole. Everything seems to be connected and the smallest events are bathed in meaning. . . .

The area of the brain that is most noticeably affected in both depression and mania is an area on the lower part of the internal surface of the prefrontal cortex . . . , the brain's emotional control center. It is exceptionally active in bouts of mania and inactive (along with other prefrontal areas) during depression.

We want to help depressed clients become active, whether it is through exercise, nutrition, or learning new ways to think and/or behave. Discerning one's life mission and what is really most important can change meaning systems and alleviate depression. Allen's work in training clients in attending behavior and social skills is one example of how we can help clients gain meaning (see Chapter 3, page 76). When clients learn to listen to someone other than themselves or to help others, they gain a new purpose, a new meaning, and they do not need to focus solely on themselves.

decision, it may be more important for the client to sort out these felt meaning constructs than the more overt facts and feelings associated with the decision.

For example, a young person may be experiencing a value conflict over career choice. Spiritual meanings may conflict with the work setting. The facts may be paraphrased accurately and the feelings about each choice duly noted, yet the underlying *meaning* of the choice may be most important. The counselor can ask, "What does each choice mean for you? What sense do you make of each?" The client's answers provide the opportunity for the counselor to reflect back the meaning, eventually leading to a decision that involves not only facts and feelings but also values and meaning.

Now that we have defined reflection of meaning, the remainder of this instructional reading section will focus on the place of this skill in different theoretical frameworks. Many counseling and therapy theories take into account issues of meaning. A few examples of how this skill undergirds the helping process follow.

Discernment: Identifying Life Mission and Goals

Listen. Listen, with intention, with love, with the "ear of the heart." Listen not only cerebrally with the intellect, but with the whole of feelings, our emotions, imaginations, and ourselves.   (de Waal, 1997, Preface)

This chapter began with the illustration of Charlis, who had experienced a heart attack and was beginning to explore the meaning of her life and seeking to determine what she should "really" be doing. Here we'd like to explore with you the concept of *discernment,* a complex term, often related to spirituality, that offers some specific questions that you may use to help clients who seriously want to examine their goals and mission in life. Moreover, discernment has broad applications to interviewing and counseling as it also describes what we do when we work with clients at deeper levels of meaning.

The word discernment comes from the Latin *discernere,* which means "to separate," "to determine," "to sort out." In a spiritual or religious sense, discernment means identifying when the spirit is at work in a situation: the spirit of God or some other spirit. Discernment is "sifting through" our interior and exterior experiences to determine their origin" (Farnham, Gill, McLean, & Ward, 1991). Moreover, the discernment process is important for all clients, regardless of their spiritual or religious orientation or lack thereof. Discernment is also a value clarification process in which the client seeks to find meaning and direction in life.

Counseling and therapy are also concerned with working through and sorting out value issues, although they tend to use a different language. Frankl (1978), for example, talks about the "unheard cry for meaning." Frankl has claimed that 85% of suicides see life as meaningless. Frankl goes on to attack an excessive focus on self and the need for transcendence and living beyond one's self. Camus (1955, p. 3) states, "There is but one truly serious problem, and that is . . . judging whether or not life is worth living."

The vision quest, usually associated with the Native American tradition, is oriented to helping youth and others find purpose and meaning in their lives. The individual often undertakes a serious outdoor experience for a period of time, understanding that the reason is to find or envision one's central life goals. Meditation or a life test is used in some cultures to help members find meaning and direction.

Box 11-3　Questions Leading Toward Discernment

We recommend that you share this list of questions with the person going through the discernment process. In addition, we encourage additional topics and questions, both from you and the client. You may find it helpful to go through this list with the client before you start, and check the questions that you both feel might be most helpful to explore. Discernment is a very personal exploration of meaning and the more the client participates, the more useful it is likely to be.

Below are example questions, strategies, or statements that may facilitate sensorimotor discovery. These questions tend to focus on the here and now and intuition.

▲ Relax, explore your body and find a feeling there that might serve as an anchor for your search. Allow yourself to build on that feeling and see where it goes.

▲ Sit quietly and allow an image (visual, auditory, kinesthetic) to build.

▲ What is your gut feeling? What are your instincts? Get in touch with your body.

▲ Discerning one's mission cannot be found solely through the intellect. What feelings occur to you at this moment?

▲ Imagine yourself talking to someone who can help you. See that person in the chair there. Speak to him or her, then move to that chair and have the person answer you. (Gestalt empty chair work)

▲ Consider drawing, painting, music, and other creative art forms. Where do they lead you?

▲ Can you recall feelings and thoughts from your childhood that might lead to a sense of direction now?

▲ What is your felt body sense of spirituality?

Within the concrete style, consider the following, relating to stories, specifics, and action.

▲ Tell me a story about that sensorimotor image above. (Or a story about any of the here-and-now experiences listed there.)

▲ Can you tell me a story that relates to your goals/vision/mission?

▲ Can you name the feelings you have in relation to your desires?

▲ What have you done in the past or are doing presently that feels especially satisfying and close to your mission?

▲ What are some blocks and impediments to your mission? What holds you back?

▲ Could we try role-playing what it would be like to act on those images?

▲ What are you going to do now that we have some sense of your mission? What actions will you take?

▲ When can you take action? What plans need to be made?

▲ Can you tell about spiritual stories that have influenced you?

For formal exploration, the following may be helpful. These focus on self-reflection.

▲ Let's go back to that original image and/or the story that goes with it. As you reflect on that experience or story, what occurs for you? (Or use any of the information gleaned from earlier sensorimotor or concrete questioning.)

(continued)

Box 11-3   (continued)

▲ Looking back on your life, what have been some of the major satisfactions? Dissatisfactions?

▲ What have you done right?

▲ What have been the peak moments and experiences of your life?

▲ What might you change if you were to face that situation again?

▲ Does a need for approval/admiration/respect relate to these goals?

▲ Do you have sense of obligation that impels you toward this vision?

▲ Most of us have multiple emotions as we face major challenges such as this. What are some of these feelings and what impact are they having on you?

▲ Are you motivated by love/zeal/a sense of morality?

▲ What are your life goals?

▲ What has been most meaningful (positive/supporting and negative/challenging) to you among your broad experience?

▲ What do you see as your mission in life?

▲ What does spirituality mean to you?

Dialectic/systemic questions place the client in larger systems and relationships—the self-in-relation.

▲ Place your sensorimotor, concrete, and formal experiences in broader context. How have various systems (family, friends, community, culture, spirituality, and significant others) related to these experiences? Think of yourself as a self-in-relation, a person-in-community.

▲ *Family.* What do you learn from your parents, grandparents, and siblings that might be helpful in your discernment process? Are they models for you that you might want to follow, or even oppose? If you now have your own family, what do you learn from them and what is the implication of your discernment for them?

▲ *Friends.* What do you learn from friends? How important are relationships to you? Recall important developmental experiences you have had with peer groups. What do you learn from them?

▲ *Community.* What people have influenced you and perhaps serve as role models? What group activities in your community may have influenced you? What would you like to do to improve your community? What important school experiences do you recall?

▲ *Cultural groupings.* What is the place of your ethnicity/race in discernment? Gender? Sexual orientation? Physical ability? Language? Socioeconomic background? Age? Life experience with trauma?

▲ *Significant other(s).* Who is your significant other? What does he or she mean to you? How does this person relate to the discernment process? What occurs to you as the gifts of relationship? The challenges?

▲ *Spiritual.* How might you want to serve? How committed are you? What is your relationship to spirituality and religion? What does your holy book say to you about this process?

_____

Discernment questions from Ivey, A., Ivey, M., Myers, J., & Sweeney, T. (2005). *Developmental Counseling and Therapy: Promoting Wellness Over the Lifespan.* Boston: Lahaska/Houghton-Mifflin. Reprinted by permission.

Visioning and finding meaning may often be more effective if issues are explored with a guide, counselor, or interviewer. Box 11-3 presents some questions generated by developmental counseling and therapy oriented to discerning life's meaning. As spirituality and religion issues often focus on meaning, you may find yourself at times exploring these issues with your client. For further reading on these issues, please review Ivey, Ivey, Myers, and Sweeney (2005).

The specific discernment questions of Box 11-3 should be considered a beginning. The client may suggest additional specific issues and questions that are important for consideration. We also recommend sharing the list of questions with the client, encouraging participation in deciding which questions and issues are most important.

Reflection of Meaning and Logotherapy

If one person were to be identified with meaning and the therapeutic process, that individual would have to be Viktor Frankl, the originator of logotherapy. Frankl (1959) has pointed out the importance of a life philosophy that enables us to transcend suffering and find meaning in our existence. He argues that our greatest human need is for a core of meaning and purpose in life.

Frankl's life is a testament to the power of meaning. He was imprisoned in the concentration camp at Auschwitz, but survived. Shortly after his liberation, he wrote his famous book *Man's Search for Meaning* (1959) within a 3-week period and it has remained a constant best seller since that time. Frankl believed that finding positive meanings in the depth of despair was vital to keeping him alive. During the darkest moments, he would focus his attention on his wife and the good things they enjoyed together; or, in the middle of extreme hunger, he would meditate on a beautiful sunset. This book will be helpful to clients facing crisis.

As you can see, Frankl was influential in our focus on wellness and the positive asset search. The recent trends toward a positive psychology and "learned optimism" (Seligman, 1998) are other examples of how an emphasis on strengths can aid the client.

Logotherapists search for positive meaning that underlies behavior, thought, and action. Dereflection is a specific strategy that logotherapy uses to uncover deeper meanings and help clients become more positive in outlook. Dereflection and modification of underlying attitudes are specific techniques that logotherapy uses to uncover meaning. Many clients "hyperreflect" (think about something too much) on the negative meaning of events in their lives and may overeat, drink to excess, or wallow in depression. They are constantly attributing a negative meaning to life.

The direct reflection of meaning may encourage such clients to continue these negative thoughts and behavior patterns. Dereflection, by contrast, seeks to help clients discover the values that lie deeper in themselves. The goal is to enable clients to think of things *other* than the negative issue and to find alternative positive meanings in the same event. The questions listed at the beginning of the instructional reading section represent first steps in helping clients dereflect and change their attitudes. The following abbreviated example illustrates this approach.

Client: I really feel at a loss. Nothing in my life makes sense right now.

Counselor: I understand that—we've talked about the issues with your partner and how sad you are. Let's shift just a bit. Could you tell me about what has been meaningful and important to you in the past? (The client shares some key supportive religious experiences from the past. The counselor draws out the stories and listens carefully.)

Counselor: (reflecting meaning) So, you found considerable meaning and value in worship and time spent quietly. You also found worth in service in the church. You drifted away because of your partner's lack of interest. And now you feel you betrayed some of your basic values. Where does this lead you in terms of a meaningful way to handle some of your present concerns?

As you may note, the process of dereflection is a special form of the positive asset search. But rather than focusing just on the concretes (spirituality, service to others, walking in the outdoors, enjoying one's friends), the counselor explores the positive meaning of these specifics. "What does spirituality mean to you?" "What sense do you make of a person who finds such joy in walking outdoors and enjoying sunsets?" "What values do you find in service to others?"

Out of the exploration of meaning may come data for restorying one's problems and even life-transformative actions. A modified dereflection exercise is included in the exercises at the end of the chapter.

Reflection of Meaning and Person-Centered Counseling

Carl Rogers's person-centered counseling (Bozarth, 1999; Rogers, 1961) is sometimes characterized as consisting of predominantly paraphrasing and reflection of feeling, at least in the early stages. However, *meaning* plays perhaps an even more important part in his overall thinking and conceptualization. The following brief excerpt illustrates how the use of reflection of meaning and reflection of feeling are closely intertwined:

Client: I have all the symptoms of the fear, even though it's something I want.

Therapist: The fear somehow hits you at the core. Is that what you mean? (Note reflection of meaning, followed by a check-out.)

Client: Somehow, fear is inside me anytime I get near my goal.

Therapist: Nearing goals reaches an issue somehow deeper for you.

This pattern of encouraging, listening to the client, and searching for deeper meanings is quite similar to the reflection of meaning discussed in this chapter. Reflection of feeling, by contrast, would work more directly with the emotions themselves. Meaning is a deeper concept and area for exploration.

Person-centered counseling is presented more fully in Chapter 14. At this point, you may want to read that material as a supplement to this brief discussion.

Reflection of Meaning and Cognitive-Behavioral Approaches

Cognitive-behavior therapy (Beck & Beck, 1995; Lazarus & Fay, 2000; Meichenbaum, 1994) has become an important area for the practice of counseling. Cognitive-behavioral theorists talk in depth about cognitive structures and internal dialogue. They are interested in overt behavior but also want to explore the underlying processes (or "inner speech") that monitor and guide more observable behavior.

A common cognitive-behavioral strategy directs clients to specific activities to change or alter thought sequences and meaning. Some examples are thought-stopping, guided imagery, and implosion. Moreover, cognitive-behavioral therapists add an additional armament of influencing skills and strategies to their work. They seek to move clients more rapidly into new patterns of cognition and meaning.

Counselor: So the reason you gamble is to avoid your own deeper feelings of self-doubt and worthlessness? Your inner speech and repeating statements to yourself seem to be "I'm no good; I'm worthless." Right?

Client: Yeah, those ideas keep running through my mind. The excitement of the racetrack helps me forget . . . for a while . . .

Counselor: Stay with that feeling now . . . Magnify that feeling of worthlessness . . . think about it. What comes to your mind?

Client: My father . . . he always said I'd turn out worthless. I guess he was right.

Counselor: He was right? Who's living your life, you or your father? Whose thoughts are you living?

Client: I guess I'm living up to his expectations.

Counselor: Down to his expectations, you mean. What self-image do you want? Right now, say to yourself, "This is my life . . . my decision . . ."

This is but one example of the many techniques that a directive cognitive-behaviorist might use to facilitate a client's change of meaning. At issue is finding the underlying thought and meaning structures that motivate life experience and then actively changing them to more positive frames of reference. The cognitive-behaviorist does not just reflect meaning, but actively works to help clients find more useful, positive meanings in their lives.

Multicultural Issues and Reflection of Meaning

Eliciting and reflecting meaning can be an important multicultural skill. Viktor Frankl generated many of his thoughts while in a German concentration camp. Frankl could not change his life situation, but he was able to change the meaning he made of it. He drew on important strengths of his Jewish tradition, which facilitated his survival and enabled him to help others.

Cultural, ethnic, religious, and gender groups all have systems of meaning that give an individual a sense of coherence and connection with others. A Christian will often make meaning of difficult situations drawing on the teachings of Jesus. Similarly, Muslim, Buddhist, and other religious groups will draw on their traditions. African Americans may draw on the strengths of Malcolm X, Martin Luther King, Jr., or on support they receive from Black churches as they deal with challenges. Women, who are often more relational, may make meaning out of relationships while men may focus more on issues of personal autonomy and completing a task effectively.

As religion plays such an important part in many people's lives, members of the dominant religion in a region or a nation may have different experiences from those of other spiritual orientations. For example, Schlosser (2003) talks of Christian privilege in North America where people of Jewish and other faiths may feel uncomfortable, even unwelcome, during Christian holidays. Anti-Semitism, anti-Islamism, anti-liberal Christianity, anti-evangelical Christianity are all possible results when spiritual and religious tolerance do not exist.

For practical purposes, in using reflection of meaning in your interview, recall the concept of focus in the preceding chapter. Your task is to focus not only on individuals but also on how they and their families make meaning. With the awareness that different groups and individuals make different meanings out of the same event, you will have an important beginning in multicultural counseling and therapy.

SUMMARY

Eliciting and reflecting meaning is a complex skill that requires you to enter the sense-making system of the client. Full exploration of life meaning requires a self-directed, verbal client willing to talk. The skill complex is most often associated with an abstract, formal-operational interviewing style. However, all of us are engaged in the process of meaning-making and trying to make sense of a confusing world. With clients who are more concrete, you will still find that eliciting and reflecting meaning is useful. But these clients may not be able to see patterns in their thinking or be as self-directed and reflective as those who think at a more complex level. You may find that the more directive approach to meaning taken by the cognitive-behavioral therapists is more useful with clients who have difficulty reflecting on themselves.

With highly verbal or resistant clients, you may find that they like to spend all their time thinking, reflecting on meaning, and thus end up intellectualizing with little or no action to change their behaviors, thoughts, or feelings. Viktor Frankl was well aware of this possible problem and encouraged his clients to take action on their meanings. Meaning that does not move into the "real world" may at times become a problem in itself.

Meanings are organizing constructs that are at the core of our being. You will find that exercises with reflection of meaning, if completed in depth, will result in your having a more comprehensive understanding of your client than is possible with most other skills. Mastering the art of understanding meaning will take more time than other skills. The exercises in this chapter are designed to assist you along the path toward this goal.

Box 11-4   Key Points

Why?	Meaning organizes life experience and often serves as a metaphor from which clients generate words, sentences, and behaviors. Clients faced with complex life decisions may make them on the basis of meaning, values, and reasons rather than on objective facts or on feelings. However, these meanings and values are often unclear to the client.
What?	Meanings may be identified through noting and observing client words and constructs that describe their values and attitudes toward important issues and people. As meaning is often implicit, it is helpful to ask questions that help clients explore and clarify meaning. For example: "What does this *mean* to you?" "What *sense* do you make of it?" "What *values* underlie your actions?"

(continued)

Box 11-4   (continued)

	"*Why* is that important to you?" "*Why?*" (by itself, used carefully) "What was your *intention* when you did that?"
How?	Meanings are reflected through the following process: 1. Begin with a sentence stem like the following: "You mean . . . ," "Could it mean that you . . . ," "Sounds like you value . . . ," or "One of the underlying reasons/intentions of your actions was. . . ." 2. Use the client's own words that describe the most important aspect of the meaning. This helps ensure that you stay within the client's frame of reference rather than using your own interpretation. 3. Add a paraphrase of the client's longer statements that captures the essence of what has been said but, again, operates primarily from the client's frame of reference. 4. Closing with a check-out such as "Is that close?" or "Am I hearing you correctly?" may be helpful. Meanings may be more complex in situations in which two values or two meanings collide. The use of questions, reflection of feelings, and so on may be required to help clients sort through meaning and value conflicts.
With whom?	Reflections of meaning are generally for more verbal clients and may be found more in counseling and therapy than in general interviewing. There can be a temptation to reflect negative meanings back to troubled clients. The task with such clients, however, is to help them find positive meanings. This may be done through questioning about situations in the client's past in which positive feelings existed and searching there for meanings that may be contrasted with the negative view. There may be no positive meaning immediately apparent, especially for a survivor of trauma. (But the simple fact of *survival* is positive meaning.) A well-timed reflection of meaning may help many clients facing extreme difficulty. It can help clarify cultural and individual differences, *if* the client is willing to share in more depth.
What else?	Meaning is not observable behavior, although it could be described as a special form of cognition that reaches the core of our being. Helping clients discern the meaning and purpose of their lives can serve as a motivator for change and provide a compass as to the direction of that change.

COMPETENCY PRACTICE EXERCISES AND PORTFOLIO OF COMPETENCE

The concepts of this chapter build on previous work. If you have solid attending and client observation skills, can use questions effectively, and can demonstrate effective use of the encourager, paraphrase, and reflection of feeling, you are well prepared for the exercises that follow.

Individual Practice

Exercise 1: Identification of Skills

Read the following client statement. Which of the following counselor responses are paraphrases (P), reflections of feeling (RF), or reflections of meaning (RM)?

I feel very sad and lonely. I thought Jose was the one for me. He's gone now. After our breakup I saw a lot of people but no one special. Jose seemed to care for me and make it easy for me. Before that I had fun, particularly with Carlos. But it seemed at the end to be just sex. It appears Jose was it; we seemed so close.

——————— "You're really hurting and feeling sad right now."

——————— "Since the break-up you've seen a lot of people, but Jose provided the most of what you wanted."

——————— "Looks like the sense of peace, caring, ease, and closeness meant an awful lot to you."

——————— "You felt really close to Jose and now are sad and lonely."

——————— "Peace, caring, and having someone special mean a lot to you. Jose represented that to you. Carlos seemed to mean mainly fun, and you found no real meaning with him. Is that close?"

List possible single-word encouragers for the same client statement in the following spaces. You will find that the use of single-word encouragers, perhaps more than any other skill, leads your client to talk more deeply about the unique meanings underlying behavior and thought. A good general rule is to search carefully for key words, repeat them, and *then* reflect meaning.

———————————————————   ———————————————————

———————————————————   ———————————————————

———————————————————   ———————————————————

———————————————————   ———————————————————

Exercise 2: Identifying Client Issues of Meaning

Affective words in the preceding client statement include *sad* and *lonely*. Some other words and brief phrases in the client statement contain elements that suggest more may be found under the surface. The following are some key words that you may have listed under possible encouragers: *the one for me, care for me, easy for me, I had fun, just sex,* and *we seemed close.* The feeling words represent the client's emotions about the current situation; the other words represent the meanings she uses to represent the world. Specifically, the client has given us a map of how she constructs the world of her relationships with men.

To identify underlying meanings for yourself, talk with a client, or someone posing as a client, observing his or her key words—especially those that tend to be repeated in different situations. Use those key words as the basis of encouragers, paraphrasing, and questioning to elicit meaning. Needless to say, this should be done with considerable sensitivity to the client and her or his needs. Record the results of your experience with this important exercise here. You will want to record *patterns* of meaning-making that seem to be basic and that may motivate many more surface behaviors, thoughts, and feelings.

———

———

_____

_____

_____

_____

_____

_____

Exercise 3: Questioning to Elicit Meanings

Assume a client comes to you and talks about an important issue in her or his life (for instance, divorce, death, retirement, a pregnant daughter). List five questions that might be useful in bringing out the meaning of the event.

1. _____

2. _____

3. _____

4. _____

5. _____

Exercise 4: Practice of Skills in Other Settings

During conversations with friends or in your own interviews, practice eliciting meaning through a combination of questioning and single-word encouragers, and then reflect the meaning back. You will often find that single-word encouragers lead people to talk about meaningful issues. Record your observations of the value of this practice here. What one thing stands out from your experience?

_____

_____

_____

_____

_____

_____

_____

_____

Exercise 5: Discernment: Examining One's Purpose and Mission

Using the suggestions of Box 11-3, work through each of the four sets of questions. You may do this by yourself, using a meditative approach and journaling. Or you may want to do this with a classmate or close acquaintance. Allow yourself time to think carefully about each area. Add questions and topics that occur to you—make this exercise fully personal.

What do you learn from this exercise about your own life and wishes?

_____

_____

_____

_____

Supplemental Exercise 6: Practice in Person-Centered Counseling

You may wish to complete a full five-stage interview emphasizing reflection of meaning using person-centered theory. If so, read the material in Chapter 14 on person-centered counseling and use the feedback form in that chapter. Can you complete a full interview listening to the client, reflecting meaning, and using a minimum number of questions?

Group Practice

Two group exercises are suggested here. The first focuses on the skill of eliciting and reflecting meaning, the second on the dereflection process as it might be used in logotherapy.

Exercise 1: Systematic Group Practice in Eliciting and Reflecting Meaning

Step 1: Divide into practice groups.
Step 2: Select a group leader.
Step 3: Assign roles for the first practice session.
▲ Client
▲ Interviewer
▲ Observer 1, who observes the client's descriptive words and key repeated words, using Feedback Form, Box 11-5.
▲ Observer 2, who notes the interviewer's behavior, using Feedback Form, Box 11-5.

Step 4: Plan.   For practice with this skill, it will be most helpful if the interview starts with the client's completing one of the following model sentences. The interview will then follow along, exploring the attitudes, values, and meanings to the client underlying the sentence.

"My thoughts about spirituality are . . ."
"My thoughts about moving from this area to another are . . ."
"The most important event of my life was . . ."
"I would like to leave to my family . . ."
"The center of my life is . . ."
"My thoughts about divorce/abortion/gay marriage are . . ."

A few alternative topics are "My closest friend," "Someone who made me feel very angry (or happy)," and "A place where I feel very comfortable and happy." Again, a decision conflict or a conflict with another person may be a good topic.

Establish the goals for the practice session. The task of the interviewer in this case is to elicit meaning from the model sentence and help the client find underlying meanings and values. The interviewer should search for key words in the client response and use those key words in questioning, encouraging, and reflecting. A useful sequence of microskills for eliciting meaning from the model sentence is (1) an open question, such as "Could you tell me more about that?" "What does that mean to you?" or "How do you make sense of that?"; (2) encouragers and paraphrases focusing on key words to help the client continue; (3) reflections of feeling to ensure that you are in touch with the client's emotions; (4) questions that relate specifically to meaning (see Box 11-2); and (5) reflecting the meaning of the event back to the client, using the framework outlined in this chapter. It is quite acceptable to have key questions and this sequence in your lap to refer to during the practice session.

Examine the basic and active mastery competencies in the Self-Assessment section and plan your interview to achieve specific goals.

Observers should study the feedback form especially carefully.

Step 5: Conduct a 5-minute practice session using the skill.
Step 6: Review the practice session and provide feedback for 10 minutes.   This may be a time that the microsupervision process includes a group discussion of the place of values in the interview. The feedback forms are useful here. It is often tempting to just talk, but you might forget to give the interviewer helpful and needed specific feedback. Take time to complete the forms before talking about the session. As always, give special attention to the mastery level achieved by the interviewer. The client can complete the Client Feedback Form of Chapter 1.

Step 7: Rotate roles.   Remember to share time equally.
Some general reminders.   This skill can be used from a variety of theoretical perspectives. It may be useful to see if an explicit or implicit theory is observable in the interviewer's behavior.

Exercise 2: Systematic Group Practice in Dereflection and Attitude Change

Follow the same steps as for Exercise 1. This exercise is intended to remind you of the power and importance of finding positive dimensions and meaning in negative situations. Your central aim here is to prevent and balance the excessive attention given to the negatives in life experience by helping people find positive dimensions and strengths in difficult, troubling situations.

The task of the client is to talk about something negative in his or her life. For example:

The breakup of a relationship
A death or illness
Loss of job or inability to find work
An accident
Betrayal by a friend

The first task of the interviewer is to draw out the person's negative experience through eliciting and reflecting meaning (as in Exercise 1). Then, however, the interviewer is to draw out a positive dimension of the experience via questioning; for example, questions for discussing an illness might be, "Could you talk about some positive experiences or something you learned at the hospital?" "What situations have you learned to value more as a result of that experience?" "What do you value even more now as a result?" These positive value statements may be reflected back to check on the accuracy of the interviewer's understanding.

With many clients, it may be useful or even wise first to ask them to tell you about life-sustaining positive events from the past. These may be spiritual, they may represent a special connection to the land, they may be the values of a friend, or they may be actual things done by the client that he or she feels proud of. This is a positive asset search from the past. To avoid the daily pain of the concentration camp, Viktor Frankl often thought about his wife. When possible, he would enjoy momentary pleasure, such as a taste of bread or a special sunset. By overlaying positive experiences on the negative, Frankl was able to survive the most difficult times.

As a final step, the interviewer may summarize the negative experience and the positive meaning statements and feed them both back to the client. (For instance, "You seem to feel that being ill and near death was your worst fright ever, but at the same time you seem to value your relationships with your children and family even more now. Does that make sense to you?") In this way the positive and negative elements in the situation may be joined in a new synthesis. Some clients find this experience useful in reframing their view, and their process of hyperreflection on the negative is balanced by the positive dimensions. This can often lead to new actions and behaviors.

Summarize here your findings about the concept of dereflection (or use the feedback form).

_____

_____

_____

_____

_____

_____

_____

_____

_____

_____

A caution: There is another, alternative way to approach this exercise. At times you will have clients who fail to see the negative in a situation; they may deny the reality of what has happened to them. This is particularly true of the survivors of childhood abuse, spousal abuse, and rape, and of those who have been held hostage. To maintain their sanity, these individuals have had to find *only* positive things in their universe; they deny the negative. Bringing out the negative meaning in such events can be therapeutic. However, you should be aware of the danger in this use of dereflection and attitude change and proceed with caution, using the direction and support of your staff and supervisor. Do not practice the negative dereflection discussed here until you have had considerable knowledge and experience both in the field and with your individual client.

Exercise 3: Discernment Practice

Take another person through the discernment procedure, working carefully with each step. Recall de Waal's statement on how to listen to the other person:

> Listen. Listen, with intention, with love, with the "ear of the heart." Listen not only cerebrally with the intellect, but with the whole of feelings, our emotions, imaginations, and ourselves.   (de Waal, 1997, Preface)

Share the list of questions and ideas with your client. Ask her or him to suggest additional questions and issues that may be missing in this list. Have the client define which questions he or she may wish to discuss. Use all your listening skills as you help the client find personal direction and meaning.

What do you and the client learn?

_____

_____

_____

_____

Box 11-5   Feedback Form: Reflecting Meaning

_____ (Date)

_____   _____
(Name of Interviewer)   (Name of Person Completing Form)

Instructions: Observer 1 completes the first part of this form, giving special attention to recording descriptive words the client associates with meaning and to key repeated words. In the second part, Observer 2 notes the interviewer's use of the reflection-of-meaning skill, giving special attention to questions that appeared to elicit meaning issues.

Part One: Client Observation
Key words/phrases:

_____

_____

_____

What are the main meaning issues of the interview?

Part Two: Interviewer Observation
List questions and reflections of meaning used by the interviewer, continuing on a separate sheet as needed.

1. _____

2. _____

3. _____

4. _____

5. _____

6. _____

Comment on the effectiveness of the reflection-of-meaning skill.

Portfolio of Competence

As you work through this list of competencies, think about how you would include the ideas related for reflection of meaning in your own Portfolio of Competence.

Use the following as a checklist to evaluate your present level of mastery. Check those dimensions that you currently feel able to do. Those that remain unchecked can serve as future goals. *Do not expect to attain intentional competence on every dimension as you work through this book.* You will find, however, that you will improve your competencies with repetition and practice.

Level 1: Identification and classification.   You will be able to differentiate this skill from the closely related skills of paraphrasing and reflection of feeling. You will be able to identify questioning sequences that facilitate client talk about meaning. You will be able to identify client words indicative of meaning issues.

❑   Ability to identify and classify the skills.
❑   Ability to identify and write questions that elicit meaning from clients.
❑   Ability to note and record key client words indicative of meaning.

Level 2: Basic competence.   You will be able to demonstrate the skills of eliciting and reflecting meaning in the interview. You will be able to demonstrate an elementary skill in dereflection.

❑   Ability to elicit and reflect meaning in a role-play interview.
❑   Ability to examine yourself and discern more fully your life direction.
❑   Ability to use dereflection and attitude change in a role-play interview.

Level 3: Intentional competence.   You will be able to use questioning skill sequences and encouragers to bring out meaning issues and then reflect meaning accurately. You will be able to use the client's main words and constructs to define meaning rather than reframing in your own words (interpretation). You will not interpret but rather facilitate the client's interpretation of experience.

❑   Are you able to use questions and encouragers to bring out meaning issues?
❑   When you reflect meaning, are you able to use the client's main words and constructs rather than your own?
❑   Are you able to reflect meaning in such a fashion that the client starts exploring meaning and value issues in more depth?
❑   In the interview, are you able to switch the focus as necessary in the conversation from meaning to feeling (via reflection of feeling or questions oriented toward feeling) or to content (via paraphrase or questions oriented toward content)?
❑   Are you able to help others discern their purpose and mission in life?
❑   When a person is hyperreflecting on the negative meaning of an event or person, are you able to find something positive in that person or event and enable the client to dereflect by focusing on the positive?

Level 4: Teaching competence.
- ❑ Ability to teach clients how to examine their own meaning systems.
- ❑ Ability to facilitate others' understanding and use of discernment questioning strategies.
- ❑ Ability to teach reflection of meaning to others.

DETERMINING YOUR OWN STYLE AND THEORY: CRITICAL SELF-REFLECTION ON REFLECTING MEANING

Meaning has been presented as a central issue in interviewing, counseling, and psychotherapy. What single idea stood out for you among all those presented in this chapter, in class, or through informal learning? What stands out for you is likely to be important as a guide toward your next steps. What are your thoughts on multicultural issues and the use of this skill? What other points in this chapter struck you as important? How might you use ideas in this chapter to begin the process of establishing your own style and theory? And, in particular, what have you learned about discernment and its relation to your own life?

_____

_____

_____

_____

_____

_____

_____

_____

REFERENCES

Barrett-Lennard, G. (1962). Dimensions of therapist response as causal factors in therapeutic change. *Psychological Monographs, 76,* 43. (Ms. No. 562)

Beck, A., & Beck, J. (1995). *Cognitive therapy: Basics and beyond.* New York: Guilford.

Bozarth, J. (1999). *Person-centered therapy: A revolutionary paradigm.* Ross-on-Wye, UK: PCCS Books.

Camus, A. (1995). *The myth of Sisyphus.* New York: Vintage Books.

Carter, R. (1999). *Mapping the mind.* Berkeley: University of California Press.

Contrada, R., Goyal, T., Cather, C., Rafalson, L., Idler, E., & Krause, T. (2004). Psychosocial factors in outcomes of heart surgery: The impact of religious involvement and depressive symptoms. *Health Psychology, 23,* 227–238.

deWaal, E. (1997). *Living with contradiction: An introduction to Benedictine spirituality.* Harrisburg, PA: Morehouse.

Farnham, S., Gill, J., McLean, R., & Ward, S. (1991) *Listening hearts.* Harrisburg, PA: Morehouse.

Fiedler, F. (1950). A comparison of therapeutic relationships in psychoanalytic, nondirective, and Adlerian therapy. *Journal of Consulting Psychology, 14,* 435–436.

Frankl, V. (1959). *Man's search for meaning.* New York: Simon & Schuster.

Frankl, V. (1978). *The unheard cry for meaning.* New York: Touchstone.

Fukuyama, M. (1990). *Multicultural and spiritual issues in counseling.* Workshop presentation for the American Counseling Association Convention, Cincinnati, March.

Ivey, A., D'Andrea, M., Ivey, M., & Simek-Morgan, L. (2002). *Theories of counseling and psychotherapy: A multicultural perspective* (5th ed.). Boston: Allyn & Bacon.

Ivey, A., Ivey, M., Myers, J., & Sweeney, T. (2005). *Developmental counseling and therapy: Promoting wellness over the lifespan.* Boston: Lahaska/Houghton-Mifflin.

Kain, C. (1992). We cannot turn our backs on AIDS. *American Counselor* (Fall).

Lazarus, A., & Fay, A. (2000). *I can if I want to.* New York: FMC Books.

Meichenbaum, D. (1994). *A clinical handbook/practical therapy manual for assessing and treating adults with post-traumatic stress disorder (PTSD).* Waterloo, Ontario: Institute Press.

Miller, S., Duncan, B., & Hubble, M. (2005). Outcome-informed clinical work. In J. Norcross & M. Goldfried (Eds.), *Handbook of psychotherapy integration* (pp. 84–104). Oxford: Oxford University Press,

Probst, R. (1996). Cognitive-behavioral therapy and the religious person. In E. Shafranski (Ed.), *Religion and the clinical practice of psychology* (pp. 391–408). Washington, DC: American Psychological Association.

Schlosser, L. (2003). Christian privilege: Breaking a sacred taboo. *Journal of Multicultural Counseling and Development, 31,* 44–51.

Rogers, C. (1961). *On becoming a person.* Boston: Houghton Mifflin.

Seligman, M. (1998). *Learned optimism: How to change your mind and your life.* New York: Pocket Books.

Wade, S., Borawski, E., Taylor, H., Drotar, D., Yeates, K., & Stancin, T. (2001). The relationship of caregiver coping to family outcomes during the initial year following pediatric traumatic injury. *Journal of Consulting and Clinical Psychology, 69,* 406–415.

ALLEN AND MARY'S THOUGHTS ABOUT CHARLIS

Eliciting and reflection of meaning is both a skill and a strategy. As a skill, it is fairly straightforward. To elicit meaning, we'd want to ask Charlis some variation of the basic question, "What does the heart attack *mean* to you, your past and future life?" As appropriate to the situation, questions such as the following can address the general issue of meaning in more detail:

"What has given you most satisfaction in your job?"
"What's *been missing* for you in your present life?"

"What do you find of *value* in your life?"

"What *sense* do you make of this heart attack and the future?"

"What things in the future will be most *meaningful* to you?"

"What is the *purpose* of your working so hard?"

"You've said that you have been wondering what God is saying to you with this trial? Could you share some of your thoughts here?"

"What would you like to leave the world as a gift?"

Questions such as these do not usually lead to concrete behavioral descriptions. They may often bring out emotions and they certainly bring out certain types of thoughts and cognitions. Typically, these thoughts are deeper in that they search for meanings and understandings. When clients explore meaning issues, the interview, almost by necessity, becomes less precise. Perhaps this is because we are struggling with defining the almost undefinable.

As part of our work with Charlis, we'd ask if she wants to examine the meaning of her life in more detail through the process of *discernment*. This is a more systematic approach to meaning and purpose defined in some detail in this chapter. If she wishes, we'd share the specific questions of discernment presented here and ask her which one's she'd like to explore. In addition, we'd ask her to think of questions and issues that are particularly important to her and we would give these special attention as we work to help her discern the meaning of her life, her work, her goals, and her mission.

Reflection of meaning as a skill looks very much like a reflection of feeling or paraphrase, but the key words "meaning," "sense," "deeper understanding," "purpose," "vision" or some related concept will be present explicitly or implicitly. "Charlis, I sense that the heart attack has led you to question some basic understandings in your life. Is that close? If so, tell me more."

It can be seen that we regard eliciting and reflecting meaning as an *opening* for the client to explore issues where there often is not a final answer, but rather a deeper awareness of the possibilities of life. At the same time, effective exploration of meaning becomes a major strategy in which you bring out client stories, past, present, *and future*. You will use all the listening, focusing, and confrontation skills to facilitate this self-examination. Yet the focus remains on meaning and finding purpose in one's life.

CHAPTER 12

Influencing Skills: Six Strategies for Change

INFLUENCING SKILLS AND STRATEGIES
REFLECTION OF MEANING
FOCUSING
CONFRONTATION
THE FIVE-STAGE INTERVIEW STRUCTURE
REFLECTION OF FEELING
ENCOURAGING, PARAPHRASING, AND SUMMARIZING
CLIENT OBSERVATION SKILLS
OPEN AND CLOSED QUESTIONS
ATTENDING BEHAVIOR
ETHICS, MULTICULTURAL COMPETENCE, AND WELLNESS

How can influencing skills help you and your clients?

Six additional influencing skills are presented in this chapter. Accompanying these skills are specific strategies that may be useful in facilitating clients' generating new stories and actions. All of these skills are oriented to helping your clients find new ways of thinking, feelings, behaving, and finding new meaning in their lives.

Mastery of the skills and strategies of microcounseling, coupled with the five-stage interview plan and the story—positive asset—restory—action model, will enable you to connect with counseling and interviewing theory—as each of the skills of this chapter relate to specific approaches to counseling and therapy.

A working definition of the key chapter concepts follows along with the function of each in the interview.

Skills and Strategies	*Definition and Function*
Interpretation/reframe	Provides the client with an alternative frame of reference from which to view life situations and generate new stories.
Logical consequences	Enables the client to look at the possible results of alternative actions.
Self-disclosure	Requires you, the interviewer, to share your own story, thoughts, or experiences *briefly*. Used carefully, self-disclosure may build a sense of equality in the session and encourage client trust and openness. If your background is substantially different from your client, self-disclosure on some of your views and life may be essential.

Skills and Strategies	*Definition and Function*
Feedback	Provides accurate data on how the client is seen by others and/or the interviewer.
Information/advice/ opinion/instruction/ suggestions	Complex of skills presents new information and ideas to the client. Career information, teaching about sexuality, results of test scores are some examples. The issue of when to provide advice is explored.
Directive	Leads the client to follow strategies and actions suggested by the interviewer, which may help in restorying or in taking concrete action on issues and problems. This is a central skill of cognitive-behavioral therapy and assertiveness training.

The interactive exercise that usually starts each chapter is on page 360. You will have an opportunity there to think about your work with influencing skills. The case of Alisia, presented in the exercise, will be discussed throughout this chapter.

INTRODUCTION: AN OVERVIEW OF INFLUENCING SKILLS

Influencing is part of all interviewing and counseling. Even if we use only attending skills with clients, we still influence what occurs in the session—being heard by another person greatly influences the way all of us think about ourselves and organize our lives. Confrontation, focusing, and reflection of meaning have been identified already as skills of interpersonal influence. This chapter adds six more skills that may be employed in the session. These six skills place more responsibility on the counselor as all of them in one way or another directly seek to change the way the client thinks or acts.

The chapter builds on the story—positive asset—restory—action framework. As this book demonstrates, we seek to draw out client stories, problems, or concerns within a positive framework of strength and possibility. We believe that one important way to communicate respect to clients is to help them tell stories of strength and say good things about themselves. A positive base often makes it easier for clients to talk about difficulties and their problems. The counselor and interviewer can then work with clients' personal strengths to attack these issues. We can restory our lives and act in new ways through being listened to by a counselor. The attending skills alone can often be all that is necessary for major life change, particularly if we build on positive assets using a wellness approach.

As you move to the direct action associated with the influencing skills, it is easy to forget the foundation of listening and empathic understanding. Carefully developed listening skills of paraphrasing, reflecting feelings, and summarizing are sometimes lost in the enthusiasm for the power of the influencing skills and strategies. Thus, we all need to step back and remember that counseling and interviewing are for the client, not for us. Ethical practice demands respect for the client and awareness of the power relationship inherent in the interview. Interviewers and counselors, by their position, have power over their clients; use this power with awareness of client needs and full client participation in the direction of the session. You may wish to read the example assertiveness interview in Chapter 14 in conjunction with this chapter, as it provides another illustration of these skills in action.

Disclosure of what you are going to do can often be helpful with clients when you use influencing skills. For example, if you are going to use a directive such as guided imagery, relaxation training, or thought-stopping, spend a moment telling the client what you are going to do and its potential benefits. The general rule is avoid surprises, although occasionally it is the very surprise that helps a client think of things differently. Think of a visit to a dentist or a physician for treatment or an examination. Medical personnel who are effective communicators will tell you what they are going to do *before* they do it and what you can expect to happen to your body. They will tell you whether it will hurt. Advance disclosure tends to build comfort and trust even when the next step of an interview or physical exam may not be comfortable.

The Strategies of Interpersonal Influence

Influencing skills such as directives, interpretation, and feedback usually are best given in the context of using active listening, being concrete and clear in your wording, and observing and checking out with your client the impact of your intervention. You need to provide a context or broader setting when using influencing skills; you need to use *strategy*. The strategic context is supplied by effective listening. You will find that influencing skills are most valuable if used sparingly. And if the first influencing skill doesn't work, move back to listening and then perhaps try a different influencing skill.

In sum, a strategic model is proposed (see Box 12-1) in which you (1) attend and listen to the client and are sure you understand where he or she is "coming from"; (2) assess that client and select an intervention that is personally relevant; and (3) observe the consequence of your interviewing lead and move from that observation to your next step.

The Confrontation Impact Scale (CIS, Chapter 9, p. 276) can be particularly useful as you work with the influencing skills. The effectiveness of your intervention

Box 12-1   The 1-2-3 Pattern of Listening, Influencing, and Observing Client Reaction

1. Listen	In any interaction with clients it is critical that you use attending, observation, and listening skills to determine the client's definition and view of the world. How does the client see, hear, feel, and represent the world through "I" statements and key descriptors for content (paraphrasing), feelings (reflection of feeling), and meaning (encouragers and reflection of meaning)?
2. Assess and influence	An influencing skill is best used after you understand clients' impressions and representations of their experience. An interpretation, self-disclosure, feedback statement, or other influencing skill provides a new frame of reference or informational base from which clients may act.
3. Check out and observe consequences via the CIS	Following the use of an influencing skill, use a check-out ("How does that seem to you?") and observe the consequences of your action. If either client verbal or nonverbal behavior, or both, become discrepant and you sense an increasing distance from you as an interviewer, return to the use of listening skills. Influencing skills, of necessity, remove you from being totally "with" clients. Consequently you must pay additional attention to your client observation skills when you attempt to use influencing skills. In determining what to do next, you will want to examine the impact of your intervention using the five levels of the Confrontation Impact Scale.

can be readily assessed. If your reframe/interpretation, feedback, or directive is effective, you will observe change in your client. If the client does not respond to an influencing skill, return to listening and understanding. Later the client may be able to use the influencing skill you introduced or you may want to try another more effective approach. Being intentional demands that you be flexible and ready to move with the client.

The degree of interpersonal influence desired in the interview varies from theory to theory. The word *influence* can be upsetting to a client-directed, formal-operational counselor, whereas many proponents of concrete behavior modification aim deliberately to work with the client to produce as much change as possible. Most theories now seem to agree on the benefit of extensive client involvement, and there is increasing evidence and acceptance of the value of changing counseling and interviewing styles to meet the unique needs of each client.

It is possible to place the skills on a rough continuum of interpersonal influence. Figure 12-1 rates the attending skills and several of the influencing skills in terms of their influence in the interview. The task of the counselor is often to open client discussion of a topic and then to close this discussion when appropriate. When a client is overly talkative or becomes upset, it may be useful to ask a closed question and then change the focus by asking an open question on another topic, depending on the goal of the interview. When an interview is moving slowly, an interpretation or directive may add content and flow.

The influencing skills are those that the interviewer uses to take greater responsibility for the interview; the listening skills allow the client to talk in greater depth and determine the direction of the session. This, of course, is an oversimplification. Even though it is generally accurate, keep in mind that an effective listening skill may be more influential than a directive or confrontation. The timing and *how* you use the skills are ultimately most important.

You may find it helpful to think of the interpersonal influence continuum from time to time during your interviews. If you feel that you are coming on too strong and the client is resisting, it may be wise to move to lower levels of influence and focus more on listening. Similarly, if the client is bogged down, the careful use of an influencing skill may help organize things and move the interview along more smoothly.

Important for the beginning and advanced interviewer is the *moderate triad* of skills: open and closed questions and focusing. Experience has shown that mastery of these skills is almost as important as attending behavior and client observation. If you can ask questions effectively and focus on varying topics, you have the ability to open and close almost any topic or issue your client presents. If a client has difficulty talking, an open question coupled with a slight change of focus may give her or him space to open up. If the topic seems inappropriate, a change of focus coupled with some closed questions will usually slow down the pace. This is particularly true when the client is more emotional than the interviewer feels able to handle. An open question may then be used to switch to another, less difficult topic. Later you can return to the challenging issue. You can achieve the same effects with the other skills, but the skills in the moderate triad are "swing" skills in terms of their influence and do not seem to disrupt the interview flow when an interview needs to change direction or a topic needs to be discussed more fully.

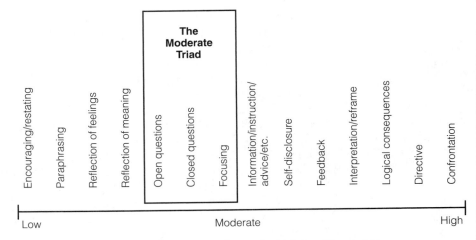

Amount of influence exerted over client talk by different skills
(All skills rest on a foundation of attending behavior and client observation.)

Figure 12-1   The interpersonal influence continuum

Let us now turn to the several skills and strategies of interpersonal influence and consider briefly how each might be of benefit to Alisia, our example client throughout this chapter.

EXAMPLE INTERVIEW AND INTRODUCTORY CHAPTER EXERCISE

As this chapter presents several skills and strategy areas, it is organized differently from the preceding chapters. Each skill discussion will contain a brief transcript of how that skill might be used in an actual interview. Alisia will be the client. The content of the interview will be consistent for all skills: Alisia has a recurring problem of lack of assertiveness, the failure to express her point of view, and the accompanying inability to obtain what she wants.

The case of Alisia is discussed throughout this chapter. Alisia, like many of us, has real difficulty expressing her wants and needs. In the following example, the counselor uses the basic listening sequence to obtain a general summary of Alisia's inability to express herself. This transcript, of course, is abbreviated and edited for clarity.

The interview began with a short rapport-establishing phase and discussion of recording the interview; it then turned to the following:

Counselor:	Could you tell me what you'd like to talk about today?
Alisia:	I couldn't do it. I simply can't express myself. I've tried many times and I can't get people to listen to me. Whether it is the boss, my lover, or the man at the garage, they all seem to run over me.
Counselor:	Run over you?
Alisia:	Yeah, I keep finding that I'm so damned accommodating that I'm always trying to get along. People like me for going along with them, but I never get what I want. Smiles, yes, but delays and nothing. I'm disgusted with myself.

Counselor:	Sounds as if you are really angry and disgusted about your inability to get what you want.
Alisia:	Right, it just goes on and on . . . I never seem to change.
Counselor:	Could you give me a specific example of the last time you had these feelings of anger and disgust? What happened? What did you say? What did they do?
Alisia:	Well, I was at a garage. I had called in for an early morning appointment. They said come in at 8:00 and so I was there on time. I had to go to a meeting at 10:00. When I checked at 9:30, they hadn't even started yet. The guy smiled and said, "Sorry, lady, we couldn't get to it." He was so smug and demeaning, but I just looked down and didn't say anything—even though I really wanted to scream. I made another appointment and my car still has that screwy, strange sound.
Counselor:	And after all this the car still isn't right. As I see you now Alisia, you seem very angry. What's happening with you as you talk to me about this?
Alisia:	Angry . . . I'm just full of it. Sexist pigs! I hurt too, deep inside. And I'm confused. (tears, but her eyes are flashing with determination) Everywhere I turn, it's there. (Alisia is expressing her emotions at this moment at the sensorimotor level, discussed in Chapter 7.)
Counselor:	I hear your anger and frustration. You're really angry and upset. You're tired of taking things as they are. The situation at the garage is just one instance of a pattern— something that repeats in various forms again and again. Have I heard you correctly? (Pattern thinking represents formal operational thought.)
Alisia:	Yes, it's a pattern. The same day as the problem at the garage, my boss started leering at me again. I'm sick and tired of it. I used to think it was my fault, but now I'm seeing that men are the problem. The garage hassles me, the boss hassles me, and my lover does the same thing in his own way.

The interview continues, and the counselor spends some time on Alisia's strengths. She uses the basic listening sequence to draw out in some detail a story of how Alisia was able to challenge a rude clerk at a store rather than let herself be run over. The counselor also discovers that Alisia has several women heroes on which she would like to model herself. After this discussion the counselor turns the focus from Alisia and the problem to the cultural/environmental context and discusses with her how matters of gender are related to Alisia's issue. Nearing the end of the interview, we hear the following:

Counselor:	So, Alisia, we've been talking for nearly an hour now. (near the end of the session) How do you put together all we've talked about?
Alisia:	As I look at it now, I realize that much of what's been happening to me is a result of sexism. I used to blame myself and try to let it go and I did what "a good girl" should do. I try to please everyone. I learned that in my family. But I'm not a girl, I'm a woman. I'm going to file harassment charges against my boss. I think I need to go with my partner for couples counseling. If we're going to stay together, something new has to evolve. There, I feel better about myself right now. The issue is how I can deal with the system. (This statement has been shortened from a much longer client–counselor exchange. Clients are not always so clear.)

Alisia has made a good start in this session. As this chapter moves on, we will see additional issues and how more active participation in the session through influencing skills can help the client.

What are some things that occur to you when working with a client such as Alisia, who has not allowed herself to express her thoughts and feelings more fully? Assuming a longer-term relationship with her, what might be your goals and plan to help her?

_____

_____

_____

_____

You may want to compare your thoughts with ours expressed at the end of this chapter on page 395.

INSTRUCTIONAL READING 1 AND EXERCISES: INTERPRETATION/REFRAMING

Interpretation/reframing is closely allied to the idea of restorying. In restorying, the client learns to talk about the problematic story in a new, more positive fashion, which will later lead to new thoughts, feelings, and actions. Likewise, when you use the microskill of interpretation/reframing, you are helping the client to look at the problem or concern from a new perspective. This new way of thinking is often central to the restorying and action process.

Interpretation/reframing is also closely allied to reflection of meaning, but in reflection of meaning, it is the client who assigns new meanings to her or his situation. In interpretation, you as interviewer take a more direct approach to the restorying process. You may draw from client observations, your own personal experience, or perhaps from some theoretical perspective as you help clients reinterpret or reframe their issues.

Focusing is another influencing skill that greatly facilitates the generation of new perspectives. By focusing on issues of gender, the counselor was able to help Alisia realize that she was not the problem, and the issue was refocused as one of sexism, a cultural/environmental/contextual issue. Alisia has reframed her issues into a new story.

Thus, you have two major alternatives to interpretation/reframing if you want to help clients generate new perspectives on their issues—focusing and reflection of meaning. However, at times you may want to be more active in the restorying process, and thus interpretation becomes a central skill.

Imagine that you had worked with Alisia over a longer period and she came to you one day upset over a troubling dream. This dream recurred frequently in childhood, and after the incident in the garage it presented itself again. Different theories would interpret the dream differently, but each would provide a new frame of reference, a new perspective for the client. For example, Alisia tells you her dream story and sums it up as follows:

Alisia: In short, I was dreaming I was walking along the cliffs with the sea raging below. I felt terribly frightened. There was a path out, but I couldn't take it. I just felt so undecided about what to do. The dream just went on and on.

Counselor: (decisional theory) You've just entered college, taken on a new job, and you have the event in the garage that kind of tipped you over. You feel almost as if you might fall off the cliff. Now you are on your own for sure, and it is frightening.

Counselor: (feminist theory) You felt frightened—I hear that. I also hear a woman who has the courage to get out on those cliffs and face the challenges. We will have to work together to help you find some support here to cope with the challenges.

Counselor: (Gestalt theory) Become that cliff. (Pause) What are you seeing and feeling, cliff? (In this example, the client takes the role of items or people in the dream. At a later point, the client might become the raging sea, and a dialogue between "sea" and "cliff" might occur.)

Counselor: (psychodynamic theory) You feel rage at your parents and you can't tell them how you really feel. It frightens you. And now you find the fellow in the garage doing much the same, but the prospect of challenging him is terrifying.

Which is the correct interpretation? Depending on the situation and context, any of these interpretations could be helpful or harmful. The first two responses deal with here-and-now reality, whereas the last two are most often associated with longer-term therapy. The value of the interpretation depends on the client's reaction to it and how she uses it over time. Think of the Confrontation Impact Scale—how does the client react to each interpretation?

All of the above provide the client with a new, alternative way to consider the situation. In short, interpretation renames or redefines "reality" from a new point of view. Sometimes just a new way of looking at an issue is enough to produce change.

Interpretation may be contrasted with the paraphrase, reflection of feeling, focusing, and reflection of meaning. In those skills the interviewer remains in the client's *own* frame of reference. In interpretation the frame of reference comes from the counselor's personal and/or theoretical constructs. The following are examples of interpretation (reframing) paired with other skills.

Client 1: (with a record of absenteeism) I'm in trouble because I missed so many days of work.

Counselor: You're really troubled and worried. (reflection of feeling)

You've been missing a lot of work and you know your boss doesn't like it. (paraphrase/restatement)

As I listen to you, I sense that the issue is your anger that your friend got the promotion and you didn't. How do you react to that? (interpretation/reframe—note that the interview goes "beyond the data" and provides a new frame of reference for viewing the situation. The check-out "How do you react to that?" enables the client to deal with the interpretation in a more open fashion.)

Client 2: (with agitation) My wife and I had another fight over sex last night. I tried to make love and she rejected me again.

Counselor: You're upset and angry. (reflection of feeling)

You had a fight after the movie and were rejected again. (paraphrase/restatement)

Sounds like you didn't take it as slowly and easily as we talked about last week and she felt forced once again. Am I close? (interpretation 1 with check-out)

Your anger with her seems parallel to the anger you discussed several sessions ago that you used to feel toward your first wife. I wonder what sense you make of that? (interpretation 2 with check-out)

The feelings of rejection really bother you. Those sad and angry feelings sound like the dream last week. Does that make sense? (interpretation 3 with check-out)

In each of these interpretations, the interviewer or counselor adds something beyond what the client has said. Many interpretive responses may be made to any client utterance, and they may vary according to the theory and personal experience of the counselor.

Interpretation has traditionally been viewed as a mystical activity in which the interviewer reaches into the depths of the client's personality to provide new insights. However, if we consider interpretation to be merely a new frame of reference, the concept becomes less formidable. Viewed in this light, the depth of a given interpretation refers to the magnitude of the discrepancy between the frame of reference from which the client is operating and the frame of reference supplied by the interviewer. For example, a client may report feeling overly upset when the boss makes a minor criticism. The interviewer, counselor, or therapist may have noted this as a constant pattern. Several interpretations could be made, with varying depth:

"You seem to react very strongly to virtually any criticism from your boss."

"You appear to have a pattern of difficulty with authority."

"You feel very unsure of yourself and need approval to validate your worth."

"Your boss represents your father, and you are repeating with him the same patterns that you experienced with your father's criticism."

Again, interpretations will vary according to the theoretical orientation of the interviewer. You may wish to examine the situations in this section and reframe them from various theoretical perspectives with your supervisor or a colleague. The successful interviewer has many alternative interpretations available and selects them according to the long-term and immediate needs of the client.

Interpretations are best given in the 1-2-3 pattern of attending carefully to the client, providing the interpretation, then checking out the client's reactions to the new frame of reference ("How does that idea come across to you?"). If an interpretation is unsuccessful, the interviewer can use data obtained from the check-out to develop another, more meaningful response, most often a return to listening skills. It should be mentioned that reflection of meaning used in combination with questioning skills often enables clients to make their own interpretations and generate their own new frames of reference. A client working effectively with an interviewer using the skill of reflection of meaning is often better able to make interpretations than a skilled counselor.

Although you can run into clients who resist counseling regardless of the skill you use, *resistance* is most often associated with interpretation and reframing—the client simply does not accept what you are saying. At a deeper level, the client may refuse counseling as well. Motivational interviewing (see Chapter 14) views resistance as one of the central issues, particularly when an interviewer is working with addictive behavior (Miller & Rollnick, 2002).

Individual Practice in Interpretation/Reframing

Interpretations provide alternative frames of reference or perspectives for events in a client's life. In the following examples, provide an attending response (question, reflection of feeling, or the like) and then write an interpretation. Include a check-out in your interpretation.

Exercise 1

I was passed over for promotion for the third time. Our company is under fire for sex discrimination, and each time a woman gets the job over me. I know it's not my fault at all, but somehow I feel inadequate.

Listening response: _____

_____

Interpretation from a psychodynamic frame of reference (i.e., an interepretation that relates present behavior to something from the past): _____

_____

_____

Interpretation from a gender frame of reference: _____

_____

_____

Interpretation from your own frame of reference in ways that are appropriate for varying clients: _____

_____

_____

Exercise 2

"I'm thinking of trying some pot. Yeah, I'm only 13, but I've been around a lot. My parents really object to it. I can't see why they do. My friends are all into it and seem to be doing fine."

Listening response: _____

_____

Reframe from a conservative frame of reference (one that opposes the use of drugs):

_____

_____

_____

Reframe from an occasional user's frame of reference: _____

_____

_____

Interpretation from your own frame of reference on this issue: _____

_____

_____

The preceding examples of interpretations and reframes are particularly good representations of value issues you will find in interviewing and counseling. What are the value issues involved in these examples and what is your personal position on these issues? Finally, how do you reconcile the importance of a client's responsibility for her or his own behavior with your position? What would you actually *do* in these situations?

_____

_____

_____

_____

_____

INSTRUCTIONAL READING 2 AND EXERCISES: LOGICAL CONSEQUENCES

The process of learning is heavily based on the consequences of actions (Dreikurs & Grey, 1968). Client actions planned for the future are likely to have consequences later on. For example, an individual may want to change jobs simply because the new one offers more pay. However, the change may disrupt family life through a move, which in turn may cause other problems. Alternatively, the same move may bring unforeseen positive consequences. Explaining to others the probable consequences of their behaviors, if done sparingly and carefully, may at times be helpful in the interview or counseling session. Better yet, help them explore future consequences through questioning and listening.

Warnings are one way to tell someone about logical consequences. They inform the other person of the negative possibilities involved in a decision or action and the consequences that may result. Warnings tend to reduce risk taking and produce conformity. They often center on *anticipation of punishment.* A teacher in a disciplinary situation wants and needs a certain minimum level of conformity, as does a correctional officer. In interviews some clients plan very risky actions without considering the full range of possible consequences. A client may have just started smoking, be considering placing an aging parent in assisted care, be thinking about dropping out of school, or be making plans for "telling the boss off." In such situations, the client may profit from being informed about the possible consequences of the action. Helping clients face potential negative consequences is not always easy. It can be useful if you ask the clients to tell you a likely negative result instead of imposing your thoughts.

Encouraging someone to take certain risks or attempt new tasks is a more positive way to introduce logical consequences. In the schoolroom some children conform too much and demonstrate little creativity. They have been so conditioned by threats and warnings and bad outcomes that they are afraid to move. In such situations the goal may be to encourage risk taking by pointing out positive outcomes of a possible decision or action. *Anticipation of rewards* is a form of logical consequences in which the individual may be asked to imagine the positive consequences and rewards of new behavior or actions. In interviews you may wish to encourage clients to be more assertive or to try some new action; anticipating positive outcomes may be helpful in this process.

Making someone aware of logical consequences may at first seem to be a type of coercion or moralizing, yet it is the rare human behavior that does not have its costs and benefits, and the counselor's task is to help the client sort out those likelihoods while working toward a decision. You may do this through the attending skill of questioning or directly through telling the client the logical consequences of a possible behavior. In providing a logical consequence response, consider the following suggestions:

1. Through listening skills, make sure you understand the situation and the way your clients understand it.
2. As clients move toward decisions, encourage them to think about possible positive and negative results of the decisions. This is often done by questioning. You may even want to develop a balance sheet showing the pluses and minuses graphically (see p. 401).
3. Provide clients with data on both the positive and negative consequences of any potential decision or action. You may ask them to tell you a possible future story of what might happen if they continue to behave as they are behaving now. If they think only in negative terms, help them think of positives. If they are thinking only in positives, prompt thinking about possible negative consequences.
4. As appropriate to the situation, provide clients with a summary of positive and negative consequences in a *nonjudgmental* manner. With many people this step is not needed; they will have made their own judgments and decisions already.
5. Let clients decide what action to take in counseling situations. In teaching or management, on the other hand, you may decide at times and actually enforce the consequences.

In counseling, showing clients logical consequences tends to be a gentle skill used to help people sort through issues more completely. It also may be used in ranking alternatives when a complex decision must be made. In management and teaching situations the same generally holds true. For disciplinary and reward issues, however, it is important to note that the power rests with the manager and teacher: they decide the *consequences*. In interviewing, the task is simply to assist the client in foreseeing consequences. In both management and teaching situations, letting the employee or student decide what action to take is critical. But people who hold power over others need to *follow through with the consequences* they warned about earlier, or their power will be lost.

You may note that the logical consequences skill works primarily within the concrete orientation and with "if . . . , then . . ." language. We describe specific behavior and then the logical concrete consequence of the behavior. For example, "We first learned, Alisia, that *if* you didn't stand up for yourself when the fellow in the garage ignored you, *then* you would continue to be ignored. If you stand up for yourself assertively, then you will get attention from him."

Once the client understands the concrete consequences of a single behavior, it then may be possible to look at patterns of behavior using the same method from the abstract style. For example, "Alisia, we now know you have a pattern of not standing up for yourself. *If* you change that pattern and become more assertive, *then* you are much more likely to feel good about yourself and handle lots of situations more to your liking." With highly concrete clients, you will often find that they will need to have the formal-operational abstract pattern of thinking illustrated with several concrete examples before they understand the commonalities of their issues.

Individual Practice in Logical Consequences

Exercise 1

Using the five steps of the logical consequences skill, briefly indicate to the client, Alisia, what the logical consequences will be for her if she continues with her lack of assertiveness.

Summarize Alisia's problem in your own words using "if . . . , then . . ." terms.

_____

_____

_____

Ask Alisia specific questions about the positive and negative consequences of continuing her behavior.

Positives: _____

_____

Negatives: _____

_____

Provide Alisia with your own feedback on the probable consequences of continuing her behavior. Use "if . . . , then . . ." language.

_____

_____

_____

Summarize the differences between the feedback just given and Alisia's view when she says she doesn't want to change (this implies the use of confrontation).

_____

_____

Encourage Alisia to make her own decision.

_____

_____

_____

Exercise 2: Logical Consequences Using Attending Skills

Through questioning skills you can encourage clients to think through the possible consequences of their actions. ("What result might you anticipate if you did that?" "What results are you obtaining right now while you continue to engage in that behavior?") However, questioning and paraphrasing the situation may not always be enough to make the client fully aware of the logical consequences of actions. Write here, for the various types of clients and situations, logical consequences statements that might help the client understand the situation more fully.

A student who is contemplating taking drugs for the first time:

_____

_____

_____

A young woman contemplating an abortion:

_____

_____

_____

A student considering taking out a loan for college:

_____

_____

_____

An executive in danger of being fired because of poor interpersonal relationships:

_____

_____

_____

A client who is consistently late in meeting you and who is often uncooperative:

_____

_____

_____

INSTRUCTIONAL READING 3 AND EXERCISES: SELF-DISCLOSURE

Should you share your own personal observations, experiences, and ideas with the client? Self-disclosure by the counselor or interviewer has been a highly controversial topic. Many theorists argue against counselors' sharing themselves openly, preferring a more distant, objective persona. However, humanistically oriented and feminist counselors have demonstrated the value of appropriate self-disclosure. Multicultural theory considers self-disclosure early in the interview as key to trust building in the long run. This seems particularly so if your background is substantially different from that of your client. For example, a young person counseling an older person needs to discuss this issue early in the session. Moreover, research reveals that clients of counselors who self-disclose report lower levels of symptom distress and like the counselor more (Barrett & Berman, 2001).

This again brings up the important issue of whether or not you have experienced the client's issues in some way. Many alcoholics are dubious about the ability of nonalcoholics to understand what is occurring for them. A client with a serious fertility problem often feels that no one can really understand her if they have not experienced her issues. Imagine that you are a heterosexual Christian Asian counselor working with a conservative Christian Latina struggling with lesbian issues. Your background is very different and truly empathizing with the client may be a challenge. Open discussion, some self-disclosure on your part, and exploration of differences may be essential. Self-disclosure of who you really are can be helpful in those situations in which you have not "been there."

It appears that self-disclosure can encourage client talk, create additional trust between counselor and client, and establish a more equal relationship in the interview. Nonetheless, not everyone agrees that this is a wise skill to include among the counselor's techniques. Some express valid concerns about the counselor's monopolizing the interview or abusing the client's rights by encouraging openness too early; they also point out that counseling and interviewing can operate successfully without any interviewer self-disclosure at all.

A self-disclosure consists of the following:

1. *Personal pronouns.* A counselor or interviewer self-disclosure inevitably involves "I" statements or self-reference using the pronouns *I, me,* and *my.*
2. *Verb for content or feeling or both.* "I think . . . ," "I feel . . . ," and "I have experienced . . ." all indicate some action on the part of the counselor.
3. *Object coupled with adverb and adjective descriptors.* "I feel happy about your being able to assert yourself more directly with your parents." "My experience of divorce was something like yours. I felt and I still feel . . ."
4. *Feeling words and expression of feeling.* These are particularly important in self-disclosure. In essence, the structure of a self-disclosure is simple: it is "self-talk" from the counselor.

Another route to effective self-disclosure is simply to tell the client your own story. For example, if you are working with a client who grew up in an alcoholic family and you yourself have had experience in your own family with alcohol, a brief sharing of your own story can be helpful. The danger of storytelling, of course, is that you can end up spending too much time on your own issues and neglect the client.

Box 12-2 When Is Self-Disclosure Appropriate?

Weijun Zhang

My good friend Carol, a European American, has had lots of experience counseling minority clients. She once told me that one of the first questions she asks her minority clients is "Do you have any questions to ask me?" which often results in a lot of self-disclosure on her part. She would answer questions not only about her attitudes toward racism, sexism, religion, and so forth, but also about her physical health and family problems. During the initial interview, as much as half of the time available could be spent on her self-disclosure.

"But is so much self-disclosure appropriate?" said a fellow student in class after I mentioned Carol's experience.

"Absolutely," I replied. We know that many minority clients come to counseling with suspicion. They tend to regard the counselor as a secret agent of society and doubt whether the counselor can really help them. Some even fear that the information they disclose might be used against them. Some questions they often have in mind about counselors are, "Where are you coming from?" "What makes you different from those racists I have encountered?" and "Do you really understand what it means to be a minority person in this society?" If you think about how widespread racism is, you might consider these questions legitimate and healthy. And unless these questions are properly answered, which requires a considerable amount of counselor self-disclosure, it is hard to expect most minority clients to trust and open up willingly.

Some cultural values held by minority clients necessitate self-disclosure from the counselor, too. Asians, for example, have a long tradition of not telling personal and family matters to "strangers" or "outsiders," in order to avoid "losing face." Thus, relative to European Americans, we tend to reveal much less of ourselves in public, especially our inner experience. A mainstream counselor may well regard openness in disclosing as a criterion for judging a person's mental health, treating those who do not display this quality as "guarded," "passive," or "paranoid." Nonetheless, traditional Asians believe that the more self-disclosure you make to a stranger, the less mature and wise you are. I have learned that many Hispanics and Native Americans feel the same way, too.

Because counseling cannot proceed without some revelation of intimate details of a client's life, what can we do about these clients who are not accustomed to self-disclosure? I have found that the most effective way is not to preach or to ask, but to model. Self-disclosure begets self-disclosure. We can't expect our minority clients do well what we are not doing ourselves in the counseling relationship, can we?

According to the guidance found in most counseling textbooks here, excessive self-disclosure by the counselor is considered unprofessional; but if we are truly aware of the different orientation of minority clientele, it seems that some unorthodox approaches are needed.

The classmate who first questioned the practice asked with a smile, "Why are you so eloquent on this topic?" I said, "Perhaps it is because I have learned this not just from textbooks, but mainly from my own experience as both a counselor and a minority person."

Mary Ivey comments: Awareness of cultural and individual differences is essential. I learned long ago that recommended "rules" and "generalizations" of counseling

(continued)

Box 12-2   (continued)

and interviewing were made to be broken as we encounter the constant uniqueness of our clients.

Storytelling can be helpful as a way of self-disclosure, and you may decide not even to mention your part in the story. Through discussing how "other people" have resolved issues, you may accomplish the same aim as self-disclosure. The goal is to find a new way to talk about issues from a different perspective.

An important dimension of self-disclosure when you work with a client who is culturally different from you is to mention this fact early in the interview. This point is perhaps made most clearly when you think about a younger person counseling an elder. It is a simple matter of respect to note the age difference and to acknowledge that one does not have life experience. Such sharing can build trust and make the elder a colleague in the interviewing process.

With gender and race, however, it will take more prudence, and it is vital that you personally are comfortable with difference. When a male works with a woman, for example, it may be useful to say, "Men don't always understand women's issues. The things you are talking about clearly relate to gender experience. I'll do my best, but if I miss something, let me know." If you are White or African American working with a person of the other race, frank disclosure of this fact at the initiation of the interview can be helpful in developing trust. "I'm Black, you're White" (or "I'm White, you're an African American"). "How is it for you working with me?" The above suggestions and statements are generalizations. Be sure that what you do or do not do is authentic to the unique person in front of you and to your own knowledge and comfort level. Also, encourage the client to ask you questions, and be willing to answer them.

Here the "1-2-3" influencing pattern is particularly important: (1) attend to your client's story; (2) assess the appropriateness of your story and share it briefly; and (3) return focus to the client—note how he or she receives your story. Making self-talk relevant to the client is a complex task involving the following issues, among others.

Genuineness.   This is a vague term, but it can be made more concrete by emphasizing that the counselor must truly and honestly have had the experience and idea. This could be termed "genuineness in relation to self," an important beginning. But the self-disclosure must also be genuine in relation to the client. For example, a client may have performance anxiety about a part in a school play. The counselor may genuinely feel anxiety about giving a lecture before 50 people. The feelings may be the same or at least similar, but true genuineness demands a synchronicity of feeling. The counselor's experience in this case is probably too distant from that of the client. Self-disclosure should be fairly close to the client's experience; for instance, the counselor might relate a situation in which he or she felt similar performance anxiety in parallel circumstances at the same age.

Timeliness.   If a client is talking smoothly about something, counseling self-disclosure is not necessary. However, if the client seems to want to talk about a topic but is having trouble, a slight, leading self-disclosure by the counselor may be helpful. Too deep and involved a self-disclosure may frighten or distance the client.

Tense. The most powerful self-disclosures are often made in the present tense ("Right now I feel _____"). However, variations in tense can also be used to strengthen or soften the power of a self-disclosure. Consider the following:

Alisia:	I am feeling really angry about the way my parents sulk when I go out on my own. They really tick me off.
Counselor:	(present tense) You're coming across really angry right now. I like that you finally are in touch with your feelings.
Counselor:	(past tense) I can share that I've had the same difficulty in expressing feelings in the past. It helped me to get in touch.
Counselor:	(future tense) This awareness of emotion can help us all be more in touch in the future. I know it will continue to aid me.

Be careful when clients say, "What do think I should do?" Clients will sometimes ask you directly for opinions and advice on what you think they should do. "What do you think I should major in?" "If you were me, would you leave this relationship?" "Should I indeed have an abortion?" Effective self-disclosure and advice can potentially be helpful in such situations, *but* it is not the first thing you need to do.

Counseling is for the client and your task is to help the client make her or his own decisions. Involving yourself too early can foster dependency and lead the client in the wrong way. Note the following exchange.

Alisia:	What do *you* think I should do about my parents?
Counselor:	I'm not in your position. First, could you explore with me a little bit more about what occurs for you? (Or another example, "What I think is not important? It's you that counts. Tell me more.")

Most often clients will then carry on with the discussion and explore the issues on their own. And, generally, when clients are carefully listened to and fully heard, they will make good decisions. If you are forced to share your thoughts when you would prefer not to, keep your comments brief and then ask the client for her or his reflections on the issue. For example, "Alisia, my own thought would be to share with my parents how pressured I feel in such situations. *Does that relate to you at all?*"

Clearly, effective self-disclosure can be a complex skill, yet it can open that interview to new dimensions of sharing and helpfulness. In a situation such as the previous example, it may be useful to help the client—here, Alisia —learn to express anger more fully. But at the same time anger must be managed, and relaxation, assertiveness training, or logical consequences may be useful. It is not usually helpful to unleash anger all at once. For men, in particular, the issue with anger may be control and management rather than expression, and a variety of techniques drawing from several theories may be needed.

Individual Practice in Self-Disclosure

The structure of a self-disclosure consists of "I" statements made up of three dimensions: (1) the personal pronoun *I* in some form; (2) a verb such as *feel, think, have experienced;* and (3) a sentence objectively describing what you think or what happened.

Exercise 1: Writing Self-Disclosure Statements

Imagine that a client comes to you. For each situation write one effective and one ineffective self-disclosure. Also, share one of your own stories briefly. Before you begin, assume that you have already listened to the client carefully and that a self-disclosure indeed might be appropriate at this moment. Nonetheless, include a check-out with each comment.

First, how would you self-disclose to Alisia after hearing her story thus far?

_____

_____

Here are two more statements that Alisia might make while talking to you. What are some possible self-disclosures where you are asked to share your thoughts. In this process, avoid advice and opinion. What from your personal history might be helpful? Unhelpful?

"My family is totally dysfunctional. You've heard my story. What do you think?"

First, provide a response that gives the client another chance to explore her own issues:

_____

_____

A self-disclosure that might be useful: _____

_____

_____

Ineffective: _____

_____

_____

"I find myself afraid and insecure in large groups. It just doesn't feel right. What should I do?"

A response that gives the client another chance to explore her own issues:

_____

_____

_____

A useful self-disclosure: _____

_____

_____

Ineffective: _____

_____

_____

INSTRUCTIONAL READING 4 AND EXERCISES: FEEDBACK

Feedback is concerned with providing clients with clear data on their performance and/or on how others may view them. It may even involve self-disclosures on your part. As indicated earlier in the Guidelines for Effective Feedback (Box 3-5 in Chapter 3), feedback is centrally concerned with the following:

To see ourselves as others see us,
To hear how others hear us,
And to be touched as we touch others . . .
These are the goals of effective feedback.

Feedback to one another is important in the process of developing skills in counseling and interviewing. The guidelines given earlier for practicing effective feedback in small-group sessions are equally critical in counseling and interviewing:

1. *The client receiving feedback should be in charge.* Feedback is more likely to be successful if the client solicits it. However, at times you as interviewer will have to determine whether the client is ready and able to hear accurate feedback. Give only as much feedback as the client can use at the moment.
2. *Feedback should focus on strengths and/or something the client can do something about.* It does little good to tell a client to change many things that are wrong. It is more effective to give feedback on positive qualities and build on strength. When you talk about negatives, they should be areas that the client can do something to change or can learn to accept as a part of reality.
3. *Feedback should be concrete and specific.* Just as with directives, it does little good to offer vague feedback. For example, "You aren't able to get along with the group" is not as helpful as "You had two arguments with Sharon that upset both of you, and now you are disagreeing strongly with Jackie. What does this mean to you?"
4. *Feedback should be relatively nonjudgmental.* Critical to being nonjudgmental are your own accepting vocal qualities and your body language. Too often feedback turns into evaluation: "You did a good job there" as compared with "I saw you relax and become more joyful as you became more assertive." Stick to the facts and specifics. Facts are friendly; judgments may or may not be.
5. *Feedback should be lean and precise.* Most people have many areas that could profit from change. However, most of us can change only one thing at a time and can hear only so much. Don't overwhelm the client. Select one or two things for feedback and save the rest for later.
6. *Check out how your feedback was received.* Just as in an effective self-disclosure or reframe, check to see how the other person reacts to feedback. "How do you react to that?" and "Does that sound close?" and "What does that mean to you?" are three examples that involve the client in feedback and will indicate whether you were heard and how useful your feedback was.

Again, note the 1-2-3 pattern: (1) attend to the client, (2) use the influencing skill appropriate to the client, and (3) check out how well the skill was received and how useful it was.

Note also that clarity and concreteness are just as important in effective feedback. Good advice and instruction both require the 1-2-3 pattern plus concreteness and clarity if they are to be meaningful to the client. These same principles apply to all the influencing skills.

Another type of feedback is more judgmental. *Praise* and certain types of *supportive statements* convey your positive judgment of the client. Negative judgments may be shown through *reprimands* and certain types of punishments. In judgmental feedback the interviewer takes a more active role.

Judgmental feedback is usually considered inappropriate in counseling situations. However, management settings, correctional institutions, schools and universities, and other face-to-face settings often require judgmental feedback. Most often judgmental feedback involves an interviewer who has some power over the life of the client or other person. (Some would argue, in fact, that nonjudgmental feedback is impossible if the interviewer has any power over the client. Thus all feedback, to some extent, could be considered judgmental.)

The suggestions for judgmental types of feedback such as praise, support, reprimands, and punishments follow the same guidelines proposed earlier in this section. Such feedback tends to be more effective when the client is in charge as much as possible and when the feedback includes some emphasis on strengths as well as weaknesses, is concrete and specific, and is lean and precise. Judgmental feedback can be delivered in a tone of voice that is nonjudgmental and factual, thereby removing some of the sting. The check-out is particularly important in judgmental feedback. It provides the client with an opportunity to react, and the interviewer has some idea of how the feedback, positive or negative, was received.

Corrective feedback may be necessary. This is a delicate balance between negative feedback and positive suggestions for the future. Here you supply ideas that might work to replace ineffective thoughts, feelings, and behaviors, but always with a positive, hopeful tone.

Let us examine some of these issues with our client Alisia.

Vague, judgmental, negative feedback	Alisia, I really don't think the way you are dealing with people who hassle you is effective. You come across as a weak person.
Concrete, nonjudgmental, positive feedback	Alisia, I sense that you tried very hard to stand up for your rights in the garage. Alisia, you have potential. May I suggest some specific things that might be helpful the next time you face that situation?
Corrective feedback	Your effort was in the right direction. You can do even more if we set up an assertiveness training session for you

Positive feedback has been described as the "the breakfast of Champions." Your positive, concrete feedback can make a major difference in clients' restorying their problems and concerns. What are your clients doing right? Let them know it!

Individual Practice With Feedback

Exercise 1: Your Experience with Feedback

We have all experienced feedback on our performance. Some of this feedback has been helpful, and some has been painful. Allow yourself to recall a positive and a

Box 12-3   Research Evidence That You Can Use: Interpretation/Reframe, Self-Disclosure, Feedback

Six skills are covered in this chapter and space does not permit exploration of them all. Interpretation has been extensively reviewed and researched by Hill and O'Brien (1999). Among her findings are that interpretations are well received by clients, even though they are used sparingly by the interviewer. She also states that the moderate depth of interpretations tends to be favored. This would suggest that the concept of reframing, particularly as it relates to reflection of meaning, is useful. The language of reframing and reflection of meaning suggest "taking another perspective" and has less of an imposition from the interviewer.

Barrett and Berman (2001) conducted the research on the therapeutic outcome of self-disclosure. In a carefully designed study, they found that clients whose therapists self-disclosed "not only reported lower levels of symptom distress, but also liked their therapist more" (p. 596). They talk about psychodynamically oriented therapists who argue that the interviewer should be a "blank slate" and a neutral observer as contrasted with humanistic therapists who have long argued for appropriate self-disclosure. We should, however, keep in mind that self-disclosure (like interpretation/reframing) needs to be used only occasionally and with sensitivity to where the client is "at the moment."

Feedback has been most investigated by group therapists. Moran, Stockton, Cline, and Teed (1998) have conducted a major review of the literature. Their findings support the definitions provided in this chapter, but they also note that early feedback facilitates the group process and that some group members have real difficulty in hearing feedback. All influencing skills need to be used sparingly or they will lose effectiveness and possibly overwhelm the client.

Influencing skills and neuropsychology

Negative emotions, located in the amygdala, bubble up and challenge us and our clients constantly.

> Clients enter therapy with their amygdalas on overdrive, anxiously fearing what the therapy relationship will hold. In the language of neurobiology, the aim is to reduce this hyperactivity and bolster the activity of the nucleus accumbens, a brain area associated with pleasure. (DeAngelis, 2005, p. 72)

We need to go to cortical functions where positive thoughts and feelings are generated so that we can deal effectively with issues and problems. For example, cognitive therapy can encourage left brain activity to gain control over negative emotions (Carter, 1999, p. 41). The influencing skills and strategies, used effectively, provide clients with specific things they can do to build more positive thoughts, feelings, and behaviors. In this way, clients can deal more effectively with their issues.

Under times of severe stress or panic, the amygdala can take over. Dimasio (2003, p. 12) also points out that we need to build positive emotions to cope with the negative. Building on wellness and strengths will enable clients to cope with major challenges. Some even speculate that practitioners will be able to tailor specific treatments to modify brain circuits through counseling, medication, meditation, or other positive interventions (Beitman & Good, 2006).

negative experience with feedback. This may have been feedback from friends, family, teachers, or a work supervisor. What do you notice personally about effective and ineffective feedback?

_____

_____

_____

_____

_____

_____

_____

_____

Exercise 2: Writing Feedback Statements

Imagine that Alisia has just said, "I'm totally lost right now. I thought I had made progress on this issue, but this last week, I totally blew it again. I'm so discouraged with myself."

Write a positive feedback statement below:

_____

_____

What might a negative feedback statement be?

_____

_____

Now try the more challenging issue of presenting corrective feedback in a positive way to Alisia.

_____

_____

Review the above feedback statements considering the six criteria of effective feedback. How many did you meet in each situation?

Here are three more client statements. Imagine that you have heard a longer story and then provide positive and negative feedback comments and evaluate them as above:

"My partner is simply impossible. So much is demanded of me. I just feel myself disappearing in the relationship. I need space."

"What do you think of me? I've done as much as I can with working through issues around my sexuality, yet others keep looking at me and talking behind my back."

"I have real difficulty with exams, no matter how hard I study. In the last test I worked really hard, read all the assignments, and thought I knew the material cold. But I still ended up with a weak C."

INSTRUCTIONAL READING 5 AND EXERCISES: INFORMATION/ADVICE/OPINION/SUGGESTION

Influencing skills can be divided into an almost infinite array of specialized topics. Beware: excessive emphasis in this area may lead you to forget attending!

All of the previously listed skills, in one form or another, are concerned with imparting information to the client. The task is to be clear, specific, and relevant to the client's world. There is no need to downplay the importance of these skills, for most forms of interviewing, counseling, and even therapy involve imparting information to clients. An important part of both feminist theory and rational-emotive behavioral therapy, for example, is directly instructing clients in the content of their theories. Relaxation trainers often teach their clients how to relax. Instructional procedures in management, medicine, and other interviewing situations are equally or more important.

Providing helpful information consists of steps closely related to those already discussed under the influencing skills:

1. Attend to the client and be sure that he or she is ready for the information or advice. As with self-disclosure, provide clients with plenty of opportunity to find their own directions.
2. Be clear, specific, concrete, and timely in your instructional procedures. The concepts already discussed under feedback, directives, and logical consequences may be especially helpful in preparing more complex programs of instruction.
3. Check out with the client that your ideas have been understood.

These skills are often used in psychological education—the direct instruction of clients in the skills of living. As you teach the skills of interviewing to clients and others, you will be needing and using the strategy of effective information giving and instruction.

The giving of information or advice might be illustrated with Alisia in the following brief examples:

Career information: Alisia, one of the issues we need to look at is your job opportunities. Notice this chart. You are working now in computer science, and we see a projected 25% increase in opportunities coming in the next decade.

Personal advice/opinion: It might be useful if you went back to the garage and got the name of the people who gave you a bad time. Just going back and writing it down can have an influence. We can talk about how you can complain effectively about your treatment.

Skill instruction: One route toward handling difficult situations such as this is to engage in attending behavior—listening to the difficult person carefully. Then using that information, we can plan a more effective response. I'll teach you some of the basics of effective listening.

Individual Practice in Information/Advice/Opinion/Suggestion

Exercise 1

What is your own experience with advice and related skills? How do you personally respond when someone gives you advice or tells you what to do? It may be useful to reflect on the skills in this area and think about positives and negatives. What occurs for you when someone gives you an opinion or advice or suggests to you what you "ought" to do? At the same time, think about how useful instruction in skills of how to do something can be. Summarize below both negative and positive thoughts you have about this area.

_____

_____

_____

_____

_____

Exercise 2: Writing Statements

Again, imagine that Alisia has just said, "I'm totally lost right now. I thought I had made progress on this issue, but this last week I totally blew it again. I'm so discouraged with myself."

Write below how you might use information, advice, instruction, opinion, or suggestion effectively. How can you involve Alisia fully in the process so that you do not foster dependency?

_____

_____

_____

_____

Repeated here are three client statements from page 378. Imagine that you have heard a longer story and then attempt to provide meaningful information, advice, instruction, opinion, or suggestion. Again, what are the potential dangers of involving yourself too much?

"My partner is simply impossible. So much is demanded of me. I just feel myself disappearing in the relationship. I need space."

"What do you think of me? I've done as much as I can with working through issues around my sexuality, yet others keep looking at me and talking behind my back."

"I have real difficulty with exams, no matter how hard I study. In the last test I worked really hard, read all the assignments, and thought I knew the material cold. But I still ended up with a weak C."

INSTRUCTIONAL READING 6 AND EXERCISES: DIRECTIVES

All influencing skills need to be situated in the client's world, which means that you first need to draw out the client's story and positive assets and wellness strengths. Directives can be useful in developing a new story, but they are especially effective in helping a client move to action. While a positive new story may be sufficient for some clients, many will profit from specific behaviors and actions offered by various directives.

Directive strategies are drawn from various counseling theories. In this section you will be introduced to some useful interviewing strategies that may lead clients to think about their stories in new ways and to apply these new ways of thinking to their daily lives. Directives are particularly important in assertiveness training interviews (see Chapter 14). Here your ability to influence clients by giving them clear and useful directions is central. You may wish to include assertiveness training as part of your examination of this important influencing skill.

A brief summary of how to give a directive follows:

1. *Involve your client as co-participant in the directive strategy.* Rather than simply telling the client what to do, be sure that you have heard her or his story, issues, and problems sufficiently. Inform the client what you are going to do and the likely result. Some theories and practitioners like to use surprises, and perhaps that is satisfactory in some situations. But as a general rule, we urge *working with,* rather than *working on,* your client.

2. *Use appropriate visual, vocal tone, verbal following, and body language.* When you use influencing skills, your attending behaviors tend to be more assertive and stronger than when you are listening. Individual and cultural difference modifies this general recommendation. With challenging clients such as an acting-out teen, you may need stronger nonverbals as you listen or provide directives. With a quieter, tentative client, it may be more appropriate to be still and tentative yourself as you share new ways of thinking about issues. We don't want to overwhelm clients.

3. *Be clear and concrete in your verbal expression.* Know what you are going to say and say it clearly and explicitly. Compare the following:

Vague:	I'd like you to go out and arrange for a test.
Concrete:	After we finish today, I'll take you out and I'll introduce you to the testing office where you will take the Strong-Campbell Interest Blank. Complete it today, and they will have the results for us to discuss during our meeting next week.
Vague:	Relax.
Concrete:	Sit quietly . . . feel the back of the chair on your shoulders . . . tighten your right hand . . . hold it tight . . . now let it relax slowly . . .

These examples illustrate the importance of indicating clearly to your client what you want to happen. Directives need to be authoritative and clear but also stated in such a way that they are in tune with the needs of the client.

4. *Check out whether your directive was heard and understood.* Just because you think you are clear doesn't mean the client understands what you said. Explicitly or implicitly check out to make sure your directive is understood. This is particularly important when a series of directives has been given. For example, "Could you repeat back to me what I have just asked you to do?" or "I suggested three things for you to do for homework this coming week. Would you summarize them to me to make sure I've been clear?"

You may wish to share these directive strategies with your clients by telling them what to expect and perhaps even why you are going to use the strategies. For example:

Counselor: Alisia, I've heard your story about how frustrated you were in the garage. I'd like to use imagery to understand the situation a bit more fully. This will require you to relax, sit back, and allow yourself to recall the situation. It often helps to visualize the specifics of the situation, what was said, and so on—almost like a movie. As we start, I'd like you to remember just one particular scene that first occurs to you. Is all this OK with you? Any questions? (Note the concreteness: the counselor has involved Alisia by providing an explanation and has checked out with her to see whether proceeding with the strategy is satisfactory to her.)

Though the example directive strategies of Box 12-4 are all presented in very brief form, with some imagination and practice—and client participation in the process—you can successfully use many of them. For more detailed presentations of these and other strategies in highly concrete form, see Ivey, D'Andrea, Ivey, and Simek-Morgan's *Counseling and Psychotherapy: A Multicultural Perspective* (2002).

The following practice exercises focus on the general structure of presenting directives. At the end of the chapter, practice exercises to implement the strategies are suggested. But before you reach the practice exercises, we'd like to expand the ideas of interpersonal influence by drawing in some example theoretical perspectives and connecting influencing skills more clearly to counseling and interviewing theory (see Box 12-4).

Individual Practice in Directives

Exercise: Trying Directive Strategies in Your Own Life

We recommend that you by yourself or with a friend test out the varying directive strategies. If you experience them yourself, you will have a better idea of their potential and how you might want to pace or time the presentation of the strategy. The following four may be useful as a beginning. Select at least one. With each of these, you may want to audiorecord the directive strategy and develop a pacing and timing that is appropriate for you. With the two guided imagery exercises, we strongly recommend that you work only with positive images. Negative imaging is potentially useful in psychotherapy but is highly inappropriate unless there is a safe context for consideration of the issues.

Guided imagery focusing on a relaxing scene. The following is a typical example of how one can help clients relax and find positives.

All of us have wellness experiences that are important for us. For example, you likely have a positive image of a place where you felt relaxed and comfortable—maybe a lakeside or mountain scene, or a snowy setting, or a quiet, special place. Take a moment now to close your eyes and relax. (pause) You may want to notice your breathing and the general feelings in your body first. Focus on that place where you felt safe, comfortable, and relaxed. Allow yourself to enter that scene. What are you seeing? . . . Hearing? . . . Feeling? . . . Perhaps there are

Box 12-4 Example Directive Strategies Used by Counselors
of Different Theoretical Orientations

	Directives are powerful skills and should be used sparingly and with caution. Several directives are described, then followed by illustrations of how that directive might be used with a person such as Alisia, our earlier example.
Specific suggestions/ instructions for action	"I suggest you try . . ." "Alisia, the next time you go to the garage, I'd like you to stand there in front of the counter, make direct eye contact with the shop head, and clearly and firmly tell him that you have a meeting at 10:00 and that you must be taken care of *now*. If he says there will be a delay, get him to make a firm time commitment. Then follow up with him firmly 15 minutes later."
	This type of clear directive is usually most effective with clients who don't have skills and need firm guidance. These are most likely too elementary for Alisia.
Paradoxical instructions	"Continue what you are doing . . ." "Do the problem behavior/thinking/action at least three times." "Alisia, I want you to notice what you're doing in terms of lack of assertiveness. Next week, at least three times, I want you deliberately to give in to other people's demands. And, at the same time, I want you to notice carefully how you are behaving and how they react."
	With clients who refuse to follow suggestions and directives, a paradoxical directive may be useful both in developing awareness and in promoting change. Alisia could profit from more awareness of her own feelings and behavior.
Imagery	"Imagine you are back in the situation. Close your eyes and describe it precisely. What are you seeing, hearing, feeling?" "What is your image of your ideal day/job/life partner?" "Alisia, you seem a bit vague about the time you gave in to your parents. Close your eyes, visualize your parents . . . (pause). . . . Now what are they saying to you? What are you saying to them? How do you feel in this process?"
	Imagery directives are often the most powerful directives and must be used with care. Images are particularly effective in helping clients experience the sensorimotor style. Additionally, many children and young people like the freedom and creativity allowed in this type of directive.
Role-play enactment	"Now return to that situation and let's play it out." "Let's role-play it again, only change the one behavior we agreed to."
	Alisia would be encouraged to take her part and play it out, with the counselor taking the role of the shop manager. (Role-playing is an especially effective technique to make the abstract concrete. It makes the client behavior clear and specific. Given that clarity, changes in concrete behavior become much easier to make. This is one of the most basic techniques used in assertiveness training.)
Gestalt empty-chair technique	"Talk to your parent as if he or she were sitting in that chair. Now go to that chair and answer as your parent would."
	This technique is similar to a role-play, but the client plays both people. The role-play concretizes the issue for both you and the client. The actual acting and movement

(continued)

Box 12-4   (continued)

	help the client to get in touch with deeper sensorimotor emotions. And, after the role-play, it is possible to use abstract formal operational analysis to consider and think about what occurred.
Gestalt nonverbal	"I note that one of your hands is in a fist; the other is open. Have the two hands talk to one another."
	This is similar to the hot-seat. Alisia would learn to be aware of her nonverbal communication and its underlying meaning.
Free association, as in psychodynamic theory	"Take that feeling and free-associate back to a childhood experience . . . to what is occurring now in your daily life." "Stay with that feeling, magnify it. Now, what flashed into your mind first?"
	This technique is best for full formal-operational clients who are self-directed. Alisia might be expected to free-associate back to a specific childhood experience that helps explain her present behavior. Many of us are repeating old parent–child interactions in the present. Paradoxically, free association works best if sensorimotor modalities of emotion are used.
Positive reframing	"Identify a negative experience, thought, or feeling. Now identify something positive in that experience and focus on that dimension. Synthesize the two." "Alisia, we've discussed the negative in the situation with your parents. What positives can you see in it? Relax . . . (pause) . . . let it flow . . ."
	This type of intervention may be expected to produce a sense of awareness, and possibly pride, in that it highlights Alisia's interpersonal sensitivity, her caring, and her flexibility. In fact, it may be these very skills that have made Alisia successful in many interpersonal situations. However, she fails to discriminate when to use the skills and when not to.
Relaxation	"Close your eyes and drift." "Tighten your forearm, very tight, now let it go." (Directions for relaxation continue.)
	Alisia may be especially tight and tense; relaxation training can be taught as a sensorimotor environmental-structuring intervention. Once Alisia is able to relax and be in control of her body, she will be better able to cope with the very real stresses she encounters daily.
Systematic desensitization	This involves (1) deep muscle relaxation, (2) construction of an anxiety hierarchy, and (3) matching objects of anxiety with relaxation. This is a more complex counseling strategy and would be useful to Alisia if she had a deep-seated anxiety problem with assertiveness. She would first learn deep muscle relaxation; then she

(continued)

some odors or tastes that come to you as well. Allow yourself to enjoy that scene in full relaxation. Enjoy it now for a moment before coming back to this room. Now, as you come back, notice your breathing (pause) and as you open your eyes, notice the room, the colors, and your surroundings. How was this experience for you?

Box 12-4   (continued)

	and her counselor would list objects of anxiety from most tension-provoking to least. At the top of the anxiety hierarchy might be speaking up for one's rights to one's parents; less anxiety-provoking might be the garage situation. Other increasingly anxiety-free situations would be generated until a state of complete calm is imagined. You would lead the client in a carefully planned session or sessions to practice relaxation under stress.
Language choice	"Change 'should' to 'want to.'" "Change 'can't' to 'won't.'" "Alisia, you say you can't do anything with the shop head. Change 'can't' to 'won't,' because you are in charge of your own behavior . . . Good . . . now say that again." Words such as *can't* imply that a situation is out of control; by changing *can't* to *won't,* Alisia is being forced to be in charge of her own behavior. This environmental-structuring approach may be particularly useful with self-directed formal-operational clients, who may need help in changing their thinking and behavior.
Thought-stopping	"We all have internal speech in which we silently talk to ourselves. Sometimes that internal 'self-talk' becomes a negative judgmental voice. I'd like to explore some of your negative thinking and self-talk. Then I want to teach you how to stop by placing a rubber band on your wrist. Every time you say something negative to yourself, snap the rubber band and say 'STOP!'"
Meditation	"Be still. Focus on one point. Relax. Concentrate on breathing. Let all thoughts slip from your mind." Alisia may prefer training in meditation as a mode of relaxation. Teaching meditation is more characteristic of a client-directed, formal-operational approach.
Family therapy communications	"Don't talk to me . . . talk to him now." "Change chairs with your wife and sit closer to your daughter . . ." You may find it helpful to bring Alisia's partner in for couples counseling. Many of the directives used individually work well when adapted to family and group counseling.
Homework	"Practice this exercise next week and report on it in the next interview." "Alisia, next week I want you to try some of the skills we've worked on in the role-play. Focus on your eye contact patterns—you've learned you tend to look down when there's the potential of conflict. Just this week, focus on this one dimension." Research increasingly shows that much learning during the interview is lost if it is not immediately transferred to daily life.

Positive images of strength. Images of people who have meant a lot to us serve as examples of strength and give us the courage to meet difficult issues. The same exercise as above can be used as you focus on a person who was important to you. As you go through this exercise, add one dimension—note where the feelings inspired by the person's image are located in your own body. Some note warm feelings in their stomach, others observe feelings of strength in their upper chest or arms—most of us can locate where the positive image is located in our body. This physical strength from others can be a resource to help cope with current problems and concerns. You will often find more power in here-and-now sensorimotor body exercises.

Spiritual and religious images.   You will also find that spiritual and religious images and experiences are very helpful to many clients. Only in recent years has the counseling and interviewing field recognized the strengths and power in this area. One route toward including spirituality as part of your practice is the community genogram, presented in the chapter on focusing. Many clients will discuss church and spirituality as an important part of their history in their community genogram.

Other positive images may be found in times when the client felt competent and "on top of things." Happy childhood memories, images of grandparents or friends, and experiences with nature are three examples of where positive images may be found. Work with the client to find positives and then draw on them when the client needs support.

Thought-stopping.   This strategy is so simple that it is easy to ignore or even forget in the many things that go on in a session. But we have consistently found that it is one of the most effective interventions a counselor or therapist can use. Please give thought-stopping your serious attention.

Many of us have a pattern of internalized negative self-talk. These stressful thoughts may be said to yourself, perhaps even several times a day, as in the following examples:

"I can't do anything right."
"I am not a good person."
"I always foul up."
"I should have done better."

Other areas for negative thoughts include guilty feelings, procrastination in which you might "overthink" the situation, worry over details, fear that things will only get worse, anger that others "never get it right," repetitively thinking about past failures, and always needing the approval of others.

Emotional thought-stopping draws on the same basics described here but is used for helping clients with depression and/or suicidal thoughts (Bernhardt, 2001). For a more detailed discussion, visit www.have-a-heart.com/depression-II.html or use the key words "emotional thought stopping, Bernhardt."

What are some of your recurrent areas of negative self-talk?

_____

_____

_____

Having identified these areas, we suggest three possibilities for thought-stopping:

1. Relax and close your eyes and imagine a situation where you will likely make the negative self-statement. Take time and let the situation evolve. When the thought comes, observe what happens and how you feel. Then tell yourself silently "STOP." In emotional thought-stopping, say it loudly—even shouting—at first.

2. Place a rubber band around your wrist and every time during the day that you find yourself thinking negatively, snap the rubber band and say "STOP!"

3. Once you have developed some understanding of how often you use negative thinking, after you say "STOP," substitute a more positive statement such as the following, which are suggested in place of the negative self-statements.

"I can do lots of things right."
"I am a capable person."
"I sometimes mess up—no one's perfect."
"I did the best I could."

4. Do this a minimum of one full day.

What are some positive self-statements that you can make to replace your own negative self-talk?

_____

_____

Supplementary Exercise: Assertiveness Training

You may wish to practice assertiveness training in conjunction with your work in the influencing skill or directives. Chapter 14 presents an example assertiveness interview, and a practice/feedback form may be found at the conclusion of that chapter.

Box 12-5   Key Points

1-2-3 pattern	In any interaction with a client, first attend to and determine the client's frame of reference, then assess her or his reaction before using your influencing skill. Finally, add a check-out to obtain the reaction to your use of the skill.
Interpersonal influence continuum	The influencing and attending skill may be classified from low to high degrees of influence. Encouragers and paraphrasing are considered relatively low in influence, whereas confrontation and directives are considered more influential.
Moderate triad	The "swing skills" of the interpersonal influence continuum are focusing and open and closed questions. They provide a framework for determining the topic of conversation while keeping a balance between influencing and attending skills.
Interpretation/reframing	The skill stresses the importance of helping clients generate a new perspective on their issues. This way of thinking may help them view problems in new ways and restructure problematic stories.
Logical consequences	This skill involves indicating the probable results of a client's action, in five steps: 1. Use listening skills to make sure you understand the situation and how it is understood by the client. 2. Encourage thinking about positive and negative consequences of a decision. 3. Provide the client with your data on the positive and negative consequences of a decision in a nonjudgmental manner. 4. Summarize the positives and negatives. 5. Let the client decide what action to take.

(continued)

Box 12-5   (continued)

Self-disclosure	Indicating your thoughts and feelings to a client constitutes self-disclosure, which necessitates the following: 1. Use personal pronouns ("I" statements). 2. Use a verb for content or feeling ("I feel . . . ," "I think . . ."). 3. Use an object coupled with adverb and adjective descriptors ("I feel happy about your being able to assert yourself . . ."). 4. Express your feelings appropriately. Self-disclosure tends to be most effective if it is genuine, timely, and phrased in the present tense. Keep your self-disclosure brief. At times, consider sharing short stories from your own life.
Feedback	Feedback entails providing accurate data on how the counselor or others view the client. Remember the following: 1. The client should be in charge. 2. Focus on strengths. 3. Be concrete and specific. 4. Be nonjudgmental. 5. Keep feedback lean and precise. 6. Check out how your feedback was received. These six guidelines are useful for all influencing skills.
Information/advice/ opinion/suggestion	These skills call for passing on information and ideas of the interviewer to the client. The "1-2-3" pattern is again critical, and the information should be clear, concrete, and timely.
Directives	Directives indicate clearly to a client what *action* to take. It is vital to involve your client with the choice of directives, even to the point of telling her or him what is to happen and what to expect as a result. Appropriately assertive body language, vocal tone, and eye contact are important, as are clear, concrete verbal expressions and checking out the degree of client participation.
What else?	Most of the influencing skills require you to be concrete and specific in your intervention, although the ability to work abstractly is particularly important in interpretation/reframing. Remember to involve your client as a co-participant as you utilize influencing skills.

COMPETENCY PRACTICE EXERCISES AND PORTFOLIO OF COMPETENCE

Influencing Skills

Small-group work with the influencing skills requires practice with each skill. The general model of small-group work is suggested, but only one skill should be used at a time. Remember to include the Client Feedback Form from Chapter 1 as part of the practice session. This is a particularly important place to practice group supervision, sharing, and feedback.

Step 1: Divide into practice groups.
Step 2: Select a leader for the group.
Step 3: Assign roles for each practice session.

▲ Client

▲ Interviewer, who will begin by drawing out the client story or issue using listening skills and then attempt one of the influencing skills. Feedback and self-disclosure may be used with other influencing skills.

▲ Observer 1, who will observe the client and complete the CIS Rating Form (see Chapter 9), deciding how much of an impact the interviewer's influencing skills have made

▲ Observer 2, who will complete the Feedback Form (Box 12-6) that follows

Step 4: Plan. In using influencing skills, the acid test of mastery is whether the client actually does what is expected (for example, does the client follow the directive given?) or responds to the feedback, self-disclosure, and so on in a positive way. For each skill, different topics are likely to be most useful. State goals you want to accomplish in each instance. Some ideas follow:

▲ *Directives.* The client may talk about a situation in which he or she was either insufficiently assertive or overly aggressive. The interviewer may first try a role-play to make the situation more specific and then may attempt one or more of the other directives to see how the client reacts.

▲ *Logical consequences.* A member of the group may present a decision he or she is about to make. The counselor can explore the negative and positive consequences of that decision.

▲ *Self-disclosure.* A member of the group may present any personal issue. The task of the counselor is to share something personal that relates to the client's concern. Again, the check-out is important to obtain client feedback.

▲ *Feedback.* A most useful approach is for the interviewer to give direct feedback to another member of the group about his or her performance in the training sessions. Alternatively, the client may talk about an issue for which the counselor will provide feedback, sharing his or her perceptions of the situation as objectively as possible. A check-out should be included.

▲ *Information/advice/other.* The interviewer may give information about a particular theory or set of facts to the group. The group gives feedback on the ability of the information-giver to be clear, specific, interesting, and helpful. We suggest that you consider teaching microskills as communication skills to your group.

▲ *Directives and specific strategies.* Select one of the strategies presented in this chapter and work through the specific steps. Involve your client in the process, and the two of you can select together the strategy that you would like to try. As part of the practice session, be sure to tell the client what to expect and the likely results.

Step 5: Conduct a 5- to 15-minute practice session using the skill. You will find it difficult to use particular influencing skills frequently, as they must be interspersed with attending skills to keep the interview going. Attempt to use the targeted skill at least twice during the practice session, however.

Step 6: Review the practice session and provide feedback for 10 to 12 minutes. Remember to stop the tape to provide adequate feedback for the counselor.

Step 7: Rotate roles.

Box 12-6   Feedback Form: Influencing Skills

_____ (Date)

_____   _____
(Name of Interviewer)   (Name of Person Completing Form)

Instructions: Observer 1 will complete the CIS Rating Form in Chapter 9. Observer 2 will complete the items below.

1. Did the interviewer use the basic listening sequence to draw out and clarify the client's story or concern? How effectively?

2. Provide nonjudgmental, factual, and specific feedback for the interviewer on the use of the specific influencing skill or directive strategy.

3. Did the interviewer check out the client's reaction to the intervention?

Portfolio of Competence

This chapter is about interpersonal influence and it covers considerable material. You cannot be expected to master all these skills until you have a fair amount of practice and experience. At this point, however, it would be helpful if you thought about the major ideas presented in this chapter and where you stand currently. Also, where would you like to go in terms of next steps?

Use the following as a checklist to summarize where you are at this point and perhaps as goals for the future. *Do not expect to attain intentional competence on every dimension as you work through this book.* You will find, however, that you will improve your competencies with repetition and practice.

Six skill areas are identified in this chapter. They are listed here with the levels of competence. Check off those areas for which you have competence so far. Be patient with yourself, as mastery of these skills will take time.

Identification/Reframe

Level 1: Identification and classification.

❏ Ability to write statements representing the skill in question and to identify and/or classify the skill when seen in an interview.

Level 2: Basic competence.

❏ Demonstration of your ability to use the skill in a role-play interview.

Level 3: Intentional competence.

❏ Demonstration of your ability to use the skill in interviews with specific impact and effects on clients.

Level 4: Teaching competence.

❏ Ability to teach the skills to others.

Logical Consequences

Level 1: Identification and classification.

❏ Ability to write statements representing the skill in question and to identify and/or classify the skill when seen in an interview.

Level 2: Basic competence.

❏ Demonstration of your ability to use the skill in a role-play interview.

Level 3: Intentional competence.

❏ Demonstration of your ability to use the skill in interviews with specific impact and effects on clients.

Level 4: Teaching competence.

❏ Ability to teach the skills to others.

Self-Disclosure

Level 1: Identification and classification.
❏ Ability to write statements representing the skill in question and to identify and/or classify the skill when seen in an interview.

Level 2: Basic competence.
❏ Demonstration of your ability to use the skill in a role-play interview.

Level 3: Intentional competence.
❏ Demonstration of your ability to use the skill in interviews with specific impact and effects on clients.

Level 4: Teaching competence.
❏ Ability to teach the skills to others.

Feedback

Level 1: Identification and classification.
❏ Ability to write statements representing the skill in question and to identify and/or classify the skill when seen in an interview.

Level 2: Basic competence.
❏ Demonstration of your ability to use the skill in a role-play interview.

Level 3: Intentional competence.
❏ Demonstration of your ability to use the skill in interviews with specific impact and effects on clients.

Level 4: Teaching competence.
❏ Ability to teach the skills to others.

Directive

Level 1: Identification and classification.
❏ Ability to write statements representing the skill in question and to identify and/or classify the skill when seen in an interview.

Level 2: Basic competence.
❏ Demonstration of your ability to use the skill in a role-play interview.

Level 3: Intentional competence.
❏ Demonstration of your ability to use the skill in interviews with specific impact and effects on clients.

Level 4: Teaching competence.
❏ Ability to teach the skills to others.

Information/Advice/Suggestion

Level 1: Identification and classification.
❑ Ability to write statements representing the skill in question and to identify and/or classify the skill when seen in an interview.

Level 2: Basic competence.
❑ Demonstration of your ability to use the skill in a role-play interview.

Level 3: Intentional competence.
❑ Demonstration of your ability to use the skill in interviews with specific impact and effects on clients.

Level 4: Teaching competence.
❑ Ability to teach the skills to others.

DETERMINING YOUR OWN STYLE AND THEORY: CRITICAL SELF-REFLECTION ON INFLUENCING SKILLS

You have encountered six skills and have had the opportunity for at least a brief introduction to each. With which of these skills do you feel most comfortable? Which might you seek to use? Which might you avoid? Perhaps most important, how do you feel about the idea of consciously influencing the direction of the interview?

What single idea stood out for you among all those presented in this chapter, in class, or through informal learning? What stands out for you is likely to be important as a guide toward your next steps. What are your thoughts on multicultural issues and the use of this skill? What other points in this chapter struck you as important? How might you use ideas in this chapter to begin the process of establishing your own style and theory?

_____

_____

_____

_____

_____

_____

Given the complexity of this chapter and the many possible goals you might set for yourself, list here three specific goals you would like to attain in the use of influencing skills within the next month.

1. _____

2. _____

3. _____

REFERENCES

Barrett, M., & Berman, J. (2001). Is psychotherapy more effective when therapists disclose information about themselves? *Journal of Consulting and Clinical Psychology, 69,* 597–603.

Beitman, G., & Good, G. (2006). *Counseling and psychotherapy essentials.* New York: Norton.

Bernhardt, S. (2001). Emotional thought stopping. www.have-a-heart.com/depression-II.html.

Carter, R. (1999). *Mapping the mind.* Berkeley: University of California Press.

Damasio, A. (2003). *Looking for Spinoza: Joy, sorrow and the feeling brain.* New York: Harvest.

DeAngelis, T. (2005, November). Where psychotherapy meets neuroscience. *Monitor on Psychology.*

Dreikurs, R., & Grey, L. (1968). *Logical consequences: A new approach to discipline.* New York: Dutton.

Hill, C., & O'Brien, K. (1999). *Helping skills.* Washington, DC: American Psychological Association.

Ivey, A., D'Andrea, M., Ivey, M., & Simek-Morgan, L. (2002). *Counseling and psychotherapy* (5th ed.). Boston: Allyn & Bacon.

Miller, W., & Rollnick, S. (2002). *Motivational interviewing: Preparing people for change.* New York: Guilford.

Moran, K., Stockton, R., Cline, R., & Teed, C. (1998). Facilitating feedback exchange in groups: Leader interventions. *Journal for Specialists in Group Work, 23,* 257–268.

ALLEN AND MARY'S THOUGHTS ABOUT ALISIA

Our orientation to the question of general direction and goals for counseling remains highly influenced by Carl Rogers. What does the client want to have happen? We really want to see the goals of counseling determined by the client. We would want to work with Alisia to ensure that *she* is determining the focus of counseling, so the listening skills of the first part of this book remain central to us.

But goal setting does not always have to be completed within a person-centered Rogerian tradition. While goals can come out of effective Rogerian work, our tendency would be to seek to make them more explicit. We would likely first draw on decisional counseling theory and practice (see the next chapter) and seek to outline with her some specific issues for exploration. And, somewhat contrary to Rogers, we would not hesitate to bring in issues that Alicia has not thought of, as you will see below. Along with goal setting, we would also keep some key questions in mind that we would raise with her. But determining whether to follow these up would be her decision.

Example possible directions we would consider for Alicia:

▲ Helping her expand her consciousness as a woman in a sometimes oppressive world. Feminist counseling theory might be useful. The emphasis on positive women models would be characteristic of our work.

▲ Encouraging her to balance negative thinking with positive assets and wellness strengths. This is one of our core values and a central part of the microskills framework and theory.

▲ Helping her become more assertive. You will see how we would do this via assertiveness training in Chapter 14.

▲ As many of these issues are related to gender, Allen would seek supervision if he worked with her and would always be aware of the possibility of referring her to a woman as the need might arise.

How do our thoughts compare with yours? Again, counseling is for the client and it is vital that we not impose our goals, thoughts, and feelings on her. We would focus on helping her make her own decisions—and here, once again, we are back to the central importance of listening!

SECTION IV

Skill Integration

What is your own style of interviewing, counseling, or psychotherapy? This section provides a framework to help you integrate the many skills and concepts of this book.

Chapter 13 presents a decisional approach to the interview in detail. Decisional counseling is a practical and integrative approach with a long history. Working from this framework will enable you to assist clients in making life decisions. You will also find that the decisional system will help you understand other theoretical orientations as you construct your own theory and orientation to the interview.

A full decisional interview transcript is presented in Chapter 13. Here you also will find information on how to plan for an interview, make case notes, and develop a long-term treatment plan. The final exercise in this chapter is the most important one in this book, suggesting that you audio- or videorecord a full interview and develop a transcript of your work. This will allow you to analyze your natural style and skills, thus facilitating your eventual development of a flowing interview.

Chapter 14 provides information on four very different orientations to interviewing and counseling—person-centered, behavioral assertiveness training, brief solution-oriented counseling, and motivational interviewing. You will find that the person-centered orientation uses virtually no questions and emphasizes listening skills. Assertiveness training uses a balance of skills but gives special attention to the use of directives, particularly the role-play. The solution-oriented interviewing and counseling approach, on the other hand, utilizes questions as the central skill, with the listening skills taking an important secondary role. Motivational interviewing has close parallels to decisional counseling, but relies even more on the skills approach of this book.

The four theoretical orientations of Chapter 14 are designed to show you the variety of ways in which the interview can be constructed using the microskills approach. We anticipate that you will find all useful in your eventual practice, but you may find that you personally favor one of the methods over the others. Each can be useful to you as you think through your own natural style and favored approach to helping others. At the same time, maintain awareness that some of your clients may prefer a different method from your first choice.

Chapter 15 provides you with the opportunity to review your work with microskills, interview structure, and various orientations to the interview. You will be asked to think about your own natural style of interviewing and plan for the future.

By the time you have completed this book and this section in particular, you will have learned how to analyze and integrate the concepts of intentional interviewing and counseling. You may aim to achieve the following:

1. Examine and analyze your own behavior and thinking in the interview and discuss your specific impact on clients.
2. Review your helping style and your sensitivity to multicultural and gender issues.
3. Define your ideas and intentions about the theory and methods you believe to be most important in counseling and interviewing.
4. Conduct a beginning full interview using five alternative theoretical/practical structures:
   ▲ *Decisional counseling.* The five stages of the interview may be constructed as a problem-solving framework. Decisions underlie most issues and skills in this theory and will be useful in many settings.
   ▲ *Person-centered counseling.* If you have completed a full interview using only the basic listening sequence, you have made a useful beginning to understand and practice this theoretical orientation. Chapter 14 adds skills of reflection of meaning, focusing, and feedback to broaden this framework and place it more in tune with Carl Rogers's original thinking.
   ▲ *Solution-oriented counseling.* The solution model is designed as a brief approach to human change. In a time of accountability, you may find this model useful in aiding clients to move to action. Questioning is the central microskill.
   ▲ *Behavioral assertiveness training.* Many of your clients will benefit from training in this strategy, regardless of your personal theoretical orientation. It is a useful supplement to help overly passive or overly aggressive clients achieve their own personal goals.
   ▲ *Motivational interviewing.* This is a relatively new approach that is rapidly gaining popularity. Developed as a therapy for work with alcoholics, it is now recognized as a highly useful strategy for all clients. You will find that your knowledge of microskills is particularly compatible with this method of helping.

The objectives above will provide you with a comprehensive overview of the interviewing and counseling process. You can use your knowledge and skills to build your own culturally intentional, culturally appropriate interviewing style.

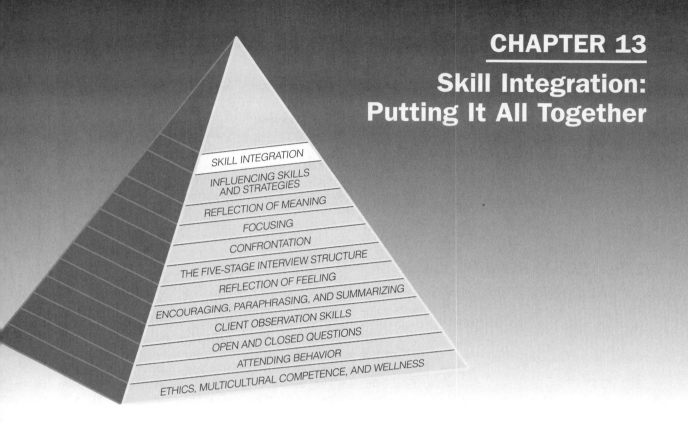

Skill Integration

Influencing Skills and Strategies

Reflection of Meaning

Focusing

Confrontation

The Five-Stage Interview Structure

Reflection of Feeling

Encouraging, Paraphrasing, and Summarizing

Client Observation Skills

Open and Closed Questions

Attending Behavior

Ethics, Multicultural Competence, and Wellness

How can skill integration help you and your clients?

Major function	The many skills, strategies, and concepts of intentional interviewing and counseling will be of no benefit to you and your clients unless you can integrate them smoothly into a naturally flowing interview and treatment plan such as the one presented in this chapter.
Secondary functions	Mastery of the concepts of this chapter results in the ability to do the following:

▲ Conduct a well-organized career interview based on a decisional model derived from trait-and-factor theory
▲ Develop long-term treatment plans for your work with a client and keep systematic interview records
▲ Utilize the concepts of this chapter to further your skills and understanding of other theoretical models of the helping process

INTRODUCTION: ANALYZING THE INTERVIEW

This chapter presents a detailed analysis of a single interview and shows how you can use a similar process to examine your own work. It is possible to take the many skills and concepts of this book and use them systematically in an entire interview. The basic objective of this chapter is to facilitate an examination of your own style. As you read this chapter, think ahead to how you will conduct your own interviews, develop your own plan for interviews, keep notes, and integrate your own ideas into a long-term treatment plan. The single practice exercise for this chapter will suggest that you audio- or videorecord an interview and use it as a basis for conducting a comprehensive analysis of your own style and its impact on clients.

INSTRUCTIONAL READING 1: DECISIONAL COUNSELING

Decisional counseling may be described as a practical model for counseling, interviewing, or psychotherapy that recognizes decision making as undergirding most—perhaps all—systems of counseling. The basic point is simple: Whether clients need to choose among careers, decide to have a child, decide to get married, or search for new ideas on how to live more effectively, *they are always making decisions.* Decisional counseling facilitates the process of decision making underlying life patterns.

Many see Benjamin Franklin as the originator of the systematic decision-making model. He suggested three stages of problem solving: (a) identify the problem clearly, (b) generate alternative answers, and (c) decide what action to take. Another term for decisional counseling is *problem-solving counseling.* The essential issue is the same regardless of the terms we use: *How can we help clients work through issues and come up with new answers?*

Decisional counseling theorists believe that all theories of counseling and therapy—behavioral, person-centered, or even psychoanalytic—deal with problem solving and decisions. The behavioral counseling process aims to teach clients how to make new decisions for assertive action or behavior control. The person-centered counselor wants to enable clients to make decisions for themselves through self-examination and self-reflection. The psychoanalytic therapist is searching for the underlying factors leading to decisions. Thus decisional counseling is a basic framework useful in many settings.

The Trait-and-Factor Legacy

The goal in decisional counseling is to facilitate decision making and to consider the many *traits and factors* underlying any single decision. Trait-and-factor theory has a long history in the counseling field, dating back to Frank Parsons's development of the Boston Vocational Bureau in 1908. Parsons pointed out that in making a vocational decision the client needs to (a) consider personal traits, abilities, skills, and interests; (b) examine the environmental factors (opportunities, job availability, location, and so on); and (c) develop "true reasoning on the relations of these two groups of facts" (Parsons, 1909/1967, p. 5). Trait-and-factor theory since that time has searched for the many dimensions that underlie "true reasoning" and decision making.

Gradually trait-and-factor theory came to be seen as limited, and new decisional and problem-solving models have arisen (Brammer & MacDonald, 2002; D'Zurilla, 1996; Egan, 2002; Janis & Mann, 1977; Ivey, D'Andrea, Ivey, & Simek-Morgan, 2002). All of those models could be described as modern reformulations of Benjamin Franklin's original model and trait-and-factor theory; however, they are based on more recent thinking and research.

The Balance Sheet and Future Diary: Bringing Emotions Into Decisional Practice

Decision making is also explored by modern neuropsychology. Emotions play a critical role in decision making for it is *how we feel* about the alternative answers and possible solutions that leads us to a decision. This approach sees decision making as involving: "(1) the facts of the problem presented, (2) the option chosen to solve it, (3) the factual outcome to the solution, and, importantly, (4) the outcome of the solution in terms of emotion and feeling" (Damasio, 2003, p. 143). This approach reminds us that decisions are related to emotional outcomes—pleasure with a decision that works and pain when it doesn't. Decisions are not just present tense; they also anticipate the future logical consequences of the decision.

An Australian, Leon Mann, developed the Balance Sheet and Future Diary to help people "take decisions" (Mann, 2001; Mann, Beswick, Allouche, & Ivey, 1989). Motivational interviewing has recently adapted these strategies and finds them helpful in facilitating decision making, particularly among those with substance abuse issues (see Chapter 14, page 472, for a sample balance sheet). The strategies are simple and straightforward but can be a powerful addition to decisional counseling. In developing a balance sheet with a client, you list all the possible solutions and rate each possibility with a " + " or " − ". Or you may use several pluses or minuses if you have several possibilities. This provides a visual map of the decisional territory.

The balance sheet is coupled with the Future Diary in which clients anticipate how they will feel emotionally about each decision. The Future Diary can be used as a homework assignment for the client. However, we believe that it is also desirable to work through the likely futures of decisions with the client in the here and now of the interview. You may wish to use guided imagery to help clients anticipate the future emotions associated with each decision.

Decisions, Problem Solving, and the Five Stages/Dimensions of the Interview

Decisional counseling concepts are basic to the five-stage/dimensions model of the interview with which we have been working throughout this book. At this point, it may be helpful to see how these five elements fit with the traditional problem-solving models as shown in Table 13-1.

Thus, decisional issues and problem-solving theory are clearly parallel. The simplest problem-solving model is defining the problem, generating or brainstorming alternatives, and then deciding among the alternatives. We have also seen that neuropsychology research follows a similar model, with special emphasis on emotion.

Applications to Counseling Theories and Multiple Settings

Decisional issues and the microskills model underlie many theoretical approaches to helping. If you have intentional competence in the microskills and the five stages/dimensions, you are well prepared to work with many different approaches to the helping process. Table 13-2 shows how the five stages might appear in differing theoretical or practical orientations.

Let us now turn to an example interview.

Table 13-1 The five stages/dimensions and traditional problem-solving models

The Five Stages/Dimensions of Intentional Interviewing and Counseling	Problem-Solving Models and the Trait-and-Factor Legacy
1. Initiating the session—rapport and structuring ("Hello; this is what might happen in this session.")	This step has often not been stressed as a separate element, but it is obviously an important part of any problem-solving attempt.
2. Gathering data—drawing out stories, concerns, problems, or issues ("What's your concern? What are your strengths or wellness resources?")	Benjamin Franklin talked of defining the problem whereas trait-and-factor theorists speak of the need to consider personal traits, abilities, skills, and interests. The environmental emphasis of trait-and-factor theory should be commended.
3. Mutual goal setting—establishing outcomes ("What do you want to happen?")	In many problem-solving approaches, this is considered part of defining the problem. Separating it out more specifically may be useful. Goal setting is often the first issue addressed in brief solution-oriented interviewing and counseling.
4. Working—exploring alternatives, confronting client incongruities and conflict, restorying ("What are we going to do about it?")	These models often use brainstorming to find alternatives and employ "true reasoning" to discover the relationship of the person to the environment.
5. Terminating—generalizing and acting on new stories ("Will you do it?")	Generalization is not always stressed but it is implicit in brainstorming and true reasoning.

Table 13-2 Estimated time spent in stages/dimensions given different theoretical/practical orientations

Stage/Dimension	Decisional Counseling	Person-Centered Theory	Assertiveness Training	Brief Solution-Oriented Counseling	Motivational Interviewing
1. Initiating the session—rapport and structuring	Medium	Some, rapport especially stressed throughout	Some	Some	Medium
2. Gathering data—drawing out stories, concerns, problems, or issues	Medium	Great	Great	Medium	Medium
3. Mutual goal setting	Medium	Little emphasis	Medium	Great, a central issue	Medium
4. Working—exploring alternatives, confronting client incongruities and conflict, restorying	Medium	Medium	Medium	Medium—Dimensions 4 and 5 combined	Medium
5. Terminating—generalizing and acting on new stories	Great	Little	Medium	See above	Great

EXAMPLE FULL INTERVIEW TRANSCRIPT: I'D LIKE TO FIND A NEW CAREER

The following interview is a role-play conducted by Allen and Mary Ivey. A written transcript was made of this interview and then Allen analyzed his behavior in the session and Mary reviewed his comments and analysis.

As in all interviews, you will find some interviewer responses that are less effective than others. We all make errors; it is our ability to learn from them and change that makes us become more effective interviewers and counselors. As you read the session and the process comments, stay alert to Allen's responses that you think could be phrased more effectively. What would *you* do differently? And what responses make sense to you?

Keep in mind as you read the example interview material that you soon may be analyzing your own interviewing style. You may wish to use the format of the interview presentation (5 columns) as a way to prepare your own interview analysis.

Planning the Session

This initial interview illustrates that career counseling is closely related to personal counseling. Note how the relationship between job change and personal issues develops during the session. Additionally, gender issues emerge as important multicultural factors that need to be considered. This shows up especially in Stage 4 of this interview.

Mary role-plays a client who is 36 years old and divorced, with two children. She has worked as a physical education teacher for a number of years. She stated in her information file completed before counseling that "I find myself bored and stymied in my present job as a PE teacher. I think it is time to look at something new. Possibly I should think about business. Sometimes I find myself a bit depressed by it all."

Given the five-stage structure of the basic interview, it is possible to develop a plan for the interview before it takes place. If you don't know where you are heading, you are less likely to help the client. This does not mean you need to impose your views or concepts on the client; rather, you can think through what you want to do to help the client achieve his or her objectives.

You are about to read the transcript of an initial interview session. However, before the interview was held, the interviewer, Allen, developed the interview plan presented in Box 13-1. This plan was developed from his study of a client file that consisted of a pre-interview questionnaire. Mary had stated in her intake form, "I'd like to do something new with my career. I'm ready for something new—but, what?" As you will note, the interview plan is oriented to help the client develop her own unique career plan and to facilitate the discussion of personal issues as well. The plan is considered a structure to help the client achieve her objectives and make her own decisions, but remains flexible enough to adapt as the interview progresses and new issues may be brought up.

Although it is, of course, tentative, the plan does illustrate that it is possible to plan a session before it happens. As it turns out, Allen's objectives were roughly realized in the session. However, if Mary had had different needs, the interview plan could have been scrapped at that time and more appropriate interventions used. Remember: An interview plan is only a plan; it is not a definitive roadmap.

Box 13-1 First Interview Plan and Objectives

	Before the first interview, study the client file and try to anticipate what issues you think will be important in the session and how you might handle them. (This plan is Allen's assessment of his forthcoming interview with Mary.)
Initiating the session— rapport/structuring	*Special issues anticipated with regard to rapport development? What structure do you have for this interview? Do you plan to use a specific theory?* Mary appears to be a verbal and active person. I note she likes swimming and physical activity. I like to run. . . . That may be a common bond. I think I'll be open about structure but keep the five stages in mind. It seems she may want to look into career choice and she may be unhappy with her present job. I should keep in mind her divorce, as this may also be a personal issue. I'll use the basic listening sequence to bring things out and probably work with a decision-making model. She's probably an abstract, formal-operational client. Therefore, in the first stages of the session, I'll need to listen to her stories and use mainly questions and reflection skills.
Gathering data— drawing out stories, concerns, problems, or issues	*What are the anticipated problems? Strengths? How do you plan to define the issues with the client? Will you emphasize behavior, thoughts, feelings, meanings?* Mary seems to be full of strengths. I'll use her wellness strengths and work on finding out what she *can* do. I'll use the basic listening sequence to bring out issues from her point of view. I'll be most interested in her thoughts about her job and her plans for the future; however, I should be flexible and watch for other issues. Mary may well bring up several issues. I'll summarize them toward the end of this phase, and we may have to list them and set priorities if there are many issues. Mainly, however, I expect an interview on career choice.
Mutual goal setting	*What is the ideal outcome? How will you elicit the idealized self or world?* I'll ask her what her fantasies and ideas are for an ideal resolution and then follow up with the basic listening sequence. I'll end by confronting and summarizing the real and the ideal. It's worked for me in the past and probably will work again. As for outcome, I'd like to see Mary defining her own direction from a range of alternatives. Even if she stays in her present job, I hope we can find that it is the best alternative for her.
Working—exploring alternatives, confronting client incongruities and conflict, restorying	*What types of alternatives should be generated? What theories would you probably use here? What specific incongruities have you noted or do you anticipate in the client?* I hope to begin this stage by summarizing her positive strengths and wellness assets. It's too early to say what the best alternatives are. However, I'd like to see several new possibilities considered. Counseling and business are indicated in her pre-interview form as two possibilities. They seem good. Are there other possibilities? The main incongruity will probably be between where she is and where she wants to go. I'll be interested in her personal life as well. How are things going since the divorce? What is it like to be a woman in a changing world? I expect to ask her questions and develop some concrete alternatives even in the first session. . . . I hope she will act on some of them following our first session. I think career testing may be useful.
Terminating—generalizing and acting on new stories	*What specific plans, if any, do you have for transfer of training? What will enable you personally to feel that the interview was worthwhile?* Drawing from the above, I'll feel satisfied if we have generated some new possibilities and can do some exploration of career alternatives after the first session. We can plan from there. I'd like it if we could generate at least one thing she can do for homework before our second session.

Table 13-3 presents the interview as it actually happened, but it is structured to present the data according to the five-stage model. The counselor-and-client verbatim transcript is supplemented by a skill-and-focus analysis of the session. The Process Comments column analyzes the effectiveness of skills throughout the interview, with special attention given to the effect of confrontations ("C") on the client's developmental change.

As you read this interview, pay special attention to Allen's interviewing style. Think about how appropriate each of his interventions is and what you might do differently. Although this book emphasizes the importance of your defining your own natural style, it is also important that you look at your style and how others would view it. Just as you might criticize or change what Allen does, what might others criticize and suggest for changes in your behavior?

The way an interviewer responds to clients, even through using listening skills, influences what he or she says next and the direction of the interview. Allen constructed the session as one addressing primarily career counseling and focused on decisions that needed to be made. Simply focusing on Mary's saying she was bored led naturally to more career discussion. If the focus had been on her "new friend" or her children, a different session might have occurred.

Moreover, a person-centered counselor might have responded very differently. For example, see Chapter 14, page 441, to discover how the interview might move rather quickly to personal counseling with a greater emphasis on reflection of feeling and reflection of meaning. A behaviorally oriented interviewer would again be expected to listen differently and lead the session in still another direction. See page 445 for an example of how a session oriented to assertiveness would be very different.

Please take a moment to move ahead to the person-centered and behavioral examples. Compare and contrast the varying styles of interviewing. Which do you prefer? The helping fields are moving rapidly toward eclectic and integrative theories and methods (Ivey, D'Andrea, Ivey, & Simek-Morgan, 2002). You will find it helpful to be equipped with skills in these orientations and to continue to learn other systems as well. We do believe that the systematic decisional counseling model and motivational interviewing underlie most, perhaps all, theories of helping and if you are skilled in this system, you have a good foundation for mastering other theories as well.

Table 13-3 The Allen and Mary five-stage interview

Skill Classifications				
Listening & Influencing	Focus	C*	Counselor and Client Conversation	Process Comments
Open question	Client		**STAGE/DIMENSION 1: Initiating the session— Rapport and structuring** 1. *Allen:* Hi, Mary. How are you today?	
	Client, interviewer		2. *Mary:* Ah . . . just fine . . . How are you?	As Mary walked in, Allen saw her hesitate and sensed some awkward-ness on her part. Note that she opens with two speech hesitations.
Information, paraphrase	Interviewer, client		3. *Allen:* Good, just fine. Nice to see you. . . . Hey, I was noting in the folder that you've done a lot of swimming.	
	Client, main theme		4. *Mary:* Oh, yeah, (smiling) . . . I like swimming; I enjoy swim-ming a lot.	Consequently, Allen decides to take a little time to develop rapport and put Mary at ease in the interview. Note that he focused on a positive aspect of Mary's past. It is often useful to build on the client's strengths even this early in the session.
Information, closed question	Main theme, interviewer, client		5. *Allen:* With this hot weather, I've been getting out. Have you been able to?	The distinction between providing information and a self-disclosure is illustrated at *Allen 5* and *Mary 6*. Allen only comments that he's been getting out, whereas Mary gives information and personal feelings as well.
	Client		6. *Mary:* Yes, I enjoy the exercise. It's good relaxation.	
Paraphrase, reflection of feeling	Client		7. *Allen:* I also saw you won quite a few awards along the way. (*Mary:* Um-hmm.) . . . You must feel awfully good about that.	Mary's nonverbal behavior is now more relaxed. Client and counselor now have more body language symmetry.
	Client		8. *Mary:* I do. I do feel very good about that. It's been lots of fun.	
Information, closed question	Main theme		9. *Allen:* Before we begin, I'd like to ask if I can tape-record this talk. I'll need your written per-mission, too. Do you mind?	Obtaining permission to tape-record interviews is essential. If the request is presented in a comfortable, easy way, most clients are glad to give permission. At times it may be useful to give the tapes to clients to take home and listen to again.
	Client		10. *Mary:* No, that's okay with me. (signs form permitting use of tape for *Intentional Interview-ing and Counseling*)	

*This column will record the presence of a confrontation.

(continued)

Table 13-3   **(continued)**

Skill Classifications				
Listening & Influencing	*Focus*	*C*	*Counselor and Client Conversation*	*Process Comments*
Information giving	Client		11. *Allen:* As we start, Mary. There are some important things to discuss. We'll have about an hour today and then we can plan for the future together. Today, I'd like to get to know you and I'll try to focus mainly on listening to your concerns. At the same time, from your file, I know that some of your issues relate to women's issues. Obviously, I'm a man and sometimes women are not always comfortable talking about certain things. I think it is important to bring this up so that you will be more likely to feel free to let me know if I seem to be "off-target" or mis-understand something. Let's follow that up by your asking me any questions you'd like around this or other matters.	Allen provides some additional structure for the session so that Mary knows what she might expect. He introduces gender dif-ferences and provides an opportu-nity for Mary to react and ask questions.
	Client, interviewer		12. *Mary:* I feel comfortable with you already. But, a couple questions. One is that I'm interesting in the counseling field as a possibility—and the other is around the issue of living with divorce and being a single parent. What can you say about those?	Mary gives the OK, but then asks two questions. She leans forward when she asks them.
Self-disclosure, open question	Interviewer, client		13. *Allen:* Well, first I'm divorced and have one child living with me while the other is in college. Of course, I'd be glad to talk about the counseling career and share some of my thoughts. What thoughts occur to you around divorce and counseling?	Keep self-disclosures brief and return the focus to the client. But be comfortable and open in that process.
	Client, main theme		14. *Mary:* That helps. Going through my divorce was the worst thing of my life. My children are so important to me.	Mary smiles, sits back, and appears to have the information she was wondering about.

(continued)

Table 13-3 (continued)

Skill Classifications				
Listening & Influencing	Focus	C	Counselor and Client Conversation	Process Comments
			Perhaps your experience with divorce will help you understand where I am coming from. Let's get started and look at what my career should be.	
Open question	Client		**STAGE/DIMENSION 2: Gathering data—Drawing out stories, concerns, problems, or issues** 15. *Allen:* Could you tell me, Mary, what you'd like to talk about today?	In this series of leads you'll find that Allen uses the basic listening sequence of open question, encourager, paraphrase, reflection of feeling, and summary, in order. Many interviewers in different settings will use the sequence or a variation to define the client's problem.
	Client, problem, others		16. *Mary:* Well . . . ah . . . I guess there's a lot that I'd like to talk about. You know, I went through . . . ah . . . a difficult divorce and it was hard on the kids and myself and . . . ah . . . we've done pretty well. We've pulled together. The kids are doing better in school and I'm doing better. I've . . . ah . . . got a new friend. (breaks eye contact) But, you know, I've been teaching for 13 years and really feel kind of bored with it. It's the same old thing over and over every day; you know . . . parts of it are okay, but lots of it I'm bored with.	As many clients do, Mary starts the session with a "laundry list" of issues. Though the last thing in a laundry list is often what a client wants to talk about, the eye-contact break at mention of her "new friend" raises an issue that should be watched for in the interview. As the session moves along, it becomes apparent that more than the career issue needs to be looked at. Mary discusses a "pattern" of boredom. This is indicative of an abstract, formaloperational cognitive/emotional style.
Encourage	Client		17. *Allen:* You say you're *bored* with it?	The key word *bored* is emphasized.
	Client, problem		18. *Mary:* Well, I'm bored, I guess . . . teaching field hockey and . . . ah . . . basketball and softball, certain of those team sports. There are certain things I like about it, though. You know, I	Note that Mary elaborates in more detail on the word *bored*. Allen used verbal underlining and gave emphasis to that word, and Mary did as most clients would: she elaborated on the meaning of the key word to her.

(continued)

Table 13-3 **(continued)**

| Skill Classifications | | | | |
Listening & Influencing	Focus	C	Counselor and Client Conversation	Process Comments
			like the dance, and you know, I like swimming—I like that. Ah . . . but . . . you know . . . I get tired of the same thing all the time. I guess I'd like to do some different things with my life.	Many times short encouragers and restatements have the effect of encouraging client exploration of meaning and elaboration on a topic. "I'd like to do some different things" is a more positive "I" statement.
Paraphrase	Client		19. *Allen:* So, Mary, if I hear you correctly, sounds like change and variety are important instead of doing the same thing all the time.	Note that this paraphrase has some dimensions of an interpretation in that Mary did not use the words *change* and *variety*. These words are the opposite of boredom and doing "the same things all the time." This paraphrase takes a small risk and is slightly additive to Mary's understanding. It is an example of the positive asset search, in that it would have been possible to hear only the negative "bored." Working on the positive suggests what *can* be done. Note her response.
	Client, family, problem		20. *Mary:* Yeah . . . I'd like to be able to do something different. But, you know, ah . . . teaching's a very secure field, and I have tenure. You know, I'm the sole support of these two daughters, but I think, I don't know what else I can do exactly. Do you see what I'm saying?	Mary, being heard, is able to move to a deeper discussion of her issues. Note that Mary continues to talk within the formal-operational orientation. She discusses patterns and generalizations. If she were primarily concrete, she would give many more linear details and tell specific stories about her issues. She continues for most of the interview in this mode of expression. She has equated "something different" with a lack of security. As the interview progresses, you will note that she associates change with risk. It is these basic meaning constructs, already apparent in the interview, that undergird many of her issues.
Reflection of feeling, followed by check-out	Client, problem	C	21. *Allen:* Looks like the security of teaching makes you feel good, but it's the boredom you associate with that security that makes you feel uncomfortable. Is that correct?	This reflection of feeling contains elements of a confrontation as well, in that the good feelings of security are contrasted with the boredom associated with teaching.

(continued)

Table 13-3 (continued)

Skill Classifications				
Listening & Influencing	Focus	C	Counselor and Client Conversation	Process Comments
	Client, problem		22. *Mary:* Yeah, you know, it's that security. I feel good being . . . you know . . . having a steady income and I have a place to be, but it's boring at the same time. You know, ah . . . I wish I knew how to go about doing something else.	Note that Mary often responds with a "Yeah" to the reflections and paraphrases before going on. Here she is wrestling with the confrontation of *Allen 21*. She adds new data, as well, in the last sentence. On the CIS, this would be a Level 3 response.
Summary, check-out	Client, family, problem	C	23. *Allen:* So, Mary, let me see if I can summarize what I've heard. Ah . . . it's been tough since the divorce, but you've gotten things together. You mentioned the kids are doing pretty well. You talked about a new relationship. *I heard you mention that.* (*Mary:* Yeah.) But the issue right now that you'd like to talk about is that . . . this feeling of boredom (*Mary:* Ummm . . .) on the job, and yet you like the security of it. But maybe you'd like to try something new. Is that the essence of it?	This summarization concludes the first attempt at problem definition in this brief interview. Allen uses Mary's own words for the main things and attempts to distill what has been said thus far in a few words. The positive asset search has been used briefly in this section ("You've gotten things together . . . kids . . . doing well"). See also other leads in this section that emphasize client strength. Mary sits forward and nods with approval throughout this summary. The confrontation of the old job with "maybe you'd like to try something new" concludes the summary. Note the check-out at the end of the summary to encourage Mary to react.
	Client, problem		24. *Mary:* That's right. That's it.	Mary again responds at Level 3 on the CIS.
Open question	Main theme		**STAGE/DIMENSION 3: Mutual goal setting** 25. *Allen:* I think at this point it might be helpful if you could define for me what some things are that might represent a more ideal situation.	In Stage 3 the goal is to find where the client might want to go in a more ideal situation. You'll note that the basic listening sequence is present in this stage, but it does not follow in order, as in the preceding stage.
	Client, problem, others		26. *Mary:* Ummm. I'm not sure. There are some things I like about my job. I certainly like interacting with the other professional people on the staff. I enjoy working with the kids. I enjoy talking with the kids.	Mary associates interacting with people as a positive aspect of her job. When she says "enjoy working with kids," her tone changes, suggesting that she doesn't enjoy it that much. But the spontaneous tone returns when she mentions "talking

(continued)

Table 13-3   (continued)

| Skill Classifications | | | | |
Listening & Influencing	Focus	C	Counselor and Client Conversation	Process Comments
			That's kind of fun. You know, it's the stuff I have to teach I'm bored with. I have done some teaching of human sexuality and drug education.	with them" and talks about teaching subjects other than team sports.
Paraphrase, open question	Client, main theme		27. *Allen:* So, would it be correct to say that some of the teaching, where you have worked with kids on content of interest to you, has been fun? What else have you enjoyed about your job?	The search here is for positive assets and things that Mary enjoys. Note the "what else?"
	Client, family, problem		28. *Mary:* Well, I must say I enjoy having summer vacations and the same vacation time that the kids have. That's a plus in the teaching field. (pause)	
Encourage			29. *Allen:* Yeah . . .	Mary found only one plus in the job. Allen probes for more data via an encourager. This type of encourager can't be classified in terms of focus.
	Client, others		30. *Mary:* You see, I like being able to . . . Oh, I know, one time I was able to do teaching of our own teachers and that was really . . . I really felt good being able to share some of my ideas with some people on the staff. I felt that was kind of neat, being able to teach other adults.	Mary brings out new data that support her earlier comment that she liked to teach when the content was of interest to her. The "I" statements here are more positive and the adjective descriptors indicate more self-assurance.
Closed question	Client, others		31. *Allen:* Do you involve yourself very much in counseling the students you have?	A closed question with a change of topic to explore other areas.
	Client, others		32. *Mary:* Well, the kids . . . you know, teaching them is a nice, comfortable environment, and kids stop in before class and after class and they talk about their boyfriends and the movies they go to and so on; I find I like that part of it too . . . about their concerns.	Mary responds to the word *counseling* again with discussion of interactions with people. It seems important to Mary that she have contact with others.

(continued)

Table 13-3 (continued)

Skill Classifications				
Listening & Influencing	Focus	C	Counselor and Client Conversation	Process Comments
Summary, closed question	Client, problem, others		33. *Allen:* So, as we've been reviewing your current job, it's the training, the drug education, some of the teaching you've done with kids on topics other than phys. ed. (*Mary:* That's right.) And getting out and doing training and other stuff with teachers . . . ah, sharing some of your expertise there. And the counseling relationships. (*Mary:* Ummmm.) Out of those things, are there fields you've thought of transferring to?	This summary attempts to bring out the main strands of the positive aspects of Mary's job. In an ongoing interview, a closed question on a relevant topic can be as facilitating as an open question. Note, however, that the interviewer still directs the flow with the closed question.
	Client, problem		34. *Mary:* Well, a lot of people in physical education go into counseling. That seems like a natural second thing. Ah . . . of course, that would require some more going to school.	Mary talks with only moderate enthusiasm about counseling. In discussing training and business, she appears more involved.
	Client, problem		Umm . . . I've also thought about doing some training for a business. Sometimes I think about moving into business . . . entirely away from education. Or even working in a college as opposed to working here in the high school. I've thought about those things, too. But I'm just not sure which one seems best for me.	Mary appears to have assets and abilities, makes many positive "I" statements, is aware of key incongruities in her life, and seems to be internally directed. She is clearly an abstract, formal-operational client. For career success, she needs to become more concrete and action-oriented.
Paraphrase, closed question	Client, problem		35. *Allen:* So the counseling field, the training field. You've thought about staying in schools and perhaps in management as well. (*Mary:* Um-hm, um-hm.) Anything else that occurs to you?	This brief paraphrase distills Mary's ideas in her own words.
	Problem		36. *Mary:* No, I think that seems about it.	
Summary, open question, eliciting meaning	Client, problem	C	37. *Allen:* Before we go further, you've talked about teaching and the security it offers. But at the same time you talk about	This summary includes confrontation and catches both content and feeling. The question at the end is directed toward issues of meaning.

(continued)

Table 13-3 (continued)

Skill Classifications				
Listening & Influencing	*Focus*	*C*	*Counselor and Client Conversation*	*Process Comments*
			boredom. You talk with excitement about business and training. How do you put this together? What does it *mean* to you?	The word *boredom* was underlined with extra vocal emphasis.
	Client, others		38. *Mary:* Uhh . . . Ah . . . If I stay in the same place, it's just more of the same. I see older teachers, and I don't want to be like them. Oh, a few have fun; most seem just *tired* to me. I don't want to end up with that.	Mary elaborates on the meaning and underlying structure of *why* she might want to avoid the occasional boredom of her job. When she talks about "ending up like that," we see deeper meanings. On the CIS, the client may again be rated as Level 3. Though considerable depth of understanding and clarity is being developed, no change has really occurred. You will find that developmental movement often is slow and arduous. Nonetheless, each confrontation moves to more complete understanding.
Encourage/ restatement	Client		39. *Allen:* You don't want to end up with that.	The key words are repeated.
	Client		40. *Mary:* Yeah, I want to do something new, more exciting. Yet my life has been so confused in the past, and it is just settling down. I'm not sure I want to risk it.	Mary moves on to talk about what she wants, and a new element—risk—is introduced. Risk may be considered Mary's opposing construct to security.
Reflection of feeling	Client		41. *Allen:* So, Mary, risk frightens you?	This reflection of feeling is tentative and said in a questioning tone. This provides an implied check-out and gives Mary room to accept it or suggest changes to clarify the feeling.
	Client, problem		42. *Mary:* Well, not really, but it does seem scary to give up all this security and stability just when I've started putting it together. It just feels strange. Yet I do want something new so that life doesn't seem so routine . . . and . . . ah . . . I think maybe I have more	Mary responds as might be predicted with a deeper exploration of feelings of fear of change. At the same time, she draws on her personal strengths to cope with all this.

(continued)

Table 13-3   (continued)

Skill Classifications				
Listening & Influencing	Focus	C	Counselor and Client Conversation	Process Comments
			talent and ability than I used to think I did.	
Reflection of meaning, check-out	Client, problem	C	43. *Allen:* So you've felt the meaning in this possible job change as an opportunity to use your *talent* and take risks in something new. This may be contrasted with the feelings of stability and certainty where you are now. But *now,* as you are, also means you may end up like some people you don't want to end up like. Am I reaching the sense of things? How does that sound?	This reflection of meaning also confronts underlying issues that impinge on Mary's decision. It contains elements of the positive asset search or positive regard as Allen verbally stresses the word *talent.*
	Client, problem		44. *Mary:* Exactly! But I hadn't touched on it that way before. I do want stability and security, but not at the price of boredom and feeling down so much of the time as I have lately. Maybe I do have what it takes to risk more.	Mary is reinterpreting her situation from a more positive frame of reference. Allen could have said the same thing via an interpretation, but reflection of meaning lets Mary come up with her own definition. This reinterpretation of Mary's meaning represents a Level 4 response on the CIS. She has a new frame of reference with which to look at herself. But this newly integrated frame is *not* problem resolution; it is a *step* toward a new way of thinking and acting. After a Level 4 or 5 response, expect a client to move back to Level 3 or even Level 2. At this point, Allen decides to move to Stage 4 of the interview. It would be possible to explore problem definition and ideal worlds in more detail. However, later interviews can take up these matters.
Feedback	Client		**STAGE/DIMENSION 4: Working—Exploring alternatives, confronting client incongruities and conflict, restorying** 45. *Allen:* Mary, from listening to you, I get the sense that you do have considerable ability.	Allen combines feedback on positive assets with some self-disclosure here and uses this lead as a transition

(continued)

Table 13-3   (continued)

Skill Classifications				
Listening & Influencing	*Focus*	*C*	*Counselor and Client Conversation*	*Process Comments*
			Specifically, you can be together in a warm, involved way with those you work with. You can describe what is important to you. You come across to me as a thoughtful, able, sensitive person. (pause)	to explore alternative actions. The emphasis here is on the positive side of Mary's experience. Allen's vocal tone communicates warmth, and he leans toward Mary in a genuine manner.
			46. *Mary:* Ummmm . . .	During the feedback, Mary at first shows signs of surprise. She sits up, then relaxes a bit, smiles and sits back in her chair as if to absorb what Allen is saying more completely. There are elements of praise in Allen's comment.
Directive, paraphrase, open question	Client, main theme		47. *Allen:* Other things for job ideas may develop as we talk . . . ah . . . I think it might be appropriate at this point to explore some alternatives you've talked about. (*Mary:* Um-hm.) The first thing you talked about was that what you liked teaching was drug education and sexuality. What else have you taught kids in that general area ?	At this point, Allen starts exploring alternatives a little more concretely and in depth. The systematic problem-solving model—define the problem, generate alternatives, and set priorities for solutions—is in his mind throughout this section. He begins with a mild directive. "What else?" keeps the discussion open.
	Client, problem		48. *Mary:* Let's see . . . The general areas I liked were human sexuality and drug education, and family life and family growth, and those kinds of things. Ah . . . sometimes communication skills.	
Closed question	Problem		49. *Allen:* Have you attended workshops on these types of things?	Closed questions oriented toward concreteness can be helpful in determining specific background important in decision making.
	Client, problem, others		50. *Mary:* I've attended a few. I kind of enjoyed them. I've enjoyed them . . . I really did. You know, I've gone to the university and taken workshops in values clarification and	Note that virtually all counselor and client comments have focused on the client and the problem. It is important to consider the client in each of your responses; too heavy an emphasis on the problem may cause you to

(continued)

Table 13-3   **(continued)**

Listening & Influencing	Focus	C	Counselor and Client Conversation	Process Comments
Skill Classifications				
			communication skills. I liked the people I met.	miss the unique person before you. At the same time, a broader focus might expand the issue and provide more understanding. Social work, for example, might emphasize the family and social context.
Reflection of feeling, information, check-out	Client, problem		51. *Allen:* Sounds like you've really enjoyed these sessions. One of the important roles in counseling, education, and business is training—for example, psychological education through teaching others skills of living and communication. How does that type of work sound to you?	Allen briefly reflects her positive feelings, then shares a short piece of occupational information. This is followed by a check-out returning the focus to Mary.
	Client, problem		52. *Mary:* I think I would enjoy that sort of thing. Um-hmmm . . . It sounds interesting.	
Paraphrase, open question	Client, problem		53. *Allen:* Sounds like you have also given a good deal of thought to . . . ah . . . extending that to training in general. How aware are you of the business field as a place to train?	Mary's background and interest in a second alternative are explored.
	Client, problem, environmental context		54. *Mary:* I don't know that much about it. You know, I worked one summer in my dad's office, so I do have an exposure to business. That's about it. They all have been saying that a lot of teachers are moving into the business field. Teaching is not too lucrative, and with all the things happening here in California and all the cutbacks, business is a better long-term possibility for teachers these days. It just seems like an intriguing possibility for me to investigate or look into. But I don't know much about it. The latest business cutbacks are scary, too.	Mary talks in considerably greater depth and with more enthusiasm when she talks about business. The important descriptive words she has used with teaching include *boring, security,* and *interpersonal interactions,* while *interest* and *excitement* were used for training and teaching psychologically oriented subjects as opposed to physical education. Now she mentions cutbacks. Business has been described with more enthusiasm and as more lucrative. We may anticipate that she will eventually associate the potential excitement of business with the negative construct of risk and the lack of summer vacations and time to be with her children.

Table 13-3 **(continued)**

Skill Classifications				
Listening & Influencing	Focus	C	Counselor and Client Conversation	Process Comments
Paraphrase, reflection of feeling	Client, problem	C	55. *Allen:* Mm-hmm, . . . so you've thought about it . . . looking into business, but you've not done too much about it yet. Neither teaching nor business is really promising now and that's a little scary.	This paraphrase is somewhat subtractive. Mary did indicate that she had summer experience with her father. How much and how did she like it? Allen missed that. The paraphrase involves a confrontation between what Mary says and her lack of doing anything extensive in terms of a search. The reflection of feeling acknowledges emotion.
	Client		56. *Mary:* That's right. I've thought about it, but . . . ah . . . I've done very little about it. That's all . . .	Mary feels a little apologetic. She talks a bit more rapidly, breaks eye contact, and her body leans back a little. Mary's response is at Level 2 on the CIS. She is only partially able to work with the issues of the confrontation.
Interpretation	Problem		57. *Allen:* And, finally, you mentioned that you have considered the counseling field as an alternative. Ah . . . what about that?	Allen has missed the boat. More exploration of business should have followed. The confrontation of thinking but absence of action is probably appropriate, but it was brought in too early. If Allen had focused on positive aspects of Mary's experience and learned more about her summer experience, the confrontation probably would have been received more easily. As this was a demonstration interview, Allen sought to move through the stages perhaps a little too fast. Also, the counseling field is an alternative, but it seems to come more from Allen than from Mary. An advantage of transcripts such as this is that one can see errors. Many of our errors arise from our own constructs and needs. This intended paraphrase is classified as an interpretation, as it comes more from Allen's frame of reference than from Mary's.
	Problem, others		58. *Mary:* Well, I've always been interested, like I said, in talking with people. People like to talk with me about all kinds of	Mary starts with some enthusiasm on this topic, but as she talks her speech rate slows and she demonstrates less energy.

(continued)

Table 13-3   (continued)

Skill Classifications				
Listening & Influencing	Focus	C	Counselor and Client Conversation	Process Comments
			things. And *that* would be interesting . . . ah . . . I think, too.	
Encourage			59. *Allen:* Um-hmmm.	
	Problem		60. *Mary:* You know, to explore that. (pause)	Said even more slowly.
Encourage			61. *Allen:* Um-hmmm. (pause)	Allen senses her change of enthusiasm, is a bit puzzled, and sits silently, encouraging her to *talk more*. When you have made an error and the client doesn't respond as you expect, return to attending skills.
	Problem		62. *Mary:* But . . . I'd have to take some *courses* . . . if I really wanted to get into it.	One reason for Mary's hesitation appears.
Interpretation/ reframe	Client, problem		63. *Allen:* So putting those three things together, it seems that you want people-oriented occupations. They are particularly interesting to you.	This is a mild interpretation, as it labels common elements in the three jobs. It could be classified also as a paraphrase. Not all skill distinctions are clear.
	Client		64. *Mary:* Definitely . . . and that's where I am most happy.	Mary has returned to a Level 3 on the CIS.
Feedback	Client, problem	C	65. *Allen:* And, Mary, as I talk I see you as . . . ah . . . coming across with a lot of enthusiasm and interest as we talk about these alternatives. I do feel you are a little less enthusiastic about returning to school. (*Mary:* Right!) I might contrast your enthusiasm about the possibilities of business and training with your feelings about education. There you talk a little more slowly and almost seem bored as you talk about it. You seem lively when you talk about business possibilities.	Allen gives Mary specific and concrete feedback about how she comes across in the interview. There is a confrontation as he contrasts her behavior when discussing two topics. Confrontation—the presentation of discrepancies or incongruity—may appear with virtually all skills of the interview. It may be used to summarize past conversation and stimulate further discussion, leading toward a resolution of the incongruity.
	Client, problem		66. *Mary:* Well, they sound kind of exciting to me, Allen. But I	Mary talks rapidly, her face flushes slightly, and she gestures with

(continued)

Table 13-3   (continued)

Skill Classifications				
Listening & Influencing	Focus	C	Counselor and Client Conversation	Process Comments
			just don't know how to go about getting into those fields or what my next steps might be. They sound very exciting to me, and I think I may have some talents in those areas I haven't even discovered yet.	enthusiasm. She meets the confrontation and seems to be willing to risk more intentionality. This, however, may still be considered a Level 3 on the CIS, although there may be movement ahead.
Feedback, information, logical consequences	Client, problem	C	67. *Allen:* Um-hmmm. Well, Mary, I can say one thing. Your enthusiasm and ability to be open and look at things is one part of your ability to do just that. As things get going in your search, you're going to find that helpful to you. Ah . . . at the same time, business and schools represent different types of lifestyles. I think I should give you a warning that if you go into the business area you're going to lose those summer vacations.	This statement combines mild feedback with logical consequences. A warning about the consequences of client action or inaction is spelled out. Mary is also confronted with some consequences of choice.
	Client, problem, others		68. *Mary:* Yeah, I know that . . . and you know, that special friend in my life—he's in education—I don't think he would like it if I was, you know, working all summer long. But business does pay a lot more, and it might have some interesting possibilities. (*Allen:* Um-hmm.) . . . It's a difficult situation.	Confrontations often result in clients presenting new concepts and facts important in life decisions that have not been discussed previously. A new problem has emerged that may need definition and exploration. Mary is still responding at Level 3 on the CIS, but Allen is obtaining a more complete picture of the problem and of the client.
Encourage/ restatement	Problem		69. *Allen:* A difficult situation?	Again, the encourager is used to find deeper meanings and more information.
	Client, problem, others		70. *Mary:* Um-hmm. I guess I'm saying that . . . I'm . . . ah . . . you know, my friend . . . I don't think he would approve or like the idea of my going to work in business and only having two weeks' vacation.	Mary has more speech hesitations and difficulties in completing a sentence here than she has anywhere in the interview. This suggests that her relationship is important to her, and her friend's attitude may be important in the final career decision.

(continued)

Table 13-3   (continued)

Skill Classifications				
Listening & Influencing	Focus	C	Counselor and Client Conversation	Process Comments
			(*Allen:* Uhhuh.) He wants me to stay in some field where I have the same vacation time I have now so we can spend that time together.	Much career counseling involves personal issues as well as career choice. Both require resolution for true client satisfaction.
Interpretation/ reframe, open question	Client, others, cultural/ environmental/ contextual	C	71. *Allen:* I hear you saying that your friend has a lot to say about your future. How does that strike you as a woman who has been independent and on your own successfully for quite a while?	Here we see the introduction of gender relations as a cultural/environmental/contextual issue. Allen's reframing of the situation offers Mary a chance to explore her relationship with Bo from a different contextual perspective.
	Others		72. *Mary:* It really is . . . well, Bo's a special person . . .	Mary's eyes brighten.
Interpretation	Client, others		73. *Allen:* And, I sense you have some reactions to his . . .	Allen interrupts, perhaps unnecessarily. It might have been wise to allow Mary to talk about her positive feelings toward Bo.
	Client, problem		74. *Mary:* Yeah, I'd like to be able to explore some of my own potential without having those restraints put on me right from the beginning.	Mary talks slowly and deliberately, with some sadness in her voice. Feelings are often expressed through intonation. Here we see the beginning of a critical gender issue. Women often feel constraints in career or personal choices, and men in this culture often place implicit or explicit restraints on critical decisions. Feminist counseling theorists argue that a male helper may be less effective with these types of problems. What are your thoughts on this issue?
Interpretation/ reframe, check-out	Client, problem, others, cultural/ environmental/ contextual	C	75. *Allen:* Um-hmm . . . In a sense he's almost placing similar constraints on you that you feel in the job in physical education. There's certain things you have to do. Is that right?	This interpretation relates the construct of boredom and the implicit constraint of being held down with the constraints of Bo. The interpretation clearly comes from Allen's frame of reference. With interpretations or helping leads from your frame of reference, the check-out of client reactions is even more important. The drawing of parallels is abstract, formal-operational in nature.

(continued)

Table 13-3 (continued)

Skill Classifications				
Listening & Influencing	*Focus*	*C*	*Counselor and Client Conversation*	*Process Comments*
	Client, problem, others		76. *Mary:* Yes, probably so. He's putting some limits on me . . . setting limits on the fields I can explore and the job possibilities I can possibly have. Setting some limits so that my schedule matches his schedule.	Mary answers quickly. It seems the interpretation was relatively accurate and helpful. One measure of the function and value of a skill is what the client does with it. Mary changes the word *constraints* to the more powerful word *limits*. Mary remains at Level 3 on the CIS, as she is still expanding on aspects of the problem.
Open question, oriented to feeling	Client		77. *Allen:* In response to that you feel . . . ? (deliberate pause, waiting for Mary to supply the feeling)	Research shows that *some* use of questions facilitates emotional expression.
	Client, problem		78. *Mary:* Ah . . . I feel I'm not at a point where I want to *limit things.* I want to see what's open, and I would like to keep things open and see what all the alternatives are. I don't want to shut off any possibility that might be really exciting for me. (*Allen:* Um-hmm.) A total lifetime of careers.	Mary determinedly emphasizes that she does not want limits.
Reflection of feeling, paraphrase	Client, problem	C	79. *Allen:* So you'd like to have a life of exciting opportunity, and you sense some limiting . . .	A brief, but important, confrontation of Bo versus career.
	Client, problem, others, cultural/ environmental/ contextual		80. *Mary:* He reminds me of my relationship with my first husband. You know, I think the reason that all fell apart was my going back to work. You know, assuming a more nontraditional role as a woman and exploring my potential as a woman rather than staying home with the children . . . ah . . . you know, sort of a similar thing happened there.	Again, the confrontation brings out important new data about Mary's present and past. Is she repeating old relationship patterns in this new relationship? The counselor should consider issues of cultural sexism as an environmental aspect of Mary's planning. This does not appear in this interview, but a broader focus on issues in the next session seems imperative. Other focus issues of possible importance include Mary's parental models, others in her life, a women's support group, the present economic climate, the attitudes of the counselor, and "we"—the immediate

(continued)

Table 13-3 (continued)

Skill Classifications				
Listening & Influencing	Focus	C	Counselor and Client Conversation	Process Comments
				relationship of Mary and Allen. Thus far he has assumed a typical Western "I" form of counseling where the emphasis is on the client. Due to the development of new, more integrated data, this could be a Level 5 response on the CIS.
Summary	Client, problem, others, cultural/ environmental/ contextual	C	81. *Allen:* There really are a variety of issues that . . . you're looking at. One of these is the whole business of a job. Another is your relationship with Bo and your desire to find your own space as an independent woman.	The interview time is waning, and Allen must plan a smooth ending and plan for the next session. He catches here the confrontation that Mary faces between work and relationship. We also should note Allen fails to pick up fully on the C/E/C focus. Many of Mary's issues relate to women's issues in a sometimes sexist world.
			82. *Mary:* (slowly) Um-hmmm . . .	Mary looks down, relaxes, and seems to go into herself.
Reflection of feeling	Client		83. *Allen:* You look a little sad as I say that.	This reflection of feeling comes from nonverbal observations and picks up on her sensorimotor reactions.
	Problem		84. *Mary:* It would be nice if the two would mesh together, but it seems like it's kind of difficult to have both things fitting together nicely.	Mary here is describing her ideal resolution. Here the interview could recycle back to Stages 2 and 3, with more careful delineation of the problem between job and personal relationships and defining the ideal resolution more fully. *A problem exists only if there is a difference between what is actually happening and what you desire to have happen.* This sentence illustrates the importance of problem definition and goal setting. Mary's response to the confrontation is 4 on the CIS. We have an important new insight, but insight is not action. She also needs to act on this awareness.
Information, directive	Problem		85. *Allen:* Well, that's something we can explore a little bit	Many clients bring up central issues just as the interview is about to end.

(continued)

Table 13-3 **(continued)**

Skill Classifications				
Listening & Influencing	Focus	C	Counselor and Client Conversation	Process Comments
			further. It seems this is an important part of the puzzle. Let's work on that next week. Would that be okay? I see our time is about up now. But it might be useful if we can think of some actions we can take between now and the next time we get together.	Allen makes the decision, difficult though it is, to stop for now and plan for more discussion later. Note that Mary is talking about her relationship mainly from an abstract, formal-operational orientation. Clients often bring up central issues late in the session.
Summary, open question	Client, problem	↓	**STAGE/DIMENSION 5:** **Terminating—Generalizing** **and acting on new stories** 86. *Allen:* We have come up so far with three things that seem to be logical: business, counseling, and training. I think it would be useful, though, if you were to take a set of career tests. (*Mary:* Uh-huh.) That will give us some additional things to check out to see if there are any additional alternatives for us to consider. How do you feel about taking tests?	Allen continues his statement and moves to Stage 5. He summarizes the career alternatives generated thus far and raises the possibility of taking a test. Note that he provides a check-out to give Mary an opportunity to make her own decision about testing.
	Client, problem		87. *Mary:* I think that's a good idea. I'm at the stage where I want to check all alternatives. I don't want *anything* to be limited. I want to think about a lot of alternatives at this stage. And I think it would be good to take some tests.	Mary approves of testing and views this as a chance to open alternatives. She verbally emphasizes the word *anything,* which may be coupled with her desire to avoid limits to her potential. Some women would argue that a female counselor is needed at this stage. A male counselor may not be sufficiently aware of women's needs to grow. Allen could unconsciously respond to Mary in the same ways she views Bo as responding to her.
Information	Client, problem, interviewer		88. *Allen:* Then another thing we can do . . . ah . . . is helpful. I have a friend at a local firm who originally used to be a coach. She's moved into personnel and training at Jones. (*Mary:* Ummm.) I can arrange	Allen suggests a concrete and specific alternative for action. Although Mary is predominantly formal-operational, she has tended to talk about issues and avoid action. This avoidance of action is also indicative of Level 3 on the confrontation

(continued)

Table 13-3 **(continued)**

Skill Classifications				
Listening & Influencing	Focus	C	Counselor and Client Conversation	Process Comments
			an appointment for you to see her. Would you like to go down and look at the possibilities there?	impact scale. Until Mary takes some form of concrete action or resolves the issue in her mind, she will remain at Level 2 or 3 on the CIS. If some action is taken on the issue during the coming week, then she will have moved at least partially to Level 4 on the CIS.
	Client, problem		89. *Mary:* Oh, I would like to do that. I'd get kind of a feel for what it's like being in a business world. I think talking with someone would be a good way to check it out.	Stated with enthusiasm. The proof of the helpfulness of the suggestion will be determined by whether she does indeed have an interview with the friend at Jones and finds it helpful in her thinking.
Feedback, open question	Client, problem		90. *Allen:* You're a person with a lot of assets. I don't have to tell you all the things that might be helpful. What other ideas do you think you might want to try during the week?	Allen recognizes he may be taking charge too much and pulls back a little. Although he is pushing Mary for action, he is now using her ideas. Too much direction and advice can make a client resistant to your efforts.
	Client, problem		91. *Mary:* What about checking into the university and ah . . . advanced degree programs? I have a bachelor's degree, but . . . maybe I should check into school and look into what it means to do more coursework.	Mary, on her own, decides to look into the university alternative. This is particularly important, as earlier indications were that she was not all that interested. Note that real generalization is usually concrete *action*.
Summary	Client, problem, cultural/ environmental/ contextual		92. *Allen:* Okay, that's something else you could look into as well. (*Mary:* Uh-huh.) So let's arrange for you then to follow up on that. I'd like to see you doing that. (*Mary:* Um-hmmmm.) And . . . ah . . . we can get together and talk again next week. You did express some concern about your relationship with your friend, Bo, ah . . . would you like to talk about that as well next week? And, as I look back on this session, one theme we haven't discussed yet is how being a woman with family	Allen is preparing to terminate the interview. Fortunately he does consider the women's issue. Probably this should have been done sooner in the session. Is this an issue with which he can help, or would you recommend referral?

(continued)

Table 13-3 **(continued)**

Skill Classifications				
Listening & Influencing	Focus	C	Counselor and Client Conversation	Process Comments
			responsibilities relates to all this. Maybe this is something to be explored next week as well?	
	Client, cultural/ environmental/ contextual		93. *Mary:* I think so, they sort of all . . . one decision influences another. You know. It all sort of needs to be discussed. And thanks for bringing up the women's issue and my children. That's important to me.	An important insight at the end. Mary realizes her career issue is more complex than she originally believed. If you were Allen's supervisor, would *you* recommend a primary emphasis on career counseling or on personal counseling in the next session? Or perhaps some combination of them both? What else would you advise him to do?
Self-disclosure	Client, interviewer		94. *Allen:* Okay. I look forward to seeing you next week, then.	
			95. *Mary:* Thank you.	

INSTRUCTIONAL READING 2: INTERVIEW TRANSCRIPT ANALYSIS AND PLANNING

Three commentaries on the interview will be presented: The first discusses taking notes on the session; the second reviews the skills used by the interviewer and their impact on the client; and the third examines the process of development in the client.

Issues in Interview Planning

The interview plan in Box 13-1 described what the interviewer planned to do in the session. If you review the box, you will find that it closely approximates what happened in the session. Although this indicates a decisional plan that was well thought through, it also indicates that other possibilities were not tested. If that plan had had different objectives, the interview would have been different. For example, if the interviewer had been oriented toward a Rogerian, person-centered style, the plan would have called for more client examination of self and feelings about self. Very different skills would have been used and the goal would have been phrased more in terms of self-understanding and self-actualization. It is important for counselors to realize that what they plan for the interview (and their own style and theory) has an impact on what actually happens.

Box 13-2   National and International Perspectives: What's Happening
With Your Client While You Are Counseling?

Robert Manthei, University of Canterbury, Christchurch, New Zealand

There is more going on in interviewing beyond what we see happening during the session. Clients are good observers of what you are doing even though they may not always tell you what they think and feel. Research shows that clients expect counseling to be shorter than do most counselors and therapists. Clients see counselors as more directive than counselors see themselves. And what the counselor sees as a good session may be seen otherwise by clients and vice versa. Counselors and clients may also vary in their perceptions of counseling effectiveness.

I conducted research studies of client and counselor experience of counseling. Among the major findings are the following:

Most people don't come for counseling immediately

By the time they see a counselor, most clients have already tried to solve their problems in a variety of ways. Talking with friends and family and trying to work it out on their own were usually tried first. Some rely on hope or deny that they have problems until they become more serious. Reading self-help books, prayer, and alcohol and drugs are among other things tried. *Implications for practice:* Ask your clients what they have tried before they came to you and find out what aspects of prior efforts seemed to have helped. Try to build on their past successes. This is an axiom of brief solution-oriented therapy (see the next chapter).

First impressions are important

That first interview sets the stage for the future, and the familiar words "relationship and rapport" are central. I found that clients generally had favorable impressions of the first sessions and viewed what happened even more positively than counselors. Sometimes sharing experience helps. One client who did not feel so positive about the first session commented, "Maybe if the counselor had gone through a similar experience of divorce and raising children, it would have helped." *Implications for practice:* Obviously, be ready for that first session. Cover the critical issues of confidentiality and legal issues in a comfortable way. Structuring and letting the client know what to expect seems important. And, empathic listening always remains central.

(continued)

Interviewing plans and notes need to be available for client review and participation. More and more agencies are involving their clients in setting up and implementing treatment plans. For example, the University of Massachusetts Behavioral Medicine Clinic (2004) has a brief one-page form that contains the following:

Patient (client) strengths/skills
Problem list (not a diagnosis)
   Problem 1
      Goal and outcome behavior
      Interventions (strategies, theory)

Box 13-2 (continued)

Counseling helps, but so do events outside of the interview	Clients attributed 68% of the improvement in their issues to counseling and 32% of improvement to things they did outside of counseling. Among the "outside" things that helped were talking and socializing more with family and friends, taking up new activities, learning relaxation, and involvement with church. *Implications for practice:* The counseling interview is important, but generalization of behavior and thought to daily life is also valuable. Homework and specific ideas for using what is learned immediately are important. Tell clients to expect this sort of change and ask them about it at the beginning of each session.
Things that clients found helpful in their counseling	Relationship variables such as warmth, understanding, and trust were important. Clients liked being listened to and being involved in making decisions about the course of counseling. Explanations and interpretations helped them see their situations in a new way; new skills such as imagery, relaxation training, and thought-stopping to eliminate negative self-talk were also useful.

All the above speak to respecting the client's ability to participate in the change process and to orchestrate their own "healing." I think it is vital that throughout counseling we tell clients what we are doing, but also ask them to share their perceptions of the session(s) with us. By being genuinely curious we can learn from and learn with more client participation.

Allen Ivey comments: Bob Manthei's comments remind me of the value of a question I like to use at the end of the interview—"What stood out for you today? What one thing seemed most helpful?" I sometimes find that what I thought was the most valuable moment is not close to what the client experienced. Asking "What was least helpful?" sometimes also brings valuable and needed feedback. The fifth stage of the interview—generalization to the real world—is what counts. And we encourage change by listening carefully and providing an array of intentional alternatives for client action. Note that counseling with Mary in this chapter's example interview will be successful only if she follows up and does something different this next week. In the next chapter, please pay special attention to the Maintaining Change and Relapse Prevention Worksheet. It can provide a useful framework to help us all maintain focus and direction.

Progress toward goal
Estimated time goal may be reached
Problems 2, 3, etc., follow the same structure

Note that the clinic list starts with positive assets of the client, thus immediately bringing a wellness approach to the treatment. The client and the counselor/ therapist both sign this sheet and review it periodically. As new issues or problems arise, they are added to the list. In addition, the clinic uses a client feedback form to review progress and determine what aspects of the session(s) were more helpful and less helpful.

The power of the counselor or interviewer to determine the session's direction cannot be denied. If you decide to conduct a career session, it usually turns out that this topic predominates. If you decide to help the person become more assertive, this too happens. And if you decide to "let the person talk and see what happens," it will occur of itself. Even though the interview may not completely follow your plan, it is your personal decision that has a strong influence on what happens. This is one reason we need interview plans, examination of our own style, and notes. In this way, we can examine ourselves and at least be aware of the action and effect we have on others.

As you might anticipate, interview notes may be structured in a fashion similar to the five-stage plan. Although not shown here, Allen's notes might have summarized issues under rapport and structure, perhaps noting his and Mary's good rapport and that they had decided to work on career issues. He should also have noted that his response to personal issues such as the relationship with Bo was limited. This was something that Allen structured into the session. His rationale might have been "One can't talk about everything. I think Bo is important, but I felt it was more critical at the time to focus on career decisions." Structure notes such as this also allow someone to supervise and work with Allen. If you were his supervisor, would you think he made the right decision? We can only really tell by examining his interview plans and notes and talking with him. We do not often have verbatim transcripts of sessions.

Notes on problem definition, Mary's goals, and the techniques and discussion of the fourth stage of the session would provide an understanding of the interview and its general function. Last but not least, notes on generalization plans should be made. You will find that you sometimes forget what happened with your clients; notes will facilitate planning for your next contact. Additionally, as we move toward increasing legal and professional accountability for our interviews, the importance of good notes cannot be overemphasized.

Develop your own a set of notes on this interview, using the five-stage model as a structure for your commentary. Include the Client Feedback Form from Chapter 1. You may find it helpful to share your notes with your client and obtain her or his impressions of what you have written.

Referral

The word "referral" appears in the interview process notes. No interviewer has all the answers and in the case of Mary, Allen thought referral to a women's group might be helpful as many of her issues are common to women looking for career change. In addition, his notes indicate the need for a referral to the university financial aid office. An important part of individual counseling is helping your clients find community resources that may facilitate their growth and development. The community genogram helps interviewers and clients think more broadly and consider appropriate referral sources. This type of referral may strengthen your relationship with the client.

A key referral issue is whether this is a case for which interviewer expertise and experience is sufficient to help the client. Allen has considerable experience in career and personal counseling and thus far the issues are within his expertise. But this is where supervision and case conferences can be helpful. Just because Allen or any other counselor thinks that he or she is working effectively may not be enough. Opening up your work to others' opinion is an important part of professional practice. Clients, of course, should be aware that you as counselor or therapist are being supervised.

If the clients' issues are clearly beyond our knowledge and experience, an appropriate referral needs to be arranged carefully. Referral at intake is fairly clear as the intake interviewer seeks to match client issues with staff expertise. Referral to another helper during an ongoing interview series is more complex. We do not want to leave our clients "hanging" with no sense of direction or fearful that their problems are too difficult. The client should be fully involved throughout the referral process and the new counselor or therapist informed of client needs. In such cases, we recommend that you maintain contact with the client as the referral process evolves, sometimes even continuing for a session or two until the change to a different counselor is complete.

Finally, sometimes the relationship simply doesn't seem to be working. In such situations, seek consultation and supervision immediately, as most often these challenges can be resolved. When you sense the relationship isn't working, focus on client goals and seek to hear the client's story completely. And ask her or him for feedback and suggestions as to what might be helpful. If referral to another interviewer becomes necessary, avoid blaming either the client or yourself.

Skills and Their Impact on the Client

Table 13-4 presents a skill summary of Allen's interview with Mary. Note the different use of skills in each stage of this interview. Allen used no influencing skills in Stages 2 or 3 (problem definition and determining outcomes) but a more balanced use of skills in the other stages. In Stage 4 Allen used both influencing skills and confrontation of incongruity and discrepancies extensively. Stage 4 could be considered the "working" phase of the interview.

In terms of a total balance of skill usage, Allen used a ratio of approximately two attending skills for every influencing skill. His focus remained primarily on the client, although the majority of his focus dimensions were dual, combining focus on Mary with focus on the problem or issue.

In examining this interview for competence levels, we find that Allen is able to identify and classify the several skills and stages of the interview. He is able to identify some of the impact of his skills on the client.

Allen also demonstrates Level 2, basic competence, which calls for the ability to use the basic listening sequence to structure an interview in five simple stages, and to employ intentional interviewing skills in an actual interview.

Level 3, intentional competence, is more difficult to assess. Let us examine this level in more detail. In terms of focus dimensions, as mentioned, Allen focused primarily on Mary and her concerns. Focus analysis is useful as it points out that he did not focus on others and the family (Bo and Mary's children, for example), although some theorists might say that family and Bo were the most important area of all. The relationship with Bo evolved with greater clarity later in the interview. At *71, 75,* and *81,* Allen brought in the cultural/environmental/contextual focus, enabling a beginning discussion of gender issues that clearly needs further work. Some would question whether a man can help a woman with these issues. Do you think that Allen could continue this topic or would you recommend referral to a woman counselor?

Table 13-4   Skill summary of Allen and Mary interview over five stages

	Listening/Attending						Focus							Influencing							C
	Open question	Closed question	Enc./restatement	Paraphrase	Reflection feeling	Reflection meaning	Summary	Client	Main theme/problem	Others	Family	Mutuality	Interviewer	C/E/C	Interpretation/reframe	Logical consequences	Self-disclosure	Feedback	Info./adv./etc.	Directive	Confrontation
STAGE/DIMENSION 1: **Initiating the session** 6 attending skills 3 influencing skills	2	2		2	1			6	2				3				1		4		
STAGE/DIMENSION 2: **Gathering data** 4 attending skills 0 influencing skills 2 confrontation skills	1			1			1	5	2		1										2
STAGE/DIMENSION 3: **Mutual goal setting** 14 attending skills 0 influencing skills 2 confrontation skills	3	3	2	2	1	1	2	8	6	2											2
STAGE/DIMENSION 4: **Working** 16 attending skills 15 influencing skills 7 confrontation skills	4	1	3	4	3		1	15	14	4				3	6	1		3	3	2	7
STAGE/DIMENSION 5: **Terminating** 5 attending skills 3 influencing skills	2						3	5	4				2	1			1	1	1		
Total: 45 attending skills 21 influencing skills 11 confrontation skills	12	6	5	9	5	1	7	39	28	6	1	0	5	4	6	1	2	4	8	2	11

We should note that Allen focused primarily on the client in the earlier phases of the interview and only in the later portions increased his emphasis on the problem focus. Thus, he does seem to have mastered to some extent the ability to focus on the person and balance that effectively with the problem focus. An ineffective interviewer might have focused early on the problem and missed the unique person completely.

In a general and perhaps more important sense, in each phase of the interview Allen fulfilled the functions suggested for each stage. In the rapport stage he was able to use client observation skills to note that Mary was somewhat tense at the beginning and then to flex and select a rapport-building exchange that enabled the client to be more comfortable in the interview. In the problem-definition stage Mary was able to identify a specific problem she wanted to resolve, which Allen summarized at *Allen 23*. In Stage 3 he again used the basic listening sequence to bring out some concrete goals (business, counseling), and through reflection of meaning *(Allen 43)* he summarized some of the key aspects of Mary's thinking about the issue.

In Stage 4 a number of incongruities were confronted, and with each confrontation Mary appeared to move a little deeper into some personal insights concerning her present and future. Note in particular *Allen 71*, which led Mary into the important area of her personal life and relationship with Bo. The slightly inaccurate confrontation at *Allen 75* fortunately included a check-out, and thus Mary was able to introduce her important construct, substituting the word *limits* for Allen's *constraints*. In the final stage Mary appeared ready and willing to take action. However, the real proof of the value of the interview will have to wait until the next meeting, which will show whether the generalization plan was indeed acted on.

Examining specific skills, we might note that Allen's open questions tended to encourage Mary to talk, that his paraphrases and reflections of feeling were often followed by Mary's saying "Yes" or "Yeah." In the important reflection of meaning at *Allen 43*, Mary responded with "Exactly! I hadn't touched on it that way before. . . ." In short, Allen does seem to be able to use the skills to produce specific results with the client.

However, as noted earlier, Allen's self-disclosures are not fully effective, and he did not demonstrate the full range of focusing that might have helped enlarge and round out the conception of Mary's problem. There is time for this, however, in ensuing interviews. The first interview is often a good time for general exploration. It might have been helpful if Allen had used the newspaper framework (who, what, when, where, how, and why) to cover the bases more completely.

In general, Mary responded primarily at Level 3 on the Confrontation Impact Scale to the 11 confrontations in this session. However, most of the confrontations seemed to enable Mary to explore her issues in more depth. A particularly successful one was at *79*, where Allen comments on Mary's desire to have a life of "exciting opportunity," but notes that she senses Bo's putting limits on her. Mary moves easily here to discuss her own personal wishes in more depth. On the Confrontation Impact Scale, she has moved here to a "5," or a new way of thinking about her issues.

The work that clients do after the interview is as important as or more important than what they do in the session with you. The real impact of the interview and the confrontations will show in the next session and in Mary's life after interviewing is completed. You may note that you can use the Confrontation Impact Scale as a way to assess the effectiveness of your interviewing and counseling over time.

You may want to think back for a moment about this session. What did you like? What advice would you give Allen? What did he do right?

Has Allen been able to produce predictable effects on Mary? "Predictable effects" refers to the functions of each of the skills and the five stages of the interview.

Box 13-3 Second Interview Plan and Objectives

<table>
<tr>
<td></td>
<td>After reviewing the preceding session, identify issues you anticipate will be important in the next session and plan how you might handle them. (This plan is Allen's assessment of his forthcoming interview with Mary.)</td>
</tr>
<tr>
<td>Initiating the session—rapport and structuring</td>
<td>*Special issues anticipated with regard to rapport development? What structure do you have for this interview? Do you plan to use a specific theory?*

Mary and I seem to have reasonable rapport. As I look at the first session, I note I did not focus on Mary's context nor did I attend to other things that might be going on in her life. It may be helpful to plan some time for general exploration *after* I follow up on the testing and her interviews with people during the week. Mary indicated an interest in talking about Bo. Two issues need to be considered at this session in addition to general exploration of her present state. I'll introduce the tests and follow that with discussion of Bo. For Bo, I think a person-centered, Rogerian method emphasizing listening skills may be helpful.</td>
</tr>
<tr>
<td>Gathering data—drawing out stories, concerns, problems, or issues</td>
<td>*What are the anticipated problems? Wellness strengths? How do you plan to define the issues with the client? Will you emphasize behavior, thoughts, feelings, meanings?*

1. Check with Mary on her career plans and how she sees her career concerns defined now. Use basic listening sequence.
2. Later in the interview, go back to the issue with Bo. Open it up with a question, then follow through with reflective listening skills. If she starts with Bo, save career issues until later. Be alert to this relationship from a women's and feminist perspective.</td>
</tr>
</table>

(continued)

Specifically, when an interviewer uses a particular skill or action, does the client *do* what is predicted by the defined function of the skill? Further, does the counselor actually accomplish the main goals of each stage of the session?

Another way to evaluate the effectiveness of an interview is in terms of the degree of freedom or choices available to Mary. In the words of a common counseling slogan, "If you don't have at least three possibilities, you don't have a choice." Mary appears to have achieved that objective in the interview, at least for the moment. In addition, the issue of her relationship with Bo has been unearthed, and this topic may open her to further counseling possibilities. Again, the question must be raised whether Allen, as a man, is the most appropriate interviewer for Mary to see. Answers to that question will vary with your personal worldview. What are your evaluations? What would you do differently?

Planning for Further Interviews

Box 13-3 contains Allen's interview plan for the second session. This plan derives from information gained in the first interview and organizes the central issues of the case, allowing for new input from Mary as the session progresses.

Box 13-3   (continued)

	3. Mary has many assets. She is bright, verbal, and successful in her job. She has good insight and is willing to take reasonable risks and explore new alternatives. These assets should be noted in our future interviews.
	4. Explore women's issues with her.
Mutual goal setting	*What is the ideal outcome? How will you bring out the idealized self or world?*
	This may not be too important in this interview. We already have her vocational goals, but they may need to be reconsidered in light of the tests, further discussion of Bo, and so on. It is possible that late in this interview or in a following session we may need to define a new outcome in which careers and her relationships are both satisfied.
Working—exploring alternatives, confronting client incongruities and conflict, restorying	*What types of alternatives should be generated? What theories would you probably use? What specific incongruities have you noted or do you anticipate in the client?*
	1. Check on results of tests and report them to Mary.
	2. Explore her reactions and consider alternative occupations.
	3. Use person-centered, Rogerian counseling and explore her issues with Bo.
	4. Relate careers to the relationship with Bo. Give special attention to confronting the differences between her needs as a "person" and Bo's needs for her. Note and consider the issue of women in a changing world. Does Mary need referral to a woman or a women's group for additional guidance? Would assertiveness training be useful?
Terminating—generalizing and acting on new stories	*What specific plans, if any, do you have for transfer of training? What will enable you personally to feel that the interview was worthwhile?*
	At the moment it seems clear that further exploration of careers outside the interview is needed. We will have to explore the relationship with Bo and determine her objectives more precisely.

In short-term counseling and interviewing, the interview plan serves as the treatment plan itself. As you move toward longer-term counseling, a more detailed treatment plan is important. There, the several problems and issues raised by the client may be outlined in much the same fashion as in the interview plan but at greater length, and priorities may be set. Similarly, an array of desired outcomes may be identified. For exploring alternatives and incongruity, several alternative interview plans and theoretical methods may be employed. A wide variety of generalization plans may be needed. Box 13-4 presents a sample treatment plan for Mary, assuming that she wishes to continue counseling over a longer period, such as 5 to 10 sessions.

Within the profession there is disagreement on the need for an interviewing plan and treatment plan. The more structured counseling theories, such as behavioral and strategic/structural, strongly urge interview and treatment plans with specific goals. Their interview and treatment plans are often more specific than those presented here. Less structured counseling theories (Gestalt, psychodynamic, Rogerian) tend not to have treatment plans, preferring to work in the moment with the client. The interview and treatment plan forms suggested here represent a midpoint that you may find helpful in thinking through your own opinion on this important issue. Share your treatment plan with your client.

Box 13-4   Long-Term Treatment Plan (Developed After Three Interviews)

Initiating the session— rapport and structuring	Mary seems to respond best when I am open. In the second interview she seemed to like and appreciate my self-disclosures, and it facilitated the assertiveness-training exercise. She responded well for a time when I tried Rogerian, person-centered methods but seemed to prefer more directiveness on my part. That may be me, or it may be her. We can talk about that during the next session. She does seem to like structured techniques more than unstructured.
Gathering data— drawing out stories, concerns, problems, or issues	The following problems have been discussed thus far. Some have been touched on only briefly. In my opinion, the more important at the moment are these: 1. The relationship with Bo: working out the issues. 2. Career choice: it now seems clear that she wants to leave the school setting, but it is quite unclear how she can handle it financially. 3. Finances are a problem we have explored only briefly. One of the attractions of the relationship with Bo is that it solves those problems. 4. Assertiveness: Mary has made a beginning but may profit from more training and practice. 5. Unpleasant dreams: Mary mentioned briefly that she had trouble sleeping this past week. I may need to explore the dream in some detail if she brings it up again. 6. How do all these issues relate to being a woman in today's society? To the solution of these problems Mary brings a considerable number of assets. She is bright, verbal, and apparently has a large number of female friends. Her hobbies and leisure-time activities seem to interest her. She has worked through her divorce, and her children are doing well.
Mutual goal setting	It seems safe to say the following: 1. Mary wants to make a decision one way or the other with Bo that makes it possible for her to be in a more satisfactory career setting. At the moment it appears that she wants to place her interests above Bo's. 2. She does want a satisfactory relationship with a man. 3. Financial planning may be necessary to help her handle her money more effectively. 4. Mary seeks to be more assertive and speak clearly for her needs. 5. Dreams have been mentioned as a problem. 6. Further attention should be given to identifying other outcomes. 7. Referral to a women's group seems necessary.
Working—exploring alternatives, confronting client incongruities and conflict, restorying	1. Career decision will continue to be worked through with trait-and-factor decision-making counseling. 2. Explore relationship with Bo via person-centered counseling as in the past, but emphasize behavioral methods to help her act more decisively. 3. Referral to the university financial aid office may be helpful. 4. Gestalt or psychodynamic dream analysis should be kept on reserve if the dream topic comes up again. Involve Mary in this decision, as it may mean more long-term counseling than she wishes. Mary has enough assets to move forward easily on her own. 5. Draw on feminist theory to inform all these issues.
Terminating—generalizing and acting on new stories	Constantly keep in mind the array of generalization techniques available so that each problem or issue discussed gets worked on outside the interview.

Box 13-5   Key Points

Decisional counseling	Decisional counseling, a modern reformulation of trait-and-factor theory, assumes that most, perhaps all, clients are involved in making decisions. By considering the many traits of the person and factors in the environment, counselor and client can arrive at a more rational and emotionally satisfying decision.
Decisional structure and alternative theories	The five-stage structure of the interview can be considered a basic decisional model underlying other theories of counseling and therapy. Once you have mastered the skills and strategies of intentional interviewing and the five-stage model, you will find that you can more easily master other theories of helping.
Interview analysis	Using the constructs of this book, it is possible to examine your own interviewing style and that of others for microskill usage, focus, structure of the interview, and the resultant effect on a client's cognitive and emotional developmental style.
Interview plan and note taking	It is possible to use the five-stage interview structure to plan your interview before you actually meet with a client. This same five-stage structure can be used as an outline for note taking after the interview is completed.
Treatment plan	A treatment plan is a long-term plan for conducting a course of interviews or counseling sessions. Using the same five-stage interview structure, it is possible to consider issues of rapport and alternative structures over time. Furthermore, one may list and prioritize key client problems and assets and list client goals to be achieved. Several alternative strategies for change may be summarized in an overall treatment plan. Finally, it is critical to develop specific plans for treatment to be generalized to daily life.

COMPETENCY PRACTICE EXERCISE AND PORTFOLIO OF COMPETENCE

Students often find that the highlight of their experience with this book is videotaping or audiotaping their own session, transcribing it, and analyzing what occurred. You will find that you have learned a great deal about what occurs in the interview and your ability to discuss what you see will be invaluable throughout your interviewing or counseling career.

Practice Exercise

There is one major practice exercise in this chapter: Prepare a paper in which you demonstrate your interviewing style, classify your behavior, and comment on your development over the term.

The following steps are suggested for this major presentation of your interviewing style.

1. Plan to conduct an interview with a member of your group, a friend, or an actual client. This interview should last at least 20 minutes (although most prefer a longer time) and should follow the basic five-stage structure. It should be an interview you are satisfied to present to others. Before you conduct the interview, be sure you have your client's permission to record the session.

Box 13-6 Interview Plan and Objectives Form

	After studying the client file before the first session or after reviewing the preceding session, complete this form indicating issues you anticipate being important in the session and how you plan to handle them.
Initiating the session—rapport and structuring	Special issues anticipated with regard to rapport development. What structure do you have for this interview? Do you plan to use a specific theory? Skill sequence?
Gathering data—drawing out stories, concerns, problems, or issues	What are the anticipated problems for this client? Wellness strengths? How do you plan to define the issues with the client? Will you emphasize behavior, thoughts, feelings, meanings? In what areas do you anticipate working on problems?
Mutual goal setting	Where do you believe this client would like to go? How will you elicit the idealized self or world? What would you like to see as the outcome?
Working—exploring alternatives, confronting client incongruities and conflict, restorying	What types of alternatives would be generated? What theories would you probably use here? What specific incongruities have you noted or do you anticipate in the client? What skills are you likely to use? Skill sequences?
Terminating—generalizing and acting on new stories	What specific plans, if any, do you have for transfer of training? What will enable you personally to feel that the interview was worthwhile?

2. Before this interview is actually held, fill in an Interview Plan and Objectives Form (Box 13-6) for the session.
3. Audiorecord or videorecord the interview.
4. Develop a transcript of the session. Place the transcript in a format similar to the one used by Allen in this chapter. Leave space for comments on the form.
5. Classify your interviewing leads by skill and focus; classify the client's focus as well.
6. Identify the specific stages of the interview as you move through them. Note that you may not always follow the order sequentially: indicate clearly that you have returned to Stage 2 from Stage 4 if that occurs.
7. Make process comments on the transcript. Use your own impressions plus the descriptive ideas and conceptual frames of this book.
8. Develop interview notes on your session using the five-stage structure of the interview.
9. Develop an interview plan for the next session (Box 13-6).
10. Develop a long-term treatment plan (see Box 13-7).

Box 13-7 Creating a Long-Term Treatment Plan

Initiating the session—rapport and structuring	How does this client develop rapport? What issues are of most comfort/discomfort? How does the client respond to structuring? At what place will structuring be most helpful?
Gathering data—drawing out stories, concerns, problems, or issues	List below, in order of importance, the several areas of client concern. Include a list of the client's wellness strengths and assets for coping with these issues.
Mutual goal setting	What are your and the client's ideal outcomes for these and other issues?
Working—exploring alternatives, confronting client incongruities and conflict, restorying	What are the client's main alternatives? What are your treatment alternatives? What are the main items of client incongruity? How might they best be confronted?
Terminating—generalizing and acting on new stories	What are the specific thoughts, feelings, and behaviors that you and the client would like to generalize to real life? Wherever possible, work with the client to set up the specific goals of generalization. Also, please see page 451 and examine the Maintaining Change and Relapse Prevention Worksheet. At times a written contract for generalization can be very helpful.
	It is often helpful to develop the treatment plan with the client so that a more egalitarian relationship is possible. Joint agreement on goals and methods can facilitate treatment.

Portfolio of Competence

After you have completed the practice exercise for this chapter, please go back to the interview you completed as you started this book, and note how your style has changed and evolved since then. What particular strengths do you note in your own work? Are you meeting your client's needs as described in Box 13-2?

As you review your own work, pay special attention to your understanding and use of cultural/environmental/contextual issues. Examine your interview from the perspective of someone from a different cultural group and gender from your own. How would he or she consider and evaluate your work?

A checklist for your portfolio has not been developed for this chapter. At this point, developing and evaluating your own interviewing style becomes the major exercise for your portfolio.

DETERMINING YOUR OWN STYLE AND THEORY: CRITICAL SELF-REFLECTION ON SKILL INTEGRATION

What single idea stood out for you among all those presented in this chapter, in class, or through informal learning? What have you learned and observed about yourself?

_____

_____

_____

_____

_____

REFERENCES

Brammer, L., & MacDonald, G. (2002). *The helping relationship* (8th ed.). Boston: Allyn and Bacon.

Damasio, A. (2003). *Looking for Spinoza: Joy, sorrow, and the feeling brain.* New York: Harcourt.

D'Zurilla, T. (1996). *Problem-solving therapy.* New York: Springer.

Egan, G. (2002). *The skilled helper* (6th ed.). Pacific Grove, CA: Brooks/Cole.

Ivey, A., D'Andrea, M., Ivey, M., & Simek-Morgan, L. (2002). *Theories of counseling and psychotherapy: A multicultural perspective* (6th ed.). Boston: Allyn and Bacon.

Janis, I., & Mann, L. (1977). *Decision making: A psychological analysis of conflict, choice, and commitment.* New York: Free Press.

Mann, L. (2001). Naturalistic decision making. *Journal of Behavioral Decision Making, 14,* 375–377.

Mann, L., Beswick, G., Allouache, P., & Ivey, M. (1989). Decision workshops for the improvement of decision making skills. *Journal of Counseling and Development, 67,* 237–243.

Parsons, F. (1967). *Choosing a vocation.* New York: Agathon. (Originally published 1909).

University of Massachusetts Memorial Medical Center, Behavioral Medicine Clinic. (2004). *Treatment plan.* (Unpublished document). Griswold Mental Health Clinic, Palmer, MA: Author.

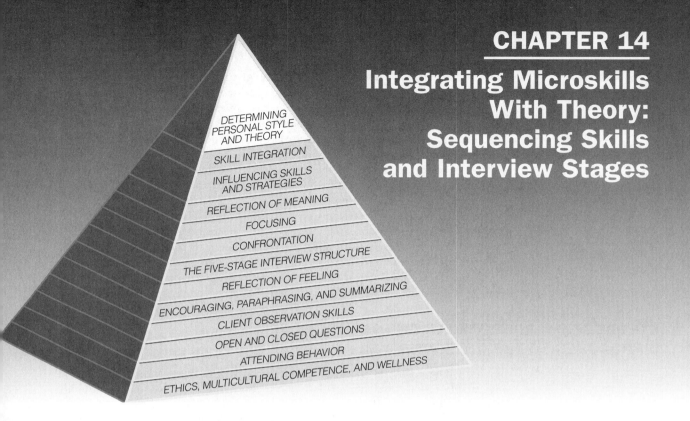

CHAPTER 14

Integrating Microskills With Theory: Sequencing Skills and Interview Stages

---

How can integrating microskills with helping theory assist you and your clients?

Major function	In this chapter, you will experience four additional important approaches to counseling—person-centered, behavioral, solution-oriented, and motivational interviewing.
	Through microskills analysis of interviews, it is possible to present combinations of skills and structure outlining key practice features of the theory or method. This will enable you to engage in a variety of interviewing styles, thus providing you and your clients with more alternatives for more intentional, effective helping. Table 14-1, presented later in the chapter, offers specific actions that you can use as an outline of each major approach.
Secondary functions	Mastery of the concepts of this chapter will enable you to

▲ Conduct a beginning person-centered interview.
▲ Engage in assertiveness training.
▲ Practice the basics of brief solution-oriented interviewing and counseling.
▲ Practice some aspects of motivational interviewing.

This chapter contains four different theories of the interview, each one of them focusing, in a different way, on a wellness approach and client positive strengths. Thus this chapter is longer than others. We recommend that you read only one section at a time. You may want to read the material on person-centered counseling together with the skill of reflection of meaning, and ideas about assertiveness training in conjunction with the skill of directives. Brief solution-oriented counseling is another form of decisional counseling but has its own style. Motivational interviewing was originally developed for work with alcohol problems, but has been expanded to general counseling issues.

Another approach is to select one or two theories of change that appeal to you and learn these more fully. Later you can return to this book to examine the other methods. Practice in each, using appropriate interview structure and the microskills, will give you a beginning understanding and some competence in each method.

Read the sections of this chapter that are appropriate for your place in your development as an interviewer and/or counselor.

INTRODUCTION: MICROSKILLS AND THEORETICAL APPROACHES TO THE INTERVIEW

We are facing a time in the helping field when we are being asked for clear and measurable results from our work. We are also being pressured to be culturally aware and come up with new ways to help clients make sense of their world. At the same time, there is a great need to maintain the traditions of the past with a focus on human dignity. These are weighty demands on us and on the helping profession.

Chapter 13 presented decisional theory as one way to organize the interview. Decisional theory has some advantages because it allows varying theoretical approaches to become part of the session. For example, you could help the client, Mary, to engage in deeper counseling about her personal issues through reflection of meaning and the person-centered approach. You could move toward action and behavioral change by making assertiveness training part of the process. Gender awareness and cultural issues could be explored. And a solution-oriented approach could be used to help identify answers in a brief counseling framework. Motivational interviewing might help develop motivation for change and action.

This chapter seeks to provide a structure to aid you in organizing your thinking and decisions about how you want to help clients. You will encounter very different ways to organize and structure the interview. You are not expected to master them all, but we believe that if you attempt one or more of the tasks in this chapter, you will be well prepared for further work in integrating theory with practice.

Theory is *not* presented in this chapter, except in the most general manner. References at the end of Chapter 15 provide suggestions for theoretical next steps. What you can expect here are some specifics that will make intuitive sense and provide you with understanding and expertise that can help you conduct your interviews more flexibly.

If you have mastered the five-stage interview structure, can use wellness and the positive asset search, you are well prepared to move to the next stage—making your own decisions about integration of skills utilizing a variety of approaches.

INSTRUCTIONAL READING AND EXAMPLE INTERVIEW 1: PERSON-CENTERED COUNSELING

A major assumption of person-centered theory is that the client is competent and ultimately self-actualizing. Skills are used to find that internal strength and resilience. A person-centered counselor is most often interested in focusing on the meaning and feelings of the client; the actual facts of the problem are considered less important. The goal is self-actualization, helping clients realize themselves more fully. Decisions may be made, but it is how the client feels about himself or herself that is most important. Therefore, the focus is much more on the person and less on the problem. Questions are considered intrusive and generally should be avoided in this orientation.

Person-centered theory is considered most appropriate for abstract, self-directed/formal-operational clients who are best able to think through their own direction. The concepts may also be helpful with other clients, and occasionally children, but a more concrete language on your part will be demanded.

To illustrate, let us take the case study of Mary from Chapter 13. Let us assume that Mary comes to us with the vocational problem presented at *Mary 16*. You have seen how the decisional counselor, Allen, used the basic listening sequence to draw out client facts and feelings and then summarized them at *Allen 23*. The problem focus was on career and decisional issues. If Mary were to see a person-centered counselor, very different things would happen.

The person-centered counselor may be expected to wait for Mary to initiate the conversational topic. After a moment's pause, Mary begins:

Mary: Well . . . ah . . . I guess there's a lot that I'd like to talk about. You know, I went through a difficult divorce and it was hard on the kids and myself and . . . ah . . . we've done pretty well. We've pulled together. The kids are doing better in school and I'm doing better. I've . . . ah . . . got a new friend. (breaks eye contact) But, you know, I've been teaching for 13 years and I'm really kind of bored with it. It's the same old thing over and over every day; you know. . . parts of it are okay, but lots of it I'm bored with.

Allen: Sounds as if *you* feel rather good about *yourself* and pulling it together, yet there are parts of *you* that feel bored and incomplete.

(This reflection of feeling includes a strong dimension of recognition and confrontation of mixed feelings. Person-centered counselors often give attention to the word *you* and emphasize it through verbal underlining [see Chapter 3] thus personalizing the interview.)

Mary: Yes, sometimes I feel confused. I know I've done well, but where do I go next? Something seems to be missing. Here I am, 36, yet alone and feeling stalemated. What does it all mean?

Allen: Mary, you say something is missing; you feel alone and stalemated when you look at *yourself* from a deeper level. There's something *missing* for you . . . (pause) . . . there's something missing that's *meaningful*.

(This is a reflection of meaning. The counselor is searching for Mary's underlying values and meanings in the belief that if she finds her true self, she will

Mary: self-actualize and solve many issues spontaneously. Here you see the client-directed/formal-operational style that expects clients to be able to solve their own dilemmas.)

Mary: Yes. . . . (pause) . . . (starts quietly crying) . . . I feel so alone. Nothing ever seems to work out. It's been so hard over these years. . . . Where am I? What should I do?

Allen: You've felt alone at the deepest level. You've had the strength and wisdom to work through many difficulties, but somehow, somewhere, something meaningful is missing for *you*. . . .

(This is a complex statement, typical of those who adopt the person-centered style. Note the reflection of feeling at the beginning, followed by feedback that points out positive assets of the client; this is characteristic of the Rogerian concept of positive regard. The final portion of the statement orients itself again to meaning: the underlying, deeply felt issues that impel us to action, often without our awareness.)

If you wish to extend your skills in person-centered theory, specifics of taking the theory into practice may be found in Ivey, D'Andrea, Ivey, and Simek-Morgan (2002). There you will find suggestions on how to include multicultural issues in this approach. You may also find it useful to read Carl Rogers's *On Becoming a Person* (1961) in which he provides the basics of his theory. Bozarth (1999) provides an update of Rogerian theory.

Specifics for structuring a person-centered interview may be found in Table 14-1. From a skills perspective, the following guidelines are suggested:

1. Seek to eliminate or minimize questions.
2. Focus almost exclusively on the client. The words *you* and *your* and the client's name are central.
3. Search for and reflect underlying meaning and consider reflection of meaning along with paraphrasing, reflection of feeling, and summarization as the basic skills.
4. Constantly use the positive asset search to help clients frame their experience in forward-moving ways.
5. Use selected influencing skills of confrontation, feedback, and self-disclosure, but sparingly.
6. Interpretation/reframes, advice, and directives are *not* part of this orientation.

Bringing Multicultural Issues Into the Person-Centered Approach

The focus on the individual client needs to remain but you can use a double focus, helping Mary see her issues in cultural context. For example, the first counselor response to Mary focuses very much on the pronoun *you*—"Sounds as if *you* feel rather good about *yourself* and pulling it together, yet there are parts of *you* that feel bored and incomplete." This reflective comment leads Mary to talk about herself in a deeper fashion. However, you can hold true to client individuality and also focus on multicultural issues. Focusing on both person and situation can enrich client experiencing. For detailed examples, see Ivey, D'Andrea, Ivey, and Simek-Morgan (2002).

Table 14-1 Five major approaches to counseling and the five-stage interview structure

Decisional Counseling: Counselor Actions	Person-Centered Counseling: Counselor Actions	Assertiveness Training: Counselor Actions	Solution-Oriented Interviewing and Counseling: Counselor Actions	Motivational Interviewing: Counselor Actions
Stage/Dimension 1. Initiating the session—Rapport and structuring ("Hello; this is what might happen in this session.") All systems give special attention to developing rapport and building a supportive alliance in a natural and personal style.				
▲ Outlines purpose of session and what client can expect. ▲ May state what to expect in each stage of the interview.	▲ Tends not to discuss structure and moves immediately to direction established by client. ▲ May subtly point out importance of allowing client to direct the session.	▲ Emphasizes importance of client participation in the session and may state the importance of the client's defining specific goals for the session. ▲ Points out that specific observable behaviors are the session focus.	▲ Clearly lets the client know what to expect—"What's your goal today?" "What has gotten better about the problem even before you got here?" ▲ Searches early for wellness strengths and positive assets.	▲ Structures session, stresses listening. "What would you like to talk about today?" ▲ Discovers motivation for change on 10-point scale. How important is it for you to change?"
Stage/Dimension 2. Gathering data—Drawing out stories, concerns, problems, or issues ("What's your concern? What are your strengths or resources?") The basic listening sequence (BLS) is central in all four approaches. Questioning skills will be used very little, if at all, by the person-centered counselor whereas they are a central solution-oriented interviewing and counseling skill.				
▲ Uses BLS to draw out facts, feelings, and organization of client's problem or decisional issue. ▲ Draws out individual and multicultural strengths.	▲ Uses listening skills to draw out client concerns with a focus on the individual client and feelings. ▲ Maintains constant emphasis on positive regard and client strengths.	▲ Uses BLS to draw out concrete behaviors in specific situations. ▲ Focuses broadly on both individual and contextual issues. ▲ Typically uses role-plays to discover behavioral specifics.	▲ Draws out client story briefly focusing on wellness and positive assets. ▲ Normalizes concerns and searches for contextual support systems. ▲ Seeks concrete examples of past successes.	▲ Uses BLS to widen awareness. "Change talk" focuses on good and bad of behavior. "What do you like about (alcohol, drugs, etc.)?" "What's the down side?" ▲ Affirms client as a person. "You handled that well."
Stage/Dimension 3. Mutual goal setting—"What do you want to happen?" Each system seeks to help the client find her or his own goals.				
▲ Uses BLS to draw out client's ideal decisional and career goals. ▲ Seeks to make them concrete and verifiable.	▲ Seeks to find client goal through listening, but even Carl Rogers has been known to ask "What would you like to see happen?" ▲ Helps client define distinction between the real self and the ideal self.	▲ Continues search for concrete goals for behavior change and may seek to define them more precisely here.	▲ Goal-setting process continues; defining objectives clearly may itself solve the problem as solutions often become apparent. ▲ Uses more questions to facilitate the process—"What are exceptions to the problem?"	▲ Explores positive and negative motivation for change on decisional balance sheet. Explores goals and values. ▲ "What would you like it to be like in the future?" "What are future consequences of change?"

(continued)

Table 14-1 (continued)

Decisional Counseling: Counselor Actions	Person-Centered Counseling: Counselor Actions	Assertiveness Training: Counselor Actions	Solution-Oriented Interviewing and Counseling: Counselor Actions	Motivational Interviewing: Counselor Actions
Stage/Dimension 4. Working—Exploring alternatives, confronting client incongruities and conflict, restorying ("What are we going to do about it?") The distinctions between the four systems become even more apparent here as each system confronts discrepancies and incongruity.				
▲ Considers the basic confrontation as between the present decisional problem and the goal. ▲ Balances influencing and listening skills. May seek to help client see impact of decision via reframing and logical consequences. ▲ May use career testing, information giving, and other strategies to facilitate decisional process.	▲ Confronts the ideal self with the real self with hope of integration. ▲ Continues to use listening skills, although reflection of meaning may become central. May engage in brief self-disclosure. ▲ Maintains little focus on problem solving while seeking to help client get a better sense of self.	▲ Considers basic confrontation as between the present behavior and the goal behavior. ▲ Repeats role-play until client is able to demonstrate the goal behavior fully. ▲ May emphasize environmental factors related to behavioral change.	▲ Often will combine Stages 4 and 5 as the emphasis here is on finding specific ways to change what occurs in the real world. ▲ Will use wellness strengths as levers to change and generalization of new thoughts, feelings, and behaviors. ▲ Expects to involve client fully in the process of brainstorming and exploring alternatives.	▲ Elaborates and affirms change talk. "Give me example of how it will be when you change." "What else can you do?" ▲ Responds to resistance. Central as many resist change. Magnify and explore discrepancies. Reframe resistance. ▲ Enhances confidence talk. Uses 10-point scale "confidence ruler." Reviews past successes.
Stage/Dimension 5. Terminating—Generalizing and acting on new stories ("Will you do it?") This is where many, perhaps even most, beginning *and* experienced human service people fail. Change that is not planned is likely not to occur.				
▲ Often prescribes some type of homework or action to follow up on the session. Techniques drawn from other theories discussed here may be used to facilitate generalization.	▲ Historically has given little attention to generalization in the belief that significant changes in attitudes, thoughts, feelings, and meanings will eventually result in major changes.	▲ Perhaps gives the most attention to generalization of any theory. Expects client to leave with a clear behavioral change plan. ▲ Often will seek to provide specific follow-up to ensure that change is maintained.	▲ Will use strategies from other orientations if there is reason to believe that the client will have difficulty in taking learning from the session to daily life.	▲ Negotiates concrete change plan. "What specifically are you going to do?" "What is the first step?" ▲ Completes change plan worksheet. Examines support system. Obtains commitment.

Gender is a multicultural issue that might enrich the person-centered discussion. Note how focusing on cultural/environmental/contextual issues as well as the client broadens the discussion and helps Mary see herself more completely.

Allen: *Mary, you* feel good about what *you* as a *woman* have done in a difficult situation. *Your* strengths as a wo*man* are many.

Mary: Exactly. I have many women friends who have supported me. I know I am not alone in this struggle. My mother's example and strength have also been important.

Listening

The way you listen can and does influence the way clients respond. Thus, it is essential that you constantly examine your own behavior in the session. Choosing to listen exclusively to "I" statements in the person-centered mode affects the way clients talk about their issues. Focusing on culture, gender, and context also affects the way they respond. Return to *Allen 17* (page 408) of the preceding chapter and examine how selective attention affected the progress of the session.

In dealing with the need to listen, you face two profound and important ethical and practical questions. How *do* you listen? How *should* you listen? The first question needs to be borne in mind constantly. The second question represents a value issue that perhaps is ultimately not answerable but that you nonetheless will be grappling with throughout your helping career. Your personal values influence *how* you listen.

INSTRUCTIONAL READING AND EXAMPLE INTERVIEW 2: ASSERTIVENESS TRAINING—A BEHAVIORAL STRATEGY

Behavioral approaches work on the assumption that changing behavior will result in more immediate change and that changes in attitudes toward self will often follow later. Assertiveness training is an important behavioral strategy to master and have available for some of your clients. It was developed in the late 1960s by behavioral psychologists (Alberti & Emmons, 2001) and has come to be one of the more widely accepted techniques for assisting clients regardless of one's theoretical orientation. For example, many feminist-oriented helpers use this method to help women become more direct and achieve their own goals. You will find that assertiveness training will often be used as part of rational-emotive behavioral therapy, that it is often needed to help career counseling clients communicate better at work or conduct better job interviews, and that even psychoanalysts may refer clients to a behavioral specialist to learn the skills of life management through assertiveness training.

Stage 1: Initiating the Session—Rapport and Structuring

The initial part of the interview will not differ from other types of counseling, as the behavioral counselor or assertiveness-training specialist also cares about people. Rapport will be established with the client, because a caring, relaxed atmosphere is essential for change.

The behaviorally oriented counselor might agree with the person-centered counselor, whose strengths are empathy and sensitivity, but would point to research suggesting that interpersonal change occurs more quickly and directly through counseling oriented toward observable behavior. Let us assume Mary begins an interview with a behavioral counselor. The behavioral counselor would tend to search for directly observable behaviors that might be identified and changed. The behavioral counselor tends to assume a concrete, coaching approach and is concerned with action and *doing,* helping clients concretely operate in their environment.

Allen: You say you are *bored,* even though you sound very busy and active. What is going on, what are *you doing* when you are bored? What's happening around you?

(Again you may wish to return to *Allen 17* for comparison purposes. The behavioral counselor is interested in knowing the concrete specifics of what is happening behaviorally in the person and in the surrounding environment. What is the client, Mary, doing, and how do others react to her behavior?)

Mary: Well, so many things are happening, between the job and the kids, that I never get time to stop and think. Something is happening all the time. I just keep going. There is always someone demanding something from me. . . . It's very frustrating not being able to *ever* have a minute.

Allen: I see; never a minute for yourself. What are some objectives you have for yourself here today?

(The counselor is aware that the session could go in many possible directions. As is typical of behavioral counselors, there is an emphasis on having Mary *participate* in that direction and goal selection. Whereas the person-centered counselor does not aim to direct the client, the behavioral counselor usually has specific objectives in mind.)

Mary: Well, I heard that you were good at helping people become more assertive, expressing themselves more . . . maybe through that I can find myself more clearly.

Allen: That sounds like a reasonable objective, Mary. If I hear you correctly, you feel pretty overwhelmed by life and are tired of living without time for yourself. You're hoping assertiveness training can be part of that process. Is that right?

In this brief excerpt, you'll have noticed an emphasis on *doing* and on *observable behavior.* You probably observed that Allen, the career counselor in Chapter 13, the person-centered counselor just presented, and this behavioral counselor all use basic listening skills to draw out the problem. Yet you may also have noticed that they focused on different things and that the interview was very different in each case. By now you should be able to see the potential value in the very different theories of helping. Mary could profit from a clearer career choice; she could also benefit from becoming more self-directed and aware of herself via person-centered work; and, in addition, she certainly could become more assertive in expressing her point of view.

Allen: Mary, we are going to try assertiveness training. As you know, this system is often helpful in enabling us to speak up more effectively and to obtain our goals, without overruling or overrunning other people. We'll discuss your situation and then role-play some ideas that may help you work through some of the issues. Is that okay?

This comment would most likely immediately precede the introduction of Stage 2. Mary would be involved in the decisional process as much as necessary. Structuring of assertiveness training or any other style of counseling and interviewing helps the client understand what is going on and may even enable the client to work with the therapist in a more mutual fashion.

Behavioral counseling, once assumed to be an imposition on the client, over time has become one of the most sophisticated methods of helping, and works actively to ensure that the interview is "person-centered" in the best sense of the term.

Stage 2: Gathering Data—Drawing Out Stories, Concerns, Problems, or Issues

(Mary has stated her general problem as never having a minute for herself and as desiring to speak for herself more clearly. The interview has set some general objectives, but the counselor wants more behavioral, observable specifics. The interview continues.)

Mary: Yes, that's right, everyone seems to run over me. I can't say no.

Allen: Could you give me a specific, recent example of a time when you didn't say no? What happened? What did you do? (Search for concreteness)

Mary: Well, I was talking to the principal today. He wants me to take on advising still another club. I'm coach of two sports now and advising the Tri-Hi-Ys. Every activity I do makes me stay after school and I get home late. And when I get home late the kids want even more from me . . . and then Bo seems to want more too! I can't say no to any of them.

Allen: You can't say no to any of them. I can see your frustration and how you are trying. Must be difficult. Let's take your example of the principal.

Mary: The principal . . . he just walks all over me. If I could start with him, maybe I could learn what to do with the others . . . maybe even Bo.

Allen: (interrupts) Okay. Rather than talk about it and analyze it to death, I'd like to see what really happens when you have to engage in a decision like that. You know, role-playing. . . . I'm going to be your principal and ask you to take on yet another responsibility. You play yourself and react to me just as you did earlier today.

(Characteristic of assertiveness training are clear directives and role-playing the actual scenario of the problem. Thus, rather than "hear about" the problem, the counselor observes actual behavior in the "here and now.")

Allen: Okay? (Mary nods in agreement.) I'm Michael, the principal, now. Mary . . . thanks for coming in. It's good to see you. You've done a great job with the swim team and the field hockey team this year. We like what we see.

Mary: (smiles) Thanks.

Allen: Mary, we've got a problem. The community is asking us to do more about drugs, and a group of parents have assembled a committee and want school participation. You have really good relations with the kids and have done a few workshops on the problems of drugs. I want you to join that committee.

Mary: (smiles, but a little weakly) Uhh . . . I'd sure like to do it. I'm beginning to see some real problems in some of my classes. But I'm simply overloaded with the teams, and my kids are getting older and need me to drive them places. I don't see how I could. . . . (hesitates, eyes downcast)

Allen: I'm glad you're interested. . . . It's only one night a week. The group wants to work with us in developing a curriculum and I want to be sure our point of view is represented. I'll call and tell them you'll do it.

Mary: (somewhat desperately, but weakly) Michael, I don't see how I can do it. . . . Sure, I'd like to help, but . . .

Allen: And you *are* a help. I'll meet with you during the week to give you additional support. I want to keep in touch with this.

Mary: (weakly) Okay.

Allen: (leaving role of principal) Is that how it is? You tend to give in rather quickly?

Mary: I'm afraid so. . . . It happens all the time. I'm just so anxious to please others.

Allen: I can understand that. Mary, what did you *do* in that role-play that got in your way? What *specifically* were you *doing* that allowed the principal to run over you?

(The brief role-play ends. The counselor then seeks to identify specific behaviors that represent lack of assertiveness. Frequently, you will find that behaviors identified in this method relate to the basics of attending behavior. As much as possible, you want the clients rather than you to identify specific behaviors. They can then define *their own* behavioral goals, which makes achievement more likely.)

Mary: Well, I certainly give in easily. I noticed I let him do most of the talking. He didn't listen to me. I felt pretty uncomfortable, I know that because I often smile when I am insecure and don't know what to do.

Allen: Uh, huh . . . and how did that uncomfortableness look? What was your body saying? Where were you looking?

Mary: I guess I look down a lot—it must all be a symptom of my wanting to please others.

(At this point, a person-centered counselor would be more interested in Mary's desire to think of others before her own needs. Rather than emphasizing concrete specifics of behavior, the counselor would aim to understand Mary's underlying emotions and meanings.)

The interview continues with further discussion of behavior specifics, much of it focusing on what Mary is doing "wrong." There is need for a positive asset search to build strengths.

Mary: It makes me very discouraged and tired.

Allen: (ignores emotions) But, Mary, let us not forget that you are doing several things right. What are they?

Mary: It all seems pretty bad to me.

Allen: First, the principal said you were doing a good job at school . . . you've apparently done a lot there. He wouldn't have selected you for this job if you weren't effective. And you couldn't do those things so well if parts of you weren't assertive. Right?

(Here we see elements of the positive asset search. Clients grow from strengths, and Mary has obvious abilities. If your clients know that you respect what they *can* do, they will have greater strength and potential for attacking their problems.)

Mary: (brightens up, smiles more hopefully) I hadn't thought of it that way. I guess I can do *some* things. (The counselor then elaborates on Mary's strengths.)

Stage 3: Mutual Goal Setting

(A concrete example of Mary's problem has been presented, with behavioral specifics that can be seen directly and even measured. Armed with an awareness of her strengths, she is now able to set up some specific goals for change.)

Allen: Given what's going on, Mary, what are some specific goals and behaviors you might want to change the next time we try that role-play?

Mary: Well, I'd like to smile less and talk more.

Allen: And what about more direct eye contact and a louder, stronger voice?

(The interview continues, and Mary and the counselor work out specific goals for change in her interview with the principal. They focus on some dimensions of her attending behavior, as just mentioned—less smiling, for example—and on Mary's assertively saying no.)

Stage 4: Working—Exploring Alternatives, Confronting Client Incongruities and Conflict, Restorying

Allen: To summarize where we are, Mary, in the past and in this interview with the principal, you gave in, and giving in is represented by smiling, looking down, and letting the other person take charge of the topic. On the other hand, your goal is to change these behaviors and take more control of your daily life. Does that sum it up?

(This is the classic confrontation statement, useful at the beginning of the fourth stage of the interview. The client's problem or past behavior is contrasted with the goal behavior. The discrepancy between the two is the issue to be resolved in this stage. At this point, the counselor is demonstrating Level 4 empathy (see Chapter 8) in that he has summarized the client's problem clearly for her—something that had not been done before and that hence may be termed additive empathy. In terms of the Confrontation Impact Scale, the goal for Mary will probably be a Level 5 resolution response: the creation of something new—change and development.)

Mary: Yeah, that seems to sum it up. What next?

Allen: Well, what we do next is another practice role-play, and we continue that until you demonstrate assertiveness with me in the situation. . . . That may take several practice sessions or it might be accomplished today. Regardless, what you take back to the school, and to your kids and Bo, is far more important than what happens here. We'll work together until you master these skills and are getting what you want. Okay, let's try another role-play. (The role-play starts.) Mary, I'd like you to take over the new drug program. The parents want your involvement.

Mary: Michael, that sounds great. I wish I could, but I have got so many things going right now, and the kids need me to drive them places. (Her vocal qualities are strong, but her eye contact is still poor.)

Allen: I can understand that. It's the ones who are busiest that you always ask. You can do it . . . you've done great workshops on drug education, and it fits with your good work in physical education.

Mary: (more weakly) No, I don't think I can. I don't want to . . . Nuts! There, you can see what I do!

Allen: Well, Mary, it was better. You did speak up stronger, and I liked the way you came up with reasons. We'll continue practicing.

(Mary and the counselor continue to practice via role-play and discussion. Gradually, Mary gets stronger and demonstrates an assertive no. This represents resolution of the discrepancy between where she was when she started the "problem"—and where she wanted to go—"the goal or outcome." She has almost reached a Level 5 response on the CIS. There are some "buts" to this rating, however, as you will see in Stage 5.)

Stage 5: Terminating—Generalizing and Acting on New Stories

Learning in the laboratory of the interview may appear to result in developmental change. However, real change occurs only in life, after the interview. Many of your clients will show considerable promise for change in the interview and then continue to behave and think in old patterns after the interview is over. Homework, planned transfer, and prevention of relapse are critical parts of effective change (Witkiewitz & Marlatt, 2004). In the following exchange, the counselor introduces Mary to relapse prevention, a systematic framework to help prevent loss of learning from the counseling session.

If you fail to include generalization and transfer of learning from your helping sessions, regardless of your theoretical decisions you will find that much of your work in helping is of little avail.

Allen: Well, this is the time, Mary, that is perhaps most important in our interview. You've certainly demonstrated that you can be more assertive and say no. The big question is whether you can generalize this to your principal and start taking charge of your own life.

Mary: Yes, and I want to do it with my kids, with some of my students, and with Bo. Everyone is running all over me.

Allen: Generally speaking, it is best to focus on one behavioral change at a time. After you have succeeded with one, you'll find the others will follow. Now, I'd like to go over the Maintaining Change and Relapse Prevention Worksheet with you. This is a way to give you some homework to ensure that you will continue to be more assertive.

The counselor hands Mary the worksheet (see Box 14-1). They work through it together; giving special emphasis to things that may come up to prevent Mary from being assertive. Research and clinical experience in counseling both reveal that this may be the most important thing you can do with clients: help them ensure that they actually *do* something different as a result of their experience in the interview. The interview closes . . .

Allen: Well, Mary, we've made good progress today. You've demonstrated that you can be assertive with a little practice. The big test will come tomorrow, with Michael. I'm sure you will maintain strong eye contact and vocal tone and will remember your very good reasons for saying no. We'll meet next week to see how it went.

Assertiveness training is useful with many types of clients. You may note that it is compatible with person-centered, solution-oriented, or decisional approaches.

Box 14-1 Maintaining Change and Relapse Prevention Worksheet:
Self-Management Strategies for Skill Retention

I. CHOOSING AN APPROPRIATE BEHAVIOR, THOUGHT, FEELING, OR SKILL TO INCREASE OR CHANGE
Describe in detail what you intend to increase or change:

_____

_____

_____

Why is it important for you to reach the above goal(s)?

_____

_____

What will you do specifically to make it happen?

_____

_____

II. RELAPSE PREVENTION STRATEGIES
A. *Strategies to help you anticipate and monitor potential difficulties: regulating stimuli*

Strategy	*Assessing Your Situation*
1. Do you understand the relapse and change process, and that there will be challenging situations when it will be difficult to engage in new behaviors?	_____
2. What are the differences between learning the behavioral skill or thought and using it in a difficult situation?	_____
3. Support network? Who can help you maintain the skill?	_____
4. High-risk situations? What kinds of people, places, or things will make retention or change especially difficult?	_____

B. *Strategies to increase rational thinking: regulating thoughts and feelings*

5. Are you aware that a slip, relaps, or mistake need only be temporary? "Relapse happens."	_____
6. What might be an unreasonable emotional response to a temporary slip or relapse?	_____
7. What can you do to think more effectively in tempting situations or after a relapse?	_____

C. *Strategies to diagnose and practice related support skills: regulating behaviors*

8. What additional support skills do you need to retain the skill? Assertiveness? Relaxation? Communication microskills?	_____

Permission to use this adaptation of the Relapse Prevention Worksheet was given by Robert Marx.

(continued)

Box 14-1 (continued)

D. Strategies to provide appropriate outcomes for behaviors: regulating consequences

9. Can you identify some probable outcomes of
succeeding with your new behavior?

   _____

   _____

10. How can you reward yourself for a job well done?
Generate specific rewards and satisfactions.

   _____

   _____

III. PREDICTING THE CIRCUMSTANCE OF THE FIRST POSSIBLE FAILURE (LAPSE)
Describe the details of how the first lapse might occur, including people, places, times, and emotional states.
This will be helpful to you in coping with the relapse when and if it comes.

_____

_____

_____

_____

The flexible interviewer or counselor may be able to utilize all four approaches with the same client. You will note that the giving of directives is particularly important in this technique and that the counselor freely used feedback on performance. Listening skills remain important, as the goal is to give the client as much power and control over the session as possible.

Bringing Multicultural Issues Into Assertiveness Training

Again focusing becomes an important skill. Mary's difficulty with the male supervisor is a common problem among women. This assertiveness training session might be enriched by bringing in the general issue of women's needing to be more assertive in the workplace. Mary then could draw on women models of assertiveness and realize that she is not alone with her issues. When the role-play occurs, the counselor could add, "How would an assertive *woman* act with this man?"

Assertiveness training has other multicultural implications beyond gender. Gay teens, for example, can profit from assertiveness training as they cope with teasing and harassment in the high school. People with disabilities can profit from assertiveness training as they seek to gain their legal rights. Parents who come from an international community in which teachers are viewed as "always right" might benefit from assertiveness training so that they learn how to express themselves more directly in U.S. and Canadian culture.

The final point above is particularly important to observe and consider. Some cultures have a more assertive style than others. For example, what may be standard assertive behavior among African Americans or European Canadians may be seen as overly aggressive by some Asian cultures. The Americans and Canadians in turn may think a traditional immigrant Asian family is too passive. What is often mistakenly interpreted as passive, however, is an effective cultural style in traditional Asian communities. Cultural differences clearly modify the appropriateness of assertiveness training with Latina women. The goals of assertiveness training in this case are sometimes in direct opposition to traditional cultural standards.

All of the above situations can profit from culturally sensitive assertiveness training. As you start this type of work, spend time finding out from your client how he or she views the words *passive, assertive,* and *aggressive.* How does your unique client define these issues in a cultural context? Then you can describe the European American model of assertiveness and how it contrasts with those of other cultures. With women who might be in danger from abuse if they become too assertive, it is important to share the potentially dangerous consequences of assertiveness. In such cases, add culturally sensitive counseling to help clients make decisions.

You will also find that assertiveness training can be useful to those who are seen by others as overly aggressive. Again, it can be helpful to consider the different perspectives on what constitutes assertiveness.

INSTRUCTIONAL READING AND EXAMPLE INTERVIEW 3: BRIEF SOLUTION-ORIENTED INTERVIEWING AND COUNSELING

Allen E. Ivey, Robert Manthei, Sandra Rigazio-DiGilio, and Mary Bradford Ivey

Brief solution-oriented interviewing and counseling originated in the work of the psychiatrist Milton Erikson, who often produced seemingly miraculous results with his clients using short-term methods. A number of authorities have examined Erikson's work, added their own thoughts, and defined some basic principles of brief therapy (e.g., Budman & Gurman, 2002; de Shazer, 1985, 1988; Hersen & Biaggio, 2000). Sklare (2004) has provided us with an excellent summary of this approach as it might be applied to children. Narrative and storytelling theories have recently been integrated with brief concepts (Monk, Winslade, & Epston, 1997). Semmler and Williams (2000) have given multicultural issues and narrative approaches special attention.

A major assumption is that clients have their own solutions readily available if we help them examine themselves and their goals. It is possible to organize much of the pragmatic work of these brief-therapy counselors within the microskills interviewing structure. In addition, you may find that several of the strategies and questions presented here will be useful regardless of the position you will eventually take on the helping process.

Whereas person-centered methods use very few or no questions, brief counseling often makes questions the central skill. We also suggest that you consider sharing the specific solution questions you will use with your volunteer and real clients, particularly in the early stages. Try using chapter questions and structure as a "worksheet" that you and the client share. As you gain experience and confidence with this method, you may wish to continue the sharing—counseling and interviewing can be more powerful and real in an egalitarian co-constructed framework.

As you begin to apply solution-oriented interviewing and counseling (SOIC), it is important to recall that building rapport and trust remains essential. We may sometimes still need to hear the client's story in detail before moving toward answers and solutions. Hearing the client's version of what has worked or has not worked is especially important as SOIC focuses so strongly on goal setting (Stage 3 of the five-stage structure). Careful attention to clients' *stories* and *their goals* is fundamental as there is a need to avoid imposing our own personal or theoretical agenda.

How long is SOIC? Anticipate one to three interviews as typical within the SOIC framework. The solution approach may be a single interview, or it may extend to as many as 5 or 10 sessions. The key word is *brief,* emphasizing solutions rather than problems.

Stage 1: Initiating the Session

As you practice SOIC, share the specific questions with your client and work on them together.

Basic questions. What has gotten better about your problem? What do you do right? How can we keep that going? The positive asset search is central.

Your own mind-set. Solution-oriented interviewing and counseling asks that you think differently about helping. Instead of focusing on defining "problems" or a long drawn-out exploration of "what happened and why," you need to focus on "solutions." This means establishing a positive expectation for both yourself and your client. Consequently, your task is to structure the session to achieve success in this important joint venture with your client.

Relationship. Traditional rapport and listening skills remain central, although you will want to use questioning skills much more as your intervention of choice. You may wish to explain to the client early on that you will be asking many questions and find out whether that is acceptable. As noted in Chapter 4 on questions, some clients and people from some cultures may be suspicious of frequent questions. If so, spend more time in explaining what you are doing and more time on listening to stories and developing trust.

Structuring the session. Letting the client know what is going to happen is especially important in SOIC. Share with the client very early in the session what you are going to do and why. For example:

> Many people can accomplish considerable progress in just a few sessions. What we are going to do here today is focus on solutions—the goals you want to achieve. Can you tell me what your goals are *for today?*

The words "for today" are important because they bring the client to the here and now and the possible, rather than leading to a lengthy attempt to resolve everything at once. Others may want more explanation. Certain issues are too large to be resolved in a few sessions. In such cases, your client may decide to work on one primary issue now and leave the others for later. Returning to counseling at some later point is not failure; rather, it shows willingness to continue to work on the many complexities we all face in daily life. And if the goals are clear, reasons for returning and motivation for change are increased. With children or adolescents, the wording may be better phrased as follows:

Jon, the teacher asked me to talk with you. Rather than talk about problems, I'd like to know how things might become better for you. Could you tell me one thing that you can do to feel better—happier about the rest of today? (Or "Before we begin, I want to know something that makes you happy. Tell me about what you like to do.")

You may find it helpful to share your question list here with your client. Important in the above is the idea that change can occur quickly, particularly if we focus on specifics of actively making things better. This type of statement also prepares the client for return at some later point. SOIC recognizes up front that we all have problems, that this is normal, and that all of us can benefit from occasional sessions with helpers. Counseling and interviewing thus become an expected, normal part of life experience.

Some useful questions. Start solution-based thinking at the very beginning of the session. Even as you listen to the client's story and/or reasons for coming to the interview, you can ask the following:

▲ Has anything changed since you decided to come to see me? Are things better in any way?
▲ What's keeping it from getting worse?
▲ Are there any exceptions in this problem? When is the problem not so much of a problem?
▲ What have you been doing to keep this issue from really dragging you down?

What else? Your client will not always respond in depth to your questions. De Shazer (1988, 1993) recommends the frequent use of "What else?" to prompt client thinking and the generation of more complete answers and solutions.

Children and adolescents (as well as adults) may initially respond negatively to your questions and even say "Nothing." Remember the importance of rapport and listening—with many clients, a sense of humor helps! With experience you will develop follow-up questions and help clients explore their issues in new ways.

Next—Stage 2 or Stage 3? If the problem was clearly defined during rapport and structuring and the goal is relatively clear, consider moving directly to Stage 3. *Do this especially if the individual is able to identify specific things that have gone better and/or times when the problem is "not a problem."* Examples of clearly defined concerns might be these:

I'd like to stop arguing so much with my partner.
My son doesn't sit still during meals.
I'd like to be able to speak up at meetings more effectively.
Our lovemaking has become too routine. I want my partner to warm up to me.
I want more challenge in my work.

Again, virtually all of these require some awareness of times when the problem is not a problem.

On the other hand, if the concern is presented vaguely, such as "My relationship is falling apart," and if the client talks about multiple issues in a confusing array and has difficulty in considering the issue of goals, exploration time in Stage 2 is clearly necessary. Solution-oriented methods work best if only one problem is in the foreground. Other issues can be dealt with later.

The decision to omit Stage 2 is not easy. Many clients need to tell you their story in detail before moving on to stating their goals. Thus, the skills of the basic listening sequence remain essential. When in doubt, it is wise to be more conservative and spend time in exploration. But even here, you can occasionally ask goal-setting questions, building a foundation for more rapid change and client involvement. Some ideas to focus problem definition follow below.

Stage 2: Gathering Data and the Positive Asset Search

Basic questions. What are the exceptions to the problem? When does the problem not occur? How do you get that more positive result to happen? What are your strengths and resources?

Being brief. You may want to tell the client that it may not be necessary to focus on all the details of the problem. Then with the client's assistance, generate a brief narrative of the problem, concern, or issue. Specifically, use the basic listening sequence to draw it out and be sure to summarize what has been said. The summary is particularly helpful in SOIC as it helps organize the session and serves as a foundation for clearer and more effective goal setting.

Many clients will describe the problem using formal/operational, abstract terms; be sure to clarify the problem/concern with concrete specifics so that the abstractions are clarified. But be sensitive to your client. As pointed out frequently in this book, some people need to tell their stories as much as or more than they need help in changing their thinking, feeling, and behavior.

Normalizing the narrative. The skill of normalizing a client's story is a particularly useful one. To accomplish this, you try to help clients see that their issues are a natural and logical result of their life situations. It is normal to have concerns and it is normal to have difficult situations. Your task is to point out to clients that while we all have issues, our concerns are solvable. Your own nonverbal behavior and confidence are part of this process.

Cultural/environmental/contextual issues such as gender, race/ethnicity, and spirituality factors may often be part of normalizing the narrative. The gay or lesbian client, for example, may begin the session by stating the problem as depression over constant harassment. As you hear this story, you note that the client is focusing on herself or himself as if the problem is internal. By focusing on cultural oppression, you help externalize and normalize the story.

Care must be taken to avoid minimizing serious concerns. An eating disorder, an abusive family history, and racial or sexual harassment are difficult issues. Normalizing the narrative is *not* stating that these are normal and expected parts of life; rather, normalizing the narrative means focusing on the idea that we all have concerns and it is indeed possible to do something about them. You can then use focusing to externalize and normalize the concern. Also, at this point, you may wish to view Box 14-2, which illustrates ethical issues that underlie SOIC. SOIC may not be sufficient with some clients and complex issues. Thus, referral is a necessary part of effective practice.

The positive asset search and wellness. As you listen to the client's story, search for strengths in the client, including positive family assets, community assets, and cultural and/or spiritual strengths. An important part of normalizing clients and situations is to help them reidentify with their strengths and power. Sometimes these assets will provide an obvious solution the client missed earlier. Chapter 3 suggested the following questions as particularly useful to draw out strengths that can be used for solutions:

▲ Considering your ethnic/racial/spiritual history, can you identify some wellness strengths, visual images, and experiences that you have now or have had in the past?
▲ Can you recall a friend or family member who represents some type of hero in the way he or she dealt with adversity? What did he do? Can you develop an image of her?
▲ We all have family strengths despite frequent family concerns. Family can include our extended family, our stepfamilies, and even those who have been special to us over time. For example, some people talk about a special teacher, a school custodian, an older person who was helpful. Tell me concretely about them and what they meant to you.

Genuinely complimenting and giving feedback to the client on specific strengths and assets is often most useful. But the clients must accept that their wellness strengths are real, or the positive asset search may seem trite and even disrespectful. With some clients and some cultural groups, direct compliments may be embarrassing. Indirect ways to compliment a client for her or his strengths include the following:

▲ How did you know that?
▲ Where did you learn that?
▲ How did you figure that out?
▲ How did you develop that strength?

These questions, are of course, useful at any point in the interview where you note client strength and resources.

Useful Stage 2 questions. It is often a good idea to share a list of solution-based questions with your client and explore them together. As you become more familiar with the ideas, remember to continue to work *with* your client, not *on* her or him.

▲ Are there times when you do not have this problem?
▲ What's different about the times when this problem does not occur?
▲ Suppose when you go to sleep tonight, a miracle happens and the concerns that brought you in here today are resolved. But since you are asleep, you don't know the miracle has happened until you wake up tomorrow; what will be different tomorrow that will tell you that a miracle has happened? (The miracle question; see de Shazer, 1988, p. 5.)

Box 14-2 Some Practical and Ethical Issues in Using Solution-Oriented Interviewing and Counseling (SOIC)

Brief solution-oriented interviewing and counseling, like any form of helping, has limits and is not appropriate for all people. When the client isn't responding, an ethical interviewer or counselor needs to face two fundamental questions:

1. What other types of counseling might assist this client?
2. Who can provide this treatment?

With question 1, you may have other skills and methods that may be useful for the client. If you do, inform the client that you together are going to examine some other approaches that may be beneficial. If you do not have other ideas that may work, then consider question 2. You may need to seek assistance from a supervisor, consultant, or more experienced colleague to help the client move to another potentially more productive counseling environment.

Evidence that a solution-oriented interviewing or counseling approach is not effective for a client might include the following:

1. The client presents with serious symptoms or problems (e.g., substance abuse, relational violence, child abuse, suicidal gestures, alcoholism, eating disorders) and does not respond to SOIC interventions. Though SOIC can frequently be helpful in working with these issues, solution methods may not always be adequate in themselves. These matters may require experienced counselors or therapists who use other approaches. Seek referral as soon as possible. (This, of course, is true for any of the skills, methods, or ideas presented in this text. Ethical interviewers or counselors are aware of both themselves and their helping methods and are ready to refer when necessary.)
2. The client is not able or willing to try the solutions generated in the session. For example, after several solutions have been successfully rehearsed in session, the client cannot or will not enact these solutions outside the counseling relationship.
3. The solutions are primarily generated by you rather than by the client. In this case the client may be unable or unwilling to construct possible solutions and may be dependent on the ideas of the counselor.

(continued)

▲ Follow up the miracle question with "How will we know the issue has been resolved?" and "What are the first steps to keep the miracle going?"

The above questions are common in solution-oriented approaches, but to be fully effective, they require follow-up and exploration. The "What else?" question is useful, as are the four questions listed above with the positive asset search. You may find that you obtain rather brief or sketchy responses to the miracle or other questions. At this point, your own skills as an interviewer become crucial—use your natural ability to expand and draw out responses.

Scaling. This method is useful in many cases of SOIC. Scaling provides a quick and useful way for you and the client to communicate regarding the current depth of the concern.

On a scale of 1 to 10, with 1 meaning the concern is fully resolved and 10 meaning that the concern almost totally overwhelms you, where would you put yourself today? (Or at this moment in the session?)

Box 14-2   (continued)

4. The client may be unable to identify exceptions to a problem from which useful goals can be constructed. This may show most clearly at Stage 3 during goal setting.

   The cues and signals may appear at any time during the sessions. Initially, failure to adequately define goals or solutions to the problem may indicate that a different interviewing style may be more appropriate. Also, solutions that are vaguely constructed and not carefully explored for positive and negative consequences may be a sign of the client's lack of commitment to the process.

   During the final stage of counseling, the client may have difficulty making the solutions real in her or his life. Sometimes clients will refuse to try out ideas that previously looked quite promising.

5. The client's context may not be receptive to certain solutions. For some clients, the interviewer needs to assess the impact that a solution might have on the client's relationship in a broader context. For example, a woman client's solution might focus on taking a much more assertive position in a marriage. If so, is there danger of abuse if the woman speaks up?

6. The client may in fact want a long-term helping relationship—meaning the discovery of self or understanding may be the client's issue. Some clients will respond more favorably to a long-term, safe relationship and may reject you or even terminate if you use only SOIC. Solution-oriented interviewing and counseling is not for everyone, nor is there any one approach, microskill, or way to structure an interview.

7. Again, we would remind you of the importance of considering family and cultural/environmental/contextual issues. There are many strengths and solutions in surroundings. Support networks may be important to help the client implement new approaches.

   This list is not meant to be exhaustive but rather to provide some ideas for you to explore as you consider the appropriateness of SOIC for you and your clients.

With children and younger adolescents, actually drawing a scale may be useful. The child can then point to where he or she is on the scale.

Use scaling periodically through the session and consider using it throughout all your interviews, regardless of theory. It will serve as a temperature gauge so that you know how clients are feeling about their problems at the moment. Scaling can be useful as an evaluation device to determine whether you and the client are in synchrony and seeing things similarly. Scaling is very helpful when you think things are going well and then find that the client sees the issue differently (or vice versa). Scaling can also let you know how the client is feeling at a deeper level. At the surface, things may seem fine, but when the client's place on the scale is revealed, it may be much different from what you think.

With experience and practice, you will want to expand your use of scaling. For example, you could have your client evaluate his or her present level of motivation for change ("How committed are you to solving the problem?"), the confidence of success ("How likely are you to succeed?"), or how he or she will deal with termination ("At what point do you feel that the problem is sufficiently resolved?").

Stage 3: Mutual Goal Setting

Basic questions. What do you want to happen? How do you cope with the problem? What have you done so far that is helpful in achieving that goal? If you don't know where you are going, you may end up somewhere else. Too much interviewing and counseling focuses on problems and fails to ask the client what he or she would consider a satisfactory solution. Often what you think is the desired answer is very different from that of the client. For this reason, SOIC recommends moving as rapidly as possible to goal setting. The most important question at this stage is the following or some variation:

> We have heard your concern (summarize again, if necessary to keep interview on track and check accuracy). . . . Now, what, specifically, do you want to happen? Be as precise as possible.

Some children and adolescents will have difficulty with goal setting, as their life experience has been focused on what people in authority want from them. Patience and setting up concrete, achievable goals are important. It is easier to complain than to set goals. Again, remember the important "What else?"

Co-constructing concrete, achievable, clear goals with the client. Be sure you negotiate specific goals that can actually be reached. Clients too often want to resolve all concerns simultaneously. Help them work toward resolving a smaller piece of the larger issue. We can think of the ripple effect—a small change can lead over time to significant differences in a client's life.

Positive asset search and wellness. As you turn the client's mind to goals, remind her or him that he or she has the resources, abilities, and past positive experiences that can make clear goals and the resolution of concerns possible. The positive ideas of Stage 2 can be applied here as well. Recall that the client must believe in these assets and that some clients have negative reactions to direct compliments. The positive asset search must be a joint cooperative endeavor.

Additional Stage 3 questions. Use the following, but strengthen them with variations on de Shazer's "What else?" question to help expand ("Can you add anything more?" "Any other thoughts?" etc.).

▲ Let's focus on the *exceptions*. Tell me about the times when the concerns are absent or seem a little less burdensome.
▲ What is different about these times?
▲ How do you get that more positive result to happen?
▲ How does it make your day go differently?
▲ What did he or she do or say when it was better?
▲ How did you get her or him to stop?
▲ How is that different from the way you usually handle it?

Some other SOIC questions follow that many find helpful. Most involve a change of pace and can add humor to the session.

▲ What do you do for fun?
▲ What would enable you to feel that life is better?
▲ Name one thing that would help.
▲ Let's take a piece of the larger concern and work on that. Okay? We can't solve it all today, but we can make a piece of it a bit better.

After you have defined the goal, the following questioning strategies are useful. Note that all are very concrete and specific. When they are combined with variations of the "What else?" question, you can obtain very specific ideas about client wishes and desires.

▲ How will your life be different?
▲ Who will be the first to notice?
▲ What will he or she do or say?
▲ How will you respond?

The above questions are especially useful when you are working with a relationship issue. Frequently, it is useful to have both members of the relationship there with you to work through SOIC.

Stages 4 and 5: Working on the Concern and Generalizing New Ideas to the Real World

Basic questions.   What have you been doing right? What do you have to keep doing so that things continue to improve? What will tell you that things are going well? How can we take what we have learned today to daily life?

SOIC combines these two stages. If you have been working effectively on identifying resources, finding exceptions to the problem, and identifying goals in the first three stages, you have already done much to brainstorm and explore solutions. Thus your goal in this stage is to solidify and organize the solutions and move toward concrete action. The issue of Stage 5 generalization remains central. We often find that interviewers and counselors spend so much time listening to problems and trying to find solutions that they may neglect work on generalization. Unless we focus on transfer of learning, the entire interview may not bring positive benefits for the client.

Particularly important here is to work on the clearly defined goals in specific manageable form. Do not try to resolve all the issues at once. Rather, work toward a "piece of the solution." Constantly focus on the idea that something can be done. Every successful idea for solution needs to have a practical use outside the session.

Thinking about change and "taking it home."   We need to change negative "I can't do anything" talk to a new conversation about change and possibility. We always need to think about how talking about change can be transferred to the real world. General guidelines from de Shazer (1985, 1993) include the following:

1. Note what the clients do that is good, useful, and effective. Find out what efforts they have been making and support their process of change. This is essentially the microskills positive asset search.
2. Note exceptions to the problem. What is going on when the problem isn't happening? Be concrete and specific in this search.
3. Promote the two above as they relate to clear, specific client goals.

Box 14-3 Can't You Be a Little Patient?

Weijun Zhang

I was once hired by an American multinational corporation to counsel its expatriate executives working in China. My first client, a middle-aged Caucasian male, had been in China for the past 3 years, functioning first as a manager of finance and then as the general manager of the joint venture. The reason he sought counseling involved his relationship with his local subordinates. "They are driving me nuts," as he initially put it.

When his eyes first met mine, I could tell right away that he had a big question mark in mind of my ability to counsel him. But after a few minutes of small talk, he seemed to be convinced that this Chinese guy was westernized enough to be his counselor. Before I realized it, he burst into a string of complaints about his sour relationship with the local managers. Apparently he trusted me, for he revealed very specific facts of his situation and made no attempt to hide his hard feelings toward his Chinese colleagues.

Being a fan of the solution-focused approach, and feeling certain of knowing the mind-sets of White male businessmen, I wasted no time in starting to intervene when I heard him say, for the third time, "They are driving me nuts." I replied, "I see that your relationship with the local managers has really deteriorated. I suppose you know that you are not the only Western expatriate who is suffering from this problem. However, let me ask you a question here. Could you recall a period of time, no matter how short it is, when your relationship with your local partner was good, or normal, or not so bitter?"

He looked very surprised upon hearing this solution-oriented question, thought for a few seconds, murmured something like, "There should be, I suppose so . . . ," and then went on criticizing his Chinese co-workers with even more vigor.

I let him continue whining for about 3 minutes before I seized another opportunity to pose another solution question: "Since the operation of this joint venture has been going on for years without interruption, is it reasonable for me to assume that there have been times when you and your local managers have communicated?"

"Yes," he answered indifferently.

(continued)

In effect de Shazer says *work on what we have already done.* If you did a good job with goals, exceptions, and other elements, the solution may already be in hand and may just need to be reemphasized.

Contract with your client for specific follow-up to determine goal attainment in the next session or by phone. SOIC represents a contract and commitment to the clients. Do not leave them at this point. Stay with them until they accomplish *their goals.*

As part of this follow-up, it is useful to assign a task that the client can use to ensure transfer from the interview. Concrete, achievable tasks, often set up in small increments, start moving the client toward significant change. Here you may consider some of the influencing skills as necessary for the individual client. Any of the dimensions of Stage 5, generalizing—including homework and relapse prevention worksheets—and other dimensions, may be added.

Box 14-3   (continued)

"Could you please give me one or two examples of this positive side of your relationship?" Reluctantly, he started to relate an incident in which he and his subordinates had had good cooperation. But when the story was barely half told, he shifted to focus on the negative and went on complaining again!

I am not a person who gives up easily. When I saw a chance to cut in on his grievance, I tried again to switch his focus to the positive side, in the hope that we could move to the goal-setting stage faster. My assumption was that since he was a busy executive in a bottom-line-oriented company, he would surely favor short-term counseling and wanted to see concrete results quickly.

Much to my surprise, he burst into anger at my attempt and began to shout at me: "Why can't you let me finish my stories? Why can't you just listen? I thought I was lucky to have finally met one Chinese who can really understand what is going on here! Why can't you be a little patient?" Seeing I was taken aback by the outburst, he added, "By the way, do you know what PRC really stands for? The People's Republic of China, isn't it? Let me tell you what. It means you have to have PATIENCE and make RELATIONSHIPS in order to earn CASH!"

This left me embarrassed. This Caucasian client of mine, after being in China for 3 years, was now teaching his Chinese counselor, who had been in the states for about twice as much time, the importance of patience and relationships! Apparently, both of us had done a good job in adjusting to the local culture, though in opposite directions.

Allen Ivey comments: Some counselors who favor SOIC may have a hidden agenda, which is to go through the first two stages as quickly as possible and avoid listening to the story of the client's concern or problem. They may wish to start the goal-setting process as quickly as possible. This can be okay as long as you are aware of your agenda and are flexible. Problems occur when you are insensitive toward clients' specific needs and impose your own wishes on them.

Though other stages may be shortened or even eliminated to achieve efficiency, Stage 1—rapport building—remains central and critical, especially in relationship-oriented cultures. In short, haste may slow your effectiveness. Very often, you have to let clients have their catharsis and time to tell their story before moving further.

The storytelling or narrative approach might add that the "writing of a new story" to describe the incident, issue, or problem is highly useful. The skill of interpretation/reframing and focusing can help to describe the problem in new ways. As part of SOIC you could ask the child, adolescent, or adult client to tell the old story from a new frame of reference. For example, the old perspective may have focused on what other people are doing to make the client's life miserable. The new story focuses on what the client has done to cope successfully with the situation. The newly developed narrative becomes the cognitive and emotional framework for behavioral change. With some clients, writing down new stories and action in journal form may be especially useful. With children, it may help to have them draw pictures of the old story and pictures of the new. White and Epston (1992) include the possibility of the counselor's writing down summaries of the clients' new possibilities and sharing them in a letter sent to the home or at the next session.

Finally, remind your client again that he or she is welcome to come back at a future time for more work on this concern or any new issues that may arise.

Bringing multicultural issues into SOIC.   As stated several times in this book, a questioning interviewing style can often be a problem if you have not built sufficient rapport and trust with your client. Thus, establishing a natural and effective rapport is perhaps even more important in this approach than it is in others. Listen to the story until the client is ready. One route toward building trust is sharing your questions and interview plan with the client. The emphasis on positives, of course, will be useful in making the solution approach culturally relevant.

Focusing on the cultural/environmental/context can be especially important. For example, you may be interviewing a woman who has experienced harassment in the workplace. If you focus solely on the problem and individual solutions, you may miss the most critical issue. The problem may be located not in the individual but in the system. If there is a family problem, it may be wise to focus on the family and not just on individual solutions.

The solution-oriented approach can easily focus on the individual, with insufficient attention to broader social issues. However, widening the focus, coupled with the emphasis on individual strengths, may make SOIC a most valuable addition to a multiculturally aware helping interview.

EXAMPLE INTERVIEW: BRIEF SOLUTION-ORIENTED COUNSELING

This demonstration of solution-oriented interviewing and counseling is a session conducted by Penny Ann John, a first-year graduate student at the University of Massachusetts, Amherst. It illustrates many of the points of this book and of SOIC. We thank Penny for permitting us to share her work with you. As you will note, she worked with a verbal client volunteer with a fairly specific concern that had some obvious solutions. It will not always be this easy and direct, but sometimes it is. For your first practice in SOIC, we suggest you find a classmate, friend, or family member to volunteer as the client. It will take some experience and practice for you to master these ideas with clients who have more complex issues or who may be resistant to the process.

We'd particularly like you to note how Penny uses the basic listening sequence and search for positive assets and wellness as a vital part of SOIC. You will find that balancing the questioning style of SOIC with listening skills generally strengthens the interview. The interview has been edited for clarity, but it remains the work of Penny. As with all interviews, Penny's work is not perfect, but it is a fine example of how the positive asset search and the focus on exceptions and solutions can make a difference in the life of volunteer and real clients. As you read this session, think how you might have handled the interview in a way that is in accord with your own natural style of helping.

Stage 1: Initiating the Session—Rapport and Structuring

Penny shared a summary statement on SOIC with the client and talked about the process initially as she began the session. Carter is a close friend of hers, so she was able to jump right in after carefully explaining what was to happen. Carter had earlier told Penny that she wanted to explore the stress she was feeling as the end of the academic term was approaching.

Penny: Carter, we talked before pretty much about what we are going to do today, which is Solution-Oriented Interviewing and Counseling, and we are supposed to take a small issue or concern for you and work through that and come up with some solutions. You said that you wanted to work on academic stress. Let us take a part of the larger issue—small parts of larger issues are often useful places to start. You've got a list of the questions just as I have here and, if you wish, add any questions I missed that you think are important.

Carter: OK.

Penny: So, to start, suppose you tell me what your goal is for today. (Note immediate focus on goal setting.)

Carter: My goal for today is for us to brainstorm and come up with ideas to manage my stress because I am feeling really stressed out. (Of course, goal definition may not always be this quick and easy. Penny has a client who verbalizes well and who "buys in" to the solution model.)

Stage 2: Gathering Data—Drawing Out Stories, Concerns, Problems, or Issues

Penny: OK, sure. What brings this topic to your mind today versus talking about this another time? (Focus on *here and now* and the reason for wanting to discuss it *now.*)

Carter: Well, I am a graduate student and it is that time of the semester. Everything is coming to what seems like crunch time and that is when I feel the most stress. There are so many things on my plate.

Penny: Right now it is stressful for you because it is coming toward the end of the semester . . . and you have a lot going on. OK? With all of this going on at the moment, what might be positive about your situation right now? (reflection of feeling—"You feel X because Y"—followed by an open question oriented to strengths and solutions already existing in the client)

Carter: Well, you know I have got to say, I have talked to a lot of people lately and they have a lot more to do . . . um hum . . . right. And I did not really realize that until I talked to them because I have been plugging along right along and doing my papers so I don't have everything to do all at once. And that made me feel a lot better.

Penny: Great. So you seem pretty organized. You seem like you are getting things done but still have some work to do, but you have been doing things right along. (positive feedback, paraphrase of problem and assets)

Carter: Yeah, I really have. I am not sure why I feel so stressed because I know I will have the time to do it and I have been doing it so far but I get, I still see the deadline at the end and it is getting closer, so it feels a little stressful.

Penny: Have there been times in the past when you have felt this type of stress but have dealt with it in a positive way? (solution-oriented question seeking exceptions to the problem and past successes)

Carter: Sure, last semester or even when I was working. There have been times when it seemed like I had a lot to do. I had a very busy job. There are things I like to do when I have time to do them. I like to go dancing. I like to be active. I like to be social and that always . . . it is a real release for me. It is like freedom. You know.

(*Penny:* Right.) And then you can forget about it for a while. (*Penny:* Uh huh.) Then it is really good and then I get rejuvenated and I can come back and do what I need to do. OK. It is just finding the time to do that.

Penny: Oh, that is really terrific. So, in the past when you have been stressed, you have gone out dancing, you have done social things, and you have done other things to keep your mind from it and then you get more energized from it also. (positive feedback, paraphrase of strength in dealing with stress)

Carter: Yeah, it does really work.

Penny: And then you are able to focus. Oh, great. What is different about the times when you don't feel stressed out? (paraphrase, positive feedback in form of compliment, question searching for positive exceptions to the problem)

Carter: One of two things. Either I am using the technique to not be stressed out by doing all the social things I need and all of the good things and fun things I enjoy or there is less to do. There is not a crunch time or a deadline time. The summer. I guess when I feel the most organized, when I feel I have things under control, I feel less stressed.

Penny: OK, so when you feel organized and have things under control, you feel less stressed. (brief summary or paraphrase/reflection of feeling)

At this point, you may want to review the interview for focus. Note that virtually every one of Penny's comments focuses both on the client and on possible solutions to the problem.

Stage 3: Mutual Goal Setting

Carter: Right, and when I finish a project and when I see that it is completed and do one thing at a time, I feel less stressed. When I try to do three things and none of them are completed but I have done all this work it is still stressful. I guess when I finish and look at it and say oh, it's done. I did this, whatever task it may be.

Penny: You feel better when things are organized and you complete your papers. When you do a part of each of your projects but don't finish any complete class project, it doesn't feel so good because you don't feel like you have completed anything. (paraphrase, reflection of feeling)

Carter: Right, and it might be more work, but it doesn't look that way because I can't check it off the list. You know?

Penny: Yeah. Say you woke up tomorrow and this stress was miraculously gone, what would it be like? What would it look like? (This is a good time for the miracle question as we have an understanding of Carter's issues and her style. The miracle question often brings out new data, often unexpected, helping us find new solutions. We may find ourselves needing to totally redefine the problem or concern with data provided by the miracle question.)

Carter: I would have everything done and I would be going on vacation.

Penny: If you get everything done, then you could be on your vacation with your boyfriend, right? (paraphrase with check-out)

Carter: Yes, exactly.

Sometimes the miracle question doesn't produce much in the way of useful data at first. In this instance it might have been more productive if Penny had been more specific and had asked, "What would you be doing differently?" Penny could also have followed up for more information on the ideal resolution, particularly as Carter said, "Yes, exactly." Penny was on track and could have asked for more concreteness. Another possibility: "Could you be more specific? What's the first thing you would notice that would be different if the stress were gone?" It takes time and practice to make the miracle question work. Also, this is a point where the client and counselor can look at questions together seeking to elaborate mutually and make the miracle question more concrete, specific, and useful.

Penny: To recap, your goal has been to brainstorm and identify ways to deal with stress. We've identified some of the strengths in dealing with stress as your organization and your ability to do one thing at a time. And it helps, as you seem to be able to take time off and forget your studies for a while and enjoy yourself. Sounds like organization of your time these next few weeks will be important. (summary)

Carter: Yes, that's it. I guess my goal is to cool down a bit as I know I can do it. Then my boyfriend and I can be off on vacation for a week.

Stages 4 and 5: Working on the Concerns and Generalizing

Penny: Let us talk about some of your strengths, your strengths in the way you can deal with this stress. (directive)

Carter: I don't know. I am pretty positive about things. I know that I will finish it and I will get everything done. I think that is the strength. It is not a question of if I will do everything; it is just as I am in the moment, things get hectic. I am not a defeatist. I know I will get my work done and I know I will graduate. I know I will and I know what I need to do . . . um hum . . . and I know the things I like to do if I could carve out the time to do them and make sure that I take that time for myself. Then it will be better. So, I think that is a strength.

Penny: Yeah. Some of your strengths are that you are positive, that you know what you need to do, you know how to do it, and you know how to get there. I also heard you say you were organized before. What are some of your strengths in other parts of your life? (summary, open question that may lead to suggestions for dealing with stress)

Carter: I think those strengths also follow through in other areas of my life. That I am a positive person, that I like to try new things, and am adventurous. I really like life and I think that has always helped me. Right. I think that is a strength and that I can do things.

Penny: You like to try new things, are positive, and enjoy life. It is interesting. . . . (paraphrase)

Carter: I just want to do everything very, very well, so sometimes that gets in the way. I want to do it perfectly or as perfectly as I possibly can. Sometimes I feel like I am not doing my best and that bothers me. Even if it is stupid stuff. Even if I know I don't have to do the paper perfectly, I still try to. Right. Sometimes I just need to give myself a break.

Penny: You have some really great strengths, and with those strengths you were able to obtain your goal today, which was to brainstorm solutions to reduce your stress. Your positive attitude and willingness to do things and being organized and a risk taker will help you work through this stress. On a scale of 1 to 10 where do you see yourself in regard to your stress level at the moment? (Positive feedback with another compliment and summarization followed by scaling. Note that the client in the next statement responds to Penny's incomplete scaling question by defining the end points herself, a sign she understands the concept well.)

Carter: It is not too bad. Let's say 10 is the most and 1 was the least. I am probably a 5. I don't think it is that bad; talking about it makes it a lot easier. Like I said, it is more when it is in the moment and I have had a crazy day. I was working all day, had my classes, I come home, and I have seven things to do and there are three messages on the answering machine and I think, I can't do everything. I probably could do most of them, and then I have to map it out and prioritize but that is hard because it is so hard to say no. Especially when you want to do fun stuff. I guess it is a 5. Giving myself a break.

Penny: It seems like you are handling your stress pretty well. (positive feedback)

Carter: Thanks, yeah, it is not too bad. It is not as bad as it seems in those moments. You know?

Penny: Um-hum. So there are times when you have a higher stress level and times when you have a lower stress level. When are times when you are a 1? (Encourage, paraphrase, question, search for exceptions. Another useful question would be "So, what do you have to do to reduce your stress from a 5 to a 4?" This later question is an important part of the scaling process, particularly as it focuses on small change rather than total resolution.)

Carter: When I am not feeling any stress?

Penny: Yeah. (Encourage.)

Carter: When I can step away from responsibilities and play with my niece or go home and be with my family or out with my friends and dancing. I am a very active person, so I feel best when I am doing something outdoors or when I am moving or exercising. (Penny's question, focused on total removal of stress, works, but with many clients the smaller change from 5 to 4 would be more manageable.)

Penny: So, you feel stress-free when you are moving, or exercising, or playing, or are social. On a scale of 1 to 10, what would be your ideal stress level at this time? (Paraphrase; the client's stress patterns are examined by scaling.)

Carter: An ideal stress level would be a 3, because you can't always be playing. A certain amount of stress is good in order to be productive.

Penny: What do you want to happen precisely? (Move toward generalization with an open question.)

Carter: I have 4 weeks left of school. I think I need to map out the next 4 weeks in how I can balance my productive time workwise and my social outlets. I have six papers left to write and 4 weeks, so that averages to about one and a half papers per week. Now that I put it in that perspective and I see it visually, it doesn't seem so bad after all.

Penny: No, it doesn't. You have time to do your work and have time for your social outlets. You seem to have a good handle on what you need to do and the amount of time you have to do it. With your strengths, it appears that you will get your work done. You are positive. You know you can get your work done. You are motivated and organized. I like the way you have put it all in perspective. (summary, positive feedback in the form of a compliment)

Carter: Yes, I feel relieved already.

Penny: Let's now generate a picture of what the next 4 weeks are going to look like for you. (directive, concretizing the plan)

Carter: Well, I have four classes and have my assistantship, which is 10 hours per week. I also have these six papers and plenty of time to exercise and I can go out a few times over the next few weekends. My stress level has reduced tremendously already.

Penny: It sounds like you are doing a lot of things right already and your future plan is very attainable. How about support networks? Do you have people around you for a support system? (positive feedback, open question focusing on support network)

Carter: Yes, I have my sister and my brother. I also have my housemates and classmates. They are the biggest support because they are going through the same thing with me. I feel like I have a good network around me.

Penny: You sure do. What positive assets in yourself can you also draw from? (Encourage, return to focus on client and strengths.)

Carter: My drive and motivation. My positive outlook and energy level.

Penny: It sounds like you have a lot to draw from. So what are you going to do differently tomorrow? (paraphrase, open question oriented toward generalization)

Carter: I will exercise first thing in the morning and get started on one of my papers. This week I have a lot of time off, so I can probably get a few papers done and I can also go for a hike or something if the weather improves.

Penny: Well, this all sounds so clear to me. You know what you need to do. You have your timeline all mapped out and have a good balance of fun and work. Will you let me know in a week or so how it is all working out? (summary, open question)

Carter: Yes, I will call you next week.

INSTRUCTIONAL READING AND EXAMPLE INTERVIEW 4: MOTIVATIONAL INTERVIEWING

Counseling alcohol abusers, drug users, and clients with other forms of addictive behavior is recognized as extremely difficult and challenging, and out of these challenges came motivational interviewing (Miller & Rollnick, 1991, 2002). Particularly at issue is motivating the client to *actually change behavior* rather than just talking about it. Thus, capturing the interest, hopes, and ultimately the motivation to change is critical. In addition, research reveals that the system works and that motivational interviewing will be useful to clients who do not themselves have substance abuse issues.

Motivational interviewing (MI) could be described as a way to integrate the ideas of this book. MI is based on microskills such as those discussed in this book plus decisional theory and practice, in particular the work of Leon Mann (2001; Mann, Beswick, Allouache, & Ivey, 1989). In addition, it uses many of the methods and questions of solution-oriented counseling, is concerned with transfer of behavior from the interview to the real world, and uses the ideas of the Maintainance of Change and Relapse Prevention Worksheet of this chapter.

The "spirit" of motivational interviewing is based on *collaboration* with the client, *evoking* positive resources and motivation for change already in this client, and affirming client *autonomy* and self-direction. MI's four general principles for practice are *to express empathy, develop discrepancy, roll with resistance, and support self-efficacy* (Miller & Rollnick, 2002, pp. 33–42). The spirit of these principles should be basically familiar to readers. *Self-efficacy* is another term for intentionality, personal agency, and general wellness.

Let us now work through the five stages of the interview as they might be played out in an actual session of motivational interviewing. When you try MI for the first time, we suggest that you have these five stages with you as a guide. Feel free to share them with your client in the spirit of collaboration.

Stage 1: Initiating the Session

Motivational interviewing tends to start the session more rapidly than we would usually recommend. Note how Farah gets "down to business" immediately. How do you react to the following?

Farah: Good morning, Jerome. Today we've got a little less than an hour. At the beginning the most important thing is that I listen to what you have to say. I'd like to know what your concerns are and what you'd like to see as a result of being with me. There will be some details I'll ask you about as well. Perhaps we could begin with your telling me what you'd like to talk about.

Needless to say, this introduction will vary with the client. MI does not stress multicultural issues and we believe that the system would profit from more attention to individual and cultural variation. Thus we would suggest that you take whatever time is necessary to build sufficient rapport and trust before starting, although many clients will be "ready to go" as with this brief introduction by Farah. However, if a true collaborative interview is to follow, it may be important in some cases to discuss differences of ethnicity/race or gender and share something of yourself.

Let us assume that Jerome comes to Farah with an issue around alcohol abuse. He shares his story and tells Farah that his family has told him that he drinks too much and needs to get help. He says that he'd like to stop. MI theory recommends that we start working on change immediately. The counselor, Farah, brings out a paper with a 10-point scale and asks Jerome to indicate how important changing his behavior is to him. Jerome rated his interest as a 7—interested in change, but not fully committed.

The rating scale is an important "hook" in MI as it immediately introduces belief in the possibility of change and the counselor also learns of the client's depth of motivation. Another valuable by-product is that the interview will have a clear focus. Change has become the central goal at the very start. This rating scale can be useful in many sessions. Consider it a tool that will be useful to you and your clients.

Stage 2: Gathering Data—Drawing Out Stories, Concerns, Problems, and Issues

"Evocation" and "change talk" are two key words within MI. Through the use of open questions, reflective listening, and summarization; the client is encouraged to provide detail and elaboration on the change issue. In particular, change talk focuses on what is positive, enjoyable, or useful in the behavior in which change is sought. This is followed by the "downside" and what goes wrong when the behavior occurs. This ultimately leads to a decisional balance as the pluses and minuses of change are discussed.

Farah: (Elaborating the positives) Jerome, what do you particularly enjoy about drinking? What would you miss if you didn't drink? What is the negative side of drinking? What doesn't work so well for you?

Jerome: (Elaborates the positives) Well, I get out of the house. Carlotta is always on me. It makes me feel good and I forget about things for awhile. . . . (He continues)

Farah: (Elaborating the negatives) What is the negative side of drinking? What doesn't work so well for you?

Jerome: Actually, I worry sometimes about getting a little too angry when I drink. I get depressed the next day, say that I'm going to stop, but . . .

Throughout the session, the counselor maintains an open, nonjudgmental approach, seeking to empathically understand the world of the client. In addition, MI uses the word *affirmation,* which is parallel to the positive asset search. The counselor is always looking for something right in the client that can be affirmed and recognized.

Farah: As I hear you, Jerome, I can see that you have thought a lot about what drinking does for and against you. I like the way you want to control yourself and your behavior more. I also heard that your children would like you to stop and I sense your caring for them.

Further change questions focus on disadvantages of the status quo ("What is likely to happen if you don't stop drinking?"), advantages of change ("How would it be at home if your drinking stopped?"), optimism about change ("What are some supports that will help you maintain change?"), and intention to change ("Would you be willing to try stopping?")

Stage 3: Mutual Goal Setting

The positive and negative motivation for change is explored on a decisional balance sheet (Mann, 2001: Mann, Beswick, Allouche, & Ivey, 1989). Their original decisional balance sheet has been adapted by MI and is a straightforward decisional procedure in which the client and counselor list together the pros and cons (costs and benefits) of changing behavior.

Jerome's balance sheet came out as follows:

Continue drinking		Stop drinking	
(+)	(−)	(+)	(−)
Get out of house	Kids unhappy	With kids more	Miss drinking buddies
Away from wife	She may leave me	Perhaps better with her	Perhaps not
Depression	Maybe happier		
Problems disappear	Boss suspects that I drink	Lose weight	Not crazy about sitting around house
	Costly	Healthier	How will I deal with tension?
	Dad died of liver disease	Live longer	
	Not getting anywhere	Job better maybe	

Two classic questions or variations help set up the final goal, "What would you like it to be like in the future?" and "What are the future consequences of change?" We should again mention that motivational interviewing has a wide array of specific questions and techniques to accomplish the aims of each part of the MI procedures.

Jerome: Well, I guess I really would like to stop drinking and the positive consequences of stopping are all there on the balance sheet. Would that it were that easy! But, I guess I do really want to try changing.

Stage 4: Working—Exploring Alternatives, Confronting Client Incongruities and Conflict, Restorying

A particular strength of MI is the way resistance is defined and made more central in the interview. First of all, resistance is the major impediment to change. For example, an alcoholic may speak of the desire to change but in truth is resistant and unmotivated to do the work that change requires—thus, the importance of increasing motivation to change. Resistance is discussed as follows (Miller & Rollnick, 2002, p. 98):

> . . . resistance arises from the interpersonal interaction between counselor and client. . . . research clearly demonstrates that a change in counseling style can directly affect client resistance, driving it upward or downward. . . .

> (Resistance) is observable client behavior that occurs within the context of treatment and represents an important sign of dissonance within the counseling process. In a way, it is a signal that the person is not keeping up with you; it is the client's way of saying, "Wait a minute; I'm not with you; I don't agree."

When this happens, it is time to *listen* and discover what is going on with the client. Rather than ignoring resistance, MI suggests that counselors need to be prepared to try several different responses until one is effective. For example, paraphrasing or reflecting the feelings that acknowledge how the client thinks and feels is often the best place to start. Shifting interview focus or reframing the discussion from another perspective may be helpful. MI even suggests agreeing with the client and pointing out that the client is in control. In short, if you are skilled in microskills, be intentional and shift your style to meet changing client needs.

The MI philosophy around resistance and listening is important for us all. And note the importance of intentionally shifting your skill or focus when you work with challenging clients. "If the first skill or strategy doesn't work, you've still got another."

One interesting way to work with resistance is magnifying discrepancies and incongruity even more than the client has stated thus far or reframing them in a new perspective.

Jerome:	Carlotta worries too much.
Farah:	(Magnifying discrepant behavior) She has no need for concern at all. You're in charge of your drinking and she need not worry about you. (Rather than argue, the counselor extends Jerome's comment to another level, showing the limitations of this thought pattern.)
Farah:	(Reframing) Do I hear you saying that you'd like her to worry less about you? You seem concerned how she thinks about you.

Confidence talk is a vital part of motivational interviewing. Behavior change takes confidence and belief in oneself. The idea here is to help the client focus on specifics of change and how he or she can deal with them effectively. The wellness and positive asset search are good places to enhance client self-efficacy and competence. This can include reviewing past successes, assessing the strength of community supports (perhaps via a community genogram) and focusing on personal strengths. Affirming the capability of the client to change is important and that positive self-talk needs to enter Jerome fully before he can be expected to make major changes in his life.

After an emphasis on strengths and confidence, Jerome might be more optimistic around change.

Jerome:	I see what you're saying. I have done some good things at a high level in my life. My wife's family has been there to help over the years and the community services you tell me about might be useful. And I keep visualizing the little ones. Just focusing on my children sometimes gives me more strength.

Motivational interviewing proposes the *confidence ruler,* another 10-point scale in which the client rates her or himself on a zero to ten scale. The self-rating could be followed by asking the client to describe in concrete language what it is like to be at various levels on that scale.

Again, recall intentionality and the need to shift styles and tactics with the resistance client. Always have something else to try—and when all else fails—*listen!*

Stage 5: Terminating—Generalizing and Acting on New Stories

Motivation interviewing uses a *change plan worksheet.* But you will find that the Maintaining Change and Relapse Prevention Worksheet of Box 14-1 covers the same territory and meets the same purpose. Taking the time to complete such a form is very worthwhile as a basis for follow-up and action from the interview.

As part of the process of planning for generalization, further positive reframing, confidence building, and listening will be essential. It is particularly important at this stage, of course, to watch for resistance.

Summary of Motivational Interviewing

First, we are obviously not going to solve challenging issues such as alcohol abuse in one easy session. But the general structure above provides a basic model as to how counseling can be effective in this area. Visit Miller and Rollnick (2002) for research on the effectiveness of this model.

And, if you complete a practice session using MI, you will see how it integrates many of the ideas proposed in this book. With the microskills you have moved from basic attending and observation to beginning mastery of some tools and theories that can serve as a foundation for your entire professional career.

Again, when in doubt, *attend and listen carefully.* Then return to the wide array of strategies you have available and continue the process.

Box 14-4   Key Points

Five approaches	Table 14-1 should be reviewed because it summarizes the structure of decisional counseling, person-centered helping, behavioral assertiveness training, solution-oriented interviewing and counseling, and motivational interviewing. Though all these approaches may be explained by their use of microskills and how the interview is structured, note that their emphases are quite different. Decisional counseling emphasizes careful listening to the story/problem/concern/challenge of the client before acting whereas the solution approach emphasizes working on the problem as quickly as possible. Person-centered helping stresses listening to the client's feelings and story in detail, and thoughts and words are central. Behavioral assertiveness training, on the other hand, seeks to work with observable concrete behavior. Solution-oriented approaches focus on finding quick answers and using many questions while motivational interviewing appears to integrate most of the ideas of this book into a single package, potentially useful for particularly challenging clients.
Multicultural issues	Each style of interviewing requires different adaptations to be meaningful in multicultural situations. Particularly helpful in this regard is the concept of focus. By focusing on the cultural/environmental/contextual dimensions, you can bring in these issues fairly easily to all helping approaches. However, you still must recognize that the aims of each approach may not be fully compatible with varying cultures. This same point, of course, should be made with the client regardless of cultural background. *Some individuals may prefer the Rogerian person-centered approach; others may want solutions and behavioral action.* Avoid stereotyping any client with prior expectations.
Intentionality	We are suggesting that the intentional interviewer and counselor will have more than one interviewing alternative available. At the same time, it is important that you select those approaches to helping that are most comfortable for you. Balancing your knowledge, skills, and interests as you counsel varying clients will be a lifetime process of learning for any helping professional.

COMPETENCY PRACTICE EXERCISES
AND PORTFOLIO OF COMPETENCE

There is one central exercise for this chapter, a special type of practice interview. Seek out a classmate, friend, or colleague who is willing to work on a single issue for a half-hour to an hour. If that person has the interest, suggest that he or she read portions of this chapter to know what to expect. Alternatively, share the key points with him or her. This is an exercise in joint discovery. You may find that sharing the interview plan with your clients is a useful thing to do in other forms of counseling and interviewing.

The best way to understand the four approaches in this chapter is to use them in conducting an interview. As you meet with your volunteer client, share the interview plan and the key points of your plan. Consider working through your first interview with the book and notes on the table; both of you can use them for reference. You may wish to use the feedback form together as a way to summarize the specific steps.

Some topics that may be amenable to all four approaches include these:

Friendship, partner, or child difficulties
Relationship problems
School- or job-related issues
Family concerns
Getting started on career planning
Anger management
Expressing feelings more openly

In short, almost any topic can be useful but the critical issue is finding a manageable part of the larger problem for a single practice session.

Feedback forms for all four approaches follow. An observer and/or the client can provide feedback for you.

Box 14-5   Feedback Form: Person-Centered Interview

_____ (Date)

_____   _____

(Name of Interviewer)   (Name of Person Completing Form)

Stage 1. Initiating the session—Rapport and structuring ("Hello; this is what might happen in this session.") How well did the interviewer establish rapport and how did he or she accomplish this objective? Was structuring the session kept at a minimum?

Stage 2. Gathering data—Drawing out stories, concerns, problems, or issues ("What's your concern? What are your strengths or resources?") Was the interviewer able to conduct the session with a minimum of questions? Was at least one positive asset or wellness strength of the client identified? Were the client's emotions connected with the story adequately explored?

Stage 3. Mutual goal setting—("What do you want to happen?") Were the client's ideal self/world and real self/world defined? How were the goals of the client expressed?

Stage 4. Exploring alternatives, confronting incongruities and conflict, restorying ("What are we going to do about it?") Was interviewer able to assist client in exploring self in more depth? Were meaning issues considered to help the client gain deeper understanding?

Stage 5. Terminating—Generalizing and acting on new stories ("Will you do it?") This tends not to be emphasized unless the client brings the issue up.

General comments on interview and skill usage:

Box 14-6   Feedback Form: Assertiveness Training

_____ (Date)

_____   _____
(Name of Interviewer)   (Name of Person Completing Form)

Stage 1. Initiating the session—Rapport and structuring ("Hello; this is what might happen in this session.") How well did the interviewer establish rapport and how did he or she accomplish this objective? Was the session clearly structured?

Stage 2. Gathering data—Drawing out stories, concerns, problems, or issues ("What's your concern? What are your strengths or resources?") Was at least one positive asset or wellness strength of the client identified? Were the client's emotions connected with the story adequately explored? Did the interviewer engage the client in a role-play so that specific behaviors could be observed?

Stage 3. Mutual goal setting—("What do you want to happen?") Were highly specific goals for client behavioral changed established? Was the client fully involved in goal settings?

Stage 4. Working—Exploring alternatives, confronting client incongruities and conflict, restorying ("What are we going to do about it?") Was a second, more successful role-play enacted (and perhaps more, as necessary)? Was specific behavioral change emphasized? Could specific changes in the client be noted?

Stage 5. Terminating—Generalizing and acting on new stories ("Will you do it?") Was the relapse prevention form completed and a specific contract for behavioral generalization formed?

General comments on the session:

Box 14-7   Feedback Form: Solution-Oriented Interviewing and Counseling

_____ (Date)

_____   _____

(Name of Interviewer)   (Name of Person Completing Form)

Stage 1. Initiating the session—Rapport and structuring ("Hello; this is what might happen in this session.") How well did the interviewer establish rapport and how did he or she accomplish this objective? Were positive assets and wellness strengths identified and preliminary goals set early in the session?

Stage 2. Gathering data—Drawing out stories, concerns, problems, or issues ("What's your concern? What are your strengths or resources?") Was at least one positive asset or strength of the client identified? How completely did the interviewer draw out the story and/or issues?

Stage 3. Mutual goal setting—("What do you want to happen?") Were the original goals of the session reviewed and were the desired outcomes of the client really clear?

Stages 4 and 5. Working, terminating and generalizing. How did the interviewer go about helping the client develop a concrete plan for action? Did the interviewer help the client plan for generalization to daily life?

General comments on the interview and skill usage:

Box 14-8   Feedback Form: Motivational Interviewing

_____ (Date)

_____   _____

(Name of Interviewer)   (Name of Person Completing Form)

Stage 1. Initiating the session—Rapport and structuring ("Hello; this is what might happen in this session.") How well did the interviewer structure the interview? How did he or she establish rapport? Was the client's motivation for change explored? Did the client complete the 10-point motivation for change score?

Stage 2. Gathering data—Drawing out stories, concerns, problems, or issues ("What's your concern? What are your strengths or resources?") Was the basic listening sequence used to widen awareness with special attention to "change talk"? Were the positives and negatives of the area for potential change explored? Did the counselor offer sufficient client affirmations?

Stage 3. Mutual goal setting—("What do you want to happen?") Were positive and negatives around change explored on the decisional balance sheet? Was there some future pacing with the client such as "What are the future consequences of change?"

Stages 4. Working—exploring alternatives, confronting client incongruities and conflict, restorying ("What are we going to do about it?") Was the counselor able to maintain a client focus on change talk? How did the counselor handle resistance (be specific)? Was the client able to shift style and try another skill or tactic if resistance was met? Was listening continued? Was a "confidence ruler" brought in and were past strengths and successes of the client reviewed as support mechanisms?

Stage 5. Terminating—generalizing and action on new stories. How did the interviewer go about helping the client develop a concrete plan for action? Was the Maintaining Change and Relapse Prevention Worksheet used? Did the interviewer really help the client plan for generalization to daily life?

General comments on the interview and skill usage:

Portfolio of Competence

A checklist for your portfolio has not been developed for this chapter. At this point, developing and evaluating your own interviewing style becomes the major exercise for your portfolio.

DETERMINING YOUR OWN STYLE AND THEORY: CRITICAL SELF-REFLECTION ON FOUR THEORETICAL ORIENTATIONS

How does the concept of theoretical orientation relate to your own developing style and theory? Which of the four approaches presented most appeals to you? Do you agree with us that decisional counseling underlies most other approaches as a basic model?

We will not ask you to assess your competence in any of these approaches as it is far too early and you will want to work further with each one. Rather, please focus your attention on your early impressions and where you think you might go next in building competence in these or other theoretical orientations.

What single idea stood out for you among all those presented in this chapter, in class, or through informal learning? What stands out for you is likely to be important as a guide toward your next step. What are your thoughts on multicultural issues and the use of this skill? What other points in this chapter struck you as important? How might you use ideas in this chapter to begin the process of establishing your own style and theory?

_____

_____

_____

_____

_____

_____

_____

REFERENCES

Alberti, R., & Emmons, M. (2001). *Your perfect right: A guide to assertiveness training* (8th ed.). San Luis Obispo, CA: Impact. (Original work published 1970)

Bozarth, J. (1999). *Person-centered therapy: A revolutionary paradigm.* Ross-on-Wye, UK: PCCS Books.

Budman, S., & Gurman, A. (2002). *Theory and practice of brief therapy.* New York: Guilford.

de Shazer, S. (1985). *Keys to solution in brief therapy.* New York: Norton.

de Shazer, S. (1988). *Clues: Investigating solutions to brief therapy.* New York: Norton.

de Shazer, S. (1993). Creative misunderstanding: There is no escape from language. In S. Gilligan & R. Price (Eds.), *Therapeutic conversations.* New York: Norton.

Hersen, M., & Biaggio, M. (Eds.). (2000). *Effective brief therapies: A clinician's guide.* New York: Academic Press.

Ivey, A., D'Andrea, M., Ivey, M., & Simek-Morgan, L. (2002). *Theories of counseling and psychotherapy: A multicultural perspective* (5th ed.) Boston: Allyn & Bacon.

Mann, L. (2001). Naturalistic decision making. *Journal of Behavioral Decision Making, 14,* 375–377

Mann, L., Beswick, G., Allouache, P., & Ivey, M. (1989). Decision workshops for the improvement of decision making skills. *Journal of Counseling and Development, 67,* 237–243.

Miller, W., & Rollnick, S. (1991). *Motivational interviewing: Preparing people to change addictive behavior.* New York: Guilford.

Miller, W., & Rollnick, S. (2002). *Motivational interviewing: Preparing people for change.* New York: Guilford.

Monk, G., Winslade, J., & Epston, D. (1997). *Narrative therapy in practice: The archaeology of hope.* San Francisco: Jossey-Bass.

Rogers, C. (1961). *On becoming a person.* Boston: Houghton Mifflin.

Semmler, P., & Williams, C. (2000). Narrative therapy: A storied context for multicultural counseling. *Journal of Multicultural Counseling and Development, 28,* 51–62.

Sklare, G. (2004). Brief counseling that works: A solution-focused approach for school counselors and administrators. Beverly Hills, CA: Corwin.

White, M., & Epston, D. (1992). *Narrative means to therapeutic ends.* New York: Norton.

Witkiewitz, K., & Marlatt, G., (2004). Relapse prevention for alcohol and drug problems. *American Psychologist, 59,* 224–235.

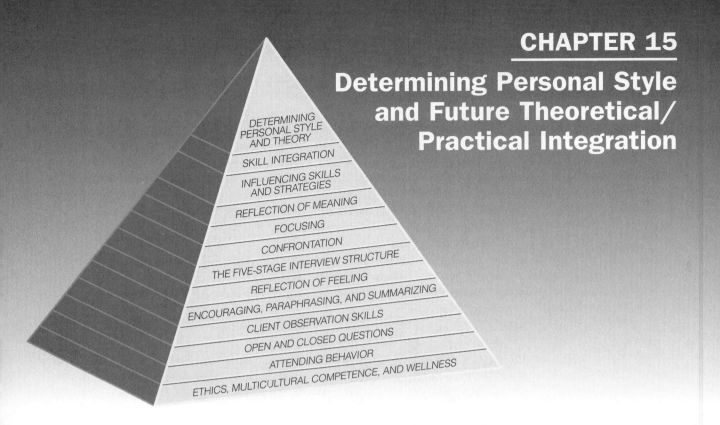

DETERMINING
PERSONAL STYLE
AND THEORY

SKILL INTEGRATION

INFLUENCING SKILLS
AND STRATEGIES

REFLECTION OF MEANING

FOCUSING

CONFRONTATION

THE FIVE-STAGE INTERVIEW STRUCTURE

REFLECTION OF FEELING

ENCOURAGING, PARAPHRASING, AND SUMMARIZING

CLIENT OBSERVATION SKILLS

OPEN AND CLOSED QUESTIONS

ATTENDING BEHAVIOR

ETHICS, MULTICULTURAL COMPETENCE, AND WELLNESS

How can determining your own personal style help you and your clients?

How do theoretical alternatives for practice relate to you and your clients?

Major function	This chapter has the following as its major purpose: You are asked to review your work with microskills and consider your own orientation to interviewing and counseling practice. You can be most effective as an interviewer, counselor, or therapist if you generate your own formulation of the helping process. What is your natural style and how does it relate to skills, interviewing structure, and alternative theoretical orientations?
Secondary functions	Developing your own personal style and theory and reflecting on the future may result in the following:

▲ A greater awareness that interviewing and counseling are not simple but incredibly complex processes requiring movement, change, and constant growth. At the same time, understanding the skills and concepts of this book provides a framework that will enable you to master this complexity. You can meet the many challenges ahead from a solid base.

▲ An awareness of the rich array of theories and concepts available to you and your clients that may be used to facilitate both your own and your clients' development.

INTRODUCTION: FINDING YOUR AUTHENTIC STYLE

Developing your own personal approach to interviewing, counseling, and therapy involves a multiplicity of factors. At this moment we are asking you to reflect on yourself, your values and personal meanings, and your skills. What are your strengths? What areas need further development? Where would you like to go?

Some counselors and therapists have developed individual styles of helping that require their clients to join them in their orientation to the world. Such individuals have found the one "true and correct" formula for counseling and therapy; clients who have difficulty with that formula are often termed "resistant" and "not ready" for counseling. These counselors do indeed have their own style, but their methods tend to be rigid and dogmatic. Such counselors and therapists can and do produce effective change, but they may be unable to serve many client populations. Missing from their orientation is an understanding of the complexity of humanity and the helping process.

Thus as you generate your own approach to the field, remember that clients may have widely different experiences from yours and may wish to head in many directions different from your own. Consequently, it is urgent that you remain aware that the skills, strategies, and theories that you favor may not be preferred by your client. If so, then you may wish to expand your skills and knowledge in areas where you are now less comfortable. Be willing to join the language and style of your client.

And as you continually expand your competence, maintain your authenticity as a person. With study, patience, and experience, you will increase your abilities to work with those who are unlike you. The opportunity to learn from clients different from us is one of the special privileges of being an interviewer, a counselor, or a therapist. You will want to expand your understanding of cultural differences including race/ethnicity, gender, spiritual/religious orientation, disability, sexual orientation, age, socioeconomic status, and other factors.

This chapter presents some structure for the evolving identification of your own personal style. To that end, the microskills hierarchy will be reviewed, but this time as a complex matrix for decision making on your part. The purpose now is to remind you just how many skills and concepts you have learned in this book. A brief map showing how these skills and concepts play themselves out in different modes of counseling and therapy should aid you in the search for your personal style.

The chapter concludes with a review of the five-stage structure of the interview, and the story—positive asset—restory—action models. This in turn leads to recommended readings for your future development in skills and theory.

INSTRUCTIONAL READING: DEFINING YOUR PERSONAL STYLE

The Microskills Hierarchy: A Summary

You have thus far been presented with 14 chapters of reading containing 36 major concepts and skill categories. Within those major divisions are more than 100 specific methods, theories, and strategies. Ideally, you will commit them all to memory and be able to draw on them immediately in practice to facilitate your clients' development and progress in the interview.

These skills, concepts, and understandings are but a beginning. We have not considered here details of personality development, testing, and the many theories of counseling and therapy you will encounter in the future. Although these skills and training concepts are used in fields as varied as medicine, management, social work, and parent education, and with many cultural groups, only a brief introduction to the multitude of applications has been provided.

How can you manage to retain and use all of this book's concepts? Most likely, you cannot at this point. But recall the story of the Samurai (page 79). With time and experience, you will develop increased understanding and expertise. As you grow as an interviewer, counselor, or therapist, you will find the ideas expressed here becoming increasingly clear, as your mastery of skills and theories will likewise continually increase.

In addition, your retaining and mastering the concepts of the microskills hierarchy may be facilitated by what is termed *chunking*. We do not learn information just in bits and pieces; we organize it into patterns. The microskills hierarchy is a pattern that can be visualized and experienced. For example, at this moment you can probably immediately recall that attending behavior has certain major concepts "chunked" under it (the "three V's + B" of culturally appropriate visuals, vocal tone, verbal following, and body language). The basic listening sequence will be easy to recall and essentially covers the first half of this book. You can probably also recall the purpose of open questions and perhaps which questions lead to which likely outcomes (for example, *how* questions lead to process and feelings).

With periodic review and experience, the many concepts will become increasingly familiar to you. If you complete a transcript examining and classifying your own interviewing style similar to the one presented in Chapter 13, the ideas of this book will become especially clear. As you must have noticed, the skills you have practiced instead of just read about are the ones you understand best and that have the most relevance for you. Reading is a useful introduction to counseling and interviewing, but the results of practice and experience will stick with you far into the future.

Table 15-1 summarizes the major concepts of the book; take some time to review your own competence levels in each of the 36 major areas and enter the results on the table. Can you do the following?

1. *Identify and classify the concept?* If the idea or behavior is present in an interview, can you label it? These concepts provide a vocabulary and communication tool with which to understand and analyze your interviewing and counseling behavior and that of others. The ability to identify concepts means most likely that you have chunked most of the major points of the skill together in your mind. You may not immediately recall all seven types of focus, but when you see an interview in progress you will probably recall which one is being used.
2. *Demonstrate basic competence?* At this level, you will be able to understand and practice the concept. Continued practice and experience forms the foundation for later intentional mastery.
3. *Demonstrate intentional competence?* Skilled interviewers, counselors, and therapists can use the microskills and the concepts of this book to produce specific, concrete effects with their clients. Please go back to the Ivey Taxonomy (Box 8-1, page 213) and review specific aspects of intentional prediction. If you reflect feelings,

Table 15-1   Self-assessment summary

Skill or Concept	Identification and classification	Basic competence	Intentional competence	Teaching competence	Evidence of Achieving Competence Level
1. Attending behavior					
2. Questioning					
3. Observation skills					
4. Encouraging					
5. Paraphrasing					
6. Summarizing					
7. Reflecting feelings					
8. Basic listening sequence					
9. Positive asset search					
10. Empathy					
11. Five stages of the interview					
12. Story—positive asset—restory—action					
13. Confrontation					
14. The Confrontation Impact Scale					
15. Focusing					
16. Reflection of meaning					
17. Noting concreteness and abstractions in self and clients					
18. Interpretation/reframe					
19. Awareness of logical consequences					
20. Self-disclosure					
21. Feedback					
22. Information/advice/opinion/ instruction/suggestion					

(continued)

Table 15-1   (continued)

Skill or Concept	Identification and classification	Basic competence	Intentional competence	Teaching competence	Evidence of Achieving Competence Level
23. Directives					
24. Analysis of the interview (Chapter 13)					
Theoretical/Practical Strategies 25. Wellness					
26. Ethics					
27. Multicultural awareness					
28. Family genogram					
29. Community genogram					
30. Decisional counseling					
31. Person-centered counseling					
32. Assertiveness training					
33. Solution-oriented interviewing and counseling					
34. Motivational Interviewing					
35. Teaching skills to clients					
36. Defining personal style and theory					

do clients actually talk more about their emotions? If you provide an interpretation, does your client see her or his situation from a new perspective? If you work through some variation of the positive asset search, does your client actually view the situation more hopefully? If you conduct a well-formed, five-stage interview, does your client's self-concept and/or developmental level change? Do the client's behavior and thinking change?

Intentional competence shows up not in your use of the skills and concepts, but rather in what your client does. To demonstrate this level of skill, you need to be able to produce *results* due to your efforts in helping.

4. *Demonstrate teaching competence?* You are not expected to master teaching all the skills and concepts of this book at this point. However, you should have learned

that you can teach attending behavior and the basic listening sequence to your clients in the interview. Some of you may have had the opportunity to take these skills and conduct teaching workshops with community volunteers, church groups, and peer counselor training programs.

For the longer term, we would suggest that you continue to think of the possibility of teaching skills to clients and their families. And if you become a professional, you most likely will find yourself teaching listening skill workshops at some point in your career.

Alternative Theories of Interviewing and Counseling

Equipped with the foundation skills of listening, observing, influencing, and structuring an interview, you are well prepared to enter the complex world of theory and practice. Soon you will be encountering the 250 or more theories competing for your attention. This final discussion is oriented to assembling what you already know, bringing it together in one place, and looking toward the future.

If you can complete the five-stage interview structured around decisions, you have a good beginning understanding of the work of many professionals. Decisional counseling has been around for a long time. Frank Parsons reminded us 100 years ago that much of our interviewing is about helping people make decisions. Some claim that decisional counseling (sometimes known as "problem-solving counseling") is the most widely practiced form of helping in current practice. The following is just a sample of places in which decisional or problem-solving counseling is practiced daily: social work, employment counseling, placement counseling, AIDS counseling, alcohol and drug counseling, school and college counseling, and in the work of community volunteers and peer helpers.

Having completed a full interview using only listening skills, you have had a beginning introduction to Carl Rogers's (1961) person-centered theory. There you likely discovered that many clients are self-directed and with a good listener, they can do much to resolve their issues on their own. Mastery of this important strategy does not make you a person-centered counselor. You will want to study Rogers's work in detail. A few additional key suggestions are presented briefly in this chapter. Consider completing an interview using *no questions* at all, a very different task from that used in solution and brief approaches.

Behavioral assertiveness training presented a very different approach to how clients make decisions about their lives. The emphasis here was on observable behavior and making decisions to change actions.

Chapter 14 also provided you with the opportunity to engage in solution-oriented approaches and the newly popular motivational interviewing. There you may have found that your listening skills and the positive asset search enabled you to conduct a very different type of decision-making interview. Questions, of course, are the central skill of these two models. You have seen five quite distinct interviewing approaches. Yet if you have mastered basic skills and strategies, you are well prepared to increase your mastery in other theoretical/practical strategies.

Box 15-1   National and International Perspectives: Using Microskills
Throughout My Professional Career

Mary V. Brodhead, Ed.D., Director, Canadian Food Inspection Agency

My first encounter with the microskills program was through my doctoral program in teacher education. I wanted to learn about counseling so that I could better reach students, particularly those at risk who often were the least likely to approach a counselor. What I learned very rapidly was that teachers often fail to listen to their students; in fact, one research study found that out of nearly 2,000 teacher comments, there was only one reflection of feeling and, of course, most comments focused on providing information. Often, even when teachers did listen, they weren't able to recognize the verbal and nonverbal cues that were reflecting the reality of their students.

After completing my degree, I entered teacher education and found that microskills lead to a more student-centered teaching. I also found that teachers who matched their students' cognitive/emotional style were the more effective (see Chapter 7). If a student is concrete, the teacher needs to provide specific examples and use concrete questions. If the student is more reflective, then formal operational strategies can be used. Bringing in emotional involvement via sensorimotor strategies enriched teaching.

I next lived on an island off Vancouver Island where my husband Dal and I worked with members of the Kwakiutl Nation. I trained teachers, but I also was involved in counseling and established an alternative program, as part of the local School Board, for youth who had dropped out of high school. I actually taught them the same interviewing skills that you are learning in this book. Needless to say, the multicultural orientation here helped sensitize me to important differences among people.

One of our first activities was a trip in the Chief's fishing boat to gather Christmas trees for the old people, a cherished tradition in the community. This was the first time most of these young people had participated in the ritual. As they delivered the trees, the recipients, in an expression of gratitude, invited them in for something to eat and began to tell stories. These old people, thrilled to have an audience, would shower attention on these alienated youth who, in response, would listen with respect. And so an upward cycle of communication began. Eventually we raised money to buy tape recorders and tried to capture these tales on tape (a common-place activity now, but quite new in those days), leading to further strengthening of the students' listening and questioning skills. In this case, attending behaviors led to an increased sense of value and respect on both sides.

After three years in British Columbia, my husband and I returned to Ottawa where I worked in a federal government employment equity program. The task

(continued)

Note the following key points as you think about the four theoretical/practical methods summarized above:

1. Different theoretical/practical systems use varying patterns of microskills. This could range from almost all the questions in the solution approach to no questions at all with the person-centered orientation.

Box 15-1   (continued)

centered around providing culturally sensitive counseling and opportunities to members of the Canadian four "equity groups" (visible minorities, Aboriginal peoples, persons with disabilities, and women in nontraditional occupations) to develop the skills needed for career development and success in the Federal Public Service.

As part of my work, I used microskills to train government officials to listen and to really hear and understand the variety of perspectives, strengths, and styles of working found within members of our multicultural workforce. Here, I also learned the importance of language and eventually became reasonably fluent in French, an essential skill for success in a bilingual, multicultural Canada. Our team developed a multicultural counseling course, used—all or in part—within 15 universities in Canada, that included many ideas presented here.

The agency where I now work is responsible for animal health, plant protection, and food safety. It is one of the "science departments," involving agriculture, fisheries, scientific research, and many other issues. Given the diversity of our workforce, my first task with my team has been to build a "culture of learning" where vast amounts of information and knowledge can be shared effectively and efficiently. Microskills are a key element in management training, and a good communication skills workshop can be vital in team building. And, of course, all our managers need to listen to and motivate those with whom they serve.

Looking back over my career, it is amazing to find that the basic microskills have been useful in my teaching, counseling, multicultural work, and governmental leadership positions. "Training as treatment" and "teaching competence" in microskills can help us all make a difference throughout our careers.

Allen and Mary Ivey comment: We asked Mary Rue Brodhead to share her experience, as we have been constantly impressed with her ability to innovate. She uses microskills as a base for many types of work. This book, of course, focuses on counseling and interviewing. But she reminds us that not all of you will be doing counseling, human relations, psychology, and social work for the rest of your lives. We think she makes an important point: Regardless of where you are personally or careerwise, it remains important that you listen.

And you will always have the potential to share what you have learned with others. We have taught microskills workshops on listening skills with elementary peer counselors, with students in church groups, medical staff, and many others. UNESCO has used the framework in Africa to teach peer counselors and conduct workshops with refugees. Managers throughout the world have been trained in microskills. Librarians, farm extension workers, and salespeople have participated in this training. We'll be interested in knowing what you are doing with these ideas 10 and 20 years from now.

2. Theoretical/practical systems tend to focus on different areas of meaning and have different stories they would tell about the interviewing, counseling, and therapeutic process.

▲ Decisional counseling focuses on making decisions about practical life issues and problems.

Table 15-2 Microskills patterns of differing approaches to the interview

Legend: ● Frequent use of skill   ◐ Common use of skill   ○ Occasional use of skill

	Decisional counseling	Person centered	Cognitive behavioral assertiveness training	Solution oriented	Motivational interviewing	Psychodynamic	Gestalt	Rational-emotive behavioral therapy	Feminist therapy	Business problem solving	Medical diagnostic interview	Traditional teaching	Student-centered teaching	Eclectic/metatheoretical
BASIC LISTENING SKILLS														
Open question	●	○	◐	●	●	◐	●	◐	◐	◐	●	◐	◐	◐
Closed question	◐	○	●	◐	◐	○	◐	◐	◐	◐	●	○	◐	◐
Encourager	●	◐	◐	◐	●	◐	◐	◐	◐	◐	○	◐	◐	◐
Paraphrase	●	●	◐	●	◐	◐	○	◐	◐	◐	◐	○	●	◐
Reflection of feeling	●	●	◐	◐	●	◐	○	◐	◐	◐	◐	○	●	◐
Summarization	◐	◐	◐	●	●	◐	○	◐	◐	◐	◐	○	●	◐
INFLUENCING SKILLS														
Reflection of meaning	◐	●	○	○	○	◐	○	◐	●	○	○	○	◐	◐
Interpretation/reframe	◐	○	○	○	●	●	●	●	◐	◐	◐	○	◐	◐
Logical consequences	◐	○	◐	○	●	○	○	●	◐	●	◐	●	◐	◐
Self-disclosure	◐	◐	○	○	◐	○	◐	○	●	◐	○	○	●	◐
Feedback	◐	◐	◐	◐	●	○	◐	◐	◐	●	◐	○	●	◐
Advice/information/ and others	◐	○	◐	○	◐	○	○	●	●	●	◐	●	●	◐
Directive	◐	○	●	◐	◐	○	●	●	◐	●	●	●	◐	◐
CONFRONTATION (Combined Skill)	◐	◐	◐	◐	●	◐	●	●	●	◐	◐	◐	◐	◐
FOCUS														
Client	●	●	●	●	●	●	●	●	◐	◐	◐	◐	●	◐
Main theme/problem	●	○	◐	●	●	◐	○	◐	◐	●	●	●	●	◐
Others	◐	○	◐	◐	◐	◐	◐	○	◐	○	○	○	◐	◐
Family	◐	○	◐	◐	◐	◐	◐	○	◐	○	◐	○	○	◐
Mutuality	○	◐	◐	◐	◐	○	○	○	◐	○	○	○	◐	◐
Counselor/interviewer	○	◐	○	○	◐	○	○	◐	◐	○	○	○	◐	◐
Cultural/ environmental context	◐	○	◐	◐	○	○	○	○	●	◐	○	○	◐	◐
ISSUE OF MEANING (Topics, key words likely to be attended to and reinforced)	Problem solving	Relationship	Behavior problem solving	Problem solving	Change	Unconscious motivation	Here-and-now behavior	Irrational ideas/logic	Problem as a "women's issue"	Problem solving	Diagnosis of illness	Information/ facts	Student ideas/ info./facts	Varies
AMOUNT OF INTERVIEWER TALK-TIME	Medium	Low	High	Medium	Medium	Low	High	High	Medium	High	High	High	Medium	Varies

LEGEND

● Frequent use of skill

◐ Common use of skill

○ Occasional use of skill

▲ Person-centered counseling emphasizes how the client understands himself or herself and makes meaning of life.

▲ Behavioral assertiveness stresses observable behaviors that can be seen directly, counted, and changed.

▲ Solution-oriented counseling uses client assets and resources to resolve issues.

▲ Motivational interviewing appears to combine all of the above in an integrated fashion. Its focus on increasing client motivation for actual change is important.

Table 15-2 illustrates once again in summary form how the microskills may be used in different approaches to the interview or group situation. The table shows that widely varying styles of helping may be understood and practiced via the microskills system.

Before you review Table 15-2, indicate below your preferred skills—the ones that you have most enjoyed and/or those that you favor and believe that you have mastered most completely. Rate each skill on a 3-point scale, with 1 representing the skills you most prefer and would like to use most often; 2, those skills that you would commonly use; and 3, those skills you would prefer to use only occasionally. You can also fill in the circles as noted in the legend at the bottom of page 490.

	MICROSKILL LEAD	Skill Preferences on 3-Point Scale			Fill in the Circle as in Table 15.2
BASIC	Open question	1	2	3	○
LISTENING	Closed question	1	2	3	○
SKILLS	Encourager	1	2	3	○
	Paraphrase	1	2	3	○
	Reflection of feeling	1	2	3	○
	Summarization	1	2	3	○
INFLUENCING	Reflection of meaning	1	2	3	○
SKILLS	Interpretation/reframe	1	2	3	○
	Logical consequences	1	2	3	○
	Self-disclosure	1	2	3	○
	Feedback	1	2	3	○
	Advice/information/and others	1	2	3	○
	Directive	1	2	3	○
	Confrontation (combined skill)	1	2	3	○
SKILL FOCUS	Client	1	2	3	○
	Main theme/problem	1	2	3	○
	Other	1	2	3	○
	Family	1	2	3	○
	Mutuality	1	2	3	○
	Counselor/interviewer	1	2	3	○
	Cultural/environmental context	1	2	3	○
MEANING	Which issues of meaning would you most like to address?				
TALK-TIME	What amount of talk-time do you believe is appropriate?		High	Medium	Low
THEORY	Which theory (theories) do you prefer?				

Key: 1 = favored skill, 2 = commonly used skill, 3 = occasionally used skill

Box 15-2   Your Natural Style and Story of Interviewing and Counseling

	The following issues are presented for you to consider as you continue the process of identifying your natural style and future theoretical/practical integration of skills and theory.
Goals	What do you want to happen for your clients as a result of their working with you? What would you *desire* for them? How are these goals similar to or different from those of decisional, solution, person-centered, and behavioral assertiveness training? What else?
Skills and strategies	You have identified your competence levels in these areas. What do you see as your special strengths? What are some of your needs for further development in the future? What else?
Cultural intentionality	With what cultural groups and special populations do you feel capable of working? What knowledge do you need to gain in the future? How aware are you of your own multicultural background? What else?
Theoretical/practical issues	What theoretical/practical story would you provide now that summarizes how you view the world of interviewing and counseling? Where next would you like to focus your efforts and interests? What else?

Then compare your preferred skill pattern with those presented there. Where are you headed in terms of theoretical preference? What other theories not presented here might you also consider and perhaps already prefer?

Your Personal Style and Future Theoretical/Practical Integration

There are two major factors to consider as you move toward identifying your own personal style and integrating the many available theories: your own personal authenticity and the needs and style of the client. Unless a skill or theory harmonizes with who you are, it will tend to be false and less effective. It is also obvious that modifying your natural style and theoretical orientation will be necessary if you are to be helpful to many clients.

In summary, remember that you are unique, and so are those whom you would serve. We all come from varying families, differing communities, and distinct views of gender, ethnic/racial, spiritual, and other multicultural issues.

If you have presented and analyzed a transcript of an interview as recommended in Chapter 13, you have an excellent beginning for understanding yourself and how clients relate to you. If you are able to engage in the five theoretical/practical methods of this book, you have at least beginning competency in issues of decisions and solutions, meaning-making, and behavioral specifics.

At this point it may be useful to summarize your own story of interviewing and counseling. Box 15-2 asks you to review your goals, your special skills, and your plans for the future. Particularly important are your plans for the future. Where are you going next?

Box 15-3 Key Points

Personal style and awareness of the client	Your arrival at the top of the microskills hierarchy offers you a time to reflect, examine your own style, and examine the appropriateness of your style for the highly diverse clients whom you will meet. This closing chapter suggests that you spend considerable time determining your own natural style. However, clients also have a natural style of being; you will want to assess and respect their styles as well. You may decide to operate from a single theoretical orientation or you may decide to offer your clients an array of skills, strategies, and theories. It is critical that you be aware of and respect both yourself and your clients as you develop a natural, integrated style of helping.
Competence with the microskills hierarchy	See Table 15-1 for a summary of the many skills and concepts in this book. Completing this table will help you develop an awareness of your skills and knowledge. Furthermore, the table can provide helpful insights into where you might want to go next in your development of skill competence.
Alternative theories of helping	This book is designed to equip you to complete five types of interviews: Rogerian or person-centered, decisional, behavioral assertiveness training, solution-oriented, and motivational interviewing. If you complete the practice exercises, you will be able to conduct an interview using all five systems. Furthermore, with an understanding of skills and the decisional structure, you will be able to engage in other theories more directly and analyze how they function for client benefit. This chapter has also specifically shown how the basic listening sequence might be used differently with the same client, depending on the theoretical orientation of the counselor. Multiple theories of interviewing and counseling can be viewed through the lens of the five-stage interview structure and the story—positive asset—restory—action model.

SUMMARY

We have come to the end of this phase of the interviewing and counseling journey. You have had the chance to learn the foundation skills and how they are structured in a variety of theoretical/practical interviews. Skills that once seemed awkward and unfamiliar are likely now automatic, and you need not think of them too often. The basic listening sequence, we hope, has become an important part of your interviewing practice and way of thinking.

The next steps are yours. Many of you will be moving on to individual theories of counseling, exploring issues of family counseling and therapy, becoming involved in the community, and learning the many aspects of professional practice. Others may find this presentation sufficient for their purposes. We have designed this book as a clear summary of the basics—a naturally skilled person can use the information here for many effective and useful helping sessions.

We have selected a few key books that will help you follow up ideas from our presentation. The books will take you in very different directions, but we have found them all helpful. Select one or two favorites to start and then expand from there. Enjoy delving more deeply into our exciting field.

We have enjoyed sharing this time with you. We have come a long way together, and we appreciate your patience. Many of the ideas for presentation in this book have come from students. We hope you will take a moment to provide us with your feedback and suggestions for the future. Again, e-mail us at info@ emicrotraining.com and say hello. This book will be constantly updated with new ideas and information. Let us grow together with new discoveries. You have joined a never-ending time of growth and development. Welcome to the field of interviewing and counseling!

SUGGESTED SUPPLEMENTARY READINGS

The literature of the field is extensive, and you will want to sample it on your own. We would like to share a few specific books that we find helpful as next steps to follow up ideas presented here. All of the materials build on the concepts of this book, but we have recommended some books that take different perspectives from our own.

Microskills

www.emicrotraining.com

Visit this web site for up-to-date information on microcounseling, microskills, and multicultural counseling and therapy. There are links to professional associations, ethics, codes, and many multicultural and professional sites.

Daniels, T., & Ivey, A. (2006). *Microcounseling* (3rd ed.). Springfield, IL: Thomas.

The theoretical and research background of microskills is presented here in detail.

Evans, D., Hearn, M., Uhlemann, M., & Ivey, A. (2004). *Essential interviewing* (6th ed.). Pacific Grove, CA: Brooks/Cole.

Microskills in a programmed text format.

Ivey, A., Gluckstern, N., & Ivey, M. (2005). *Basic attending skills* (4th ed.). North Amherst, MA: Microtraining Associates.

Perhaps the most suitable book for beginners and those who would teach others microskills. Supporting videotapes are available (www.emicrotraining.com).

Ivey, A., Gluckstern, M., & Ivey, M. (1997). *Basic influencing skills* (3rd ed.). North Amherst, MA: Microtraining Associates.

More data on the influencing skills. Supporting videotapes are available.

Theories of Interviewing and Counseling With a Multicultural Orientation

Ivey, A., D'Andrea, M., Ivey, M., & Simek-Morgan, L. (2002). *Counseling and psychotherapy: A multicultural perspective* (5th ed.). Boston: Allyn & Bacon.

The major theories are reviewed, with special attention to multicultural issues. Includes many applied exercises to take theory into practice.

Thomas, R. (2000). *Multicultural counseling and human development theories: 25 theoretical perspectives.* Springfield, IL: Charles C Thomas.

Multiple orientations are presented in a comprehensive fashion.

Suggestions for Follow-Up on Specific Theories

Decisional Counseling

Brammer, L., & McDonald, G. (2004). *The helping relationship: Process and skills* (8th ed.). Boston: Allyn & Bacon.

An articulate discussion of skills and the helping process with a strong decisional flavor.

D'Zurilla, T. (1999). *Problem-solving therapy* (2nd ed.). New York: Springer.

An entire counseling model derived from decision making.

Parsons, F. (1967). *Choosing a vocation.* New York: Agathon. (Originally published 1909)

It is well worth a trip to your library to read Parsons's work. You will find that much of his thinking is still up to date and relevant.

Solution-Oriented Interviewing and Counseling

Connell, B. (2005). *Solution-focused therapy.* Beverly Hills, CA: Sage.

The basics of brief counseling in brief form.

Sklare, G. (2004). *Brief counseling that works* (2nd ed.). Beverly Hills, CA: Corwin.

School-focused and clear. Also available—videotape of a real interview with a child (www.emicrotraining.com)

Person-Centered Counseling and Humanistic Theory

Frankl, V. (1959). *Man's search for meaning.* New York: Pocket Books. (Originally published 1946)

This is one of the most memorable books you will ever read. It describes Frankl's survival in German concentration camps through finding personal meaning. You will find that referring your future clients to this book is an excellent counseling and therapeutic tool in itself.

Rogers, C. (1961). *On becoming a person.* Boston: Houghton Mifflin.

The classic book by the originator of person-centered counseling.

Cognitive-Behavioral Counseling

Alberti, R., & Emmons, M. (2001). *Your perfect right: A guide to assertiveness training* (8th ed.). San Luis Obispo, CA: Impact. (Originally published 1970)

The classic book by the originators.

Davis, M., Eshelman, E., & McKay, M. (2000). *The relaxation and stress reduction workbook* (5th ed.). Oakland, CA: New Harbinger.

One of many books on cognitive and behavioral counseling. This is clear, direct, and easily translatable into microskill approaches to the interview.

Dobson, K. (Ed.). (2002). *Handbook of cognitive-behavioral therapies* (2nd ed.). New York: Guilford.

A comprehensive and specific presentation of key skills, strategies, and theories.

Wolpe, J., & Lazarus, A. (1966). *Behavior therapy techniques.* New York: Pergamon.

Old, but in many ways still the best and worth searching for in the library.

Multicultural Counseling and Therapy

Sue, D. W., Carter, R., Casas, M., Fouad, N., Ivey, A., Jensen, M., LaFromboise, T., Manese, J., Ponterotto, J., & Vazquez-Nuttall, E. (1998). *Multicultural counseling competencies.* Beverly Hills, CA: Sage.

The most comprehensive coverage of the necessary skills and competencies in the multicultural area.

Sue, D. W., Ivey, A., & Pedersen, P. (1999). *A theory of multicultural counseling and therapy.* Pacific Grove, CA: Brooks/Cole.

A general theory of MCT with many implications for practice.

Sue, D. W., & Sue, D. (2003). *Counseling the culturally diverse* (4th ed.). New York: Wiley.

The classic of the field. This book helped launch a movement.

Integrative/Eclectic Orientations

Ivey, A., Ivey, M., Myers, J., & Sweeney, T. (2005). *Developmental counseling and therapy: Promoting wellness over the lifespan.* Boston: Lahaska/Houghton-Mifflin.

Ivey, A. (2000/1986). *Developmental therapy.* Framingham, MA: Microtraining.

Two books that offer skill integration as well as a developmental theory. Useful follow-up from microskill training, particularly due to the focus on sensorimotor, concrete, formal, and dialectic/systemic strategies.

Lazarus, A. (1989). *Brief but comprehensive psychotherapy: The multimodal way.* New York: Springer.

This is still the basic model for multimodal therapy. You will find the BASIC-ID model useful in conceptualizing broad treatment plans.

Psychodynamic Theory

Bowlby, J. (1988). *A secure base: Parent–child attachment and healthy human development.* New York: Basic Books.

Bowlby has become the most researched figure in the psychodynamic framework. His work on human attachment will be the foundation for psychodynamic practice in the near future.

Miller, A. (1981). *The drama of the gifted child.* New York: Basic Books.

One route to understanding psychodynamic theory is looking at your own experience. This book reads well and can make a difference in the way you think about yourself as a counselor. Our students often thank us for referring them to Alice Miller.

Appendix: Glossary of Terms for Brain Areas Discussed in Research Portions of This Book

The glossary presents definitions and functions of areas of the brain discussed in this book, which has emphasized neuropsychology research and issues most closely related to counseling and interviewing. Needless to say, many important areas of the brain and research implications have not been covered here. Consider the ones discussed in the book a brief introduction to an area that will likely become more clearly defined and more central your work in the years to come.

For practical purposes in counseling and therapy, neuropsychological research has the following major implications:

1. In general, research on the brain supports much of our thinking about the practice of interviewing and human change.
2. Expect wide variation in brain functioning of each individual, just as we expect wide individual differences in personality and cultural backgrounds.
3. Negative emotions are embedded deeply in the brain and positive emotions seem to be required to keep the negative under some control. The positive asset search and wellness appear to be key to enabling a counseling and therapy based on strengths.
4. Selective attention, key to the foundation skill of attending behavior and other skills as well, has been clearly identified as a vital part of brain functioning.
5. Empathy appears to have a neurological base but can be modified by life experience. Human survival has depended on awareness of others.
6. *Neuroplasticity* is the term given to evidence that the brain can grow and learn at any age. Expect effective interviewing, counseling, and therapy to make a difference in functioning for both the client and you.

Amygdala A key part of the limbic system within the brain stem that has an important part in emotion and motivation. Traumatic and fear memories are located here and may override positive emotions.

Brain stem One of three major sections of the brain. Near the spinal cord, it controls involuntary actions. It is the seat of arousal, attention, and much of emotion.

Cerebellum One of three major sections of the brain. Important for balance and motor coordination, it also participates in language processing and selective attention.

Cerebrum Largest of the three major sections of the brain, the site of intelligence, learning, and judgment with functions in many areas such as language, memory, personality, and vision. The cerebrum is divided into left and right hemispheres. There are four regions (lobes), named for the skull bones that cover them—*frontal, parietal, temporal,* and *occipital.*

Cortex The brain's external layer—a dense layer of cells. The *neocortex* is the brain's outermost layer. The *prefrontal cortex* is where much of the brain's integrative functions occur such as decision making and attending. This portion is also considered important in positive and more complex culturally related emotions. The *visual cortex* is concerned with vision, the *anterior cingulate cortex* with touch.

Hippocampus Important for learning and memory retrieval.

Left hemisphere Controls the right side of the body and may often be the dominant hemisphere. Speech and language are located here, although speech may be located on the right for some left-handed people. This part of the brain helps analyze details and is active in analytical concepts such as math, logic, and speech. Positive emotions such as happiness and joy are located here. Evidence suggests that experiencing positive emotions requires negative emotions to be absent or at least controlled; therefore, building a positive frame of wellness seems particularly relevant to personal growth.

Limbic system (amygdala, hypothalamus, hippocampus formation, thalamus, and other modules) Controls emotional experience, particularly fear and other negative emotions. In addition, the sense of smell is located here and may explain the power of odor in emotional and sexual experience.

Medulla oblongata Controls heart rate, respiration, and other basic functions. The *reticular activating system* is located here.

Neurons Nerve cells that make up the central nervous system; they store and transmit information.

Neuroplasticity The ability of the brain to develop new connections/neurons throughout the lifespan in response to new situations.

Neuropsychology ". . . the study of relations between brain function and behavior" (Kolb & Wishaw, 2003).

Pons The link between the cortex and the cerebellum.

Reticular activating system Neural system in *medulla* that controls overall nervous system arousal in the cortex. Attentiveness, sleep, and wakefulness rise here.

Right hemisphere Controls the left side of the body. Visual information and abstract ideas are processed here. This hemisphere synthesizes observations into a general impression. It is associated with creativity and artistic talent as well as positive emotions.

Thalamus Relay station in the limbic system and part of the reticular activating system. "Focus is brought about by the thalamus, which operates rather like a spotlight, turning to shine on the stimulus" (Carter, 1999, p. 186).

Further Reading and Study on Neuropsychology

The field is obviously changing rapidly and you will want to keep abreast of new developments. At this time, we have found the following particularly useful.

Carter, R. (1999). *Mapping the mind.* Berkeley: University of California Press.
Despite the book's age, its clear outline of the brain itself and the relevant research it provides offer the best overall background for beginners. Watch for updates of this important book.

Damasio, A. (2003). *Looking for Spinoza: Joy, sorrow, and the feeling brain.* New York: Harcourt.
A literary approach, this book is particularly relevant to counselors and therapists.

Kolb, B., & Whishaw, I. (2003). *Fundamentals of human neuropsychology* (5th ed.). New York: Worth.

Considered one of the best textbooks in the field. Again, watch for the latest edition.

Restak, R. (2003). *The new brain.* New York: Rodale.

Written in a popular style, this book is another excellent introduction to neuropsychology and the brain.

Name Index

Subject Index

TO THE OWNER OF THIS BOOK:

I hope that you have found *Intentional Interviewing and Counseling,* Sixth Edition, useful. So that this book can be improved in a future edition, would you take the time to complete this sheet and return it? Thank you.

School and address:_____

Department:_____

Instructor's name:_____

1. What I like most about this book is:_____

_____

_____

2. What I like least about this book is:

_____

_____

3. My general reaction to this book is:

_____

_____

4. The name of the course in which I used this book is:

_____

5. Were all of the chapters of the book assigned for you to read?_____

   If not, which ones weren't?_____

6. In the space below, or on a separate sheet of paper, please write specific suggestions for improving this book and anything else you'd care to share about your experience in using this book.

_____

_____

_____

THOMSON

BROOKS/COLE

BUSINESS REPLY MAIL
FIRST-CLASS MAIL   PERMIT NO. 102   MONTEREY CA

POSTAGE WILL BE PAID BY ADDRESSEE

Attn: Social Work Editor

BrooksCole/Thomson Learning
60 Garden Ct Ste 205
Monterey CA 93940-9967

FOLD HERE

OPTIONAL:

Your name:_____ Date: _____

May we quote you, either in promotion for *Intentional Interviewing and Counseling,* Sixth Edition, or in future publishing ventures?

Yes: _____   No: _____

Sincerely yours,

Allen Ivey and Mary Bradford Ivey